Handbook of Advanced Semiconductor Field Effect Transistors

Scrivener Publishing
100 Cummings Center, Suite 541J
Beverly, MA 01915-6106

Publishers at Scrivener
Martin Scrivener (martin@scrivenerpublishing.com)
Phillip Carmical (pcarmical@scrivenerpublishing.com)

Handbook of Advanced Semiconductor Field Effect Transistors

Edited by

Ekta Goel

Archana Pandey

Shiromani Balmukund Rahi

and

Arun Samuel

Scrivener Publishing

WILEY

This edition first published 2026 by John Wiley & Sons, Inc., 111 River Street, Hoboken, NJ 07030, USA and Scrivener Publishing LLC, 100 Cummings Center, Suite 541J, Beverly, MA 01915, USA

For more information about Scrivener publications please visit www.scrivenerpublishing.com.

Wiley Global Headquarters
111 River Street, Hoboken, NJ 07030, USA

For details of our global editorial offices, customer services, and more information about Wiley products visit us at www.wiley.com.

The manufacturer's authorized representative according to the EU General Product Safety Regulation is Wiley-VCH GmbH, Boschstr. 12, 69469 Weinheim, Germany, e-mail: Product_Safety@wiley.com.

Library of Congress Cataloging-in-Publication Data

ISBN 9781394412570

Front cover image of transistor courtesy of Wikimedia Commons
Cover image: Generated with AI using Adobe Firefly
Cover design by Russell Richardson

Set in size of 11pt and Minion Pro by Manila Typesetting Company, Makati, Philippines

Printed in the USA

10 9 8 7 6 5 4 3 2 1

Contents

Preface

The transistor has become one of the major significant breakthroughs in the electronics field. New techniques of circuit design, miniaturization, and reliability are now emerging, due to the utilization of the transistor and other semiconductor devices. Latest design trends for military equipment have been predominantly towards the use of these devices for the active elements. The purpose of this handbook is to provide the semiconductor players with a reference of reliable, well-designed examples of contemporary devices circuits suitable for use. The brief description about included chapter has been summaries as followings'

In Chapter 1, the field of semiconductor reliability analysis and modeling and provides a detailed examination of the methods and tools used to forecast and improve the durability and efficiency of semiconductor devices. The rapid advancement of semiconductor technology has significantly increased the complexity and functionality of these devices, necessitating the development of sophisticated reliability analysis techniques to meet the changing demands of the industry. The chapter begins with an introduction to the basic principles of semiconductor reliability, including the definitions and importance of reliability metrics and failure mechanisms specific to semiconductor devices. The following sections offer a thorough review of cutting-edge reliability modeling techniques. In this chapter, there are comprehensive discussions on the application of these models in practical settings, underscoring their relevance and effectiveness in forecasting device failures. The particular emphasis is placed on accelerated life testing and its importance in reliability prediction as well as on emerging trends and challenges in the field, such as those introduced by new materials and miniaturization. By merging theoretical knowledge with practical examples, this chapter to provide readers with the knowledge required to understand and address the complexities of semiconductor reliability, enhance their ability to reduce risks, and prolong the functional lifespan of these crucial components.

Chapter 2 describes, the scaling challenges encountered in the nanometer regime with standard single gate bulk MOSFETs are significant because of stern short-channel effects, leading to an exponential rise in the off current and heightened susceptibility to process variations. The double-gate FinFET has emerged as the preferred option among multi-gate transistor architectures, owing to the self-alignment of its two gates, and it also has fabrication steps similar to the existing standard CMOS technology. As the prevalence of embedded systems and mobile devices increases, power consumption has become a paramount concern in contemporary microprocessor designs. Novel latches and flip-flops based on IG-FinFETs, which consume less power and have smaller layout areas, are introduced. To address the challenges associated with the continuous scaling of MOSFETs, fin field-effect transistors (FinFETs) have surfaced as viable alternatives. This enables the ongoing scaling and manufacturing of integrated circuits (ICs) for the creation of smart devices, thereby facilitating smart environments and integrated ecosystems. This chapter provides a concise overview of the fundamental features and operating principles of the FinFETs. It delves into the scaling constraints of mainstream MOSFETs beyond the 22-nm scale owing to short-channel effects (SCEs) and explores scalable alternative FinFET devices for various VLSI circuit design applications. In addition, this chapter highlights the advantages of multiple-gate ultra-thin-body FinFET devices. The chapter concludes by addressing the major challenges in the FinFET process, device, and circuit design.

In Chapter 3, the demand for compact, efficient, and low-power electronics has grown exponentially over the last few decades. However, conventional field-effect transistors (FETs) face subthreshold swing (SS) limitations. Thus, modern integrated circuits (ICs) with billions of transistors exhibit high overall power consumption. The negative-capacitance ferroelectric FET (NC-FeFETs) is among the many transistor concepts devised to solve this issue. The NC-FeFET structure is realized by incorporating a ferroelectric material into the FET gate oxide. This modification amplifies the internal voltage near the device channel region, resulting in a steeper SS with a high I_{ON}/I_{OFF} ratio, thus reducing power consumption. In this chapter, the fundamentals of NC-FeFETs are discussed in detail. The first section introduces the NC-FeFET and discusses its evolution and operation. The chapter discusses NC-FeFET modeling approaches, applications, performance optimization, and challenges. A comparative analysis of the NC-FeFET with other transistors is also presented. Finally, the chapter concludes with future prospects and research trends highlighting the possibilities of NC-FeFETs as future devices.

Chapter 4 focused on relentless drive towards miniaturization in micro- and nanoelectronics systems has underscored the criticality of power consumption management. This pursuit encounters a formidable obstacle known as Boltzmann tyranny, wherein the generation of a significant drain current necessitates a minimum gate voltage of approximately 60 mV. This imposes a substantial limitation on the scalability of supply voltage in Ultra-Large-Scale Integration circuits, hindering their ability to adapt to the diminishing dimensions of traditional transistors. In response to this pressing challenge, the negative capacitance field-effect transistor has emerged as a promising solution. The NCFET represents a ground-breaking advancement in the field of low-power electronics, offering a promising avenue for overcoming the fundamental limitations of traditional transistor technologies. This novel transistor leverages the unique property of negative capacitance to achieve sub thermal swing and enhance the performance of electronic devices, thereby enabling unprecedented reductions in power consumption. The NCFET operates based on the principle of exploiting ferroelectric materials to induce negative capacitance, resulting in an intrinsic voltage amplification effect. This breakthrough technology effectively addresses the power-efficiency challenges faced by conventional transistors, allowing for the realization of ultralow-power electronic devices without compromising performance metrics. This chapter explores the underlying physics, engineering aspects, design, and simulation methodology of NCFET, highlighting their probable application in circuits. The NCFET has emerged as a transformative technology, promising a paradigm shift towards ultralow-power electronics, highlighting their significance in the ongoing quest for energy-efficient and high-performance electronic devices.

In Chapter 5, Tunnel Field-Effect Transistors have emerged as a promising technology for low power electronic applications due to their steep subthreshold swing and proficiency to operate at lower supply voltages compared to conventional MOSFETs. This chapter provides a comprehensive overview of TFETs, focusing on their fundamental principles, designs, and fabrication techniques. It explores the unique quantum tunneling mechanism that enables TFETs to attain superior performance in terms of power efficiency and scaling. The chapter also discusses the application of TFETs in several domains, including digital electronics, analog and mixed-signal circuits, biosensors, energy harvesting, and communication systems. Furthermore, the potential of TFETs in emerging technologies, such as wearable and implantable medical devices, is discussed, showcasing their role in the future of low power electronics. Future trends and ongoing research efforts aimed at overcoming current challenges and enhancing TFET capabilities are also examined. This comprehensive overview provides valuable insights for researchers, engineers, and industry professionals interested in harnessing the advantages of TFET technology for next-generation electronic applications.

In Chapter 6, MOSFETs have been used in the integrated circuits for a long time. The device dimensions are shrinking continuously due to scaling. A reduction in transistor dimensions is essential for high chip density, low power, increased speed of operation, reduced chip cost, and overall enhanced performance. However, the transistor's smaller size results in numerous undesirable short channel effects , including hot carrier effect, velocity saturation, drain induced barrier lowering , and reduced gate control over the channel. But for highly scaled devices, the performance of FinFETs is compromised due to various short channel effects which start degrading the device performance. A promising alternative to FinFETs in the near future is the Gate All Around FET (GAA FET), which has excellent short channel performance. GAA FETs have gates on all four sides, in contrast to FinFETs, which have gates around the channel on three sides only. Because of this, the GAA FET has better gate control, has higher drive current, higher I_{ON}/I_{OFF}, and is less susceptible to process fluctuations. Also, the short channel performance of GAA FET is better than as compared to FinFET. Multi-gate transistors, such as FinFET and GAA FET, are extremely complex to fabricate. For instance, it is challenging to accurately adjust the fin's height, width, and inclination angle. In the same way, it is challenging to etch the gate oxide uniformly. Another problem is the oxide and channel contact surface roughness, which can lead to scattering. Another significant problem that might deteriorate the electrical properties of the device is the random positioning of dopants. The very limited number of dopants for narrow gate lengths makes them difficult to manage during construction. The performance of the device may be deteriorated by variations in the threshold voltage brought on by variations in the dopant concentration over various areas of the channel. Because of this, the threshold voltage varies greatly at higher channel doping values and less at lower channel doping values. Dopant concentrations can range from very low in the center to quite high on the edges. This is because the average dopant concentration might change for very tiny gate length values due to the discrete nature of dopant atoms. Therefore, it is crucial to study the impact of process variations due to modifications in device dimensions during the fabrication process. The impact of process variations on device performance is covered in this chapter. A detailed discussion has been done regarding the impact of varying the radius, gate oxide thickness, gate length, and channel doping of GAA FETs. For each of the aforementioned variations, the percentage change in the drive current and threshold voltage has been examined. Changes in oxide thickness, nanowire radius, gate length, channel doping, and nanowire thickness do not significantly affect the drive current or threshold voltage. The sensitivity of oxide thickness and gate length is significantly lower than that of the GAA FET's channel and radius doping. The proposed device structure is therefore appropriate for uses where a high degree of immunity to process fluctuations is needed.

Chapter 7 focuses on the growing need for high-performance, energy-efficient electronics, this study investigates Tunnel Field-Effect Transistors as a possible substitute for conventional MOSFETs. TFETs use band-to-band tunneling to obtain lower operating voltages and improved performance in contrast to conventional MOSFETs, which have drawbacks such as short-channel effects and excessive power consumption. The advantages of TFETs in terms of power efficiency and scalability are highlighted in this study's examination of their historical developments, operating principles, and device topologies. It also addresses the difficulties in fabricating TFETs, such as material compatibility and technology integration. This research emphasizes how TFETs can revolutionize semiconductor technology by assessing their performance parameters and investigating diverse applications, especially in low-power and high-speed devices. Creative solutions are crucial to overcome fabrication obstacles and enhance the overall device performance in contemporary electronic systems.

In Chapter 8, tunnel field effect transistors , characterized by gated p-i-n structures, are renowned for their rapid switching capabilities owing to their band-to-band tunneling mechanism, which makes them capable of achieving a low subthreshold swing (SS). Extensively researched as potential substitutes for MOSFETs, particularly for low-voltage and low-power applications. The numerous benefits of TFETs extend their utility across sensing applications including biosensing, image sensors, and temperature sensing. This study primarily focuses on exploring TFETs' optical sensitivity. Optical sensing has wide-ranging applications in medical diagnostics, chemical-mixing configurations, military equipment, and energy harvesting. Recently, optical sensors based on FETs have garnered significant attention owing to their compatibility with integrated circuits in low-power applications. In this study, we explore an extended source double-gate TFET (ES-DGTFET) photosensor. The device facilitates lateral tunneling as well as vertical band-to band tunneling owing to the extended source. As a result, the device was capable of achieving a high ON current and low SS. The upper gate of the ES-DGTFET was replaced by an optical gate that was responsible for the generation of photovoltage upon excitation by light of a specific wavelength and intensity. To investigate the performance of the

TFET photosensor, different parameters, such as photocurrent, response, and quantum efficiency, were observed for the proposed device. The photosensor is capable of achieving a current of 5.08×10^{-6}A/μm, responsivity of approximately 105, and quantum efficiency of more than 50% for an incident light of wavelength 450 nm and intensity of 0.8 W/cm^2.

Chapter 9 describes, the development of field-effect transistor (FET) sensors has undergone significant progress in recent years, owing in large part to the emergence of two-dimensional (2D) materials. The 2D materials, such as graphene, transition metal dichalcogenides (TMDs), and black phosphorus, have unique electrical, mechanical, and chemical properties that make them ideal candidates for sensing applications. This chapter provides an overview of the ongoing progress in FET sensors using 2D materials and explains their major standards, creation procedures, and different applications. This chapter begins with an introduction to 2D material-based FET sensors and their working principles. In next section interesting properties and applications of 2D material will be explained, for example, graphene, transition metal dichalcogenides (TMDs), and dark phosphorus, which make them reasonable for sensor applications. It further investigates the procedure utilized for the fabrication of FET sensors. An overview of the sensing mechanisms used will be explained further. This chapter highlights the challenges and future scope of 2D material-based FET sensors.

Chapter 10 described on study of a two-dimensional FET (2D-FET) works with a channel made of MoS_2, $MoSe_2$, and $MoTe_2$ and different dielectric materials. Materials such as molybdenum disulfide (MoS_2), molybdenum diselenide ($MoSe_2$), and molybdenum ditelluride ($MoTe_2$) belong to the transition metal dichalcogenide (TMD) family and are known for their fascinating electronic properties. Transition metal dichalcogenides like MoS_2 have emerged as promising candidates, boasting satisfactory bandgaps, thermal stability, high carrier mobility, and compatibility with silicon CMOS processes. Their ability to reduce scattering events, electrical insulating qualities, and suitability for 2D semiconductor channels make them excellent choices for low-power setups. The ideal dielectric, however, must also consider the particular needs of the device, including its operating environment, targeted performance indicators, and 2D semiconductor channel material. These materials have showcased immense potential in device applications, offering superior electrostatics and resilience against short-channel effects compared to traditional silicon channels. From the classic Metal-Oxide-Semiconductor FET (MOSFET) to the evolution of dual-gate FETs, tri-gate devices, and the current pinnacle, Gate-All-Around Field-Effect Transistors (GAA FETs), each technological era has presented unique opportunities and challenges. These materials have different band gaps, carrier mobilities, and intrinsic properties that make them ideal candidates for 2D transistor channels. The I_{ON}/I_{OFF} graph is a critical metric for evaluating the performance of Field-Effect Transistors (FETs), providing insights into the on-state (I_{ON}) and off-state (I_{OFF}) current characteristics. In scenarios involving a HfO_2 spacer, MoS_2 consistently demonstrates superior performance compared to alternative 2D materials such as $MoTe_2$ and $MoSe_2$. The I_{ON}/I_{OFF} graph for MoS_2 FETs typically exhibits a more favorable ratio, emphasizing a higher on-state current coupled with a lower off-state current. This advantageous combination is attributed to MoS_2's unique electronic properties, including its favorable band structure and superior carrier mobility. The HfO_2 spacer enhances the gate control over the channel, contributing to efficient modulation of the FET's conductivity. In contrast, $MoTe_2$ and $MoSe_2$ may exhibit comparatively higher off-state currents or reduced on-state currents due to factors like lower carrier mobility or less effective gating. The I_{ON}/I_{OFF} graph underscores MoS_2's prowess as a promising material for FET applications with HfO_2 spacers, highlighting its potential for high-performance electronic devices. In this study, we investigate the influence of different dielectric materials on the performance of 2D FETs with MoS_2, $MoSe_2$, and $MoTe_2$ as channel materials, with a strong focus on high-speed operation. Our study analyzes performance metrics like on-current (I_{ON}), off-current (I_{OFF}), and I_{ON}/I_{OFF} ratio which will be more when compared with Si-based FETs. Notably, the choice of dielectric significantly influences these metrics, with high-k dielectrics and HfO_2 often surpassing SiO_2.

Chapter 11 describe the scaling down the metal oxide semiconductor field-effect transistors (MOSFETs) to achieve exhibiting superior behavior and integration density is accompanied by various challenges. The leakage current with a reduction in gate dielectric and tunneling effects compromises device reliability. The integrating MOSFETs with other semiconductor technologies poses challenges in terms of compatibility and performance optimization. Therefore, addressing these issues is essential for advancing the MOSFET technology and ensuring its viability in future electronic systems. One non-classical device, the high electron mobility transistors (HEMTs), has several advantages over MOSFETs because it exhibits significantly higher electron mobility, enabling faster electron transport and improved high-frequency performance with higher switching speeds. The HEMTs excel in high-power applications because of their high current-carrying capacity, making them ideal for power amplifiers in RF and microwave circuits. Additionally, HEMTs offer low-noise figures, especially in the microwave and mm-wave frequency domains, making them suitable for low noise amplifiers in communication systems and sensitive measurements. The monolithic integration capability of HEMTs allows them to be easily unified with other semiconductor devices, thereby enabling the design of complex and high-performance integrated circuits. Furthermore, HEMTs based on materials such as Gallium Nitride (GaN) exhibit good radiation tolerance, making them suitable for space- and high-energy applications. Metal-oxide-semiconductor high electron mobility transistors (MOSHEMTs) combine the advantages of both MOSFETs and HEMTs by offering improved gate control, allowing precise modulation of the channel and better control over device characteristics such as threshold voltage and transconductance. They also reduce gate leakage, enhancing the power efficiency by utilizing the gate oxide layer as a barrier against current leakage. Additionally, the gate oxide layer in MOSHEMTs enhances device reliability by protecting against hot-carrier effects and prolonging the device lifespan. MOSHEMTs can be seamlessly integrated with Complementary MOS (CMOS) technology, enabling complex systems that leverage the strengths of both MOSFETs and HEMTs, such as power amplifiers with digital control circuits. Furthermore, MOSHEMTs exhibit enhanced performance in terms of power handling capability, speed, and linearity, which is attributed to the enhanced channel modulation enabled by the MOS structure. This chapter presents the MOSHEMT structure, working principle, and its use in different application domains, particularly for biosensing.

Chapter 12, explains the general concepts of quantum computing and digital twins using technological terminology. Specifically, considering the development of modern semiconductor technology and quantum computing technology, this chapter explores the interconnected domains of quantum computing, digital twins, and semiconductor development (especially non-memory semiconductor development), analyzing their combined influence on technological progress. This chapter investigates the basics and fundamentals of quantum computing, the progression and utilization of digital twins in diverse industries, and the essential role of emerging semiconductor technologies in making them operational. Through a multidisciplinary approach, this chapter reveals the synergies, obstacles, and potential uses in various applications, illuminating the considerable impact of these technologies across a wide range of fields. Furthermore, we address ethical concerns and outline future research directions and collaborative opportunities, offering informative insights into the technological landscape of the future.

Chapter 13 has continued advancements in the field of CMOS technology have led to aggressive scaling of device dimensions with the goal of boosting the chip integration density. However, it has become increasingly hard to keep pace with Moore's law due to short channel effects that worsens with scaling. To address these challenges, the semiconductor industry has explored various solutions. One of the most promising solutions was adopting alternative device architectures like FinFETs which are associated with double or triple gates. These multiple gates enhanced the control exercised by the gate over the channel thus mitigating the severity of several second order effects referred as short channel effects. With adoption of FinFET Technology, it became a norm of packing smaller devices with more features on a chip. This has led to a greater focus on ensuring these devices perform efficiently within the limits of their power supply. This becomes critical for areas like mobile computing, IoT devices, biosensors and etc where reliable operation is to be ensured with lower than nominal supply voltages. There is a lot of emphasis today on low voltage circuit design. The FinFETs have exhibited superior subthreshold characteristics when compared to planar devices Moreover, it has been established that the minimum value to which the supply voltage can be scaled in a CMOS inverter realized with FinFET devices is quite small. This voltage is well below 36mV, which is the fundamental limit for planar devices at room temperature. Owing to these, FinFETs present themselves as an excellent candidate for low voltage and low power circuit designs. However simply replacing the planar devices with FinFETs is not the right approach for optimized performance of circuits at low voltages. There are several device-circuit interaction issues arising that need to be addressed in a comprehensive manner before finalizing the circuit design. FinFETs have quantized widths due to discrete number of fins and circuits cannot be designed for arbitrary widths like in conventional CMOS technology. By virtue of its structure, FinFETs inherently have source and drain extension (SE and DE) regions. The carrier density in these regions is bias dependent. This results in several interesting device level phenomenon. One critical observation with FinFETs based circuits is the dependence of effective capacitances on circuit parameters like supply voltage of operation, the transition time of input signals and the fanout the circuit. It has also been reported that saturation phenomenon in FinFETs is quite distinct when compared with planar devices. This significantly affects the performance of circuits designed from the analog perspective as well. A systematic methodology for FinFET based low voltage circuit design is required for meeting performance criteria be it a digital circuit or an analog circuit. This chapter first presents an in-depth review of the existing literature on low voltage circuit design using FinFETs. It focuses on first reviewing the critical device-circuit co-design challenges in low voltage domain for FinFETs.

After developing this understanding, we review the existing methodologies used to design low voltage circuits using FinFETs in the literature.

Chapter 14 describes the reversible logic gates are crucial in quantum computing and low-power digital circuit designs because they uniquely conserve information. Conventional combinational circuits face issues such as information loss and energy inefficiency, which reversible computing addresses by ensuring that input bits are maintained in the output, thereby improving the overall efficiency. This study explores the construction and utilization of a full adder employing gates such as the Peres, Toffoli, and Feynman gates. The proposed solution is both energy-efficient and low-power, aiming to minimize ancilla inputs, quantum cost, and garbage outputs. Multipliers, critical for arithmetic operations, digital signal processing, and other computing tasks, require effective design and implementation to enhance circuit performance. Vedic multipliers, based on ancient Indian mathematical principles have been particularly noted for their efficiency and speed. This study examines and compares 8-bit Vedic multiplier architectures that use different adder structures, such as the Carry Lookahead Adder and the Ripple Carry Adder. The goal is to evaluate the performance, delay, and logic levels of these multipliers relative to the existing designs. The proposed adder design demonstrates a 75% reduction in garbage count, while the multiplier shows a 57% improvement in delay compared with previous study. The validation was performed using the Vivado 2022.2.

Chapter 15 describe the increasing demand for real-time digital signal processing applications has called for the development of fast and efficient parallel prefix adder (PPA) designs. This paper examines various PPA architectures created for 64-bit applications, as well as their performance compared to traditional adders. All design entities were verified using Xilinx ISE, and the synthesis results were subsequently analyzed. It was observed that the 64-bit Pipelined Parallel prefix Brent-Kung adder (PPPBKA) exhibited the highest operating speed amongst the tested adders, including regular and parallel prefix adders. Furthermore, the power delay product (PDP) of the Pipelined Brent Kung adder was found to be lower than those of the other adders. The simulation and synthesis were conducted using Xilinx ISE.

Chapter 16 describes the rapidly evolving landscape of Internet of Things, efficient communication protocols play a pivotal role. One critical aspect is the baud rate, which determines the rate at which data can be transmitted over a communication channel. Traditional UART systems require pre-agreed baud rates. Once set, changing the baud rate during communication is difficult. Because of this, irrespective of the state of the receiver, the amount of data being transmitted will be same. This may lead to either data loss or inefficient use of the receiver. In this method of design, that is, when the baud rate is prefixed, the power consumption and delay are high. When compared among the different baud rates like 9600, 19200, 38400, the power consumption is 0.452W and the delay is 2.194ns when baud rate is prefixed to 9600 bps. The area consumption is up to 20 LUT's. The results are more efficient when baud rate is 9600 bps, when compared with others. This project introduces an innovative solution to mitigate the aforementioned limitation. It involves the design of a high-performance adaptive baud rate generator, which dynamically adjusts the baud rate based on the state of the receiver. In the presented design, the power consumed and delay are 0.181W and 0.766ns respectively, which are reduced significantly when compared to the existing design. These achieved parameters establish the adaptive baud rate generator as well-suited for applications demanding low power consumption and fast communication.

Chapter 17 described, today's world, healthcare costs are rising due to an aging population, while governments struggle to allocate sufficient funds. As a result, patients face significant challenges in accessing medical services, leading to distress and hardship. This chapter explores key advancements in medical and biomedical sensing using VLSI technology. With continuous progress in VLSI, its integration into biomedical applications is set to further transform healthcare delivery and enhance patient outcomes.

Chapter 18 describes the biosensors have revolutionized various fields by providing rapid, sensitive, and portable means of detecting biological analytes. They find applications in healthcare, environmental monitoring, food safety, and biodefense, among others. As traditional biosensors often face challenges related to power consumption and limited operational lifetimes, the incorporation of self-powering mechanisms becomes a transformative solution. In recent years, a significant focus has been directed towards the development of self-powered biosensor FETs, which incorporates energy harvesting mechanisms to enable autonomous operation without the need for external power sources. This chapter provides a comprehensive overview of the need of FET biosensors, working principle, need of energy harvesting, latest advancements in self-powered biosensor FET technology, fabrication techniques and applications.

Chapter 19 introduces and explores a number of aspects pertaining to tunnel field effect transistors for dielectric-modulated label-free biosensors. An overview of DM TFET biosensors has been given in brief. Principles and design perceive are highlighted in the discussion of the dielectric modulation idea in TFETs. The incorporation of implanted nanogaps into

TFET geometries using computer-aided design (TCAD) tool and suitable physics-based simulation models will be explored. Through TCAD simulations, the sensitivity of gate-engineered TFET is investigated under several conditions. Lastly, the state of the proposed V-TFET is pointed out, and an overview map showing the sensitivities of the major efforts in DM label-free biosensors is presented.

In Chapter 20, it describes the advancements in sensing and detection technologies have significantly impacted various sectors, including the food industry, medical field, pharmaceuticals, defense, and environmental monitoring. Recent progress in micro-electromechanical systems (MEMS) has enabled the miniaturization of sensor technology, allowing the integration of transducing devices and signal processing systems onto a single platform. The MEMS-based sensors are celebrated for their small size, biocompatibility, electrical nature, ease of fabrication, low cost, low power consumption, ultrahigh sensitivity, fast detection speed, high accuracy, and real-time analysis. These advantages make MEMS sensors suitable for applications such as bioimplants, including retinal and neural implants. Common transducing techniques used in MEMS sensors include piezoelectric, piezoresistive, electrostatic, electromagnetic, electrothermal, and optical methods. However, these techniques often suffer from low output responses. To address this issue, the integration of MEMS sensors with field-effect transistors (FETs) has been proposed, enhancing the performance of MEMS-FET sensors. This integration allows changes in the output parameters of the FET (such as mobility, threshold voltage, on/off ratio, and source-drain currents) in response to MEMS inputs, enabling precise detection of various substances. This chapter provides a detailed review of MEMS-FET sensors, starting with an in-depth discussion of the advantages of MEMS sensors over traditional sensors. It then explores the operating principles, advantages, disadvantages, and limitations of various transduction technologies. The chapter continues with a description of different configurations, classifications, and applications of MEMS sensors. The focus will be on the working principle of MEMS-FET sensors, detailing the methodology of coupling FET with MEMS sensors. Advances in FETs and MEMS structures and their compatibility will be discussed. Special attention will be given to various types of MEMS-FET sensors, including force, accelerometer, pressure, and gas sensors, and their applications in fields, such as aerospace, navigation, unmanned aerial vehicles, explosive impact detection, and biomedical applications. Finally, the chapter will highlight future perspectives in the field of MEMS-FET sensors, providing a comprehensive overview from the fundamentals of sensor technology to advanced MEMS-FET sensor development.

Chapter 21: This work examines key factors in memory design, such as the operating system, application size, and operation diversity, which influence memory usage and its type (virtual or real). Traditional dynamic random access memory (DRAM), using one MOSFET and one capacitor, can lead to space inefficiencies for multi-bit storage. To counter this, row and column logic are placed outside the memory array. Activating a row line selects the entire row, requiring a buffer for longer lines. On the other hand, control logic latches address signals, with the column address determining which bits to access once a row is selected. This chapter outlines the circuits needed for RAM implementation. A microarchitectural framework for protocol controller design is presented, breaking down memory functions into fundamental activities for effective protocol execution. Moreover, we introduced two memory designs: reduced-area memory (area: 1/8 and 1/4) and high-density memory (density: 4/16 bits/cell) using conventional DRAM unit cells based on the soft-error principles. We presented the design objectives and concept of the new memory designs, showing that the new memory designs are cost-effective without bringing in large-area overhead for error correction code (ECC) or error localization. We have found that, without the use of any proprietary tools, a useful ECC current waveform function library has been developed to be the building blocks of exploratory computer-aided design (CAD) for DRAM data sensing.

In Chapter 22, the reliability issues of interface trap charges (ITCs) on 4H-SiC (4H–silicon carbide) Trench metal-oxide-semiconductor field-effect transistor (4H-SiC-T-MOSFET) have been discussed in terms of analog parameters. Reliability issues in 4H-SiC-T-MOSFET related to trap charges primarily stem from the interface between the gate oxide and the silicon carbide semiconductor. Trap charges at this interface can lead to threshold voltage shifts, degradation of device performance, and long-term reliability issues. The density of interface traps at the SiC/SiO_2 interface is a critical parameter affecting device reliability. High trap densities can lead to increased interface state charges, which can adversely affect device performance and reliability. Trap charges at the interface can cause threshold voltage shifts, leading to variations in device characteristics and performance. Threshold voltage instability can affect circuit operation and reliability over time. Therefore, while trap charges at the SiC/SiO_2 interface can pose reliability challenges for 4H-SiC-T-MOSFET, ongoing research and development efforts focus on mitigating these issues through improved materials, processing techniques, and device design strategies. By addressing interface trap–related reliability concerns, 4H-SiC-T-MOSFET can achieve enhanced long-term reliability and performance for various high-power and high-temperature applications.

1

Semiconductor Reliability Analysis and Modeling

Reinhard S. Park

Kyunghee University, Seoul, Republic of Korea

Abstract

This chapter discusses the field of semiconductor reliability analysis and modeling and provides a detailed examination of the methods and tools used to forecast and improve the durability and efficiency of semiconductor devices. The rapid advancement of semiconductor technology has significantly increased the complexity and functionality of these devices, necessitating the development of sophisticated reliability analysis techniques to meet the changing demands of the industry. The chapter begins with an introduction to the basic principles of semiconductor reliability, including the definitions and importance of reliability metrics and failure mechanisms specific to semiconductor devices. The following sections offer a thorough review of cutting-edge reliability modeling techniques. In this chapter, there are comprehensive discussions on the application of these models in practical settings, underscoring their relevance and effectiveness in forecasting device failures. Particular emphasis is placed on accelerated life testing and its importance in reliability prediction as well as on emerging trends and challenges in the field, such as those introduced by new materials and miniaturization. By merging theoretical knowledge with practical examples, this chapter seeks to provide readers with the knowledge required to understand and address the complexities of semiconductor reliability, enhance their ability to reduce risks, and prolong the functional lifespan of these crucial components.

***Keywords*:** Semiconductor reliability, semiconductor reliability modeling, semiconductor analysis, reliability improvement

Email: reinhardpark@protonmail.com

Ekta Goel, Archana Pandey, Shiromani Balmukund Rahi and Arun Samuel (eds.) Handbook of Advanced Semiconductor Field Effect Transistors, (1–18) © 2026 Scrivener Publishing LLC

1.1 Introduction

Recently, there has been rapid development and advancement in the field of semiconductor [1–6]. In the rapidly evolving field of technology, ensuring the reliability of semiconductor devices is crucial [7–10]. Semiconductors are fundamental to modern electronics, and their failure can cause significant disruptions across various applications, from consumer electronics to critical sectors, such as aerospace and medical devices. This chapter provides a thorough examination of the semiconductor reliability analysis and modeling. This chapter explores the essential principles of semiconductor functionality, significance of reliability, and techniques used to predict and improve the longevity and performance of semiconductor devices.

1.2 History and Fundamental of Semiconductors

Semiconductors are materials that exhibit electrical conductivity between that of conductors and insulators. The development of semiconductor technology began in the early 20th century, with significant milestones such as the invention of the transistor in 1947 by John Bardeen, Walter Brattain, and William Shockley. This invention laid the groundwork for modern electronics and revolutionized the field of solid-state physics.

Silicon is the most commonly used semiconductor material because of its abundance and cost-effectiveness, and other materials such as germanium, gallium arsenide, and silicon carbide are also employed for specific applications because of their unique electrical properties. These materials were selected based on factors such as the electron mobility, bandgap energy, and thermal conductivity.

To appreciate the importance of the reliability of semiconductors, it is vital to understand their basic operating principles. Semiconductors are materials with electrical conductivities that fall between those of conductors (such as metals) and insulators (such as ceramics). This unique property makes them ideal for controlling the electrical currents in devices such as transistors, diodes, and integrated circuits.

A core concept in semiconductor technology is doping, which involves adding impurities to pure semiconductor materials, such as silicon, to modify their electrical properties. Doping creates two types of semiconductor materials: n-type, which has an excess of electrons and p-type, which has an excess of holes (positive charge carriers). When n- and p-type materials are combined, they form a p–n junction, which is crucial for the operation of many semiconductor devices.

In a p–n junction, electrons from the n-type material recombine with holes from the p-type material at the junction interface, creating a depletion region in which no charge carriers are present. This depletion region acts as a barrier to the current flow. However, when a voltage is applied across the p–n junction, the barrier can be overcome, allowing current to flow. This principle underlies the operation of the diodes, transistors, and other semiconductor devices.

1.3 Brief Overview of Semiconductor Fabrication

The fabrication of semiconductor devices is an intricate and tightly regulated process that converts raw materials into highly functional electronic components. This process comprises multiple detailed steps, each of which requires meticulous execution to guarantee the reliability and performance of the final product. The key stages of semiconductor fabrication are summarized below.

1.3.1 Wafer Preparation

The process begins with the production of silicon wafers that act as substrates for semiconductor devices. High-purity silicon was extracted from quartz sand and melted to form a large cylindrical crystal known as a boule. The boule was sliced into thin wafers using a diamond saw. These wafers were polished to achieve a smooth and flat surface, which is essential for subsequent processing steps. Any imperfections at this stage can negatively affect the performance of the final semiconductor devices.

1.3.2 Photolithography

Photolithography is a critical step in defining geometric patterns of semiconductor devices on wafers. A light-sensitive photoresist was applied to the wafer surface and a mask containing the desired pattern was placed over it. Ultraviolet light was then shone through the mask, transferring the pattern onto the photoresist. The exposed areas of the photoresist were developed, leaving behind a precise pattern that guided the subsequent processing steps on the wafer.

1.3.3 Etching

Etching involves removing material from the wafer surface to create desired structures, such as transistor gates and interconnects. There are two primary types of etching: wet and dry etching. Wet etching uses liquid chemicals to dissolve the exposed material, whereas dry uses employ plasma to etch away the material. The choice of the etching method depends on the specific requirements of the fabricated device.

1.3.4 Doping

Doping introduces impurities into specific regions of a wafer to alter its electrical properties. This step is essential for creating p- or n-type regions, which are critical for the formation of p–n junctions. The two common doping techniques are ion implantation and diffusion. In ion implantation, the ions of the dopant material are accelerated and embedded into the wafer. During diffusion, the wafer is exposed to dopant gas at high temperatures, allowing dopant atoms to diffuse into the silicon lattice.

1.3.5 Deposition

During the fabrication process, various materials such as insulating layers, conductive interconnects, and protective coatings are deposited on the wafer. Techniques such as chemical vapor deposition (CVD), physical vapor deposition (PVD), and atomic layer deposition (ALD) are used to apply these materials to thin, uniform layers. Each deposition step must be precisely controlled to ensure correct thickness and uniformity, which are critical for device performance.

1.3.6 Planarization

As multiple layers are added to the wafer, the surface becomes uneven. Planarization was used to smooth the wafer surface, ensuring a flat and uniform surface for the subsequent layers. Chemical mechanical planarization (CMP) is a common technique that combines chemical etching with mechanical polishing to achieve desired surface flatness.

1.3.7 Metallization

Metallization (also known as the metal process) involves creating electrical interconnections between different regions of the device. Metals, such as aluminum, copper, and tungsten, were deposited onto the wafer to form these interconnects. The metal layers were patterned by photolithography and etching to create the desired circuit pathways. The integrity of these interconnects is crucial to the device's overall electrical performance and reliability.

1.3.8 Packaging

Once the devices are fabricated on the wafer, they must be separated and encased in protective packages. The wafer was cut into individual chips, each containing one or more semiconductor device. These chips are then mounted onto a substrate and encapsulated in a protective package that protects them from physical damage, moisture, and other environmental factors. The packaging process also involves creating external connections, such as solder balls or pins, that allow the device to be integrated into electronic systems.

1.3.9 Testing

Throughout the fabrication process, rigorous testing was conducted to ensure the quality and reliability of semiconductor devices. The initial testing was performed at the wafer level to identify and discard defective devices. After packaging, further testing was performed to verify the electrical performance and functionality of the final product. Reliability tests, such as accelerated life testing, can help predict the long-term performance of devices under various conditions.

1.4 Definition and Explanation of Bathtub Curve

Figure 1.1 illustrates the concept of a bathtub curve. The bathtub curve is a graphical tool widely used in reliability engineering to depict the failure rate of a product over time [11–14]. The curve is named for its shape, which resembles a bathtub, featuring high failure rates at the beginning and end of the product's lifecycle, with a low steady failure rate in between. Understanding bathtub curves is crucial for predicting and enhancing the reliability and longevity of semiconductor devices. A curve typically consists of three distinct phases: infant mortality, normal life, and wear-out. As shown in Figure 1.1, a bathtub curve consists of three main parts: 'infant mortality phase,' 'normal life phase,' and 'wear-out phase.'

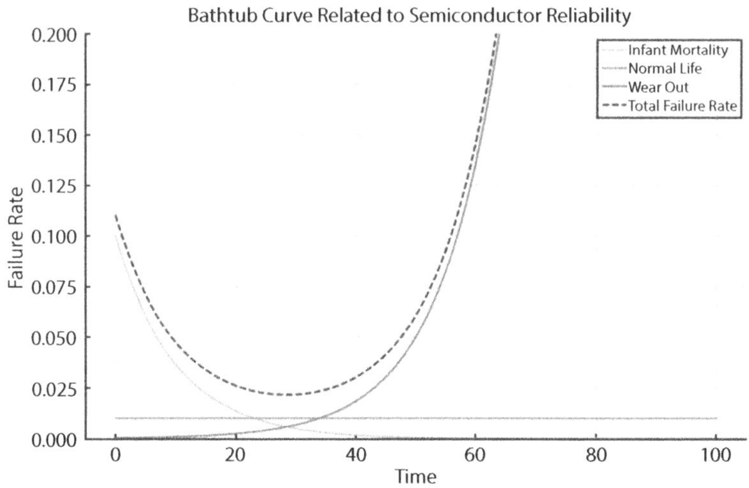

Figure 1.1 Illustration explaining the concept of 'Bathtub Curve' related to semiconductor reliability.

1.4.1 Infant Mortality Phase

The first phase of the bathtub curve, known as the infant mortality period or early failure period, is marked by a high initial failure rate, which rapidly decreases over time. Failures during this stage are often caused by manufacturing defects, design flaws, or early life stresses that reveal product

Table 1.1 Summary of common reasons of infant mortality.

Cause of infant mortality	Description
Manufacturing Defects	Imperfections introduced during production, such as contamination, improper assembly, or defects in raw materials, can lead to early failures.
Design Flaws	Errors in the semiconductor device design, such as inadequate thermal management or incorrect component specifications, can cause early-life failures.
Early-Life Stresses	Environmental stresses, such as temperature fluctuations, mechanical vibrations, and electrical surges, can exacerbate inherent weaknesses and lead to early failures.

weaknesses. The common causes of infant mortality in semiconductor devices are summarized in Table 1.1.

To mitigate infant mortality, manufacturers often use burn-in testing, in which devices are subjected to elevated stress levels for a certain period to identify and eliminate defective units before they reach customers.

1.4.2 Normal Life Phase

The second phase of the bathtub curve, known as the normal life period, useful life phase, or random failure period, is characterized by a relatively low and constant failure rate. Most semiconductor devices operate reliably during this stage and failures are typically random and rare. The factors contributing to the random failures in this phase are listed in Table 1.2.

Maintaining reliability during the normal life phase involves ensuring stable operating conditions, regular maintenance, and monitoring for signs of potential issues.

1.4.3 Wear-Out Phase

The final phase of the bathtub curve is the wear-out period, during which the failure rate increases again as the product reaches the end of its useful life. Failures during this stage are typically due to the gradual degradation of materials and components owing to prolonged use. The common wear-out mechanisms in semiconductor devices are listed in Table 1.3 [15–19].

Table 1.2 Factors leading to random failures in normal life phase.

Type	Description
Environmental Factors	Variations in operating conditions, such as temperature, humidity, and electromagnetic interference, can lead to sporadic failures.
Random Defects	Unpredictable and isolated defects that were not detected during manufacturing can occasionally cause failures.
Usage Patterns	The way the semiconductor device is used, including the frequency and intensity of operation, can influence the likelihood of random failures.

Table 1.3 Wear-out mechanisms in semiconductor devices.

Concept	Description
Electromigration	The movement of metal atoms caused by high current densities, leading to the formation of voids and hillocks that can interrupt electrical connections. Figure 1.2 visually shows the concept and possible result of electromigration [15–19].
Thermal Cycling	Repeated temperature fluctuations cause expansion and contraction of materials, leading to fatigue and failure of interconnects and solder joints.
Oxide Breakdown	The gradual deterioration of the insulating oxide layer in transistors, leading to leakage currents and eventual failure.

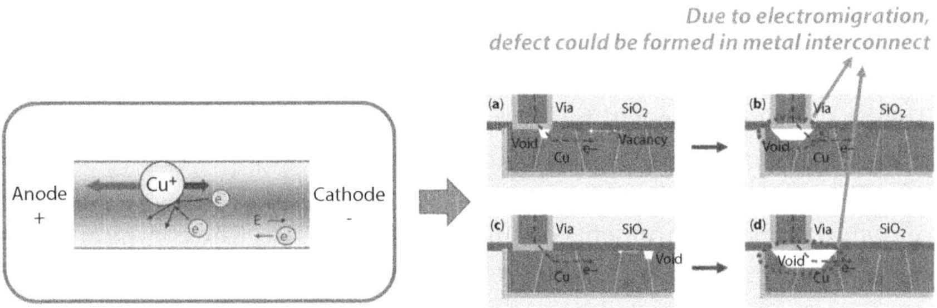

Figure 1.2 Figure explaining the concept and possible result of electromigration.

To address wear-out failures, reliability engineers may implement planned obsolescence strategies where devices are replaced or refurbished before significant failures occur. Additionally, they may use accelerated life testing to predict wear-out behavior and establish maintenance schedules.

1.4.4 Applications in Semiconductor Reliability

Bathtub curves are a valuable tool for reliability engineers in the semiconductor industry. By understanding and applying this model, engineers can predict the reliability and lifespan of semiconductor devices and develop strategies to mitigate the failures in each phase.

- During the Infant Mortality Phase: Rigorous testing and quality control measures are implemented to identify and eliminate defective units before they reach the market.
- During the Normal Life Phase: Stable operating conditions and regular maintenance are ensured to sustain a reliable operation.
- During the Wear-Out Phase: Replacement schedules and maintenance plans are established based on accelerated life-testing data to prevent unexpected failures.

1.5 Failure Mechanisms in Semiconductor

Failure mechanisms are critical factors that affect the performance and reliability of semiconductor devices. These failures often arise from the complex interactions of electrical, thermal, and mechanical stresses within the device. The failure mechanisms of semiconductors can be classified as listed in Table 1.4.

1.6 Failure Mechanism Modeling and Prediction

Accurately modeling and prediction of failure mechanisms are essential to ensure the reliability of semiconductor devices. This process involves utilizing theoretical models, empirical data, and simulation tools to understand and forecast how and when the devices fail. Various failure mechanism models have been developed, some of which are presented in Table 1.5.

'Accelerated-life testing' is widely used for life testing. The lifetime of a semiconductor can be evaluated by accelerating the semiconductor with temperature or external stress. Accelerated-life testing involves exposing semiconductor devices to elevated stress levels, such as higher temperatures or voltages, to induce failures more rapidly than under normal operating conditions. The data obtained from these tests were used to create models that predict the performance of the devices under typical conditions. Accelerated life testing helps pinpoint weaknesses and enhance device reliability.

Table 1.4 Type of failure mechanism.

Type	Description
Hot Carrier Injection (HCI)	Hot carrier injection occurs when high-energy carriers (electrons or holes) gain sufficient kinetic energy to break through the potential barrier at the silicon-oxide interface. These hot carriers can cause damage to the gate oxide, leading to shifts in the threshold voltage and degradation of transistor performance over time.
Time-Dependent Dielectric Breakdown (TDDB)	The TDDB is a failure mechanism in which the insulating properties of the gate dielectric degrade over time owing to prolonged electric field stress. This degradation eventually leads to a breakdown of the dielectric material, resulting in a short circuit between the gate and channel.
Negative Bias Temperature Instability (NBTI)	NBTI affects the p-channel MOSFETs when a negative bias is applied to the gate at elevated temperatures. This stress causes the threshold voltage to shift, leading to reduced drive current and slower transistor performance. NBTI is a significant reliability concern for modern semiconductor devices.
Stress-Induced Leakage Current (SILC)	An SILC is an increase in the leakage current through the gate dielectric caused by electrical stress. This phenomenon is particularly problematic in devices with ultrathin gate oxides, where even minor increases in leakage current can significantly impact device performance and power consumption.
Soft Errors	Soft errors are temporary changes in the state of a semiconductor device caused by external factors, such as cosmic rays or alpha particles. These errors do not cause permanent damage, but can disrupt the operation of digital circuits, leading to data corruption and system crashes.

Table 1.5 Failure mechanism models.

Model type	Description
Physics-Based Models	Physics-based models apply fundamental physical principles to describe failure mechanisms. For instance, electromigration models account for the diffusion of metal atoms under high current densities, whereas thermal cycling models consider the thermal expansion and contraction of materials. These models offer insights into the root causes of failure and help predict the lifespan of devices.
Empirical Models	Empirical models rely on experimental data and statistical analyses. They used historical failure data to establish correlations between operating conditions and failure rates. Weibull and log-normal distributions are frequently employed to model the time-to-failure data and predict the reliability of semiconductor devices.
Simulation Tools	Simulation tools such as finite element analysis (FEA) and computational fluid dynamics (CFD) are utilized to model the physical and thermal behavior of semiconductor devices. These tools aid engineers in visualizing the stress distributions, temperature gradients, and other critical factors that influence failure mechanisms. Engineers can identify potential failure points and design more robust devices by simulating various scenarios.

1.7 Design for Reliability Strategies for Semiconductor

Design for Reliability (DfR) is a critical approach in the development of semiconductor devices, ensuring that products perform as intended and have long lifespans under various operating conditions. The DfR involves a set of strategies and practices aimed at anticipating and mitigating potential failure mechanisms during the design phase, thereby enhancing the overall reliability of semiconductor devices.

1.7.1 Early Failure Analysis

The core principle of DfR is identifying and addressing potential failure modes early in the design process. This includes a comprehensive failure mode and effects analysis (FMEA), in which designers systematically assess each component and process to determine possible failure points and their consequences. By identifying issues early, corrective actions can be implemented before a product moves into manufacturing.

1.7.2 Robust Design Techniques

Robust design techniques aim to create semiconductor devices that are resilient to variations in the manufacturing processes and operating environments. This can be achieved through tolerance design, ensuring that the components and materials can withstand deviations without compromising performance. Techniques such as the design of experiments (DoE) and Taguchi methods are used to optimize the design parameters for maximum reliability.

1.7.3 Accelerated Testing and Modeling

Accelerated life testing (ALT) and accelerated stress testing (AST) are essential components of the DfR. These tests expose semiconductor devices to extreme conditions to quickly identify the potential failure mechanisms. Data from these tests are used to develop predictive models, such as the Arrhenius equation for thermal acceleration and the Coffin-Manson model for thermal cycling, which help estimate the device's lifespan under normal conditions.

1.7.4 Material Selection and Characterization

The selection of appropriate materials is crucial for ensuring the reliability of semiconductor devices. Properties such as the thermal conductivity, mechanical strength, and electrical characteristics must be carefully evaluated. Advanced characterization techniques, including scanning electron microscopy (SEM) and X-ray diffraction (XRD), have been used to analyze material behavior under various conditions. The choice of material affects the device's resistance to thermal stress, electromigration, and other degradation mechanisms.

1.7.5 Redundancy and Fault Tolerance

Incorporating redundancy and fault tolerance into semiconductor designs can significantly enhance their reliability. Redundancy involves adding extra components or pathways that can be taken over in the case of a failure. Fault-tolerant design strategies, such as error-correcting codes (ECC) and built-in self-test (BIST) circuits, enable the device to function correctly even when some components fail. These approaches are particularly important in mission-critical applications such as aerospace and medical devices.

1.7.6 Thermal Management

In semiconductor design, effective thermal management is vital for maintaining the reliability of the semiconductor devices. Excessive heat can accelerate failure mechanisms such as electromigration and thermal cycling. In addition, excessive heat can degrade electron mobility, thereby decreasing the performance of transistors [20–26]. From transistor level to circuit levels, several approaches have been developed to address these issue and improve thermal characteristics of semiconductor [27–30]. DfR strategies include designing efficient heat sinks (including transistor, circuit, and system levels), optimizing the layout for better heat dissipation, and using advanced cooling techniques, such as liquid cooling and thermoelectric cooling [23, 31–38]. Thermal simulations were also conducted to predict the temperature distribution and identify hotspots.

1.7.7 Process Control and Monitoring

The implementation of stringent process control and monitoring throughout the manufacturing process is essential to ensure that semiconductor devices meet reliability standards. Statistical process control (SPC) techniques are used to monitor the key parameters and detect variations that can lead to defects. Real-time monitoring and feedback systems enable immediate corrective actions, thereby reducing the likelihood of reliability issues in the final product.

1.7.8 Environmental Stress Screening

Environmental stress screening involves exposing semiconductor devices to conditions beyond the normal operating levels to precipitate latent defects. ESS tests can include thermal cycling, vibration, and humidity exposure with the aim of identifying and eliminating weak units before they reach the field. This screening process ensures that only the most robust devices are deployed.

1.7.9 Reliability-Centered Maintenance

The RCM is a proactive maintenance strategy that focuses on preserving the inherent reliability of semiconductor devices. It involves regularly scheduled inspections, predictive maintenance based on condition monitoring, and the timely replacement of components before failure. RCM strategies help extend the operational life of semiconductor devices and reduce the downtime due to unexpected failures.

Additionally, DfR is an ongoing process that evolves with technological advancements and feedback from field performance. It is essential to implement a feedback loop in which data from field failures are analyzed and used to improve future designs. Lessons learned from real-world applications can lead to refinements in the design, materials, and processes, further enhancing the reliability of semiconductor devices.

1.8 Conclusion

So far, we have investigated the concepts, principles, and applications of semiconductor reliability. With the help of the bathtub curve, it is possible to properly analyze the lifetime of the semiconductor. Semiconductor breakdown can be suppressed and semiconductor reliability can be improved by utilizing the bathtub curve concept and the principle of semiconductor breakdown. Accelerated-life testing is one of the most valuable strategies for testing the reliability of semiconductors. Overall, with the advancement of semiconductor testing techniques and new materials, it might be possible to improve semiconductor reliability and design for high semiconductor reliability.

References

1. Nakazawa, T. and Kulkarni, D.V., Anomaly Detection and Segmentation for Wafer Defect Patterns Using Deep Convolutional Encoder–Decoder Neural Network Architectures in Semiconductor Manufacturing. *IEEE Trans. Semicond. Manuf.*, 32, 2, 250–256, May 2019. doi: 10.1109/TSM.2019.2897690.

2. Yasuda, S., Tanaka, T., Kitabata, M., Jisaki, Y., Chamber and Recipe-Independent FDC Indicator in High-Mix Semiconductor Manufacturing. *IEEE Trans. Semicond. Manuf.*, 34, 3, 301–306, Aug. 2021. doi: 10.1109/TSM.2021.3075255.
3. Shenai, K., Dudley, M., Davis, R.F., Rugged Electrical Power Switching in Semiconductors: A Systems Approach. *Proc. IEEE*, 102, 1, 35–52, Jan. 2014. doi: 10.1109/JPROC.2013.2278616.
4. Li, X., *et al.*, Thermal Imbalance Among Paralleling Chips in Power Modules and the Impact From Traction Inverter System View. *IEEE Trans. Ind. Electron.*, 70, 3, 2231–2240, March 2023. doi: 10.1109/TIE.2022.3159920.
5. Nag, J., *et al.*, Non-Contact, Sub-Surface Detection of Alloy Segregation in Back-End of Line Copper Dual-Damascene Structures. *IEEE Trans. Semicond. Manuf.*, 28, 4, 469–473, Nov. 2015. doi: 10.1109/TSM.2015.2480037.
6. Imoto, K., Nakai, T., Ike, T., Haruki, K., Sato, Y., A CNN-Based Transfer Learning Method for Defect Classification in Semiconductor Manufacturing. *IEEE Trans. Semicond. Manuf.*, 32, 4, 455–459, Nov. 2019. doi: 10.1109/TSM.2019.2941752.
7. Chien, W.-T.K. and Hao, F., An Extended Building-In Reliability Methodology on Evaluating SRAM Reliability by Wafer-Level Reliability Systems. *IEEE Trans. Device Mater. Reliab.*, 20, 1, 106–118, March 2020. doi: 10.1109/TDMR.2020.2964999.
8. Chien, W.-T.K., Chung, A., Kuo, W., Fast Semiconductor Reliability Assessments Using SPRT. *IEEE Trans. Reliab.*, 68, 2, 526–538, June 2019. doi: 10.1109/TR.2018.2869597.
9. Siddiqui, J., Ortega, J., Albus, B., On the relationship between semiconductor manufacturing volume, yield, and reliability, in: *2017 IEEE International Reliability Physics Symposium (IRPS)*, Monterey, CA, USA, pp. SR-1.1–SR-1.5, 2017, doi: 10.1109/IRPS.2017.7936409.
10. Hromádka, A. and Martínek, Z., Usability Assessment of Mathematical Models of the Bathtub Curve, in: *2019 20th International Scientific Conference on Electric Power Engineering (EPE)*, Kouty nad Desnou, Czech Republic, pp. 1–5, 2019, doi: 10.1109/EPE.2019.8778149.
11. Wondmagegnehu, E.T., Navarro, J., Hernandez, P.J., Bathtub shaped failure rates from mixtures: a practical point of view. *IEEE Trans. Reliab.*, 54, 2, 270–275, June 2005. doi: 10.1109/TR.2005.847271.
12. Hofmeister, J., Pena, W., Goodman, D., Curti, C., Condition-Based-Events Life Curve: Conceptual View to Support Fault Management of Complex Systems of Overlapped, Distributed Events, in: *2022 IEEE International Conference on Prognostics and Health Management (ICPHM)Detroit (Romulus), MI, USA*, pp. 48–51, 2022, doi: 10.1109/ICPHM53196.2022.9815575.
13. Jopling, J., Risk Management Informed by an Uncertain Bathtub Curve (Invited), in: *2023 IEEE International Reliability Physics Symposium (IRPS)*, Monterey, CA, USA, pp. 1–8, 2023, doi: 10.1109/IRPS48203.2023.10117694.

14. Kudo, H., *et al.*, Demonstration of high electromigration resistance of enhanced sub-2 micron-scale Cu redistribution layer for advanced fine-pitch packaging, in: *2017 IEEE CPMT Symposium Japan (ICSJ)*, Kyoto, Japan, pp. 5–8, 2017, doi: 10.1109/ICSJ.2017.8240055.
15. Posser, G., de Paris, L., Mishra, V., Jain, P., Reis, R., Sapatnekar, S.S., Reducing the signal Electromigration effects on different logic gates by cell layout optimization, in: *2015 IEEE 6th Latin American Symposium on Circuits & Systems (LASCAS)*, Montevideo, Uruguay, pp. 1–4, 2015, doi: 10.1109/LASCAS.2015.7250429.
16. Mohanasundaram, S.M., Pratap, R., Ghosh, A., Tuning the Sensitivity of a Metal-Based Piezoresistive Sensor Using Electromigration. *J. Microelectromech. Syst.*, 21, 6, 1276–1278, Dec. 2012. doi: 10.1109/JMEMS.2012.2211579.
17. Liang, C.-L., *et al.*, Electromigration Failure Study of a Fine-pitch 2μm/2μm L/S Cu Redistribution Line Embedded in Polyimide for Advanced High-density Fan-out Packaging, in: *2020 IEEE 70th Electronic Components and Technology Conference (ECTC)*, Orlando, FL, USA, pp. 361–366, 2020, doi: 10.1109/ECTC32862.2020.00065.
18. Posser, G., Mishra, V., Reis, R., Sapatnekar, S.S., Analyzing the electromigration effects on different metal layers and different wire lengths, in: *2014 21st IEEE International Conference on Electronics, Circuits and Systems (ICECS)*, Marseille, France, pp. 682–685, 2014, doi: 10.1109/ICECS.2014.7050077.
19. Kumar, N., Kaushik, P.K., Gupta, A., Singh, P., Self-Heating Effects (SHEs) in Gate-All-Around FETs with Horizontally Stacked Multiple Junctionless Nanowires, in: *2022 IEEE Delhi Section Conference (DELCON)*, New Delhi, India, pp. 1–4, 2022, doi: 10.1109/DELCON54057.2022.9753115.
20. Rathore, S., *et al.*, Self-Heating Aware Threshold Voltage Modulation Conforming to Process and Ambient Temperature Variation for Reliable Nanosheet FET, in: *2023 IEEE International Reliability Physics Symposium (IRPS)*, Monterey, CA, USA, pp. 1–5, 2023, doi: 10.1109/IRPS48203.2023.10117918.
21. Jaisawal, R.K., *et al.*, Self-Heating and Interface Traps Assisted Early Aging Revelation and Reliability Analysis of Negative Capacitance FinFET, in: *2023 7th IEEE Electron Devices Technology & Manufacturing Conference (EDTM)*, Seoul, Korea, Republic of, pp. 1–3, 2023, doi: 10.1109/EDTM55494.2023.10103127.
22. Song, Y.S., Kim, S., Kim, J.H., Kim, G., Lee, J.-H., Choi, W.Y., Enhancement of Thermal Characteristics and On-Current in GAA MOSFET by Utilizing Al2O3-Based Dual-κ Spacer Structure. *IEEE Trans. Electron Devices*, 70, 1, 343–348, Jan. 2023. doi: 10.1109/TED.2022.3223321.
23. Chang, C.W., Liu, S.E., Lin, B.L., Chiu, C.C., Lee, Y.-H., Wu, K., Thermal behavior of self-heating effect in FinFET devices acting on back-end interconnects, in: *2015 IEEE International Reliability Physics Symposium*, Monterey, CA, USA, pp. 2F.6.1–2F.6.5, 2015, doi: 10.1109/IRPS.2015.7112696.

24. Tayal, S., *et al.*, Incorporating Bottom-Up Approach Into Device/ Circuit Co-Design for SRAM-Based Cache Memory Applications. *IEEE Trans. Electron Devices*, 69, 11, 6127–6132, Nov. 2022. doi: 10.1109/ TED.2022.3210070.
25. Kim, K.Y., *et al.*, Reliable High-Voltage Drain-Extended FinFET With Thermoelectric Improvement. *IEEE Trans. Electron Devices*, 69, 11, 5985–5990, Nov. 2022. doi: 10.1109/TED.2022.3209141.
26. Umana-Membreno, G.A., Akhavan, N.D., Antoszewski, J., Faraone, L., Cristoloveanu, S., Thermal broadening of the electron mobility distribution in FD-SOI MOSFETs, in: *2020 Joint International EUROSOI Workshop and International Conference on Ultimate Integration on Silicon (EUROSOI-ULIS)*, Caen, France, pp. 1–4, 2020, doi: 10.1109/ EUROSOI-ULIS49407.2020.9365363.
27. Song, Y.S., Tayal, S., Rahi, S.B., Kim, J.H., Upadhyay, A.K., Park, XXXB.-G., Thermal-Aware IC Chip Design by Combining High Thermal Conductivity Materials and GAA MOSFET, in: *2022 5th International Conference on Circuits, Systems and Simulation (ICCSS)*, Nanjing, China, pp. 135–140, 2022, doi: 10.1109/ICCSS55260.2022.9802341.
28. Pan, S., Feng, S., Zheng, X., He, X., Li, X., Bai, K., Effects of Temperature and Bias Voltage on Electron Transport Properties in GaN High-Electron-Mobility Transistors. *IEEE Trans. Device Mater. Reliab.*, 21, 4, 494–499, Dec. 2021. doi: 10.1109/TDMR.2021.3109088.
29. Chang, T.-S., Lu, T.Y., Chao, XXXT.-S., Temperature Dependence of Electron Mobility on Strained nMOSFETs Fabricated by Strain-Gate Engineering. *IEEE Electron Device Lett.*, 33, 7, 931–933, July 2012. doi: 10.1109/ LED.2012.2194982.
30. Song, Y.S., Kim, J.H., Kim, G., Kim, H.-M., Kim, S., Park, XXXB. -G., Improvement in Self-Heating Characteristic by Incorporating Hetero-Gate-Dielectric in Gate-All-Around MOSFETs. *IEEE J. Electron Devices Soc.*, 9, 36–41, 2021. doi: 10.1109/JEDS.2020.3038391.
31. Song, Y.S., *et al.*, Electrical and thermal performances of omega-shaped-gate nanowire field effect transistors for low power operation. *J. Nanosci. Nanotechnol.*, 20, 7, 4092–4096, July 2020.
32. Kang, S.J., *et al.*, Investigation of Self-Heating Effects in Vertically Stacked GAA MOSFET With Wrap-Around Contact. *IEEE Trans. Electron Devices*, 69, 3, 910–914, March 2022. doi: 10.1109/TED.2022.3140283.
33. Song, Y.S., *et al.*, Improvement of self-heating effect in Ge vertically stacked GAA nanowire pMOSFET by utilizing Al2O3 for high-performance logic device and electrical/thermal co-design. *Jpn. J. Appl. Phys.*, 60, 1–9, Mar. 2021. https://iopscience.iop.org/article/10.35848/1347-4065/abec5c/pdf.
34. Aoki, H., Sakairi, H., Kuroda, N., Nakamura, Y., Chikamatsu, K., Nakahara, K., AlGaN/GaN MIS HEMT Modeling of Frequency Dispersion and Self-Heating Effects, in: *2018 IEEE International Symposium on Radio-Frequency*

Integration Technology (RFIT), Melbourne, VIC, Australia, pp. 1–3, 2018, doi: 10.1109/RFIT.2018.8524087.
35. Ramezani, Z. and Orouji, A.A., Improving Self-Heating Effect and Maximum Power Density in SOI MESFETs by Using the Hole's Well Under Channel. *IEEE Trans. Electron Devices*, 61, 10, 3570–3573, Oct. 2014. doi: 10.1109/TED.2014.2352317.
36. Chohan T. *et al.*, Implication of Self-Heating Effect on Device Reliability Characterization of Multi-Finger n-MOSFETs on 22FDSOI, in: *IEEE Transactions on Device and Materials Reliability*, 22, 3, 387–395, Sept. 2022, doi: 10.1109/TDMR.2022.3183630.
37. Song, Y.S., *et al.*, Reliability improvement of self-heating effect, hot-carrier injection, and on-current variation by electrical/thermal co-design. *Solid State Electron.*, 197, 108436, Nov. 2022.
38. Rathore, S., Jaisawal, R.K., Kondekar, P.N., Bagga, N., Design Optimization of Three-Stacked Nanosheet FET From Self-Heating Effects Perspective. *IEEE Trans. Device Mater. Reliab.*, 22, 3, 396–402, Sept. 2022. doi: 10.1109/TDMR.2022.3181672.

2

Unveiling the Potential of FinFETs: An Alternative Paradigm to MOSFET

Nitish Vashishth*, **Neha Goel and R. K. Yadav**

Department of Electronics & Communication Engineering, Raj Kumar Goel Institute of Technology, Ghaziabad, India

Abstract

The scaling challenges encountered in the nanometer regime with standard single-gate bulk MOSFETs are significant because of stern short-channel effects (SCE), leading to an exponential rise in the off current and heightened susceptibility to process variations. The double-gate FinFET has emerged as the preferred option among multi-gate transistor architectures, owing to the self-alignment of its two gates, and it also has fabrication steps similar to the existing standard CMOS technology.

As the prevalence of embedded systems and mobile devices increases, power consumption has become a paramount concern in contemporary microprocessor designs. Novel latches and flip-flops based on IG-FinFETs, which consume less power and have smaller layout areas, are introduced. To address the challenges associated with the continuous scaling of MOSFETs, fin field-effect transistors (FinFETs) have surfaced as viable alternatives. This enables the ongoing scaling and manufacturing of integrated circuits (ICs) for the creation of smart devices, thereby facilitating smart environments and integrated ecosystems.

This chapter provides a concise overview of the fundamental features and operating principles of the FinFETs. It delves into the scaling constraints of mainstream MOSFETs beyond the 22-nm scale owing to short-channel effects (SCEs) and explores scalable alternative FinFET devices for various VLSI circuit design applications. In addition, this chapter highlights the advantages of multiple-gate ultrathin-body FinFET devices. The chapter concludes by addressing the major challenges in the FinFET process, device, and circuit design.

**Corresponding author*: nitish.vashishth@gmail.com; ORCID: orcid.org/0000-0003-4058-3071
Neha Goel: ORCID: orcid.org/0000-0002-7189-6305
R. K. Yadav: ORCID: https://orcid.org/0000-0003-4628-0688

Ekta Goel, Archana Pandey, Shiromani Balmukund Rahi and Arun Samuel (eds.) Handbook of Advanced Semiconductor Field Effect Transistors, (19–42) © 2026 Scrivener Publishing LLC

***Keywords*:** FinFET, SCE, multi gate transistor, ICs, VLSI, SCEs

2.1 Introduction to Transistor Technology

In the modern world, integrated circuits have revolutionized the field of electronics. They are used in almost every aspect of life, ranging from simple handheld devices to ultrafast microcomputer chips. The core of ICs is the transistor technology, which has evolved over time. Transistors can be used as switches and as sophisticated amplifiers and have been extensively used in digital electronic circuits. Transistor technology consumes less power than vintage vacuum tube technology. Furthermore, transistors do not have any moving parts, i.e., solid state devices. As the size of transistors decreases, they find their usage in a variety of applications owing to their very large-scale integration. Metal oxide semiconductor field effect transistors (MOSFETs) are fundamental building blocks of VLSI. As the transistor size is scaled down, device miniaturization becomes possible for extremely large-scale integrated (ELSI) circuits. However, with every scaled-down process, the feature size of the MOSFET was reduced. Owing to scaling, the short-channel effect (SCE) has become a prominent issue. To address this unavoidable issue, researchers have attempted to modify the structure and material used between the gate and the channel and have developed a 3-D FET structure as compared to an earlier 2-D FET structure [1]. FinFET technology has undergone rapid development and advancement across various chip manufacturers. FinFET technology was adopted by INTEL for the core I-series processors. However, as with any mature technology, we encounter some practical hurdles in development. The unique structure of FinFETs, along with the decreased gate distances, poses challenges in its manufacturing process. Consequently, gate-all-around (GAA) technology has emerged as a potential solution to these challenges.

2.1.1 Introduction to FinFET

In 1989, Hisamato *et al.* developed a new type of transistor called the DELTA transistor, which they abbreviated as Depleted Lean Channel Transistor. This transistor made use of a new gate structure and an ultrathin silicon-on-insulator layer. The formation of bulk single-crystal SOI was achieved by selective oxidation. In a DELTA transistor, the channel forms on a surface that is vertical; therefore, the width of the channel depends on the Si-island's height [2]. This is the first report of a FinFET structure.

FinFETs have become popular over the past decade because of their superior short-channel performance compared to MOS technology [3].

2.1.2 FinFET Structure

The FinFET channel, sometimes referred to as the fin, is vertical, whereas the planar MOSFET channel is horizontal, as shown in Figure 2.1. Therefore, the FinFET's width (*W*) is determined by the channel's height (H_{FIN}). This gives rise to a unique characteristic of FinFETs called width quantization which states that the FinFET width must be a multiple of channel's height, so that FinFET's width (*W*) can be enhanced by using a number of fins. Thus, it is impossible to obtain arbitrary FinFET widths. While offering more flexibility, smaller fin heights can result in a larger silicon area, because they have many fins. Taller fins, on the other hand, reduce the silicon footprint, but have the risk of causing structural instability. Process engineers often set the fin height, which should not be more than four times (4×) the fin thickness [4, 5].

FinFETs are generally fabricated on silicon-on-insulator (SOI) wafers; however, they have also been widely used in conventional bulk wafers [6–8]. Double-gate SOI MOSFETs have also been explored [39, 40]. Figure 2.2 shows the implementation of FinFETs on conventional bulk and SOI wafers. In bulk-based FinFETs, all fins share a common silicon bulk (body), whereas in SOI-based FinFETs, the fins are physically isolated. Because of the easy migration from conventional bulk MOSFET to bulk FinFETs, some manufacturers prefer the bulk technology. Nevertheless, FinFETs fabricated on both wafers are roughly equal in terms of performance, production cost, and throughput.

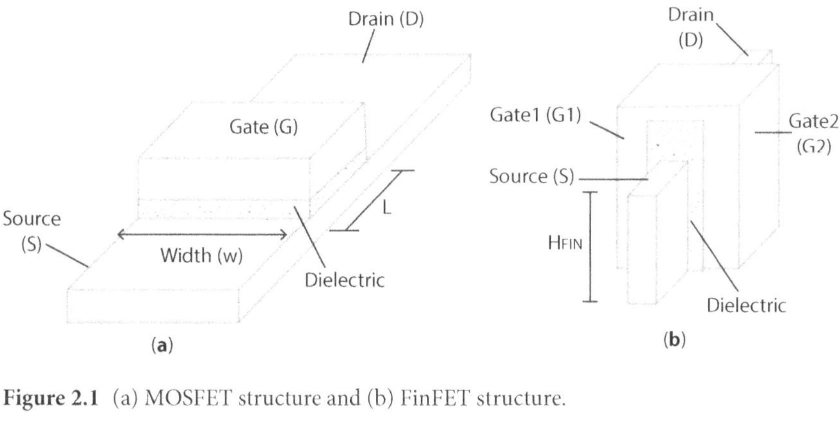

Figure 2.1 (a) MOSFET structure and (b) FinFET structure.

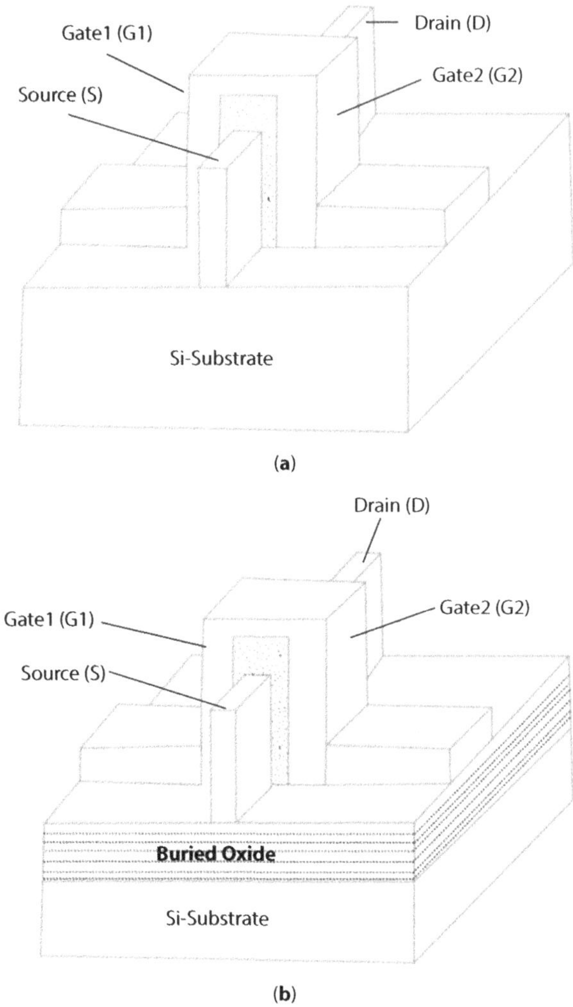

Figure 2.2 (a) Bulk FinFET and (b) SOI FinFET.

Trigate FET is a type of FinFET which has a third gate on the top of fin. In 2012, Trigate FETs were introduced by Intel at the 22 nm node in the Ivy-Bridge processor [9, 10]. The FinFET and a Trigate FET are shown in Figure 2.3.

In Trigate FETs, the thickness of the dielectric on top of the fin is lowered to accommodate the third gate. The thickness of the fin additionally increased the channel width because of the third gate. Trigate FETs therefore have a small width advantage over FinFETs. Because of the greater

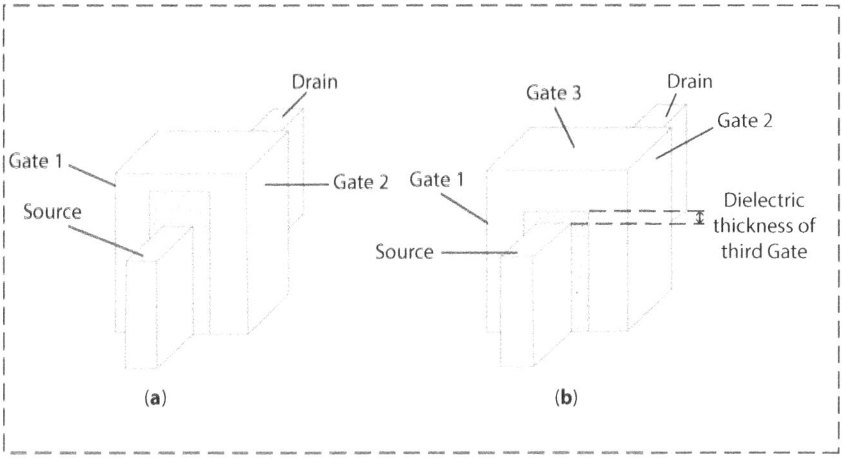

Figure 2.3 Comparison of (a) FinFET and (b) trigate structure.

current conduction at the top surface, trigate FETs also have lower gate-source capacitance than FinFETs; however, this benefit is offset by the higher parasitic resistance [11]. In the long run, Trigate FETs are inferior to FinFETs because they exhibit a high parasitic resistance. Yang and Fossum compared these both types and concluded that Trigate FETs no doubt have a little bit more body thickness than FinFETs, but the latter has better short channel performance. In addition, to achieve the same SCE profile performance, Trigate FETs are associated with the disadvantage of a large layout area [12].

2.1.3 Types of FinFET

FinFETs are mainly classified into two types: Shorted Gate (SG) FinFETs and Independent Gate (IG) FinFETs, commonly referred to as three-terminal (3-T) and four-terminal (4-T) FinFETs. Shorted Gate FinFETs justify their name, as gates at the front and back sides are physically shorted in SG-FinFETs, but in the case of IG-FinFETs, both gates are independent (Figure 2.4). This implies that both gates simultaneously control the channel operation in the SG-FinFETs. Therefore, Shorted Gate-FinFETs offer high magnitude on-current as well as off-current (sub-threshold current) compared to Independent Gate FinFETs. However, the IG-FinFET has its own advantages, *viz.*, it offers the application of various voltages and signals to gates as both gates are physically isolated. This makes it possible to linearly modulate the front-gate threshold voltage using the back-gate bias.

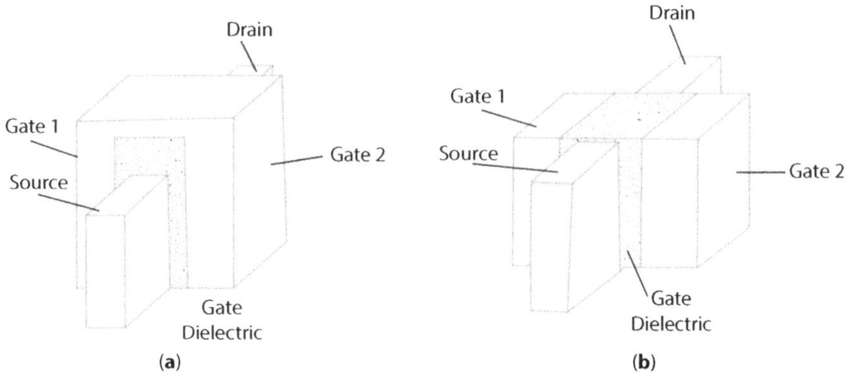

Figure 2.4 Structure of (a) shorted gate-FinFET and (b) independent gate-FinFET.

However, because two distinct gate contacts must be placed, IG FinFETs have the disadvantage of a large-area requirement. Generally, FinFETs with both gates have the same work function (Φ). However, the work functions can also be altered. This results in asymmetries in device parameters and a new type of SG-FinFET came to existence which is known as ASG-FinFETs (Asymmetric Gate Function SG-FinFETs) (Figure 2.5) [13, 14].

In asymmetric SG-FinFETs, the two gates exhibit different doping profiles. Asymmetric SG-FinFETs show good short-channel characteristics and a significantly lower off-current than SG-FinFETs but slightly higher on-current than SG-FinFETs [15].

FinFETs have also been investigated for other asymmetries in addition to gate work function asymmetry. As the channel length is indirectly

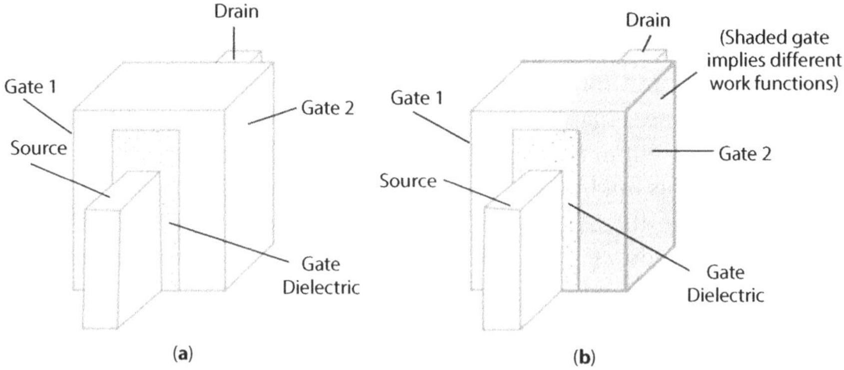

Figure 2.5 (a) Shorted gate-FinFET and (b) asymmetric SG-FinFET structure.

increased in the case of Asymmetric Drain Spacer Extended-FinFETs, they can result in better characteristics under a short-channel constraint [16]. However, it has the major disadvantage of a greater area requirement for the layout. In addition, because of this asymmetry, the source and drain are not interchangeable, in a conventional CMOS. The additional underlap causes asymmetry in the source-to-drain and drain-to-source currents, which has an impact on the performance of FinFET pass transistors. Asymmetric drain-source doped (AD) FinFETs were explored in [17], where the doping concentrations of drain and source doping were varied. Due to the low doping profile of the drain region, the reduced electric fields in the asymmetric drain-source-doped FinFETs led to better SCEs. There have also been proposals for FinFETs exhibiting an asymmetric silicon dioxide thickness (AT_{ox}) [18, 19]. These FinFETs exhibit good subthreshold slopes. Variable V_{ths} are also enabled in this situation by using IG FinFET (or 4T FinFET). An etching technique boosted by ion bombardment could be used to generate this asymmetry. Finally, research has also been conducted on asymmetric fin-height FinFETs [20, 21]. Taller pFinFETs can compensate for the intrinsic mobility imbalance between charge carriers because the width of the FinFET channel is directly correlated with the fin height.

Bhoj and Jha examined shorted gate, independent gate, and asymmetric SG-FinFETs under temperature fluctuations and discovered that at a higher temperature, although the off-current declines for all three FinFETs, Asymmetric SG-FinFETs still remain the best, with a multifold advantage over SG-FinFETs [15].

2.1.4 Symbols of FinFETs

Figure 2.6 shows the symbols/schematics of various shorted gate (SG) and independent gate (IG) FinFETs.

Figure 2.7 shows the symbols/schematics of various asymmetric shorted gate nFinFET and pFinFET.

2.2 Implementation of Inverters and NAND Gates Using FinFETs

Bhoj and Jha conducted a detailed examination and comparison of shorted, independent, and asymmetric SG FinFET-based NOT gates (inverters) and 2-input NAND gates (NAND2) [15]. The inverters and NAND2 gates

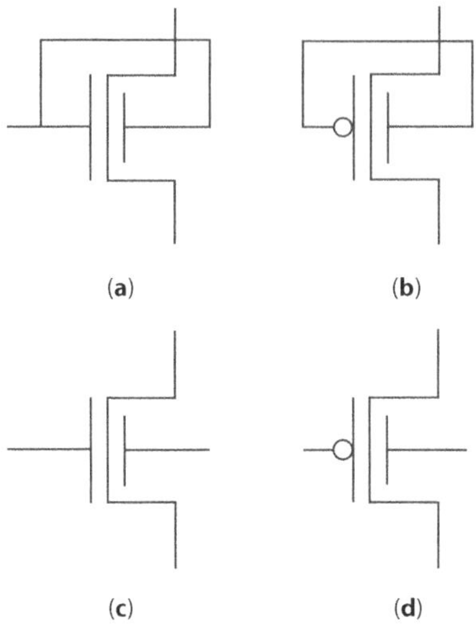

Figure 2.6 Symbol of (a) SG-nFinFET, (b) SG-pFinFET, (c) IG-nFinFET, and (d) IG-pFinFET.

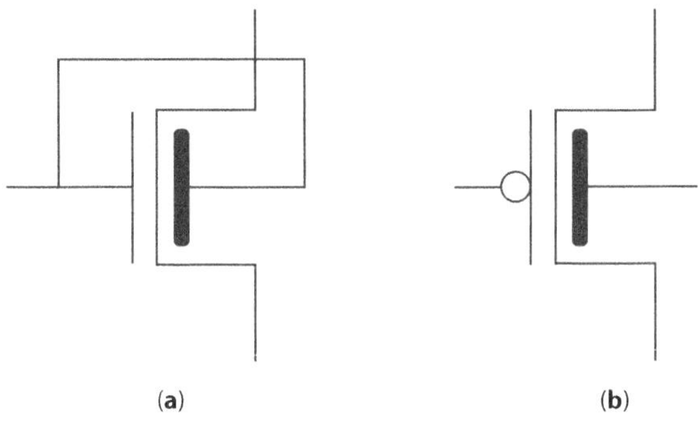

Figure 2.7 Symbols of asymmetric SG- (a) nFinFET and (b) pFinFET.

alone can be used to construct logic networks. This makes them the most essential logic library building block.

2.2.1 SG/IG-FinFET Based Inverters

An inverter can be implemented in four different ways depending on the how shorted gate (SG) and independent gate-FinFETs. These are identified as shorted gate (SG), low-power (LP) inverter, independent gate nFinFET-based (IGn), and independent gate pFinFET-based (IGp) inverters. Figure 2.8 depicts the schematic diagrams of these inverters.

The shorted gate inverter contained one nFinFET and one pFinFET. The layout area requirement of the SG inverter is the lowest among the four

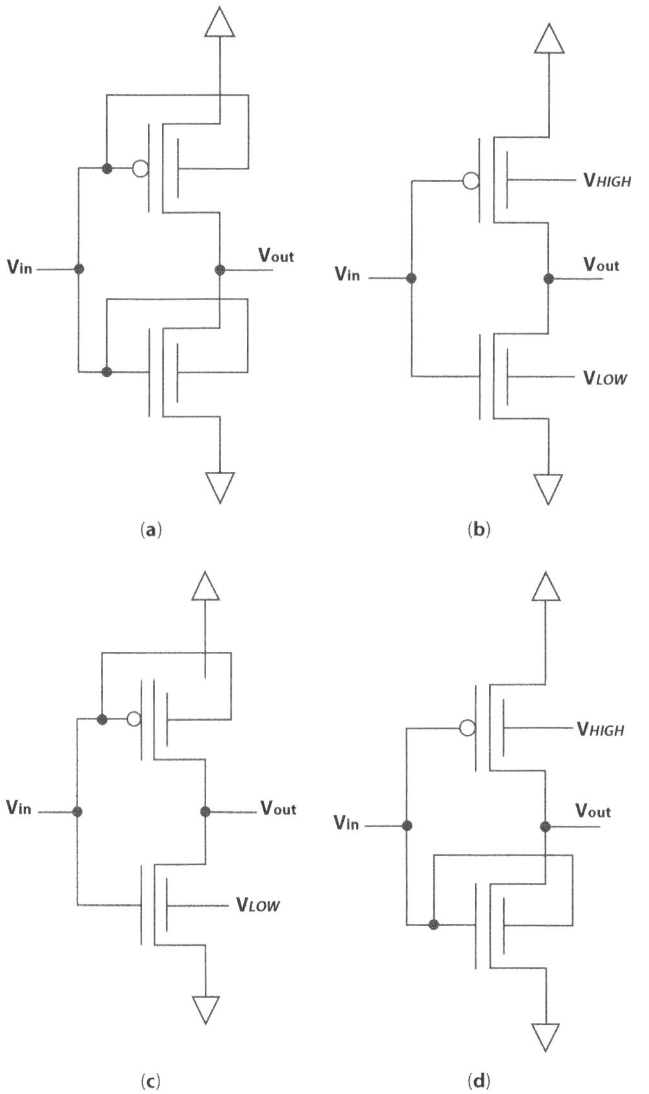

Figure 2.8 (a) Shorted gate (SG), (b) low power (LP), (c) IGn, and (d) IGp inverter schematics.

Table 2.1 Comparison of various FinFET inverters [15].

Topology	**SG**	**LP**	**IGn**	**IGp**
Area	1	1.36	1.36	1.36
I_{off} (Avg.)	20.92	1	2.75	19.25
T*pd*	1	3.67	1.67	2.92

inverters, as it uses only SG-FinFETs. This is not the case in the other three configurations; at least one independent gate FinFET is used.

The gate at the backside of an independent gate FinFET is connected to two different signals: V_{HIGH} and V_{LOW}. By providing suitable biasing at the backside gate of the independent gate FinFET, a prominent reduction in the off current (I_{off}) is achieved. Additionally, independent gates require more complex layouts, resulting in a 36% area overhead when compared with shorted gate inverters of double the size.

The propagation delay, area, and leakage current of different inverters are compared in Table 2.1. It is obvious that Shorted Gate inverter has the lowest propagation delay (*Tpd*) and area, but it leaks far more leakage (off) current as compared to a low-power inverter. However, the performance of low-power inverters is poor in terms of area and propagation latency. However, given its intermediate area, latency, and leakage, the independent gate IGn inverter appears promising.

2.2.2 SG/IG-FinFET Based 2-Input NAND Gates

Similar to inverter configurations, 2-input NAND gates have shorted gate or low-power topologies, where all transistors are shorted gates or independent gate-FinFETs. Combining shorted gate and independent gate

Table 2.2 Comparison of various 2-input NAND gates [15].

Topology	**SG**	**LP**	**MT**	**IG**	**IG2**	**XT**	**XT2**
Area	1	1.27	1.27	1	1	1.27	1
I_{off} (Avg.)	18.4	1	7.0	18.4	7.73	18.13	7.73
Tpd (Toggle A)	1	4.13	3.8	1.6	2.08	3.20	1.47
Tpd (Toggle B)	1	4.50	3.88	1.69	2.02	3.58	1.38
Tpd (Toggle AB)	1	3.48	3.09	1	1.55	2.38	1.55

FinFETs is more feasible with a 2-input NAND gate than with an inverter because it contains more transistors. This resulted in the following additional configurations: XT, IG2, MT, IG2, and XT2. The propagation delay, area and leakage current of different NAND gates are compared in Table 2.2. Figure 2.9 displays schematics of the SG, LP, and MT 2-input NAND

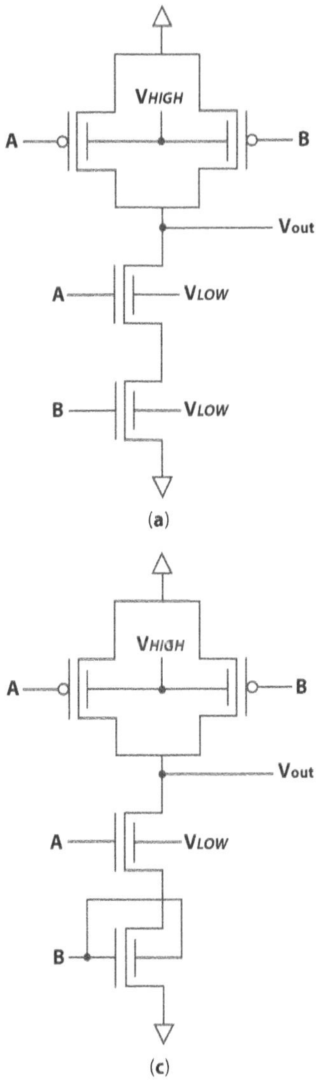

Figure 2.9 2-Input NAND gate in (a) LP, (b) SG (shorted gate), and (c) mixed terminal (MT).

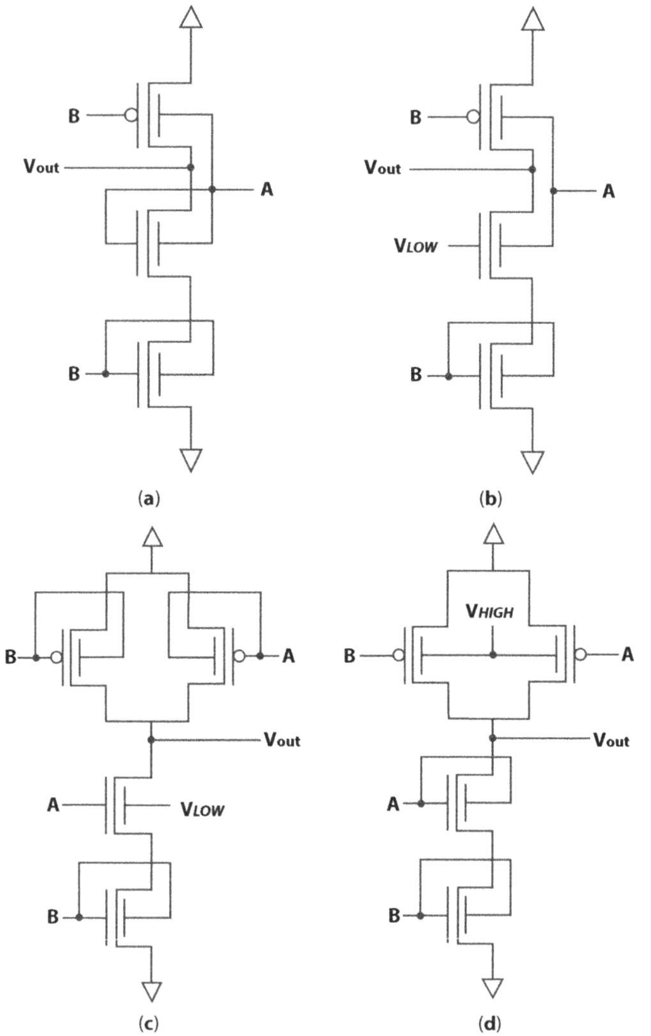

Figure 2.10 2-Input NAND gate in (a) IG, (b) IG2, (c) XT2, and (d) XT configurations.

gates. Figure 2.10 shows schematic designs for the IG, IG2, XT, and XT2 configurations of 2-input NAND gates.

2.2.3 Asymmetric Shorted Gate FinFET Based Logic Gates

Bhoj and Jha studied inverters and 2-input NAND gates and asymmetric SG-FinFETs [15]. Any SG-FinFET logic gate's schematics or layouts can be

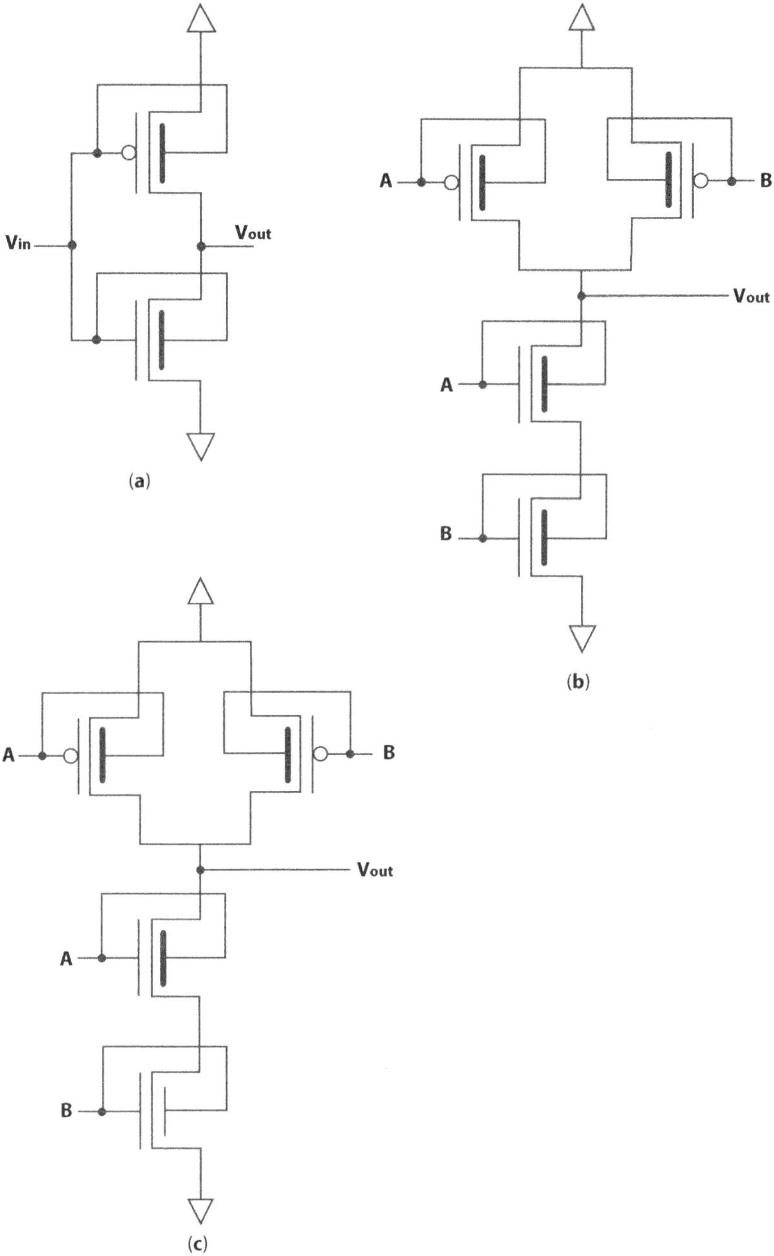

Figure 2.11 Asymmetric SG-FinFET logic gates: (a) inverter, (b) 2-input NAND gate, and (c) SG-NAND2 gate.

changed to an asymmetric SG-FinFET logic gate, without requiring more layout space, as shown in Figure 2.11. Therefore, the use of ASG-FinFETs affects propagation delay and leakage. Leakage-delay trade-offs are made possible by preserving a portion of the SG-FinFETs in the Shorted Gate NAND2 circuit, as shown in Figure 2.11(c). Pure shorted gates offer a reduced propagation delay and pure ASG gates offer less leakage.

2.3 Implementation of Latches and Flip-Flops Using FinFETs

Transmission gate (TG) and half swing (HS) latches and flip-flops using SG, IG, and Asymmetric SG FinFETs have also been researched [15, 22], as shown in Figures 2.12 and 2.13. By adding an independent gate FinFET inverter to the feedback path of all Shorted Gate Transmission Gate (TG) latch in Figure 2.12(a), an IG latch was presented in [22].

The Independent Gate-FinFETs in the inverter (I3) are made less wide than the drive inverter (I1) by appropriately reverse-biasing the back gates. Consequently, a safe write operation can be ensured without the requirement for the drive inverter to be large, as is typically the case. Compared to the traditional SG latch, the independent gate latch has the advantage of reduced leakage power and small-area requirements at nominal process corners with nearly no deterioration in setup time or propagation delay. Comparing Independent Gate flip-flops with Transmission Gate flip-flops, similar gains in power and area were achieved (Figure 2.13(a)).

Similar trade-offs were noted by Bhoj *et al.* after introducing Asymmetric SG-FinFETs in the TG and HS locks. All three latch inverters (I1 to I3) have ASG-FinFETs installed, which produce minimal leakage power and maximum propagation delay. Similarly, to the IG latch, ASG-FinFETs are introduced only in I3. Although the propagation latency increased by approximately 30%, this new arrangement reduced the leakage power by approximately 50%. Due to the potential to reduce the size of I1 and retain the required write stability, this arrangement also saves space. The outcomes for the ASG flip-flops were comparable. Inverters (I1 and I2) and nFinFETs (M1 to M4) that combine SG, IG, and ASG-FinFETs produce a variety of half-swing-latches (as shown in Figure 2.12(b)) and flip-flops (as shown in Figure 2.13(b)), similar to the TG latches and flip-flop. However, this reduction in leakage power comes at the cost of twice as much propagation delay as that predicted for the all-ASG arrangement. The fascinating configuration that is produced using Asymmetric SG-FinFETs in M2/M4

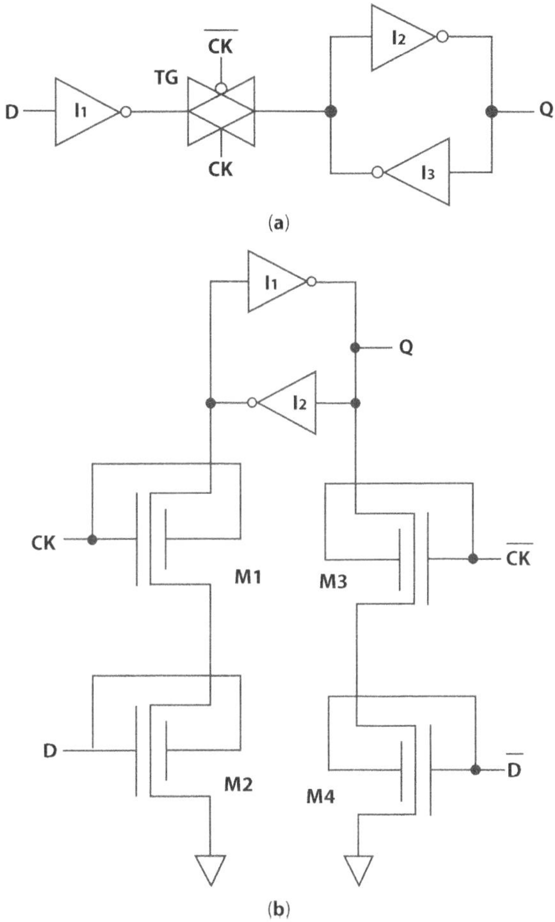

Figure 2.12 FinFET latches: (a) TG (transmission gate) and (b) HS (half-swing).

only results in a 20% greater reduction in leakage power, at a trivial expense of not more than 5% in latency. For HS flip-flops, similar outcomes were seen.

2.4 Implementation of SRAM Using FinFETs

Static RAM is an essential part of the modern microprocessor on-chip cache. SRAMs typically cover more than 50% of the die area in modern multicore processors [23]. Statistics are a major hindrance to SRAM performance because SRAMs are constructed using transistors with the

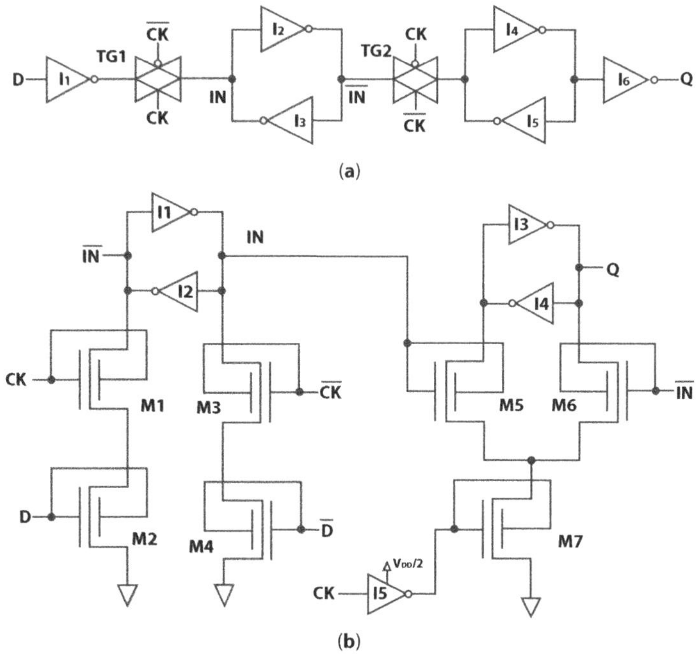

Figure 2.13 (a) TG flip-flop and (b) HS flip-flop schematic.

smallest dimensions available at a technology node (to maximize memory capacity). Mismatches in transistor strengths and threshold voltage resulting from RDF and other process variables plague the deeply scaled SRAMs that are constructed on top of planar MOSFETs. SRAMs also soak up the majority of the leakage power on the chip owing to the lengthy idle times of their massive memory arrays.

SRAMs also soak up the majority of the leakage power on the chip due to their massive memory arrays' lengthy idle times. In the last 10 years, many studies have been conducted on six-transistor (6T) FinFET SRAMs (see Figure 2.14) with the goal of reducing the leakage power and addressing the bitcell variability that has increased [16, 17, 24–30].

Utilizing a combination of shorted gate, independent gate, and asymmetric SG-FinFETs, relatively novel static-RAM bitcell topologies have been proposed [27, 28, 31, 32]. FinFET SRAMs are divided into the following groups in [27]: The three types of shorted-gate configurations are as follows: (a) vanilla, in which all FinFETs are of Shorted Gate type; (b) independent gate, here one or more Shorted Gate-FinFETs are replaced with Independent Gate-FinFETs; and (c) several work-function shorted

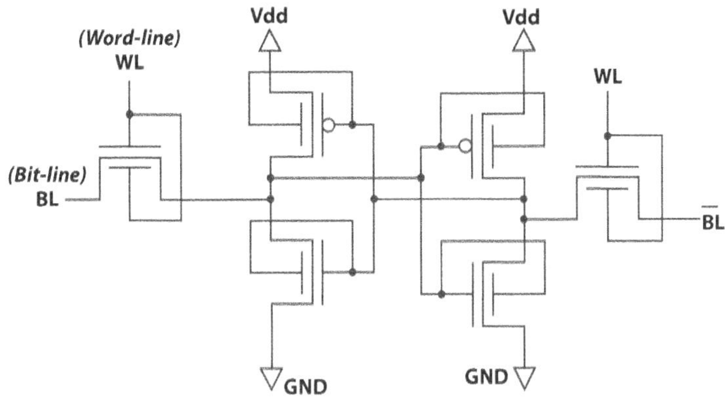

Figure 2.14 6-Transistor SRAM cell using FinFET.

gate, which involves replacing one or more Shorted Gate-FinFETs with Asymmetric SG-FinFETs.

2.5 Implementation of DRAM Using FinFETs

DRAM offers fewer area requirements for storing a single bit of information. Because of their large space advantage over SRAMs, single-transistor (1T) dynamic random-access memories have historically been utilized in both on-chip caches and off-chip main memories. A floating body cell (FBC), commonly referred to as a capacitorless 1T-DRAM, was proposed. Compared with traditional embedded DRAMs, this DRAM requires less space and a simpler manufacturing method [33–35]. The threshold voltage shift caused by the majority carrier concentration in the floating body is the basis of its operation. However, planar PDSOI 1T-DRAMs' retention properties deteriorate owing to the increased impurity concentration required to inhibit the SCEs. These scalability problems with 1T-DRAMs can be solved using double-gate FinFET DRAMs (DG FinDRAM) [36–38].

2.6 Challenges and Limitations of FinFET Technology

The development of Fin Field-Effect Transistor (FinFET) technology has significantly impacted the production of semiconductors, especially integrated circuits (ICs) and microprocessors. However, like any technology, it has drawbacks and restrictions on its own:

(a) Manufacturing Complexity: Compared to planar MOS transistors, the manufacturing of FinFETs requires precise control, thereby making its manufacturing a complex process.
(b) Production Cost: Increased manufacturing complexity can lead to higher production costs compared with planar transistor technology.
(c) Design Challenges: The development of circuits using FinFET technology necessitates a thorough understanding of the peculiar properties and behaviors of these semiconductors. To fully benefit from the FinFET technology, circuit designers must optimize their designs, which can take more time and money.
(d) Reliability: Reliability problems with FinFETs include hot-carrier injection, time-dependent dielectric breakdown (TDDB), and negative bias temperature instability (NBTI). These issues may affect the durability and robustness of FinFET-based devices.
(e) Scalability Concern: To continue growing beyond FinFETs, new materials and technologies may be needed when transistor dimensions approach atomic sizes.
(f) Power Dissipation: FinFETs still exhibit issues with power dissipation, particularly in high-performance applications. Controlling power dissipation is essential, particularly in data centers and mobile devices, where energy efficiency is vital.
(g) Thermal Management: FinFET devices' higher integration density and faster switching rates allow them to produce more heat than planar transistors. Reliable operation and prevention of thermally induced performance deterioration necessitate the use of effective thermal-management technologies.
(h) Compatibility With Other Technologies: FinFET technology may be difficult to integrate with other new technologies, such as 3D integration or novel materials, because of compatibility problems and variances in manufacturing processes.

FinFET technology is still the mainstay of semiconductor manufacturing despite these obstacles and restrictions, allowing improvements in performance, power efficiency, and integration density in a variety of electronic products. Pushing the limits of semiconductor technology and furthering the development of FinFET technology will require addressing these issues through ongoing research and innovation.

2.7 Potential Future Developments in FinFET Technology

The objective of prospective FinFET technological advancements is to surmount the existing constraints and augment semiconductor device performance, power efficiency, and integration density. Possible advancements in the future are as follows:

(a) Gate-All-Around (GAA) FinFETs: With a gate that entirely encircles the fin structure, GAA FinFETs are the next development in FinFET technology. Compared with conventional FinFETs, GAA FinFETs have superior electrostatic control and lower leakage currents, which may allow for further scaling and enhanced performance.
(b) Novel Materials: The performance and dependability of FinFETs can be improved by investigating new materials for fin and gate stacks. The transistor performance can be enhanced while the power consumption is decreased by including materials with increased carrier mobility or superior gate dielectrics.
(c) Vertical FinFETs: Vertical FinFETs can provide higher packing densities and scalability by positioning the fins vertically rather than horizontally. The goal of research on vertical FinFET structures is to enable higher levels of integration in future semiconductor devices by overcoming the drawbacks of horizontal FinFETs.
(d) Beyond CMOS: New FinFET-based devices with improved functionality and efficiency may result from investigating other device designs and computing paradigms outside the conventional Complementary Metal-Oxide-Semiconductor (CMOS) paradigm. Tunnel FETs (TFETs),

spintronics-based devices, and memristors are some examples.

(e) Integration with Nanotechnology: New avenues for device downsizing and performance improvement can be created by combining FinFET technology with nanoscale materials and structures. It is possible to investigate methods such as self-assembly and nanowire integration to further push the boundaries of the FinFET technology.

(f) Applications of Quantum Computing: The development of quantum computing devices may benefit from the application of FinFET technology. FinFET fabrication techniques can be used to create scalable quantum computing systems through research on semiconductor-based qubits and hybrid quantum-classical architectures.

(g) Heterogeneous Integration: Highly integrated and adaptable semiconductor devices for a range of applications can be created by combining FinFET technology with other cutting-edge technologies, including 3D integration, photonics, and the heterogeneous integration of diverse materials and components.

(h) Low-Power Design Techniques: Creating cutting-edge circuit topologies and low-power design strategies specifically suited for FinFET technology will help further increase battery life and energy efficiency in mobile devices, the Internet of Things, and other power-constrained applications.

(i) Manufacturing Process Innovations: The scalability and accuracy of FinFET production can be further improved by ongoing improvements in semiconductor manufacturing processes, including directed self-assembly (DSA), extreme ultraviolet lithography (EUV), and sophisticated patterning techniques.

(j) The long-term reliability and longevity of semiconductor devices can be increased by addressing the reliability issues related to FinFETs through enhanced device design, materials engineering, and reliable manufacturing processes.

To meet the growing demands of the semiconductor industry for faster, more efficient, and more integrated electronic devices, future development of FinFET technology is likely to involve a combination of material innovation, device architecture exploration, manufacturing process advancements, and system-level optimization.

2.8 Conclusion

In this chapter, we discuss the problem of MOS technology with reduced dimensions. We discuss the advantages of FinFET devices over conventional planar MOS transistors. FinFET characteristics and trade-offs among shorted gates, independent gates, and asymmetric SG FinFETs are explored in this chapter. We designed various digital circuits using FinFETs, such as inverters, 2-input NAND gates, latches, flip-flops, and SRAM memory cells. The tradeoffs between the leakage power and propagation delay can occur at any stage of the design hierarchy and have been observed. This chapter also summarizes the challenges and limitations of the FinFET technology. Finally, the chapter concludes with potential future developments in the FinFET technology.

References

1. Veeraraghavan, S. and Fossum, J.G., Short-channel effects in SOI MOSFETs. *IEEE Trans. Electron Devices*, 36, 3, 522–528, 1989.
2. Hisamoto, D., Kaga, T., Kawamoto, Y., Takeda, E., A fully depleted lean-channel transistor (DELTA)—a novel vertical ultra thin SOI MOSFET, in: *Proceedings of the International Electron Devices Meeting (IEDM '89)*, Washington, DC, USA, pp. 833–836, December 1989.
3. Nowak, E.J., Aller, I., Ludwig, T., *et al.*, Turning silicon on its edge [double gate CMOS/FinFET technology]. *IEEE Circuits Devices Mag.*, 20, 1, 20–31, 2004.
4. Alioto, M., Comparative evaluation of layout density in 3T, 4T, and MT FinFET standard cells. *IEEE Trans. Very Large Scale Integr. VLSI Syst.*, 19, 5, 751–762, 2011.
5. Collaert, N., Demand, M., Ferain, I., *et al.*, Tall triple-gate devices with TiN/HfO2 gate stack, in: *Proceedings of the Symposium on VLSI Technology*, pp. 108–109, June 2005.
6. Park, T.-S., Cho, H.J., Choe, J.D., *et al.*, Characteristics of the full CMOS SRAM cell using body-tied TG MOSFETs (Bulk FinFETS). *IEEE Trans. Electron Devices*, 53, 3, 481–487, 2006.
7. Kawasaki, H., Okano, K., Kaneko, A., *et al.*, Embedded bulk FinFET SRAM cell technology with planar FET peripheral circuit for hp32 nm node and beyond, in: *Proceedings of the Symposium on VLSI Technology (VLSIT '06)*, pp. 70–71, June 2006.
8. Kim, S.Y. and Lee, J.H., Hot carrier-induced degradation in bulk FinFETs. *IEEE Electron Device Lett.*, 26, 8, 566–568, 2005.

9. Auth, C., 22-nm fully-depleted tri-gate CMOS transistors, in: *Proceedings of the IEEE Custom Integrated Circuits Conference (CICC '12)*, San Jose, Calif, USA, pp. 1–6, September 2012.
10. Markoff, J., *Intel increases transistor speed by building upward*, The New York Times Company, New York, USA, May 2011, http://www.nytimes.com/2011/05/05/science/05chip.html.
11. Lin, C.-H., Chang, J., Guillorn, M., Bryant, A., Oldiges, P., Haensch, W., Non-planar device architecture for 15 nm node: FinFET or trigate?, in: *Proceedings of the IEEE International Silicon on Insulator Conference (SOI '10)*, pp. 1–2, October 2010.
12. Yang, J.-W. and Fossum, J.G., On the feasibility of nanoscale triple-gate CMOS transistors. *IEEE Trans. Electron Devices*, 52, 6, 1159–1164, 2005.
13. Kedzierski, J., Fried, D.M., Nowak, E.J., *et al.*, High-performance symmetric-gate and CMOS-compatible Vt asymmetric-gate FinFET devices, in: *Proceedings of the IEEE International Electron Devices Meeting (IEDM '01)*, pp. 437–440, December 2001.
14. Mathew, L., Sadd, M., White, B.E., *et al.*, FinFET with isolated n+ and p+ gate regions strapped with metal and polysilicon, in: *Proceedings of the IEEE International SOI Conference Proceedings*, pp. 109–110, October 2003.
15. Bhoj, A.N. and Jha, N.K., Design of logic gates and flip-flops in high-performance FinFET technology. *IEEE Trans. Very Large Scale Integr. VLSI Syst.*, 21, 11, 1975–1988, 2013.
16. Goel, A., Gupta, S.K., Roy, K., Asymmetric drain spacer extension (ADSE) FinFETs for low-power and robust SRAMs. *IEEE Trans. Electron Devices*, 58, 2, 296–308, 2011.
17. Moradi, F., Gupta, S.K., Panagopoulos, G., Wisland, D.T., Mahmoodi, H., Roy, K., Asymmetrically doped FinFETs for low-power robust SRAMs. *IEEE Trans. Electron Devices*, 58, 12, 4241–4249, 2011.
18. Masahara, M., Surdeanu, R., Witters, L., *et al.*, Demonstration of asymmetric gateoxide thickness four-terminal FinFETs having flexible threshold voltage and good subthreshold slope. *IEEE Electron Device Lett.*, 28, 3, 217–219, 2007.
19. Masahara, M., Surdeanu, R., Witters, L., *et al.*, Demonstration of asymmetric gate oxide thickness 4-terminal FinFETs, in: *Proceedings of the IEEE International Silicon on Insulator Conference (SOI '06)*, pp. 165–166, October 2006.
20. Sachid, A.B. and Hu, C., Denser and more stable SRAM using FinFETs with multiple fin heights. *IEEE Trans. Electron Devices*, 59, 8, 2037–2041, 2012.
21. Liu, Y., Matsukawa, T., Endo, K., *et al.*, Advanced FinFET CMOS technology: TiN-Gate, fin-height control and asymmetric gate insulator thickness 4T-FinFETs, in: *Proceedings of the International Electron Devices Meeting (IEDM '06)*, San Francisco, Calif, USA, pp. 1–4, December 2006.
22. Tawfik, S.A. and Kursun, V., Characterization of new static independent-gate-biased FinFET latches and flip-flops under process variations, in:

Proceedings of the 9th International Symposium on Quality Electronic Design (ISQED '08), San Jose, Calif, USA, pp. 311–316, March 2008.
23. Patt, Y.N., Patel, S.J., Evers, M., Friendly, D.H., Stark, J., One billion transistors, one uniprocessor, one chip. *Computer*, 30, 9, 51–57, 1997.
24. Bansal, A., Mukhopadhyay, S., Roy, K., Device-optimization technique for robust and low-power FinFET SRAM design in NanoScale era. *IEEE Trans. Electron Devices*, 54, 6, 1409–1419, 2007.
25. Bhoj, A.N. and Joshi, R.V., Transport-analysis-based 3-D TCAD capacitance extraction for sub-32-nm SRAM structures. *IEEE Electron Device Lett.*, 33, 2, 158–160, 2012.
26. Bhoj, A.N., Joshi, R.V., Jha, N.K., Efficient methodologies for 3-D TCAD modeling of emerging devices and circuits. *IEEE Trans. Comput.-Aided Des. Integr. Circuits Syst.*, 32, 1, 47–58, 2013.
27. Bhoj, A.N. and Jha, N.K., Parasitics-aware design of symmetric and asymmetric gate-workfunction FinFET SRAMs. *IEEE Trans. Very Large Scale Integr. Syst.*, 22, 3, 548–561, 2014.
28. Endo, K., O'Uchi, S.-I., Matsukawa, T., Liu, Y., Masahara, M., Independent double-gate FinFET SRAM technology, in: *Proceedings of the 4th IEEE International Nanoelectronics Conference (INEC '11)*, pp. 1–2, June 2011.
29. Joshi, R., Kim, K., Kanj, R., FinFET SRAM design, in: *Proceedings of the 23rd International Conference on VLSI Design (VLSID '10)*, Bangalore, India, pp. 440–445, January 2010.
30. Joshi, R.V., Kim, K., Williams, R.Q., Nowak, E.J., Chuang, C.-T., A high-performance, low leakage, and stable SRAM row-based back-gate biasing scheme in FinFET technology, in: *Proceedings of the 20th International Conference on VLSI Design held jointly with 6th International Conference on Embedded Systems (VLSID '07)*, pp. 665–670, January 2007.
31. Carlson, A., Guo, Z., Balasubramanian, S., Zlatanovici, R., Liu, T.J.K., Nikolic, B., SRAM read/write margin enhancements using FinFETs. *IEEE Trans. Very Large Scale Integr. VLSI Syst.*, 18, 6, 887–900, 2010.
32. Tawfik, S.A., Liu, Z., Kursun, V., Independent-gate and tiedgate FinFET SRAM circuits: design guidelines for reduced area and enhanced stability, in: *Proceedings of the 19th International Conference on Microelectronics (ICM '07)*, Cairo, Egypt, pp. 171–174, December 2007.
33. Yoshida, E. and Tanaka, T., A design of a capacitorless 1T-DRAM cell using gate-induced drain leakage (GIDL) current for lowpower and high-speed embedded memory, in: *Proceedings of the IEEE International Electron Devices Meeting*, Washington, DC, USA, pp. 3761–3764, December 2003.
34. Tran, L.C., Challenges of DRAM and flash scaling—potentials in advanced emerging memory devices, in: *Proceedings of the 7th International Conference on Solid-State and Integrated Circuits Technology Proceedings (ICSICT '04)*, vol. 1, pp. 668–672, October 2004.

35. Bhoj, A.N. and Jha, N.K., Gated-diode FinFET DRAMs: device and circuit design-considerations. *ACM J. Emerg. Technol. Comput. Syst.*, 6, 4, 12:1–12:32, 2010.
36. Tanaka, T., Yoshida, E., Miyashita, T., Scalability study on a capacitor-less 1T-DRAM: from single-gate PD-SOI to doublegate FinDRAM, in: *Proceedings of the IEEE International Electron Devices Meeting*, pp. 919–922, December 2004.
37. Bawedin, M., Cristoloveanu, S., Flandre, D., A capacitorless 1T-DRAM on SOI based on dynamic coupling and double-gate operation. *IEEE Electron Device Lett.*, 29, 7, 795–798, 2008.
38. Yoshida, E., Miyashita, T., Tanaka, T., A study of highly scalable DG-FinDRAM. *IEEE Electron Device Lett.*, 26, 9, 655–657, 2005.
39. Goel, N., Design device for subthreshold slope in DG fully depleted SOI MOSFET. *Int. J. Nano Electron. Phys. Ukraine*, 9, 1, 01022(4pp), 2017.
40. Goel, N., Performance of Double Gate SOI MOSFET. *Int. J. Electron. Eng.*, 4, 1, 57–59, 2012.

3

Prospects of Negative-Capacitance Ferroelectric Field-Effect Transistors in Low-Power Electronics and Beyond

Ningombam Ajit Kumar[1]*, Khuraijam Nelson Singh[2], Sisira Hawaibam[3], Sushmita Dandeliya[2] and Sonal Agrawal[4]

[1]Department of Electronics and Communication Engineering, G H Patel College of Engineering & Technology (GCET), The Charutar Vidya Mandal (CVM) University, Gujarat, India
[2]Department of Electronics and Communication Engineering, Indian Institute of Information Technology, Nagpur, Maharashtra, India
[3]Department of Electronics and Communication Engineering, National Institute of Technology, Imphal, Manipur, India
[4]Advanced Materials Research Group, Department of Engineering Sciences, ABV-Indian Institute of Information Technology and Management, Gwalior, M.P., India

Abstract
The demand for compact, efficient, and low-power electronics has grown exponentially over the last few decades. However, conventional field-effect transistors (FETs) face subthreshold swing (SS) limitations. Thus, modern integrated circuits (ICs) with billions of transistors exhibit high overall power consumption. The negative-capacitance ferroelectric FET (NC FeFET) is among the many transistor concepts devised to solve this issue. The NC-FeFET structure is realized by incorporating a ferroelectric material into the FET gate oxide. This modification amplifies the internal voltage near the device channel region, resulting in a steeper SS with a high I_{ON}/I_{OFF} ratio, thus reducing power consumption. In this chapter, the fundamentals of NC-FeFETs are discussed in detail. The first section introduces the NC-FeFET and discusses its evolution and operation. The chapter then discusses NC-FeFET modeling approaches, applications, performance optimization, and challenges. A comparative analysis of the NC-FeFET with other transistors is also presented. Finally, the chapter concludes with future prospects and research trends highlighting the possibilities of NC-FeFETs as future devices.

**Corresponding author*: ningombama@gmail.com

Ekta Goel, Archana Pandey, Shiromani Balmukund Rahi and Arun Samuel (eds.) Handbook of Advanced Semiconductor Field Effect Transistors, (43–72)

Keywords: Ferroelectric, low power, MOSFET, negative capacitance, sub-threshold swing

3.1 Introduction

Conventional MOSFETs are considered the backbone of modern electronics. However, the conventional MOSFET's sub-threshold swing (SS) limitation, also known as the Boltzmann limit, is 60 mV/dec at 300 K. This has limited voltage downscaling in MOSFETs. This obstruction in downscaling the gate voltage results in increased power consumption in modern electronics, consisting of transistor counts in billions. To mitigate this limitation, the concept of introducing a thin layer of ferroelectric material in the gate stack of a field-effect transistor (FET) to produce a negative capacitance (NC) has been introduced [1]. This amplifies the internal voltage while maintaining the integrity of the transport phenomena. The ferroelectric material amplified the internal voltage near the device channel. This allows the achievement of a sub-60 mV/dec SS, resulting in low power consumption [2, 3]. Due to its steep switching characteristic and high I_{ON}/I_{OFF} ratio, the NC-FeFET is a potential device that can replace conventional MOSFET. The structures of the NC-FET and NC-FeFET are the same. The only difference is in their V–I characteristics. The NC-FeFET exhibits hysteresis V–I characteristics, whereas the NCFET does not, as shown in Figure 3.1 [4]. Because of this hysteresis, NC-FeFETs have applications in memory devices. According to the quasi-static NC model, the NCFET overall gate capacitance is positive, proving that the ferroelectric insulator stack for a single state provides a stable NC state. Meanwhile, for the NC-FeFET, the overall gate capacitance is negative. Thus, the I–V characteristics lie between the two states, resulting in hysteresis.

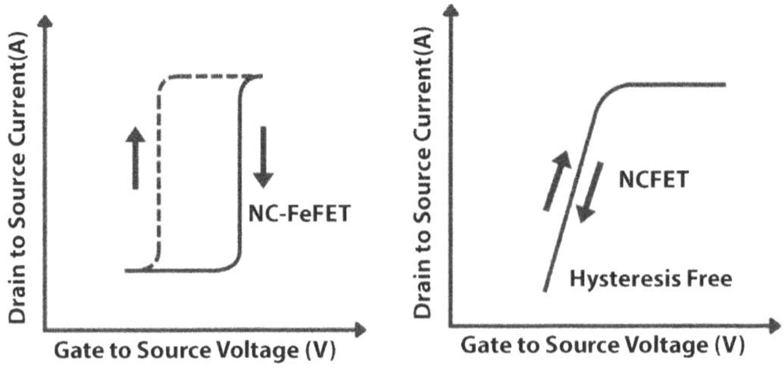

Figure 3.1 Characteristic of NC-FeFET and NCFET.

In this chapter, we provided a comprehensive overview of the NC-FeFET technology, including its evolution in research and significant milestones. The fundamental concept of NC, its working principle, and materials used to achieve NC are discussed. The chapter also highlights the modeling approach and investigates the application of the NC-FeFET, focusing on its potential applicability in low-power electronics and memory devices. Approaches for enhancing device performance and challenges such as stability, material, compatibility, and fabrication process are discussed. This chapter also presents a basic comparative analysis of NC-FeFETs with various transistor technologies, such as conventional MOSFET and tunneling field-effect transistors (TFETs), to provide a deeper insight into its advantages and limitations. Finally, the chapter presents the future prospects and trends foreseeing the possible future of NC-FeFETs in research and technological changes.

3.1.1 Overview of Negative Capacitance Ferroelectric FET Technology

With the ever-increasing demand for a more convenient lifestyle, the demand for compact and powerful devices has surged. Supporting this idea, in 1966, Moore predicted that the number of transistors in an IC would increase by approximately double every two years. Researchers and the semiconductor industry have maintained this scaling trend ever since. Nevertheless, in the 2020s, some claimed that the law was dead, while others claimed the opposite. In transistor scaling, as the channel length decreases, the power per switching event decreases and the transistor density in the IC increases. However, as the channel length decreases to a length comparable to the depletion width of the source and drain junctions, undesirable short-channel effects (SCEs) occur. Some of these effects include punch-through, drain-induced barrier lowering, surface scattering, the hot electron effect, velocity saturation, and impact ionization [5–8].

Over the last five decades, technology has made many achievements driven by the continuous scaling of complementary metal oxide semiconductor (CMOS) technology by providing dense and fast integration of transistors. The need for better performance and integration in almost all devices has accelerated scaling. As a result, today's transistors, which are 20 times faster, account for less than 1% of the total size compared to transistors five decades ago [9]. It increases the chip's power density and overall power consumption. It also allows the modification of various very large-scale integration (VLSI) design parameters to achieve an IC with high performance. Consequently, CMOS has become the leading technology in ICs owing to its scalability, compact size, and high performance.

According to Dennard's scaling law, the thickness of the gate oxide decreases as the channel length decreases. This hinders further scaling at a certain oxide thickness. As scaling increases, the capacitance and overall power consumption decreased [5]. However, scaling the supply voltage comes with the issue of scaling the threshold voltage (V_{TH}) to maintain the same overdrive voltage. If both parameters are reduced, the driving capability is sustained, but the leakage current increases owing to the reduction in V_{TH}. Therefore, excessive voltage scaling is difficult to achieve. Moreover, the SS worsens with scaling. Therefore, a transistor with a larger SS requires a larger gate voltage to turn ON the transistor. Consequently, the inability or limit to reduce the supply voltage hinders the reduction in the overall power consumption [10, 11].

A large SS restricts the rate at which the drain current increases with reference to a gate voltage of 60 mV/dec. The SS of the transistor is generally defined as,

$$SS = \frac{\partial V_G}{\partial \log_{10} I} = \left(\frac{\partial V_G}{\partial \Psi_S}\right)\left(\frac{\partial \Psi_S}{\partial \log_{10} I}\right) \tag{3.1}$$

where, Ψ_S, V_G, and I defines the channel surface potential, gate voltage, and drain current. In Eq. (3.1), $\frac{\partial V_G}{\partial \Psi_S}$ defines the control of gate voltage over the surface potential. While the second term $\frac{\partial \Psi_S}{\partial \log_{10} I}$ reflects current variation through the bending of the surface energy band. In the sub-threshold region of MOSFET, diffusion currents dominate the channel conduction resulting in [12],

$$\frac{\partial \Psi_S}{\partial \log_{10} I} = 60mV / dec. \tag{3.2}$$

Moreover, for traditional MOSFET $\frac{\partial V_G}{\partial \Psi_S}$ is greater than or equal to 1, resulting in a minimum theoretical SS of 60 mV/dec.

Two approaches have been proposed to achieve sub-60 mV/dec SS. The first approach lowers the SS by reducing the second term $\frac{\partial \Psi_S}{\partial \log_{10} I}$ in Eq. (3.1). They consist of devices such as TFETs [13–15], punch-through impact ionization MOSs (PIMOSs) [16–18], feedback FET (FB-FET) [19, 20], and suspended-gate MOSFET (SG-MOSFETs) [21, 22]. TFETs reduce the SS by their electron tunneling effect in the source-channel and drain-channel junctions [14]. PIMOSs amplify system nonlinearity to control the breakdown voltage in the p-i-n structure. The FB-FET achieves steep switching using a positive-feedback mechanism. SG-MOSFET achieve low SS by adding a suspended metal gate [21]. However, the practical applications of these devices have been limited owing to their complicated fabrication processes and reliability issues. The second approach decreases the first term $\frac{\partial V_G}{\partial \Psi_S}$ of Eq. (3.1) to a value of less than 1. The author of [23] proposes incorporating ferroelectric layer incorporated in the gate stack of the NC-FeFET to achieve a steeper SS. This results in low power dissipation, high ON current (I_{ON}), and low OFF current (I_{OFF}) [2, 24]. NC-FeFETs provide a simple symmetrical structure, better compatibility with the present fabrication process, and nonvolatile computing [25, 26]. In addition, it can achieve better electrical performance by combining it with other FETs, such as TFETs and FinFETs. Thus, the NC-FeFET is a potential energy-efficient semiconductor device.

3.1.2 Research Evolution of Negative Capacitance FET and its Significant Milestones

The NC-FeFET is a future device that can achieve sub-60 mv/dec at 300 K. In 2008, Salahuddin and Datta first reported that ferroelectric materials could be used to lower the SS below the Boltzmann limit in FET [23]. Later, in 2011, Khan described NC effects in a nanoscale ferroelectric-dielectric heterostructure [27]. They observed that ferroelectric $Pb(Zr_{0.2}Ti_{0.8})O_3$, when bilayered with $SrTiO_3$, exhibited a composite capacitance greater than the constituent $SrTiO_3$ capacitance. This indicates the effectiveness of $Pb(Zr_{0.2}Ti_{0.8})O_3$ as NC. They used Landua's mean field theory in their study and observed that temperature effectively tunes the NC parameter and enhances the dielectric heterostructure capacitance at high temperatures.

In 2012, a non-hysteretic NCFET was proposed, which provides a sub-30 mV/dec SS by implementing body profile engineering [28]. This approach enhances the performance of the SS by employing a thin channel layer over the bottom layer, which is heavily doped. This amplifies the effect of NC voltage. It achieves an I_{OFF} of 10 pA/μm and an I_{ON} of 0.3 mA/μm at a supply voltage (V_{DD}) of 0.3 V. These results were obtained with a 100 nm gate length, excluding any increment in mobility due to strain. Nevertheless, incorporating a thin semiconductor epitaxial layer into the conductor improves the device's performance.

Two-dimensional (2D) transition metal dichalcogenides, such as MoS_2, are a suitable option for a channel material owing to their extremely thin thickness and improved stability of channel capacitance [29]. Combined with the NC effect, this resulted in a significant improvement in the SS over a broad range of currents. The authors in [29] fabricated 2D NCFETs using MoS_2 and hafnium zirconium oxide ($HfZrO_2$ or HZO) as the channel and the ferroelectric materials, respectively. These materials are suitable for use in the CMOS technology. This allows consistent and long-lasting switching at a rate of less than 60 mV/dec. This makes them a viable option for transistors that can be scaled down in size and operate at low voltage. Prior to this, a 2D channel NCFET with a polymeric ferroelectric was demonstrated [30]. However, this approach led to the development of an unstable and incompatible CMOS, although it enhanced low-voltage switching.

In 2013, the author in [31] proposed a transistor using a combination of an ultrathin body of Si to suppress the SCEs and NC to lower the SS below 60 mV/dec. This NCFET, while excluding the mobility enhancement by strain, achieved an I_{ON} of 250 μA/μm, I_{OFF} of 250 pA/μm, and 21 mV/dec SS with 0.3V V_{DD}. In [32], an NCFET with a gate oxide of 1.8 nm thickness Zr-doped hafnium dioxide (HfO_2) was fabricated in an FDSOI wafer. This demonstrated a hysteresis-free operation. Compared with the conventional Si-HfO_2 stacked gate oxide, it also exhibited a steeper SS with a tenfold reduction in I_{OFF} and constant I_{ON} for a channel length of 30 nm. Furthermore, NCFET offers a higher I_{ON} at a constant V_{DD} at the matched I_{OFF}.

Several experiments have been performed by modifying the NC-FeFET channel with group III–V compounds. In [33], the author presented a simple model of an NC-MOSFET with as monolayer MoS_2 (molybdenum disulfide). The model accurately describes the negative output differential conductance (NDC) effects and reverse DIBL for a sub-10 nm NC-MOSFET. A low supply voltage also shows improvement in SS owing to the electrostatic SCEs, resulting in a linear increment of the reverse DIBL and NDC. Similarly, other low-dimensional materials, such as tungsten diselenide (WSe_2) and carbon nanotubes, can also be used as channels in NCFETs to improve their performance [34, 35]. An ultra-thin body (UTB) NCFET with Si and Ge as the channel is investigated, and the SS was compared [36]. To avoid hysteresis, the concept of design space is utilized. This shows that designing an NCFET with a channel made from a high-permittivity material is more suitable because it increases the channel capacitance and drain coupling. Thus, germanium on insulator-based NCFETs lowers the SS compared to silicon-based NCFETs because of its higher channel capacitance and drain coupling.

In 2016, Khandelwal *et al.* presented a compact model for NC-FinFETs for ultralow-power applications [37]. The study showed that NC in the FinFET structure improves the inverter delay by 11 times compared to a 14 nm ITRS FinFET. In the same year, a SPICE model for NC-FeFETs using the Landau–Khalatnikov (L–K) equation was presented [38]. However, this model does not consider the effects of any non-idealities. Subsequently, an improved compact model for NC-FinFETs using the LK equation was introduced, incorporating trap charges [39].

In 2016, Khan analyzed the impact of leaky ferroelectric materials on the stabilization of NC [40]. The leakage was defined by a ferroelectric capacitor in conjunction with a parallel resistor. Thus, it provides an extra route for the dissipation of charges, thereby compromising the overall performance. The solution to the aforementioned issue can be achieved by meticulous selection of the metal work functions [41]. Additionally, the authors asserted that properly engineered leaky NCFET can initiate NC at a lower gate voltage. At the same time, work function engineering allows it to have a higher subthreshold slope and drain current.

In 2019, an undoped HfO_2-based NCFET was examined by varying its ferroelectric thickness and studying its effect on device performance [42]. A resistive load inverter was designed to demonstrate that when the ferroelectric material thickness increased within a controlled range, the noise margin improved, and power dissipation was reduced. In [43], the author constructed and analyzed a doping-less ring NCFET utilizing charge plasma. This design offers an SS that is much lower than that of conventional TFETs and a high I_{ON}. Its excellent analog performance with linearity makes it well suited for radio-frequency (RF) and space applications.

3.2 The Fundamentals of Negative Capacitance Ferroelectric FET

3.2.1 Concept of Negative Capacitance

NC in the field of semiconductors is a fascinating phenomenon currently being widely studied owing to its potential to transform the energy consumption and performance of various devices. It was first observed in thin-film amorphous semiconductor chalcogenides in the 1960s [44]. Later, it was reported in numerous semiconductor devices, such as pn-junction [45], solar cells [46], Schottky diodes [47], and quantum devices [48]. However, it took several decades for researchers to understand that almost all NC occurrences are strongly associated with the transient process,

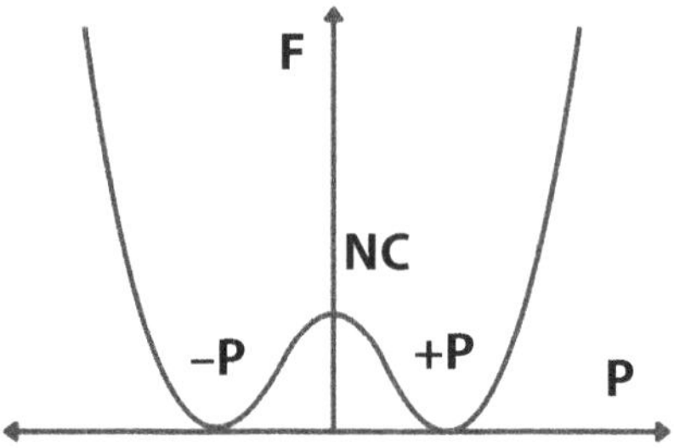

Figure 3.2 Free energy Vs polarization.

providing a theoretical explanation and fundamental physical basis for the experimental evidence connected to NC.

The NC effect in ferroelectric materials is generally explained using Landau's theory of phase transitions [49]. Landau's analysis of the ferroelectric characteristics around the phase transition is shown in Figure 3.2 [50]. This suggests that there is an NC region that may be sustained by the interaction between a typical dielectric and ferroelectric material between the two stable states, +P and −P. To control the NC effect, it is necessary to precisely match the ferroelectric and dielectric layers in transistors.

The forces operating on the material must be considered when discussing switching dynamics. When there is no electric field, every ferroelectric unit cell is an electronic dipole surrounded by adjacent dipoles [51]. The lowest energy state in a single crystal is the condition in which every dipole orientation direction is the same. When a field is applied, the dipoles start automatically aligning. The net polarization is determined by the total +P and −P domains of the material. Single-dipole switching is one of the fastest electrical switching processes currently in use. However, in a multicrystalline structure, not all the dipoles with randomly oriented films will switch at the same time. It takes a limited amount of time to change the net polarization from +P to −P, or vice versa.

Jonscher *et al.* proposed a method to analyze NC by time-domain behavior under a step function bias, where the current first decreases and continuously increases until it eventually decays to zero [53]. The interaction between transitory NC and the negative curvature of the free-energy profile during polarization switching has been demonstrated in recent research [54, 55]. Furthermore, NC exhibit a time-dependent switching nature [56].

Small-signal capacitance can be determined by computing the inverse of the second derivative of the potential energy (U) with charge (Q). Hence, NC is often utilized as a shorthand for negative differential capacitance. The ferroelectric material capacitance is typically denoted as:

$$C_F = \left(\frac{\partial^2 U}{\partial Q_F^2} \right)^{-1} \tag{3.3}$$

where, U and Q_F are the potential energy and ferroelectric polarization charge.

3.2.2 Working Principle, Operation, and its Advantages

Capacitance can be defined as the inverse of the double derivative of the potential with respect to the charge, as shown in Eq. (3.3). Figure 3.3 shows the mechanism of NC formation. When the gate voltage is 0 V, the ferroelectric polarization points upward, whereas it is inverted when the gate voltage is greater than the coercive field (E_c), as depicted in Figures 3.3a,b. The energy landscape of the ferroelectric exhibits an inverted parabolic curve around charge $Q = 0$, contrary to that of an ordinary capacitor. This curvature, as seen in Figure 3.3c, results in NC. This indicates that only a slight modification in the gate voltage results in a substantial variation in the channel surface potential. However, achieving NC with a ferroelectric material mostly depends on its capability to transition from a local minimum state to a nonequilibrium state.

Figures 3.3c, d depict the changes in the energy distribution before and after the polarization inversion. In the non-equilibrium state, the capacitance has a negative value without offering stability to the system. The device stability can be attained by changing the CMOS gate stack. The options for connecting the ferroelectric and dielectric layers are either by connecting them directly or by using a metal connection between them. The overall capacitance increased as the NC effect developed, and this NC effect significantly enhanced the magnitude of the internal voltage, V_{in}. In addition, the incorporation of capacitance in series improves the overall stability and surface potential (ψ_s) of the channel, allowing a larger voltage modification of the applied gate voltage.

Recent studies have documented the rapid switching properties of NCFETs achieved with a carbon nanotube channel with silicon (Si) and zirconium (Zr) doped ferroelectric layers [57]. The operating point in a ferroelectric capacitor can be identified by setting the polarization charge density equal to the channel charge density. The intersection point between the load line of the channel charge and the polarization–electric field (P–E) curve defines the static operating point. If the curve exhibits only one intersection point in the NC area, the ferroelectric MOS transistor works as an NCFET.

Hysteresis in NCFETs results from leakage current via the dielectric layer in the device [11]. This triggers a ferroelectric depolarization field, which results in the formation of charge traps inside the ferroelectric film. Nevertheless, the hysteresis experienced by a ferroelectric capacitor can be prevented by altering its ferroelectric thickness. In the work done by Kim *et al.* [11], the properties of the reference materials were optimized for static operating points. The thickness of these materials differs depending on the $\psi_s - V_G$ relation of the MOS capacitor in the NCFET under the influence of a positive ferroelectric voltage [3]. Within the 3–5 nm range, ψ_s increased without hysteresis. The amplification achieved is remarkable considering the ferroelectric material's extremely small thickness, which is only a few

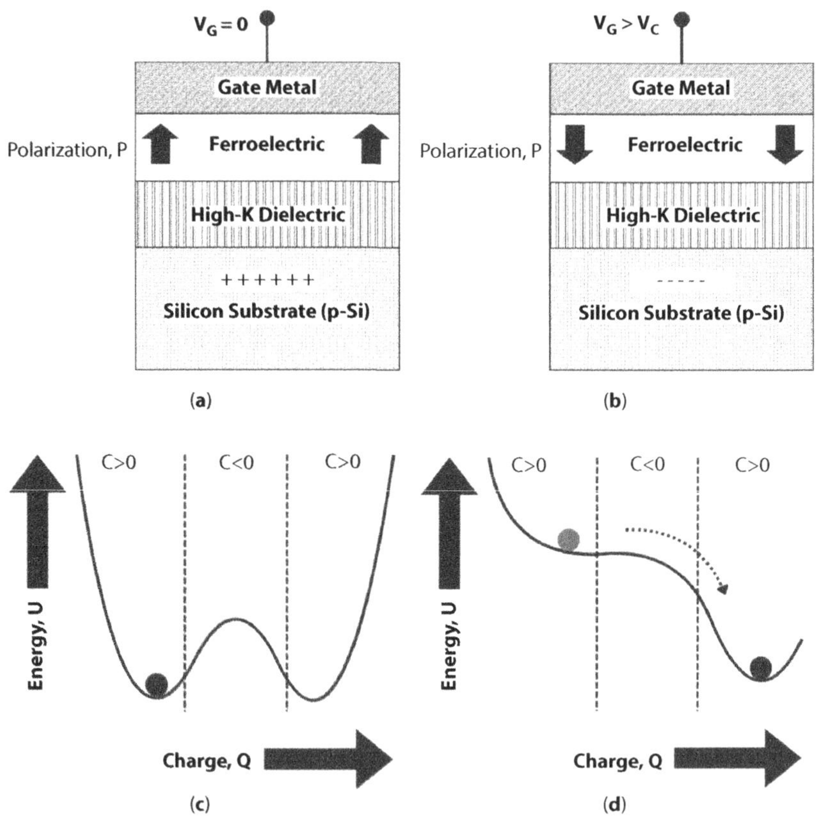

Figure 3.3 NC-FeFET working operation (a) zero biased ferroelectric polarization, (b) ferroelectric polarization for $V_G > V_C$, (c) energy landspace of ferroelectric capacitor before polarization inversion, and (d) energy landspace of ferroelectric capacitor after polarization inversion.

nanometers. Furthermore, a small voltage (0.2 V) is sufficient to magnify ψ_s, enabling operation with a minimal supply voltage. It is important to mention that when the ferroelectric thickness is very large, the static operation point cannot be properly adjusted. This results in a hysteresis effect on the ψ_s–V_G curve [58]. Despite the existence of hysteresis, NCFET shows considerable potential for operating inside the SS limit. Regarding NC,

$$\frac{\partial V_G}{\partial(\Psi_S)} = \left(1 + \frac{C_S}{C_{OX}}\right) \tag{3.4}$$

In Eq. (3.4), when the C_{OX} is negative, $\frac{\partial V_G}{\partial(\Psi_S)}$ is less than 1. For an NC-FeFET, the C_{OX} is negative, which theoretically makes the SS less than 60 mV/dec. For such devices, the fabrication process can be performed according to the existing fabrication process by including an oxide layer with an NC effect. This provides a significant advantage to the NC-FeFET.

The working principle of NC is its ability to counterbalance the positive capacitance of a device. This results in a modified gate capacitance (C_G). So, gate capacitance can be defined by,

$$C_G = \left(C_F^{-1} + C_{FET}^{-1}\right)^{-1} \tag{3.5}$$

Eq. (3.5) results in an SS of less than 60 mV/dec. Figure 3.4 shows the capacitor-equivalent model of the NC-FeFET. This shows that the ferroelectric capacitor (C_F) and channel capacitor are connected in series, representing a ferroelectric FET capacitor.

As shown in Figure 3.4a, the C_{FET} capacitor functions as a positive capacitance. When the total gate capacitance becomes positive, it can stabilize the NC condition in the ferroelectric. This condition specifies that the magnitude of $|C_F|$ must exceed that of C_{FET} to achieve a stable NC condition of the ferroelectric in the NC-FeFET. Furthermore, the applied voltages can have an impact on $|C_F|$ and C_{FET}. When the above conditions are not met for any combination of V_G and V_D, NC-FeFETs exhibit instability, resulting in hysteresis in their electrical properties. To prevent such instability, it is necessary to restrict the ferroelectric thickness because the coefficient of $|C_F|$ is directly related to the ferroelectric material thickness (T_F). Moreover, the value of $|C_F|$ is also influenced by E_c and the remnant polarization (P_r) of the ferroelectric material. When a ferroelectric material exhibits high P_r

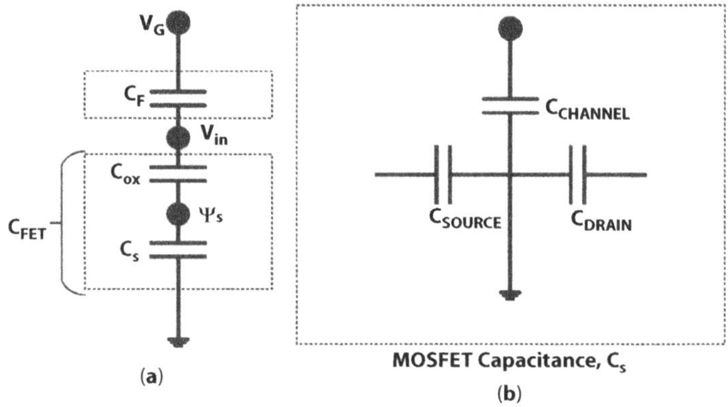

Figure 3.4 (a) Capacitance circuit model of NC-FeFET and (b) equivalent capacitance of substrate.

and low E_c, $|C_F|$ increases. This indicates that the absence of hysteresis is consistent over the entire range of operating voltages. Furthermore, a low T_F, high P_r, and low E_c result in decreased capacitance matching between C_F and C_{FET}. The interface of the ferroelectric oxide has an internal voltage that is higher than the gate voltage, resulting in a decrease in SS below 60 mV/dec. Consequently, compared with conventional transistors, the device allows for a higher current with a smaller supply voltage.

3.2.3 Various Structures and Materials Employed for Achieving Negative Capacitance

The three main structures of NCFET are metal-ferroelectric-semiconductor (MFS), metal-ferroelectric-insulator-semiconductor (MFIS), and metal-ferroelectric-metal-insulator-semiconductor (MFMIS). Figure 3.5 show their structural diagrams and equivalent capacitances. The NC-FeFET structures shown in Figure 3.5 are identical to those of the MOSFET structures. The only difference is the presence of an extra-thin layer of ferroelectric material, or ferroelectric material replacing the oxide layer. The MFS is a basic NC-FeFET formed by replacing the oxide layer of an FET with a ferroelectric material to achieve NC. MFIS incorporates an insulator layer to enhance the performance of ferroelectric materials, such as barium titanate and lead zirconate titanate. Despite their crystal imperfections on Ge and Si substrates, they provide the advantage of direct integration into the gate stack. The MFMIS structure consists of an insulator layer and a

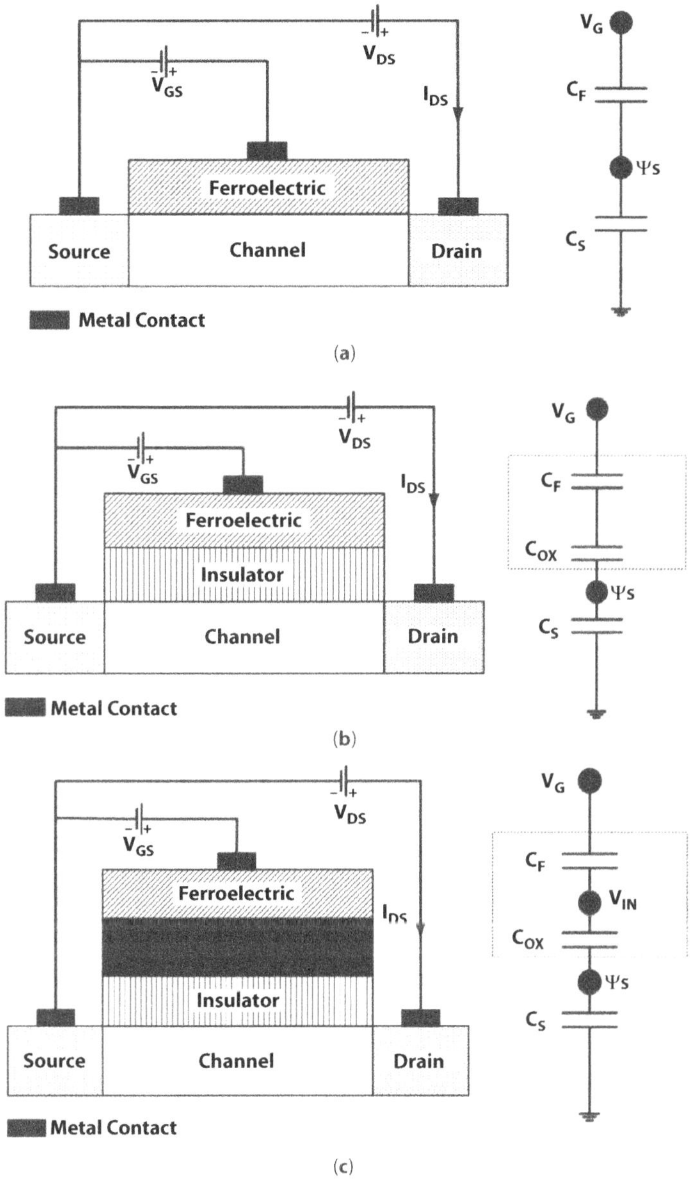

Figure 3.5 (a) Structure of MFS, and its capacitance circuit model, (b) Structure of MFIS, and its capacitance circuit model, and (c) structure of MFMIS, and its capacitance circuit model.

ferroelectric layer, with an interior metal layer between them. This uniform arrangement ensures that the ferroelectric polarization remains constant along the channel. This uniformity is maintained because the top electrode and inner metal layer had a consistent potential between them.

NC-FeFETs generally use three main types of materials: organic ferroelectric materials, inorganic hafnium dioxide ferroelectric materials, and inorganic perovskite ferroelectric materials. The polyvinylidene fluoride-trifluoroethylene copolymer (P(VDF-TrFE)) is a widely recognized organic ferroelectric material with excellent ferroelectric properties. It includes a high remnant Pr and a comparatively low Ec. Organic ferroelectric materials have features such as flexibility, low processing temperatures, and compatibility with flexible substrates, making them appropriate for specific applications such as flexible electronics. It also presents challenges related to the stability, interface compatibility, and long-term reliability of electronic devices.

Hafnium dioxide (HfO_2) is an inorganic substance that is currently being thoroughly investigated due to its ferroelectric characteristics. Doping HfO_2 with dopants such as yttrium (Y), silicon (Si), zirconium (Zr), or aluminum (Al) can enhance the ferroelectric properties of HfO_2. The incorporation of HfO_2-based ferroelectric materials based on doped HfO_2 in NC-FeFET devices is particularly fascinating because of their stable integration into the existing semiconductor fabrication process. Due to its strong dielectric characteristics, HfO_2 is advantageous for achieving a substantial charge storage capacity. In addition, inorganic ferroelectric materials have superior scalability and long-term durability compared with their organic counterparts. This makes them suitable for use in high-density memory systems. By appropriately controlling the material composition, doping levels, and deposition processes, the ferroelectric characteristics of HfO_2-based NC-FeFETs can be enhanced, leading to superior performance.

Lead zirconate titanate (PZT) and barium titanate (BTO) are inorganic perovskite-type ferroelectric materials. PZT is well known for its high Curie temperature and robust ferroelectric characteristics. Perovskite ferroelectrics exhibit multifunctional characteristics, including piezoelectricity, pyroelectricity, and ferroelectricity, making them viable devices for applications beyond NC-FeFETs. However, the production of perovskite ferroelectric materials involves a complex fabrication process. Accurate optimization during integration with semiconductor devices is also required to guarantee compatibility and dependability.

Numerous factors, including device performance specifications, processing compatibility, scalability, and environmental considerations, determine the selection of ferroelectric materials for integration into NC-FeFETs. Current research, trends and development efforts are focused on improving these factors and integrating them into future electronic applications.

3.3 Modeling

The modeling comprises both analytical and simulation components. This allows researchers to comprehend the device's behavior and aid in optimizing it for various uses in future electronics.

3.3.1 Modeling Approach

Two fundamental techniques are used in the modeling of NC-FeFETs: the L–K model and Miller's model [59]. Both models are frequently used to describe the polarization and voltage phenomenon across ferroelectric materials. The L–K equation, which is a dynamic derivation of the Landau–Devonshire theory, has drawn more interest than the Miller model. The L-K model is a theoretical framework that clearly explains the dynamics of ferroelectric materials, mainly the polarization switching phenomenon. The L–K model is implemented in NC-FeFET modeling to precisely depict the behavior of the ferroelectric layer in reaction to external electric fields and its influence on device functioning. This explains the temporal fluctuations in the polarization of the ferroelectric material induced by an external electric field. It considers phenomena such as the movement of domain walls and the reversal of polarization.

Researchers have implemented the L–K model in conjunction with semiconductor device physics models to effectively predict the interaction between the ferroelectric layer and channel in NC-FeFETs. This implementation enabled the researcher to analyze the device's behavior in different biasing scenarios. The Miller model is an approximation model that is widely implemented in the field of electronics, particularly in the amplifier design. This clearly explains the influence of parasitic capacitance on device performance. In the NC-FeFET, the Miller model is used to account for the impact of parasitic capacitances on its operation, including the effects of the ferroelectric layer's capacitance. In this model, the interaction between the input and output terminals of an NC-FeFET device increases the Miller effect, which is the effect of the ferroelectric layer's NC on the transistor's effective capacitance. This in turn affects the voltage transfer characteristics and overall performance of the device. L–K equation describe the ferroelectric material as below,

$$\tau\frac{\partial P}{\partial t}+\frac{\partial G}{\partial P}=0 \tag{3.6}$$

where τ, P, and G define the response time of polarization, polarization charge per unit, and Gibbs free energy, respectively, which is derived by using the ferroelectric anisotropy structure as Equation 3.7,

$$G(P,E_{ex})=\alpha P^2+\beta P^4+\gamma P^6-E_{ex}P \tag{3.7}$$

where, α, β, γ, and E_{ex} denote the Landau parameters, and external applied electric field. From Eqs. 3.6 and 3.7 gives,

$$E_{ex}=2\alpha P^2+4\beta P^4+6\gamma P^6+\tau\frac{\partial P}{\partial t} \tag{3.8}$$

For thermal equilibrium system, $\frac{\partial P}{\partial t}=0$; so, Eq. (3.8) becomes,

$$E_{ex}=2\alpha P^2+4\beta P^4+6\gamma P^6 \tag{3.9}$$

Based on the precise experimental values of landau parameters for various ferroelectric materials, the P–E curve is developed. Surprisingly, $\frac{\partial P}{\partial E}<0$ term serves as crucial enabling voltage amplification. As V_G increases, the voltage drop at ferroelectric falls due to $\frac{\partial P}{\partial E}<0$, and these amplifies the surface potential naturally.

As shown in the flowchart in Figure 3.6, the NC-FeFET can be simulated in a technology computer-aided design (TCAD) by combining a one-dimensional (1D) L–K equation [48, 50]. Subsequently, the intrinsic MOSFET voltage drops across the gate, V_{DP} and channel charge distribution, Q_D can be determined by solving the Poisson–Schrodinger equation. Furthermore, the ferroelectric field, E, and polarization charge density (P) are determined. When V_{DP}–Q_D and E–P match ($|V_{DP}|=|E|\cdot T_F$ and $|Q_D|=|P|$), the appropriate set of device structural and bias parameters of the NC-FeFET are said to be obtained. Otherwise, these two parameters are changed until the matching condition is established.

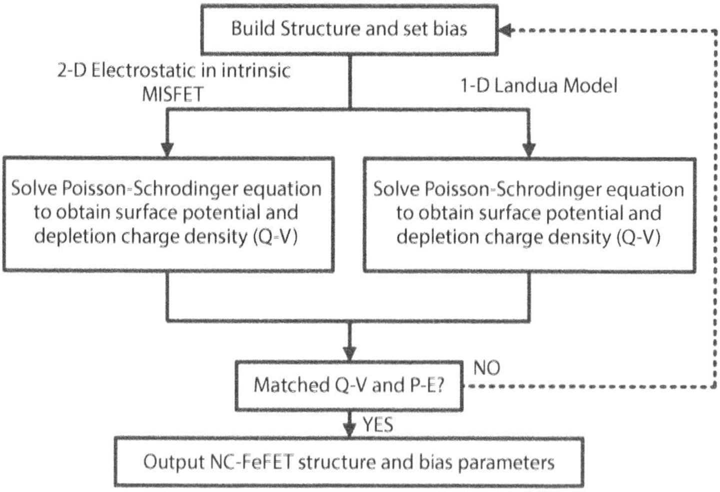

Figure 3.6 Simulation flowchart of basic NC-FeFET [54].

3.4 Applications

Considering the major advancements in semiconductor technology, NC-FeFETs offer a promising path towards low-power electronics, novel memory devices, and other applications. A brief discussion of some significant uses of NC-FeFETs in low-power devices in this investigation, as well as their potential for application in memory devices and other applications, is presented.

3.4.1 Low-Power Electronics

The achievement of low-power electronics using NC-FeFETs is possible because of their potential to realize ultralow-power NC-FeFETs. NC-FeFETs enable the creation of low-power digital and logic circuits that exhibit remarkable energy efficiency in the field of digital electronics. Due to their unique NC effect, they exhibit steeper subthreshold slopes, resulting in a significant reduction in the leakage currents and fast switching. NC-FeFET-based logic circuits have strong functionality while consuming low power due to their operation at lower supply voltages. NC-FeFETs offer substantial advantages for ultra-low-power applications in nonvolatile memory. They exhibit strong compatibility with static random-access memory (SRAM) [60] and ferroelectric random-access memory (FeRAM) [61], resulting in reduced energy consumption for writing and enhanced endurance. The steep switching property of the NCFET allows it to be used as an SRAM device. The NCFET-based SRAM bit cell is a favorable option for achieving a high yield at a near-threshold voltage of V_{DD} [60]. The implementation of the NC-FeFET-based nonvolatile memory macro in a nonvolatile processor demonstrated a 27% increase in forward progress compared with that of FeRAM [61]. The NC effect of NC-FeFETs decreases power consumption during write and read operations, making them well-suited for energy-efficient nonvolatile memory (NVM) solutions.

NC-FeFETs also play a crucial role in the development of energy-efficient neuromorphic computer architectures [62]. These topologies leverage the sharp subthreshold slope of NC-FeFETs to create spiking neural networks (SNNs) and synaptic devices that mimic the structure and function of the human brain, while minimizing energy consumption. This enables the development of artificial intelligence systems with unparalleled efficiencies and processing capacities. The integration of NC-FeFETs into energy harvesting systems [63] and Internet-of-Things (IoT) devices [55] significantly enhances the energy efficiency and prolongs the battery life. Compared to their CMOS counterparts, NC-FeFET logic designs have demonstrated substantial enhancements. It was observed that the NC-FeFET logic leads to savings of more than 54%, 68%, 85%, and 95% in power, delay, and energy, respectively [64]. NC-FeFETs operate at reduced power-supply voltages and exhibit decreased leakage currents. These devices, such as wireless sensor networks and wearable technologies, are ideal for IoT applications that require energy conservation. In regions with limited resources, the energy-efficient characteristics of these devices enable sustainable operation and extended battery longevity.

3.4.2 Potential Application in Memory Devices and Beyond

NC-FeFETs exhibit significant potential to become the future of NVM systems. Owing to their NC characteristics and low power consumption, they offer substantial improvements in FeRAM, phase-change memory (PCM), and resistive random-access memory (RRAM) [60–64]. NVM devices based on NC-FeFET technology will offer superior performance, reliability, and energy efficiency compared to conventional memory technologies. This presents futuristic opportunities for significant advancements in data storage and retrieval [60, 61].

NC-FeFETs can be integrated into emerging sectors such as reconfigurable electronic devices, analogue circuits, and hardware security. In addition to their employment in memory devices, NC-FeFETs can be used in analog-to-digital converters (ADCs) and voltage regulators due to their enhanced performance and energy efficiency. In addition, they can also be utilized in hardware security applications such as physically unclonable functions (PUFs) and random number generators [62–64].

3.5 Performance Optimization and Challenges

3.5.1 Approaches for Enhancing Performance

The performance of the NC-FeFET can be enhanced by designing its device parameters to achieve a steep SS, and high I_{ON}/I_{OFF} ratio with desirable hysteresis. Some methods to enhance the NC-FeFET performance are as follows:

a) Gate stacks: At low voltage, the I–V characteristics of metal ferroelectric insulator semiconductor FET (MFISFET) and metal ferroelectric metal insulator semiconductor FET (MFMISFET) show no changes in its I–V curve. However, when V_D is high, the MFISFET shows a better result in terms of I_{ON} and SS than the MFISFET. Hence, at higher drain voltages, the MFISFET is less prone to a mismatch between ferroelectric capacitance and insulator capacitance.

b) Ferroelectric material thickness: When the thickness (T_F) of the ferroelectric material increases, the SS steepness increases and the hysteresis becomes unacceptable. Thus, there is always a trade-off between SS and hysteresis. A basic NC-FeFET (MFSFET), is a hysteresis-free ferroelectric material $T_F < \frac{1}{2|\alpha|C_S}$. This implies that the ferroelectric capacitance C_F is larger than that of C_S and a stable positive capacitance is achieved with a hysteresis-free curve. When $T_F \geq 6$ nm, the subthreshold slope became steeper with hysteresis.

c) Material parameters: The ferroelectric material parameters E_C and P_r can be accurately adjusted to significantly change the I–V characteristics of the NC-FeFET. For a basic NC-FeFET (MFSFET), when the ferroelectric material parameters E_C and P_r change, they are reflected in its I_{ON} current and SS. For drain voltage $V_D = 0.2$ V, when E_C and P_r are 0.6 mV/cm and 25 μC/cm^2 gives a higher switching ratio in NC-FeFET than conventional FET.

d) Doping concentration: The subthreshold slope of the NC-FeFET can be improved using a certain silicon doping concentration. For an NC-FeFET with a ferroelectric layer of $Sr_{0.8}Bi_{2.2}Ta_2O_9$, when the silicon doping concentration is increased from 10^{17} to 10^{20} at $V_D = 0.2$ V, the subthreshold slope becomes steeper with increasing drain current [65].
e) Temperature effect: When the temperature increased, the effect of the ferroelectric material decreased and became weak. This results in a decrease in the influence of the channel surface potential on the gate voltage. When the external temperature of an NC-FeFET (MFSFET) increased from 290 K to 380 K, the SS increased consistently and caused narrowing of the NC area of the ferroelectric material [65].
f) High-K gate dielectric: The oxide layer can no longer support the voltage drop when the device channel reaches the sub-nanometer range. SiO_2 is used as the oxide layer in a typical NC-FeFET. High-K dielectric oxides can be used as a replacement for the oxide layer to decrease the leakage current and enhance the device's stability and electrical performance.
g) Hysteresis: In the NC-FeFET, the phenomenon of hysteresis is due to the presence of leakage current through its dielectric layer. A depolarization field is created due to this leakage current within the ferroelectric material, resulting in the formation of charge traps. Hysteresis can be restricted by several means such as lowering V_D, reducing oxide thickness, using intrinsic Si or Ge as channel to improve channel capacitance, and precise ferroelectric material parameters. NC-FeFETs with hysteresis are used in memory devices and flip-flops.

3.5.2 Challenges

Every new technology presents numerous challenges. Similarly, NC-FeFETs have their own fair share of challenges and are discussed here. Although NC-FeFETs can be integrated with the current semiconductor fabrication process, device manufacturing without compromising yield and device performance is a challenge. The NC effect of ferroelectric materials depends on the type of material, material parameters, ferroelectric thickness, channel doping, and temperature, as discussed in Section 3.5.1. Optimizing device performance with all such parameters to make it compatible with the present CMOS technology or complex logic or memory circuits will be a significant challenge. Additionally, the environmental factor affecting the device's performance and lifespan to maintain the NC effect for a long time is critical for achieving reliable operation in future electronics. Interdisciplinary efforts in device physics, material science, circuit design, and manufacturing technologies are required to overcome these challenges of NC-FeFET.

3.6 Comparative Analysis with Other Transistor Technologies

3.6.1 NC-FeFET vs MOSFET

A quantitative investigation of the subthreshold slope of the NC-FeFET and conventional MOSFET is shown in Figure 3.7. As shown in the figure, the NC-FeFET has a steeper SS than the conventional MOSFET. Due to this steeper SS, the NC-FeFET has an efficient switching speed with low power consumption in comparison with the conventional MOSFET. The on-state current I_{ON} for both transistors is almost similar to the high off-current I_{OFF} in the MOSFET, making the I_{ON}/I_{OFF} ratio of the NC-FeFET larger than that of the conventional MOSFET. Simultaneously, a high I_{OFF} is advantageous in specific applications where a fast switching speed is required. It also makes conventional MOSFET prone to high leakage current, high-power consumption, heat dissipation, and reliability concerns, particularly in low-power applications.

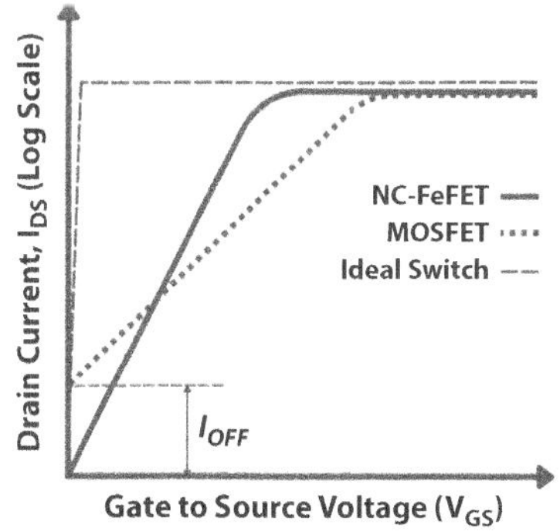

Figure 3.7 Transfer characteristic of NC-FeFET vs MOSFET.

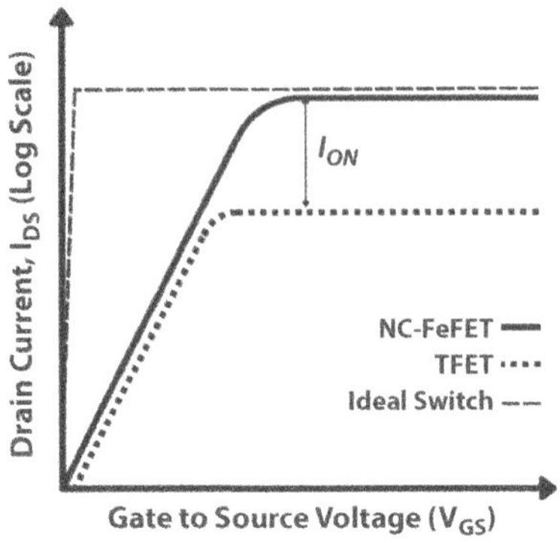

Figure 3.8 Transfer characteristic of NC-FeFET vs TFET.

3.6.2 NC-FeFET vs TFET

A quantitative investigation of the subthreshold slopes of the NC-FeFET and conventional MOSFET is shown in Figure 3.8. As illustrated in the figure, both structures have steep SS, making them suitable for low-power applications and high-speed digital circuits. The NC-FeFET has a high I_{ON} and low I_{OFF}, which suggests a reduced leakage current, whereas the TFET has low I_{ON} and I_{OFF} currents. Reducing the I_{OFF} effectively reduces the dynamic power dissipation. However, the reduction of I_{OFF} by methods such as increasing the bandgap results in lowering I_{ON}, and a low I_{ON} is not favorable for achieving quick switching. This affects the I_{ON}/I_{OFF} ratio and lowers gm more than the normal value, which is an important parameter in designing digital circuits. The lower value of g_m also indicates that TFET have low cut-off frequency and gain bandwidth products, limiting TFETs application in high-frequency applications [66, 67].

3.7 Future Prospects and Trends

Over the last and a half decades, development efforts have been made to advance the NC-FeFET to enable it for ultralow-power applications in electronics. Despite the advancement and futuristic potential shown by NC-FeFETs, integrating them into real-world applications is still in its early stages. To fully unlock the potential of NC-FeFETs, researchers must address challenges such as device reliability, scalability, manufacturing compatibility, and circuit integration. Ferroelectrics on HfO_2 have shown CMOS compatibility for future development. Other ferroelectric materials, such as P(VDF-TrFE), encounter significant fabrication challenges in the present high-temperature fabrication process. Similarly, perovskite ferroelectric materials face compatibility issues due to the hydrogen-annealing process. Another limitation is the low bandgap of the perovskite ferroelectric materials. Hence, finding potential ferroelectric materials compatible with the CMOS technology should be the focus of research [66].

Following the Samsung foundry strategic trajectory, it is on the verge of producing 2 nm and 1.4 nm node transistors, including the gate-all-around field effect transistor (GAAFET) architecture. NC GAAFET, as a high-performance, ultralow-power device, is a promising device for sustaining Moore's law to 0.5 nm node. Certain challenges must be addressed to achieve ultralow-power NC devices. It is necessary to develop an ultra-thin ferroelectric material that will keep the NC effect intact while optimizing the ferroelectric and MOS capacitance and identifying thermally sustainable ferroelectric materials. In addition, after reaching the device scaling limit, 2D materials are likely to replace Si and Ge. Thus, it is important to overcome the challenges of developing scalable and systematic techniques for ferroelectric materials in 2D materials. Many studies have also reported on the use of NC-FeFETs in various fields of IoT, wearable devices, memory technology, electronic devices, quantum computing, and artificial intelligence. However, practical applications have not yet been achieved. Thus, NC-FeFET devices are on the right track to support a technologically sustainable, advanced future.

The future of NC-FeFETs is on the verge of significant transformation in the field of low-power electronics and beyond. NC-FeFETs exhibit an SS<60 mV/dec, a high I_{ON}/I_{OFF} ratio, and the ability to alter hysteresis. This provides the potential to decrease the required supply voltage and power loss in the device. Consequently, they are suitable options for low-power electronic and memory devices. Nevertheless, NC-FeFET devices encounter various challenges, including integration complexities, manufacturing line corruption with heavy metals, and the presence of traditionally thick ferroelectrics. Traditional ferroelectric materials such as PZT and BTO are not suitable for contemporary CMOS technology. Therefore, extensive research is necessary to enhance the efficiency of NCFET by focusing on channel engineering, investigating various structural engineering methods, exploring different ferroelectric materials, and optimizing their parameters. These efforts have the potential to facilitate the swift implementation of NCFET-based devices in electronics in the coming years. As NC-FeFET devices move from being theoretical notions to implementable devices, they have the potential to completely change the way the present computing approaches for energy-efficient operation. Nevertheless, to fully utilize the revolutionary potential of NCFETs, overcoming the obstacles associated with NC-FeFET stability, CMOS compatibility, and scalability is crucial. As research advances and more industries adopt NC-FeFETs, they will play a crucial role in shaping a future that is both sustainable and technologically advanced.

3.8 Summary

This chapter discusses the core concept of the NC-FeFET, including its working principle and characteristics. The potential of the NC-FeFET as a future low-power and fast-switching device has been explored through various research findings and recent trends in the semiconductor industry. The chapter also highlights the modeling approach and investigates the application of the NC-FeFET, focusing on its potential applicability in low-power electronics and memory devices. Approaches for enhancing device performance and challenges such as stability, material compatibility, and fabrication process are also discussed. A comparative analysis of the NC-FeFET with various transistor technologies has been presented to provide a deeper insight. Finally, the chapter presents the future prospects and trends foreseeing the possible future of NC-FeFETs in research and technological changes.

References

1. Salvatore, G.A., Bouvet, D., Ionescu, A.M., Demonstration of subthrehold swing smaller than 60mV/decade in Fe-FET with P(VDF-TrFE)/SiO2 gate stack, in: *2008 IEEE International Electron Devices Meeting*, pp. 1–4, IEEE, San Francisco, CA, USA, Dec. 2008, doi: 10.1109/IEDM.2008.4796642.
2. Khan, A.I., Yeung, C.W., Hu, C., Salahuddin, S., Ferroelectric negative capacitance MOSFET: Capacitance tuning & antiferroelectric operation, in: *2011 International Electron Devices Meeting*, pp. 11.3.1–11.3.4, IEEE, Washington, DC, USA, Dec. 2011, doi: 10.1109/IEDM.2011.6131532.
3. Malvika, B.C. and Mummaneni, K., A Review on a Negative Capacitance Field-Effect Transistor for Low-Power Applications. *J. Electron. Mater.*, 51, 3, 923–937, Mar. 2022. doi: 10.1007/s11664-021-09384-8.
4. Valasa, S., Kotha, V.R., Vadthiya, N., Beyond Moore's law – A critical review of advancements in negative capacitance field effect transistors: A revolution in next-generation electronics. *Mater. Sci. Semicond. Process.*, 173, 108116, Apr. 2024. doi: 10.1016/j.mssp.2024.108116.
5. Dennard, R.H., Gaensslen, F.H., Yu, H.-N., Rideout, V.L., Bassous, E., LeBlanc, A.R., Design of ion-implanted MOSFET's with very small physical

dimensions. *IEEE J. Solid-State Circuits*, 9, 5, 256–268, Oct. 1974. doi: 10.1109/JSSC.1974.1050511.

6. Kumar, N.A., Singh, A.D., Singh, N.B., Examining the Short Channel Characteristic and Performance of Triple Material Double Gate Silicon-on-Nothing Metal Oxide Semiconductor Field Effect Transistors with Grading Channel Concentration. *J. Nanoelectron. Optoelectron.*, 14, 12, 1672–1679, Dec. 2019. doi: 10.1166/jno.2019.2667.
7. Ajit Kumar, N. and Nelson Singh, K., Analysis of Channel Grading on Triple Material Double Gate Stack Oxide SON MOSFET, in: *Advances in Communication, Devices and Networking*, Lecture Notes in Electrical Engineering, vol. 1037, Dhar, S., Do, D.-T., Sur, S.N., Liu, C.-M. (Eds.), Singapore: Springer Nature Singapore, 2023, pp. 43–49. doi: 10.1007/978-981-99-1983-3_5.
8. Kumar, N.A., Singh, A.D., Singh, N.B., Partially Depleted and Fully Depleted Silicon on Insulator: A Comparative study using TCAD. *Int. J. Electron.*, 6, 3, 3, 2017.
9. Isaac, R.D., Reaching the limits of CMOS technology, in: *IEEE 7th Topical Meeting on Electrical Performance of Electronic Packaging (Cat. No. 98TH8370)*, IEEE, West Point, NY, USA, p. 3, 1998, doi: 10.1109/EPEP.1998.733476.
10. Kish, L.B., End of Moore's law: thermal (noise) death of integration in micro and nano electronics. *Phys. Lett. A*, 305, 3–4, 144–149, Dec. 2002. doi: 10.1016/S0375-9601(02)01365-8.
11. Kim, Y. J., Park, M. H., Jeon, W., Kim, H. J., Moon, T., Lee, Y. H., Kim, K. D., Hyun, S. D., Hwang, C. S., Interfacial charge-induced polarization switching in Al2O3/Pb(Zr,Ti)O3 bi-layer. *J. Appl. Phys.*, 118, 22, 224105, 14 December 2015, doi: 10.1063/1.4937544.
12. Van Vliet, K.M., Linear response theory revisited. II. The master equation approach. *J. Math. Phys.*, 20, 12, 2573–2595, Dec. 1979. doi: 10.1063/1.524020.
13. Boucart, K. and Ionescu, A.M., Double-Gate Tunnel FET With High-k Gate Dielectric. *IEEE Trans. Electron Devices*, 54, 7, 1725–1733, Jul. 2007. doi: 10.1109/TED.2007.899389.
14. Khatami, Y. and Banerjee, K., Steep Subthreshold Slope n- and p-Type Tunnel-FET Devices for Low-Power and Energy-Efficient Digital Circuits. *IEEE Trans. Electron Devices*, 56, 11, 2752–2761, Nov. 2009. doi: 10.1109/TED.2009.2030831.
15. Fiori, G. and Iannaccone, G., Ultralow-Voltage Bilayer Graphene Tunnel FET. *IEEE Electron Device Lett.*, 30, 10, 1096–1098, Oct. 2009. doi: 10.1109/LED.2009.2028248.
16. Moselund, K.E., Bouvet, D., Pott, V., Meinen, C., Kayal, M., Ionescu, A.M., Punch-through impact ionization MOSFET (PIMOS): From device principle to applications. *Solid State Electron.*, 52, 9, 1336–1344, Sep. 2008. doi: 10.1016/j.sse.2008.04.021.

17. Hui, K., Hu, C., George, P., Ko, P.K., Impact ionization in GaAs MESFETs. *IEEE Electron Device Lett.*, 11, 3, 113–115, Mar. 1990. doi: 10.1109/55.46951.
18. Bude, J.D., Gate current by impact ionization feedback in sub-micron MOSFET technologies, in: *1995 Symposium on VLSI Technology. Digest of Technical Papers*, Japan Soc. Appl. Phys, Kyoto, Japan, pp. 101–102, 1995, doi: 10.1109/VLSIT.1995.520877.
19. Fulkerson, D.E., Feedback FET logic: a robust, high-speed low-power GaAs logic family. *IEEE J. Solid-State Circuits*, 26, 1, 70–74, Jan. 1991. doi: 10.1109/4.65714.
20. Padilla, A., Yeung, C.W., Shin, C., Hu, C., Liu, T.-J.K., Feedback FET: A novel transistor exhibiting steep switching behavior at low bias voltages, in: *2008 IEEE International Electron Devices Meeting*, IEEE, San Francisco, CA, USA, pp. 1–4, Dec. 2008, doi: 10.1109/IEDM.2008.4796643.
21. Ionescu, A.M., *et al.*, Modeling and design of a low-voltage SOI suspended-gate MOSFET (SG-MOSFET) with a metal-over-gate architecture, in: *Proceedings International Symposium on Quality Electronic Design*, IEEE Comput. Soc, San Jose, CA, USA, pp. 496–501, 2002, doi: 10.1109/ISQED.2002.996794.
22. Akarvardar, K., *et al.*, Analytical Modeling of the Suspended-Gate FET and Design Insights for Low-Power Logic. *IEEE Trans. Electron Devices*, 55, 1, 48–59, Jan. 2008. doi: 10.1109/TED.2007.911070.
23. Salahuddin, S. and Datta, S., Use of Negative Capacitance to Provide Voltage Amplification for Low Power Nanoscale Devices. *Nano Lett.*, 8, 2, 405–410, Feb. 2008. doi: 10.1021/nl071804g.
24. Yeung, C.W., Khan, A.I., Salahuddin, S., Hu, C., Device design considerations for ultra-thin body non-hysteretic negative capacitance FETs, in: *2013 Third Berkeley Symposium on Energy Efficient Electronic Systems (E3S)*, IEEE, Berkeley, CA, USA, pp. 1–2, Oct. 2013, doi: 10.1109/E3S.2013.6705876.
25. George, S., *et al.*, Device Circuit Co Design of FEFET Based Logic for Low Voltage Processors, in: *2016 IEEE Computer Society Annual Symposium on VLSI (ISVLSI)*, IEEE, Pittsburgh, PA, USA, pp. 649–654, Jul. 2016, doi: 10.1109/ISVLSI.2016.116.
26. Li, X., *et al.*, Enabling Energy-Efficient Nonvolatile Computing With Negative Capacitance FET. *IEEE Trans. Electron Devices*, 64, 8, 3452–3458, Aug. 2017. doi: 10.1109/TED.2017.2716338.
27. Islam Khan, A., *et al.*, Experimental evidence of ferroelectric negative capacitance in nanoscale heterostructures. *Appl. Phys. Lett.*, 99, 11, 113501, Sep. 2011. doi: 10.1063/1.3634072.
28. Yeung, C.W., Khan, A.I., Cheng, J.Y., Salahuddin, S., Hu, C.M., Non-hysteretic negative capacitance FET with Sub- 30mV/dec swing over 106X current range and ION of 0.3mA/μm without strain enhancement at 0.3V VDD, in: *Natl. Yang Ming Chiao Tung Univ. Acad. Hub*, 2012.
29. McGuire, F.A., Lin, Y.-C., Rayner, B., Franklin, A.D., MoS2 negative capacitance FETs with CMOS-compatible hafnium zirconium oxide, in: *2017 75th*

Annual Device Research Conference (DRC), IEEE, South Bend, IN, USA, pp. 1–2, Jun. 2017, doi: 10.1109/DRC.2017.7999478.

30. McGuire, F.A., Cheng, Z., Price, K., Franklin, A.D., Sub-60 mV/decade switching in 2D negative capacitance field-effect transistors with integrated ferroelectric polymer. *Appl. Phys. Lett.*, 109, 9, 093101, Aug. 2016. doi: 10.1063/1.4961108.
31. Chun Wing Yeung, A.I., Sarker, A., Salahuddin, S., Hu, C., Low power negative capacitance FETs for future quantum-well body technology, in: *2013 International Symposium on VLSI Technology, Systems and Application (VLSI-TSA)*, IEEE, Hsinchu, pp. 1–2, Apr. 2013, doi: 10.1109/VLSI-TSA.2013.6545648.
32. Kwon, D., *et al.*, Negative Capacitance FET With 1.8-nm-Thick Zr-Doped HfO_2 Oxide. *IEEE Electron Device Lett.*, 40, 6, 993–996, Jun. 2019. doi: 10.1109/LED.2019.2912413.
33. Dong, Z. and Guo, J., A Simple Model of Negative Capacitance FET With Electrostatic Short Channel Effects. *IEEE Trans. Electron Devices*, 64, 7, 2927–2934, Jul. 2017. doi: 10.1109/TED.2017.2706182.
34. Srimani, T., *et al.*, Negative Capacitance Carbon Nanotube FETs. *IEEE Electron Device Lett.*, 39, 2, 304–307, Feb. 2018. doi: 10.1109/LED.2017.2781901.
35. Si, M., Jiang, C., Chung, W., Du, Y., Alam, M.A., Ye, P.D., Steep-Slope WSe_2 Negative Capacitance Field-Effect Transistor. *Nano Lett.*, 18, 6, 3682–3687, Jun. 2018. doi: 10.1021/acs.nanolett.8b00816.
36. Lee, H.-P., Yu, C.-L., You, W.-X., Su, P., Investigation and comparison of design space for ultra-thin-body GeOI/SOI negative capacitance FETs, in: *2017 International Symposium on VLSI Technology, Systems and Application (VLSI-TSA)*, IEEE, Hsinchu, Taiwan, pp. 1–2, Apr. 2017, doi: 10.1109/VLSI-TSA.2017.7942460.
37. Khandelwal, S., Khan, A.I., Duarte, J.P., Sachid, A.B., Salahuddin, S., Hu, C., Circuit performance analysis of negative capacitance FinFETs, in: *2016 IEEE Symposium on VLSI Technology*, IEEE, Honolulu, HI, USA, pp. 1–2, Jun. 2016, doi: 10.1109/VLSIT.2016.7573446.
38. Aziz, A., Ghosh, S., Datta, S., Gupta, S., Physics-Based Circuit-Compatible SPICE Model for Ferroelectric Transistors. *IEEE Electron Device Lett.*, 37, 6, 1–1, June 2016. doi: 10.1109/LED.2016.2558149.
39. Khandelwal, S., Duarte, J.P., Khan, A.I., Salahuddin, S., Hu, C., Impact of Parasitic Capacitance and Ferroelectric Parameters on Negative Capacitance FinFET Characteristics. *IEEE Electron Device Lett.*, 38, 1, 142–144, Jan. 2017. doi: 10.1109/LED.2016.2628349.
40. Khan, A.I., Radhakrishna, U., Chatterjee, K., Salahuddin, S., Antoniadis, D.A., Negative Capacitance Behavior in a Leaky Ferroelectric. *IEEE Trans. Electron Devices*, 63, 11, 4416–4422, Nov. 2016. doi: 10.1109/TED.2016.2612656.
41. Khan, A.I., Radhakrishna, U., Salahuddin, S., Antoniadis, D., Work Function Engineering for Performance Improvement in Leaky Negative Capacitance

FETs. *IEEE Electron Device Lett.*, 38, 9, 1335–1338, Sep. 2017. doi: 10.1109/LED.2017.2733382.
42. Awadhiya, B., Kondekar, P.N., Meshram, A.D., Effect of Ferroelectric Thickness Variation in Undoped HfO2-Based Negative-Capacitance Field-Effect Transistor. *J. Electron. Mater.*, 48, 10, 6762–6770, Oct. 2019. doi: 10.1007/s11664-019-07483-1.
43. Gupta, A.K., Raman, A., Kumar, N., Charge-Plasma-Based Negative Capacitance Ring-FET: Design, Investigation and Reliability Analysis. *J. Electron. Mater.*, 49, 8, 4852–4863, Aug. 2020. doi: 10.1007/s11664-020-08205-8.
44. Vogel, R. and Walsh, P.J., Negative Capacitance in Amorphous Semiconductor Chalcogenide Thin Films. *Appl. Phys. Lett.*, 14, 7, 216–218, Apr. 1969. doi: 10.1063/1.1652784.
45. Jones, B.K., Santana, J., McPherson, M., Negative capacitance effects in semiconductor diodes. *Solid State Commun.*, 107, 2, 47–50, May 1998. doi: 10.1016/S0038-1098(98)00162-8.
46. Neupane, G.R., Bamidele, M., Yeddu, V., Kim, D.Y., Hari, P., Negative capacitance and hysteresis in encapsulated MAPbI3 and lead–tin (Pb–Sn) perovskite solar cells. *J. Mater. Res.*, 37, 7, 1357–1372, Apr. 2022. doi: 10.1557/s43578-022-00540-2.
47. Bilkan, C., Gümüş, A., Altındal, S., The source of negative capacitance and anomalous peak in the forward bias capacitance-voltage in Cr/p-si Schottky barrier diodes (SBDs). *Mater. Sci. Semicond. Process.*, 39, 484–491, Nov. 2015. doi: 10.1016/j.mssp.2015.05.044.
48. Wang, L., *et al.*, Negative Quantum Capacitance Induced by Midgap States in Single-layer Graphene. *Sci. Rep.*, 3, 1, 2041, Jun. 2013. doi: 10.1038/srep02041.
49. Hoffmann, M., Ravindran, P.V., Khan, A.I., Why Do Ferroelectrics Exhibit Negative Capacitance? *Materials*, 12, 22, 3743, Nov. 2019. doi: 10.3390/ma12223743.
50. Chandra, P. and Littlewood, P.B., A Landau Primer for Ferroelectrics, in: *Physics of Ferroelectrics*, Topics in Applied Physics, vol. 105, pp. 69–116, Springer Berlin Heidelberg, Berlin, Heidelberg, 2007, doi: 10.1007/978-3-540-34591-6_3.
51. Alam, M.N.K., Roussel, P., Heyns, M., Van Houdt, J., Positive non-linear capacitance: the origin of the steep subthreshold-slope in ferroelectric FETs. *Sci. Rep.*, 9, 1, 14957, Oct. 2019. doi: 10.1038/s41598-019-51237-2.
52. Dragoman, M., Aldrigo, M., Dragoman, D., Iordanescu, S., Dinescu, A., Modreanu, M., The Rise of Ferroelectricity at Nanoscale: Nanoelectronics is rediscovering the ferroelectricity. *IEEE Nanotechnol. Mag.*, 15, 5, 8–19, Oct. 2021. doi: 10.1109/MNANO.2021.3098217.
53. Jonscher, A.K., The physical origin of negative capacitance. *J. Chem. Soc., Faraday Trans. 2*, 82, 1, 75, 1986. doi: 10.1039/f29868200075.
54. Tu, L., Wang, X., Wang, J., Meng, X., Chu, J., Ferroelectric Negative Capacitance Field Effect Transistor. *Adv. Electron. Mater.*, 4, 11, 1800231, Nov. 2018. doi: 10.1002/aelm.201800231.

55. Chang, S.-C., Avci, U.E., Nikonov, D.E., Manipatruni, S., Young, I.A., Physical Origin of Transient Negative Capacitance in a Ferroelectric Capacitor. *Phys. Rev. Appl.*, 9, 1, 014010, Jan. 2018. doi: 10.1103/PhysRevApplied.9.014010.
56. Sharma, P., Zhang, J., Ni, K., Datta, S., Time-Resolved Measurement of Negative Capacitance. *IEEE Electron Device Lett.*, 39, 2, 272–275, Feb. 2018. doi: 10.1109/LED.2017.2782261.
57. Qin, L., *et al.*, Recent Developments in Negative Capacitance Gate-All-Around Field Effect Transistors: A Review. *IEEE Access*, 11, 14028–14042, 2023. doi: 10.1109/ACCESS.2023.3243697.
58. Kobayashi, M. and Hiramoto, T., On device design for steep-slope negative-capacitance field-effect-transistor operating at sub-0.2V supply voltage with ferroelectric HfO2 thin film. *AIP Adv.*, 6, 2, 025113, Feb. 2016. doi: 10.1063/1.4942427.
59. Saha, A.K., Datta, S., Gupta, S.K., 'Negative capacitance' in resistor-ferroelectric and ferroelectric-dielectric networks: Apparent or intrinsic? *J. Appl. Phys.*, 123, 10, 105102, Mar. 2018. doi: 10.1063/1.5016152.
60. Hong, Y., Choi, Y., Shin, C., NCFET-Based 6-T SRAM: Yield Estimation Based on Variation-Aware Sensitivity. *IEEE J. Electron Devices Soc.*, 8, 182–188, 2020. doi: 10.1109/JEDS.2020.2973966.
61. George, S., *et al.*, Nonvolatile memory design based on ferroelectric FETs, in: *Proceedings of the 53rd Annual Design Automation Conference*, ACM, Austin Texas, pp. 1–6, Jun. 2016, doi: 10.1145/2897937.2898050.
62. Li, X., *et al.*, Advancing Nonvolatile Computing With Nonvolatile NCFET Latches and Flip-Flops. *IEEE Trans. Circuits Syst. Regul. Pap.*, 64, 11, 2907–2919, Nov. 2017. doi: 10.1109/TCSI.2017.2702741.
63. George, S., Aziz, A., Li, X., Sampson, J., Datta, S., Gupta, S.K., Narayanan, V., NCFET Based Logic for Energy Harvesting Systems. *SRC Techcon.*, 1–4, 2015. https://nics.ee.tsinghua.edu.cn/people/lixq/resources/SRC_TechCON2015-NCFET.pdf.
64. Sanapala, K., Satyanarayana, S.V.V., Sakthivel, R., Near-zero computing using NCFET for IoT applications. *Int. J. Intell. Enterp.*, 8, 2/3, 288, 2021. doi: 10.1504/IJIE.2021.114514.
65. Xiao, Y.G., *et al.*, An improved model for the surface potential and drain current in negative capacitance field effect transistors. *RSC Adv.*, 6, 105, 103210–103214, 2016. doi: 10.1039/C6RA21955A.
66. Ionescu, A.M., *et al.*, Ultra low power: Emerging devices and their benefits for integrated circuits, in: *2011 International Electron Devices Meeting*, pp. 16.1.1–16.1.4, IEEE, Washington, DC, USA, Dec. 2011, doi: 10.1109/IEDM.2011.6131563.
67. Wu, C., Huang, R., Huang, Q., Wang, C., Wang, J., Wang, Y., An Analytical Surface Potential Model Accounting for the Dual-Modulation Effects in Tunnel FETs. *IEEE Trans. Electron Devices*, 61, 8, 2690–2696, Aug. 2014. doi: 10.1109/TED.2014.2329372.

4

Unleashing the Potential of Negative Capacitance Field Effect Transistors: A Paradigm Shift in Low-Power Electronics

Malvika[1]*, Jagritee Talukdar[2], Bijit Choudhuri[1] and Kavicharan Mummaneni[1]

[1]Department of Electronics and Communication Engineering, National Institute of Technology Silchar, Assam, India
[2]Department of Electrical Engineering, Indian Institute of Technology Bombay, Mumbai, India

Abstract

The relentless drive towards miniaturization in micro- and nanoelectronic systems has underscored the criticality of power consumption management. This pursuit encounters a formidable obstacle known as Boltzmann tyranny, wherein the generation of a significant drain current necessitates a minimum gate voltage of approximately 60 mV. This imposes a substantial limitation on the scalability of supply voltage in Ultra-Large-Scale Integration (ULSI) circuits, hindering their ability to adapt to the diminishing dimensions of traditional transistors. In response to this pressing challenge, the negative capacitance field-effect transistor (NCFET) has emerged as a promising solution. NCFET represents a ground-breaking advancement in the field of low-power electronics, offering a promising avenue for overcoming the fundamental limitations of traditional transistor technologies. This novel transistor leverages the unique property of negative capacitance to achieve subthermal swing and enhance the performance of electronic devices, thereby enabling unprecedented reductions in power consumption. The NCFET operates based on the principle of exploiting ferroelectric materials to induce negative capacitance, resulting in an intrinsic voltage amplification effect. This breakthrough technology effectively addresses the power-efficiency challenges faced by conventional transistors, allowing for the realization of ultralow-power electronic devices without compromising performance metrics. This chapter explores the underlying physics, engineering aspects, design, and simulation methodology

**Corresponding author*: malvikathakur16@gmail.com

Ekta Goel, Archana Pandey, Shiromani Balmukund Rahi and Arun Samuel (eds.) Handbook of Advanced Semiconductor Field Effect Transistors (73–88) © 2026 Scrivener Publishing LLC

of NCFET, highlighting their probable application in circuits. The NCFET has emerged as a transformative technology, promising a paradigm shift towards ultralow-power electronics, highlighting their significance in the ongoing quest for energy-efficient and high-performance electronic devices.

Keywords: Polarization, negative capacitance, ferroelectric, capacitance matching, NCFET

4.1 Introduction

The semiconductor electronics industry has seen extraordinary progress throughout the 20th century, with the continuous reinvention of the underlying device technologies over time. The constraints on self-heating and power consumption have driven these reinventions. In the 1950s, bipolar transistors became more widespread as the vacuum tubes approached their maximum capacity in terms of power dissipation. The 1960s introduced complementary metal oxide semiconductors (CMOSs), which gave bipolar transistors a serious power efficiency defeat. With rapid advancements in semiconductor device technology, transistor feature sizes have significantly decreased, and metal oxide semiconductor field-effect transistors (MOSFETs) touched the 100 nm node by 2002. However, simply reducing the channel size is no longer sufficient to meet consumer needs, prompting the need for novel materials and structural designs [1].

At the 90 nm node, stress engineering was established to foster carrier mobility in the channel area [2]. To reduce leakage currents resulting from the downscaling of the SiO_2 oxide layer, high-k metal gate (HKMG) technology was employed at the 45 nm node. This technology consists of a metal gate and high-k dielectric layer [3]. However, in deep-submicrometer regions, decreasing the channel lengths, gate oxide layer, and threshold voltages (V_{th}) resulted in higher leakage currents, which significantly increased the power dissipation in CMOS circuits. Researchers have proposed several different process- and circuit-level solutions to address the problem of leakage in submicrometer regimes. At the process-technology level, halo doping and retrograde channel doping are utilized to scale down the length of the channel and enhance the drive current without increasing the OFF current. At the circuit level, techniques such as scaling of the supply voltage (V_{DD}), dynamic threshold voltage, multiple V_{th}, and stacking of transistor methods are employed to curtail leakage current issues [4]. Severe problems with traditional bulk planar FETs surfaced as MOSFET feature sizes reached the 22 nm node. Due to the larger subthreshold swing

(SS) and higher leakage currents caused by short-channel effects (SCEs), the increased power dissipation prompted researchers to investigate new structural devices, including double-gate MOSFET, ultra-thin body SOIs (UTB-SOIs), silicon-on-insulators (SOIs), and tri-gate FinFETs [5]. However, these devices must overcome the significant SCE-related difficulties. Up to the 7 nm node, FinFETs have been shown to be efficient; nevertheless, additional downscaling has led to significant SCEs, elevated V_{th} fluctuations, parasitic capacitances, and other performance deteriorations [6]. Subsequently, Gate all-around (GAA) technology, which offers improved gate-to-channel controllability by entirely casing the gate, was established to construct a more vigorous structure with diminished SCEs. GAA transistors can scale down to the 3 nm node and improve performance by over 30% compared with FinFETs [7]. Nonetheless, as transistor feature sizes dropped below 10 nm, degradation in the subthreshold region became unavoidable owing to SCEs, causing increases in SS and V_{th}.

Transistor feature sizes are expected to decrease to 1 nm by 2033 according to the 2021 International Roadmap for Devices and Systems (IRDS), with SS reduced to 40 mV/decade to satisfy the demand for low-power devices [8]. Even if the global semiconductor sector has achieved a spectacular $500 billion in turnover, to maintain this progress, low voltage or steep slope transistors must be developed in order to reduce power dissipation and improve MOSFET scalability. The lowering of the supply voltage does not keep up with the feature size scaling of MOS transistors. The SS of MOSFETs in the subthreshold region at room temperature is 60 mV/decade because of their thermionic emission transport method, which adversely affects power efficiency [9]. Therefore, to further reduce transistor feature sizes, it is imperative to find structures that limit their SS. Within this framework, multiple steep-slope device topologies (SS <60 mV/dec) have been proposed, including avalanche transistors, tunnel FETs (TFETs), impact ionization MOSFETs (IMOSFETs), nano-electro-mechanical field-effect transistors (NEMFETs), and Negative Capacitance FETs (NCFETs) [10–13]. The evolution of these transistors is shown in Figure 4.1. While these devices can attain an SS of less than 60 mV/decade, they are still hindered by several technical constraints. Several important features of tunneling transistors include their low ON current (I_{ON}), such as asymmetric source and drain operations, and lower tunneling probability. Avalanche transistors pose a significant challenge owing to their excessively high breakdown voltages, which compromise the device stability. Nano-electro-mechanical principal transistors are currently not widely used due to their complicated fabrication processes and slow switching speeds, and they remain in the developmental stage.

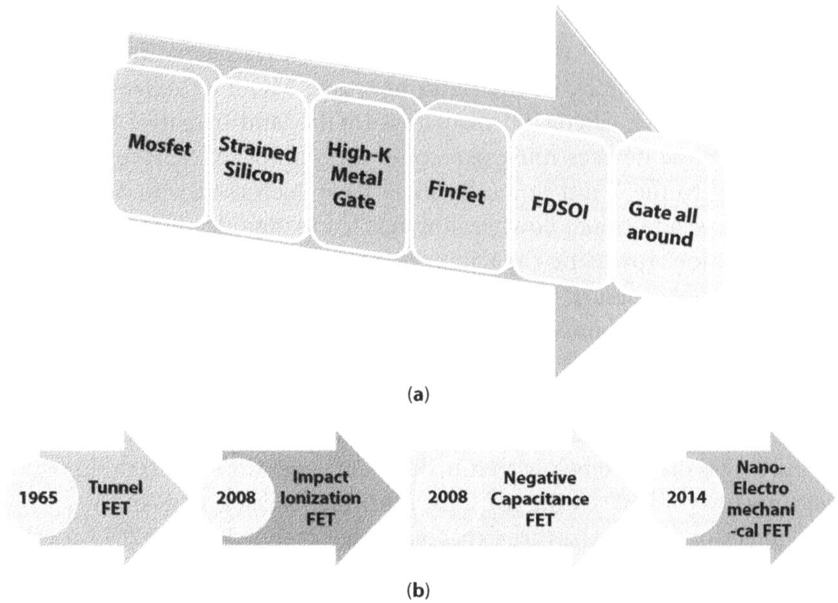

Figure 4.1 (a) Progression of transistors and (b) development of steep slope transistors.

Recently, NCFETs have gained attention owing to their simplified structure and reduced manufacturing complexity compared with MOSFETs. NCFETs incorporate a ferroelectric (FE) layer in the gate stack, which functions as negative capacitance. This FE layer amplifies the surface potential of the transistor, allowing it to overcome the limit of 60 mV/decade at 300 K. The NCFET concept was first introduced by Salahuddin and Datta in 2008. Since then, NCFETs have become prominent owing to their excellent steep switching and superior I_{ON}. A significant milestone was achieved when Global Foundries demonstrated the NCFET fabrication utilizing their advanced 14 nm FinFET technology. This demonstrates the compatibility of the NCFET technology with existing CMOS fabrication methods, which is a crucial factor for the adoption of new devices in the semiconductor industry. Practically, NCFETs enable transistors to attain a similar I_{ON} that is devoid of increasing the OFF current (I_{OFF}) at a lower V_{DD} owing to their steeper SS. Consequently, processors using the NC technique can maintain equivalent efficiency to conventional FETs while operating at lower supply voltages, leading to significant power savings. Their compatibility with CMOS current processes, low power dissipation, steep slope features, feasibility in fabrication, and symmetrical operation of source-drain makes NCFETs a favorable option for upcoming nanoscale and low-power applications. Researchers are continually proposing new materials, technologies, and advanced structures to enhance the performance of NCFETs further.

4.1.1 Operating Mechanism of Negative Capacitance FET

To attain an appropriate performance in MOSFETs, one must maintain a high ratio between the drain current (I_D) in the ON and OFF states to obtain large I_{ON} and a lower I_{OFF}. Boltzmann's limit of 60 mV/decade limits the gate voltage (V_G) to change the drain current by 10 factors. Operating MOS devices at lower voltages can significantly reduce the power dissipation. However, the Boltzmann Tyranny limits the ability to function in extremely scaled devices at lower voltages. The SS is the important parameter that limits the V_{DD} and is expressed as

$$SS = \frac{\partial V_G}{\partial(\log I_D)} = \frac{\partial V_G}{\partial \psi_s} \bullet \frac{\partial \psi_s}{\partial(\log I_D)} \tag{4.1}$$

$$\frac{\partial V_G}{\partial \psi_s} = \left(1 + \frac{C_S}{C_{OX}}\right) = Body\ factor \tag{4.2}$$

$$\frac{\partial \psi_s}{\partial(\log I_D)} = 2.3 \bullet \frac{K_B T}{q} = \frac{60mV}{decade} = Transport\ factor \tag{4.3}$$

where SS is subthreshold swing, ψ_S is surface potential, C_S is semiconductor capacitance, C_{OX} is overall oxide capacitance, q is the charge, T is the temperature and K_B is Boltzmann constant. The term in Eq. (4.2) is known as the body factor and Eq. (4.3), is the transport factor. Lowering either body factor, transport factor, or both would significantly reduce SS. For both planar and non-planar MOS devices, the transport factor cannot be reduced below 60 mV/decade at 300 K, owing to the Boltzmann Tyranny. From Eq. (4.2), the body factor <1 is feasible only when C_{OX} has a negative value, i.e., C_{OX} <0. Integrating an NC layer into a MOSFET involves the addition of an FE layer to the gate stack, as shown in Figure 4.2. This layer amplifies the internal voltage and maintains the device performance at significantly lower supply voltages. To comprehend the voltage amplification competence of FE materials, a device capacitance model depicted in

Figure 4.2 can be considered in which C_{FE} (capacitance of ferroelectric) is serially coupled with oxide capacitance.

As per the rule of voltage divider, amplification factor (η) is defined as

$$\eta = \frac{\partial V_{internal}}{\partial V_G} = \frac{C_{fe}}{C_{fe} + C_{OX}} \tag{4.4}$$

where C_{fe} is the ferroelectric capacitance, which is lesser than 0 owing to the negative capacitance provided by the ferroelectric material that leads η greater than 1.

The SS presented in Eq. (4.1) can be defined in terms of negative capacitance as:

$$SS_{fe} = \frac{\partial V_G}{\partial(\log I_{Ds})} = \frac{\partial V_G}{\partial V_{internal}} \bullet \frac{\partial V_{internal}}{\partial(\log I_{DS})} \tag{4.5}$$

$$SS_{fe} = \frac{SS}{\eta} \tag{4.6}$$

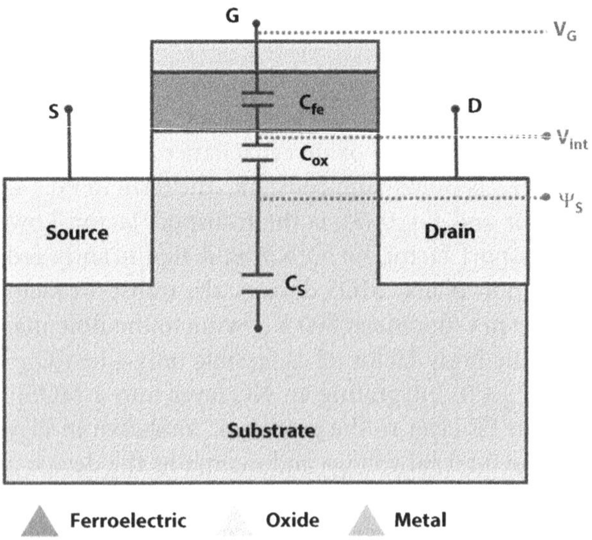

Figure 4.2 Structure of NCFET.

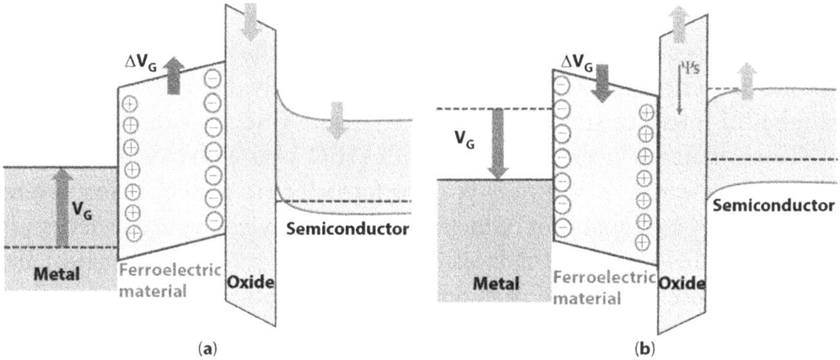

Figure 4.3 Band diagram of NCFET (a) $V_G < V_{FB}$ and (b) $V_G > V_{FB}$.

where SS_{fe} is a subthreshold swing of NCFET. The SS_{fe} value is always lesser than SS of a traditional FET. Figure 4.3(a) illustrates the band bending diagram for NCFETs when the V_G is lesser than flat band voltage (V_{FB}), causing band bending upward near the boundary of oxide and semiconductor. As V_G increases, the band near the interface of oxide and semiconductor begins to bend downward, as illustrated in Figure 4.3(b). The field generated across the oxide layer makes the ψ_S larger than the applied V_G ($\partial\psi_S / \partial V_G > 1$). This indicates that ψ_S is amplified relative to V_G. This amplification in ψ_S, results in additional charges in the channel.

4.2 A Brief Survey

The incorporation of the NC phenomenon into transistors is achieved by integrating a ferroelectric material layer into the gate stack. This feature provides the NC impact with the significant benefit of strong compatibility with diverse transistor architectures. The NC effect can be easily applied to many transistor designs, owing to its versatility. To date, numerous transistor configurations have been developed that utilize the NC effect to improve device performance. The configurations included planar NCFET, GAA NCFET, 2D-NCFET, and NC FinFETs. Significantly, these concepts have not only been proposed but also confirmed by experimental demonstrations. One notable accomplishment of these developments is the achievement of a sub-60 mV/decade SS and improved I_{ON}, demonstrating substantial performance enhancements. Additionally, the NC effect can be used in other semiconductor channel materials, including 2D materials, such as molybdenum disulfide, tungsten diselenide, and black phosphorus

(MoS_2, WSe_2, and BP). Nevertheless, it is crucial to acknowledge that the III–V group-based NCFETs have not yet attained the requisite sub-60 mV/decade SS performance. This challenge is primarily due to the substantial number of interface states between the III and V semiconductor and ferroelectric material layers. Hafnium oxide (HfO_2)-based ferroelectric materials are preferred over other types of ferroelectric materials for several reasons. They are capable of reducing the size of devices, which is necessary to increase the number of devices in current integrated circuits (ICs) and allow for more innovative memory and logic applications. Furthermore, ferroelectric materials based on HfO_2 exhibit a high level of compatibility with contemporary semiconductor fabrication techniques, enabling effortless incorporation into the current CMOS technology. They reduced I_{OFF} because of their greater dielectric constants. Furthermore, they provide improved stability and durability, as they are less vulnerable to external influences, such as temperature and radiation. This guarantees consistent operation of the device over an extended period.

Rusu *et al.* conducted an initial experimental proof of NCFET with sub-60 mV/decade by employing a polymer ferroelectric as the dielectric in the gate stack [14]. To create a p-MOSFET, they utilized n-bulk silicon as the channel material. This resulted in an average SS ranging from 51 to 59 mV/decade. They asserted that the surface potential was magnified when the ferroelectric entered the S-shaped portion of the polarization-electric field (P–E) curve. In addition, the hysteresis loop exhibited a small S shape at the coercive voltage (V_C) of the ferroelectric layer, which contradicts the original QSNC theory [15, 16]. The sub-60 mV/decade regime, in practical scenarios, is restricted to a narrow range of drain currents, often ranging from picoamperes to nanoamperes. This range is inadequate for devices operating at low voltages. To meet the requirements of lower operating voltage targets, which necessitate a lower threshold voltage and an I_{ON} equivalent to the most advanced silicon field-effect transistors (FETs), it is essential to consider the range of current in the steep-slope regime when determining a realistic figure of merit for devices with SS <60 mV/decade [17]. To obtain a low operating voltage for FETs, it is recommended that the current should be at least 1–10 μA/μm. Jo and Shin linked an FE capacitor to a p-channel bulk MOSFET and discovered an SS values below 60 mV/decade, in the current range of pA/μm to nA/μm [18]. Using this particular setup, they observed a SS of 48 mV/decade which is notably superior to that of an identical device with an SS of 110 mV/decade when an FE capacitor is not present.

The initial demonstrations employed channel lengths of a considerable magnitude to validate the NC effect. Recently, efforts have been made to combine FE capacitors with traditional CMOS transistors such as FDSOI and FinFET architectures [19]. Of these options, FinFETs have become a preferred alternative because of their exceptional performance in the sub-60 mV/decade regime [20–22]. The NC-FinFET, which combines the FinFET and negative capacitance state, was first presented in 2015, and its performance was confirmed by experimental validation, achieving an SS of 55 mV/decade [23]. NC-FinFETs provide numerous benefits such as simplified manufacturing, seamless integration, improved current driving abilities, and superior management of SCEs [24–27]. The two main architectures for NC-FinFETs are metal ferroelectric insulator semiconductor (MFIS) and metal ferroelectric metal insulator semiconductor (MFMIS) structures. The primary difference lies in the inclusion of an inner metal in the MFMIS configuration. However, in contemporary CMOS devices featuring channel lengths below 20 nm, the inner metal is unable to fit inside the gate trench. This constraint restricts the operation of NCFETs because of problems related to charge injection and accumulation. The Landau framework, which encompasses many domains, is adjusted to demonstrate that the presence of a floating metal gate can cause leakage of the FE material, leading to instability in NC and oscillations in the threshold voltage [28]. This assertion is supported by a model that is based on surface potential and specifically designed for a gate-all-around (GAA) NCFET of the MFIS type. Although it may be difficult to independently design both the semiconductor channel and FE, the MFIS-based FET is still the preferred choice [29–32].

4.3 Simulation Strategy of NCFET and its Application in Circuit

This section described the design and simulation of a single-gate NCFET (SG-NCFET) structure. In addition, the SG-NCFET device-based inverter has been investigated for low-power-switching applications.

4.3.1 Device Structure

The device structure of the SG-NCFET is illustrated in Figure 4.4(a). The gate was made of aluminum, with a thickness of 5 nm. The gate oxide was silicon dioxide with 0.6 nm thickness. Silicon-doped ferroelectric (Si: HfO_2) materials are used in the gate stack. The purpose of incorporating ferroelectric materials is to attain NC properties, which improve the subthreshold

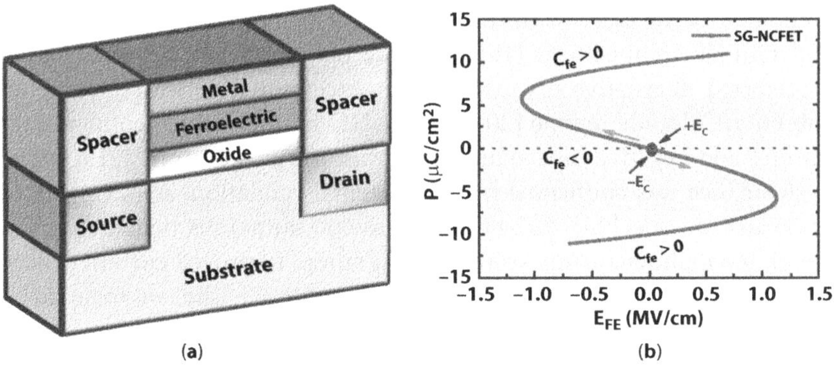

Figure 4.4 (a) 3D representation of SG-NCFET and (b) polarization and electric field curve (S-curve) of SG-NCFET.

swing and enable NCFETs to be used in various low-power applications [33]. An oxide layer was introduced between the ferroelectric material and silicon to diminish the lattice mismatch. The structure maintains a stabilized region in the Landau energy landscape of the FE model when an appropriate V_G is applied. The correlation between the polarization charge density and the electric field in the FE material was described using the Landau equation. Moreover, the device includes spacers, which help to increase the ON current. The increase in the outer fringing field with the spacer dielectric constant causes an increase in the threshold voltage.

4.3.2 Simulation Methodology

Simulations were performed using Synopsys Sentaurus TCAD tool software [34]. Fermi–Dirac statistics is used to handle large doping at the source and drain. The ENormal mobility and doping-dependent mobility models were integrated to account for the effects of doping concentration. The dynamic nonlocal path band-to-band model accompanied by the Shockley–Read–Hall recombination model was enabled. The ferroelectric and drift-diffusion models were activated to support the ferroelectric characteristics and carrier transport, respectively, during the simulation. To achieve the necessary results, the 1-dimensional Landau equation and 2-dimensional electrostatics were calculated independently. The investigation of the polarization and electric field (P–E) performance of FE was implemented using the dynamic LK theory equation. The formulas in Ref. [35] was applied to calculate crucial ferroelectric parameters, including the remanent polarization (P_r), coercive field (E_C), saturation polarization (P_S), and various parameters such as alpha (α), gamma (γ), and beta (β). Figure 4.4(b) displays the simulated P–E curve (S curve) of the SG-NCFET.

4.4 Analysis of Device Performance and its Application as an Inverter

In this section, the static analysis and SG-NCFET based inverter are discussed. The transfer characteristics of the SG-NCFET device are shown in Figure 4.5(a). The I_{ON}, I_{OFF}, V_{th}, I_{ON}/I_{OFF}, and SS of the device are 1.35 mA/μm, 1.9×10^{-11} A/μm, 0.39 V, 0.71×10^{8}, and 36.2 mV/decade, respectively as presented in Figure 4.5(b). The design of SG-NCFET based inverter is presented in Figure 4.6(a). The NCFET technology exhibits better performance than the traditional FET, despite the scaled supply voltages. This is due to the supplementary ferroelectric layer, which increases the total capacitance and improves device functionality.

With the incorporation of C_{fe}, internal voltage amplification occurs, which further allows the NCFETs to attain a larger ON current. Therefore, the NCFET device must maintain an appropriate FE thickness to develop a potent inverter with low energy usage. The SG-NCFET based inverter was designed, which utilizes a complementary NCFET device. Simulations of the proposed inverter were performed for a single input cycle, as shown in Figure 4.6(b). The inverter performance was investigated at a supply voltage of 1 V with a load capacitance (C_L) of 0.42 fF. Figure 4.6(c) depicts

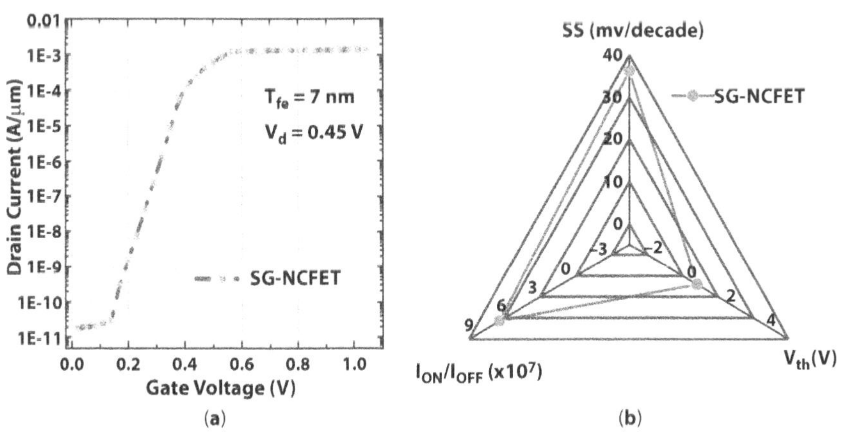

Figure 4.5 (a) I_D-V_G characteristics of SG-NCFET and (b) radar chart for the SS, V_{th}, and I_{ON}/I_{OFF} of SG-NCFET.

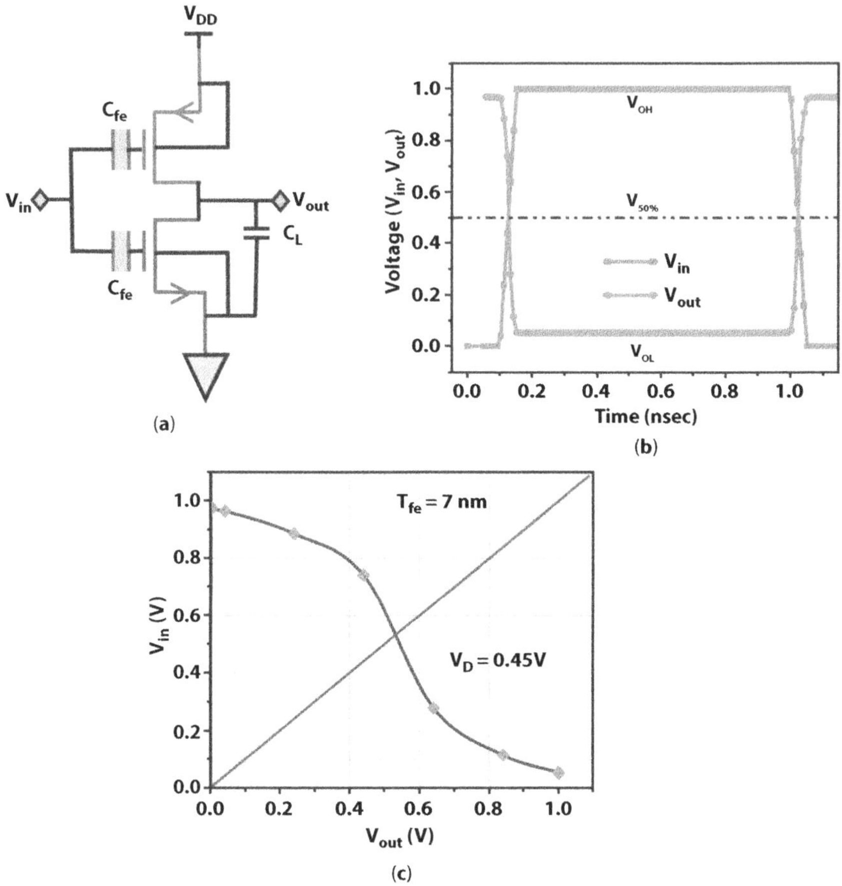

Figure 4.6 (a) Representation of SG-NCFET based inverter, (b) transient simulation of SG-NCFET inverter, and (c) VTC of SG-NCFET inverter.

the voltage transfer curve (VTC) of the SG-NCFET based inverter. The NCFET-based inverter exhibits a fast transition that can be utilized in digital circuits for low-power applications and enhanced switching activity.

4.5 Conclusion

In the last 15 years, significant efforts have been made to enhance NCFETs to reduce the energy dissipation for low-power applications. Despite considerable progress and evident potential, the practical implementation of these devices remains in the early stages, requiring further advancements.

In the current material landscape, HfO_2-based FE stands out as the only viable candidate for future NCFET development because of its compatibility with CMOS technology. Other materials, such as $BaTiO_3$, P(VDF–TrFE), and PZT, face significant integration challenges with modern semiconductor processes. The advancement of ultrathin ferroelectric materials has ensured that their properties are maintained at the sub-nanometer level. This optimization allows for precise matching of the capacitance between the FE material and CMOS, resulting in a steep SS without any hysteresis. In addition, this development addresses the issue of significant parasitic capacitance in nanoscale NCFETs. Furthermore, it identifies a thermally stable FE material that holds promise for future commercialization. Research has highlighted the significance of NCFETs for IoT, optoelectronics, photodetectors, and wearable applications; however, several application areas remain underexplored. These include integrating NCFETs in digital circuits, memory arrays, signal processing, and high-performance processor applications; exploring the impact of NCFETs on non-volatile memory applications and energy consumption; and investigating the possibility of using NCFETs in neuromorphic computing to accelerate energy-efficient AI algorithms and machine learning tasks. Additionally, it aims to assess the suitability of NCFETs in circuits, computing paradigms such as post-CMOS technologies, and quantum computing, and how the NC effect can improve the control of qubits and decrease power dissipation in devices. The future of NCFET is on the verge of a significant change in low-power computer electronics and other fields. As these devices transition from conceptual concepts to practical implementations, they possess the potential to revolutionize energy-efficient computing. Furthermore, in this work, a comprehensive design of NCFET and its application in switching are presented, along with the simulation methodology. Nevertheless, to fully exploit the revolutionary potential of NCFETs, it is imperative to address the obstacles associated with their stability, scalability, and CMOS compatibility. With the advancement of research and increasing acceptance by industries, NCFETs have been poised to revolutionize the future by promoting sustainability and technological advancement.

References

1. Skotnicki, T., Hutchby, J.A., King, T.J., Wong, H.S.P., Boeuf, F., The end of CMOS scaling: toward the introduction of new materials and structural changes to improve MOSFET performance. *IEEE Circ. Dev. Mag.*, 21, 1, 16–26, 2005. https://doi.org/10.1109/MCD.2005.1388765.

2. Chu, M., Sun, Y., Aghoram, U., Thompson, S.E., Strain: a solution for higher carrier mobility in nanoscale mosfets. *Annu. Rev. Mater. Res.*, 39, 203–229, 2009. https://doi.org/10.1146/annurev-matsci-082908-145312.
3. Chudzik, M., *et al.*, High-performance high-κ/metal gates for 45nm CMOS and beyond with gate-first processing, in: *Dig. Tech. Pap. - Symp. VLSI Technol*, pp. 194–195, 2007, https://doi.org/10.1109/VLSIT.2007.4339689.
4. Roy, K., Mukhopadhyay, S., Mahmoodi-Meimand, H., Leakage current mechanisms and leakage reduction techniques in deep-submicrometer CMOS circuits. *Proc. IEEE*, 91, 2, 305–327, 2003. https://doi.org/10.1109/JPROC.2002.808156.
5. Colinge, J.P., The SOI MOSFET: from single gate to multigate, in: *FinFETs and Other Multi-Gate Transistors*, pp. 1–48, 2008, https://doi.org/10.1007/978-0-387-71752- 4_1.
6. Razavieh, A., Zeitzoff, P., Nowak, E.J., Challenges and limitations of CMOS scaling for FinFET and beyond architectures. *IEEE Trans. Nanotechnol.*, 18, 999–1004, 2019. https://doi.org/10.1109/TNANO.2019.2942456.
7. Valasa, S., Tayal, S., Thoutam, L.R., Ajayan, J., Bhattacharya, S., A critical review on performance, reliability, and fabrication challenges in nanosheet FET for future analog/digital IC applications. *Micro Nanostruct.*, 170, 207374, 2022. https://doi. org/10.1016/j.micrna.2022.207374.
8. https://irds.ieee.org/editions/2021/beyond-cmos.
9. Sakurai, T., Perspectives of low-power VLSI's. *IEICE Trans. Electron.*, E87-C, 4, 429–436, 2004.
10. Gopalakrishnan, K., Griffin, P.B., Plummer, J.D., Impact ionization MOS (I-MOS)- Part I: device and circuit simulations. *IEEE Trans. Electron. Dev.*, 52, 1, 69–76, 2005. [Online]. Available, http://ieeexplore.ieee.org/lpdocs/epic03/wrapper. htm?arnumber=1372710.
11. Sarkar, D., *et al.*, A subthermionic tunnel field-effect transistor with an atomically thin channel. *Nature*, 526, 7571, 91–95, 2015. https://doi.org/10.1038/nature15387.
12. Kam, H., Lee, D.T., Howe, R.T., King, T.J., A new Nano-electro-mechanical Field Effect Transistor (NEMFET) design for low-power electronics. *Tech. Dig. - Int. Electron Devices Meet. IEDM*, 2005, 463–466, 2005. https://doi.org/10.1109/iedm.2005.1609380.
13. Salahuddin, S. and Datta, S., Use of negative capacitance to provide voltage amplification for low power nanoscale devices. *Nano Lett.*, 8, 2, 405–410, 2008. https://doi.org/10.1021/nl071804g.
14. Rusu, A., Salvatore, G.A., Jiménez, D., Ionescu, A.M., Metalferroelectric-meta-oxide-semiconductor field effect transistor with sub-60mV/decade subthreshold swing and internal voltage amplification, in: *Tech. Dig. Int. Electron Devices Meeting (IEDM)*, pp. 1–4, 2010, doi: 10.1109/IEDM.2010.5703374.
15. Íñiguez, J., Zubko, P., Luk'yanchuk, I., Cano, A., Ferroelectric negative capacitance. *Nat. Rev. Mater.*, 4, 4, 243–256, 2019.

16. Cao, W. and Banerjee, K., Is negative capacitance FET a steep-slope logic switch? *Nat. Commun.*, 11, 1, 1–8, Jan. 2020. doi: 10.1038/s41467-019-13797-9.
17. Nazir, G., Rehman, A., Park, S.-J., Energy-efficient tunneling field-effect transistors for low-power device applications: Challenges and opportunities. *ACS Appl. Mater. Interfaces*, 12, 42, 47127–47163, 2020.
18. Jo, J. and Shin, C., Negative capacitance field effect transistor with hysteresis-free sub-60-mV/decade switching. *IEEE Electron Device Lett.*, 37, 3, 245–248, Mar. 2016.
19. You, W.-X. and Su, P., Intrinsic difference between 2-D negative capacitance FETs with semiconductor-on-insulator and doublegate structures. *IEEE Trans. Electron Devices*, 65, 10, 4196–4201, Oct. 2018. doi: 10.1109/TED.2018.2866125.
20. Hisamoto, D., *et al.*, FinFET-a self-aligned double-gate MOSFET scalable to 20 nm. *IEEE Trans. Electron Devices*, 47, 12, 2320–2325, Dec. 2000.
21. Yu, B., *et al.*, FinFET scaling to 10 nm gate length, in: *Dig. Int. Electron Devices Meeting*, pp. 251–254, 2002.
22. Ieong, M., Doris, B., Kedzierski, J., Rim, K., Yang, M., Silicon device scaling to the sub-10-nm regime. *Science*, 306, 5704, 2057–2060, Dec. 2004. doi: 10.1126/SCIENCE.1100731.
23. Li, K.S., *et al.*, Sub-60mV-swing negative-capacitance FinFET without hysteresis, in: *Tech. Dig. Int. Electron Devices Meeting (IEDM)*, pp. 1–4, Feb. 2015, doi: 10.1109/IEDM.2015.7409760.
24. Khan, A.I., *et al.*, Negative capacitance in short-channel FinFETs externally connected to an epitaxial ferroelectric capacitor. *IEEE Electron Device Lett.*, 37, 1, 111–114, Jan. 2016. doi: 10.1109/LED.2015.2501319.
25. Ko, E., Lee, J.W., Shin, C., Negative capacitance FinFET with sub-20-mV/decade subthreshold slope and minimal hysteresis of 0.48 v. *IEEE Electron Device Lett.*, 38, 4, 418–421, Apr. 2017. doi: 10.1109/LED.2017.2672967.
26. Markov, I.L., Limits on fundamental limits to computation. *Nature*, 512, 7513, 147–154, 2014.
27. Pahwa, G., Dutta, T., Agarwal, A., Chauhan, Y.S., Designing energy efficient and hysteresis free negative capacitance FinFET with negative DIBL and 3.5X ION using compact modeling approach, in: *Proc. ESSCIRC Conf. 42nd Eur. Solid-State Circuits Conf.*, pp. 49–54, 2016, doi: 10.1109/ESSCIRC.2016.7598240.
28. Pahwa, G., Agarwal, A., Chauhan, Y.S., Numerical investigation of short-channel effects in negative capacitance MFIS and MFMIS transistors: Subthreshold behavior. *IEEE Trans. Electron Devices*, 65, 11, 5130–5136, Nov. 2018. doi: 10.1109/TED.2018.2870519.
29. Amrouch, H., Pahwa, G., Gaidhane, A.D., Henkel, J., Chauhan, Y.S., Negative capacitance transistor to address the fundamental limitations in technology scaling: Processor performance. *IEEE Access*, 6, 52754–52765, 2018. doi: 10.1109/ACCESS.2018.2870916.

30. Pahwa, G., Dutta, T., Agarwal, A., Chauhan, Y.S., Compact model for ferroelectric negative capacitance transistor with MFIS structure. *IEEE Trans. Electron Devices*, 64, 3, 1366–1374, Mar. 2017. doi: 10.1109/TED.2017.2654066.
31. Kobayashi, M., A perspective on steep-subthreshold-slope negative capacitance field-effect transistor. *Appl. Phys. Exp.*, 11, 11, 110101(1-20), Nov. 2018. Art. no. 110101, doi: 10.1109/TED.2017.2654066.
32. Malvika, C,.B. and Mummaneni, K., A Review on a Negative Capacitance Field-Effect Transistor for Low-Power Applications. *J. Electron. Mater.*, 51, 923–937, 2022. https://doi.org/10.1007/s11664-021-09384-8.
33. Malvika, C,.B. and Mummaneni, K., A new pocket-doped NCFET for low power applications: Impact of ferroelectric and oxide thickness on its performance. *Micro Nanostruct.*, 169, 207360, 2022. https://doi.org/10.1016/j.micrna.2022.207360.
34. Sentaurus, *Sentaurus sDevice 2022*, Simulation, pp. 1–12, 2022. (Manuals, Synopsis, Mountain View). Available online: https://www.synopsys.com/manufacturing/tcad/device-simulation/sentaurus-device.html.
35. Ricinschi, D., Harnagea, C., Papusoi, C., Mitoseriu, L., Tura, V., Okuyama, M., Analysis of ferroelectric switching in finite media as a Landau-type phase transition. *J. Phy. Condens. Matt.*, 10, 477–492, 1999. 10.1088/0953-8984/10/2/026.

5

The Future of Low Power Electronics: Tunnel Field-Effect Transistors

Sourav Das[1], Ekta Goel[2] and Kunal Singh[1*]

[1]*Solid State Electronics and VLSI Lab, Department of ECE, NIT Jamshedpur, Jharkhand, India*
[2]*National Institute of Technology, Warangal, India*

Abstract
Tunnel Field-Effect Transistors (TFETs) have emerged as a promising technology for low power electronic applications due to their steep subthreshold swing and proficiency to operate at lower supply voltages compared to conventional MOSFETs. This chapter provides a comprehensive overview of TFETs, focusing on their fundamental principles, designs, and fabrication techniques. It explores the unique quantum tunneling mechanism that enables TFETs to attain superior performance in terms of power efficiency and scaling. This chapter also discusses the application of TFETs in several domains, including digital electronics, analog and mixed-signal circuits, biosensors, energy harvesting, and communication systems. Furthermore, the potential of TFETs in emerging technologies, such as wearable and implantable medical devices, is discussed, showcasing their role in the future of low power electronics. Future trends and ongoing research efforts aimed at overcoming current challenges and enhancing TFET capabilities are also examined. This comprehensive overview provides valuable insights for researchers, engineers, and industry professionals interested in harnessing the advantages of TFET technology for next-generation electronic applications.

Keywords: TFET, quantum tunneling, low power electronics, band-to-band tunneling, steep subthreshold swing, analog circuits, digital circuits, next-generation electronics

**Corresponding author*: kunalsingh.ece@nitjsr.ac.in

Ekta Goel, Archana Pandey, Shiromani Balmukund Rahi and Arun Samuel (eds.) Handbook of Advanced Semiconductor Field Effect Transistors, (89–106) © 2026 Scrivener Publishing LLC

5.1 Introduction

Tunnel Field-Effect Transistors are a specific type of semiconductor device that operates based on quantum tunneling phenomena, offering significant advantages in low-power applications, in addition to traditional MOSFETs. Unlike MOSFETs, which trust thermal emission over a potential barrier, TFETs utilize quantum mechanical tunneling through a narrow-bandgap semiconductor material between drain and source terminals. The key advantage of TFETs is their steep subthreshold swing, typically below 60 mV/decade, which allows them to switch on and off more sharply and operate effectively at lower supply voltages. This characteristic reduces power consumption and enhances energy efficiency, making TFETs particularly suitable for applications that demand extended battery life, such as IoT devices, wearable electronics, and medical implants. TFETs are also explored for their potential in analog and mixed-signal applications where minimizing leakage current and achieving high signal fidelity are crucial. Additionally, their sensitivity to low voltages makes them promising candidates for biosensing applications that offer precise and energy-efficient detection of biological molecules. Continued research aims to optimize the TFET performance, explore new materials, and advance fabrication techniques to unlock their full potential in next-generation electronics, contributing to the advancement of more capable and sustainable electronic devices. Compared to the lateral TFET, the V-TFET offers several advantages, one of which is a steeper subthreshold slope as tunneling is directly regulated by the gate voltage. The steep subthreshold slope makes it possible to operate at a low power and OFF current. When the gate voltage strikes a particular threshold, tunneling arises vertically to the Si-oxide interface, directly regulates tunneling. As a result, this device may have a smaller SS and a greater ON-current than the lateral gated p-i-n construction [1, 2].

5.2 Fundamental Principles of TFET Operation

We continue this chapter by discussing the design and working principles of the TFETs. We describe the origins and evolution of some TFETs and their electrical characteristics. We established an approximation relating the electrical and physical features of the TFETs and the tunneling current. This will make it easier for readers to modify the device parameters to maximize TFETs' electrical properties. This chapter concludes by highlighting the benefits and drawbacks of TFETs and providing a brief explanation of why they might replace traditional MOSFETs in future applications. We discuss a few applications using TFETs and the proposed device structure of vertical-based TFET with significant advantages and disadvantages of TFETs in analog applications.

5.2.1 Device Structure

The TFET has a simply gated reverse-biased p-i-n structure. Figure 5.1 depicts the basic operation of n-type and p-type TFET. Specific doping of the source and drain is the fundamental characteristic of a TFET that distinguishes it from other electrical devices. In TFET, different types of doping are used in the source and drain terminals.

The source of an n-type TFET is doped p+, and the drain is doped n+, while the source of a p-type TFET is doped n+, and the drain is doped p+. A lightly doped or intrinsic p- or n-type semiconductor forms a channel. A layer of dielectric differs from the channel of the gate electrode, similar to a traditional MOSFET. The n- and p-type TFET biasing strategies are shown in Figure 5.1. The drain and gate electrodes of an n-type TFET are given a positive voltage, while the source is grounded. For a p-type material, a negative is applied in the gate and drain contacts, and the source is kept at ground. For n-type TFET, when the dominant carriers in the channel are electrons, p-type TFET holes in the channel are considered as the dominant carriers. The source and drain of an n-type TFET are where the electrons enter and exit the channel, respectively. Through the source and drain, the holes in the p-type TFET flowed into and out of the channel, respectively. Band-to-band tunneling (BTBT) is the mechanism by which dominant carriers reach the channel.

The TFET structures depicted in Figure 5.1 have many shortcomings, the most serious of which is the exceedingly small ON current. Therefore, to address these issues with the TFET's basic implementation, researchers have developed and created TFETs in a variety of other structural configurations. For this reason, we chose the DGTFET, which has a simplified construction and can be used to quickly establish an understanding of TFET.

A cross-sectional view of an n-type DG TFET is shown in Figure 5.2 [3]. It should be mentioned that several researchers have employed structures similar to the DGTFET in both theoretical and experimental studies [4–7].

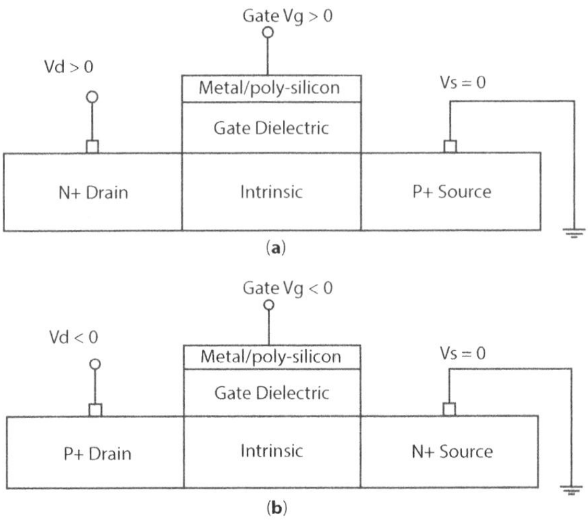

Figure 5.1 A simple configuration of a TFET: (a) n-type TFET and (b) p-type TFET [3].

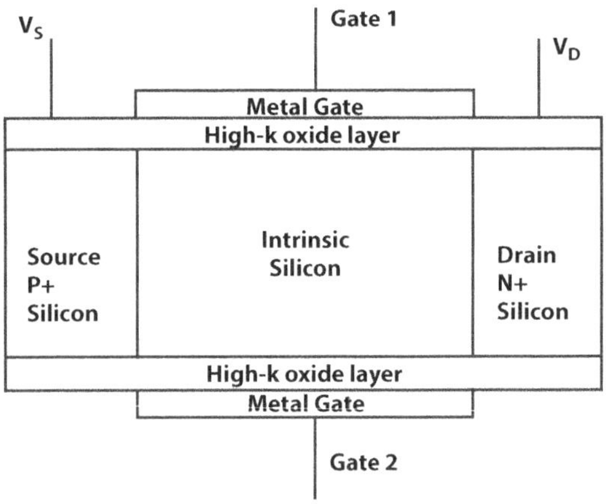

Figure 5.2 2-D sectional view of a DGTFET [3].

5.2.2 Operation

Figure 5.3 displays the band illustrations of an n-type TFET in both the ON and OFF states. When the applied gate voltage remains zero it is considering OFF-state of the TFET. The conduction band in the channel is above the valence band of the source. Consequently, BTBT is prevented in the OFF state with very little drain current. An increase in gate voltage causes the conduction band in the channel to be pushed downward and

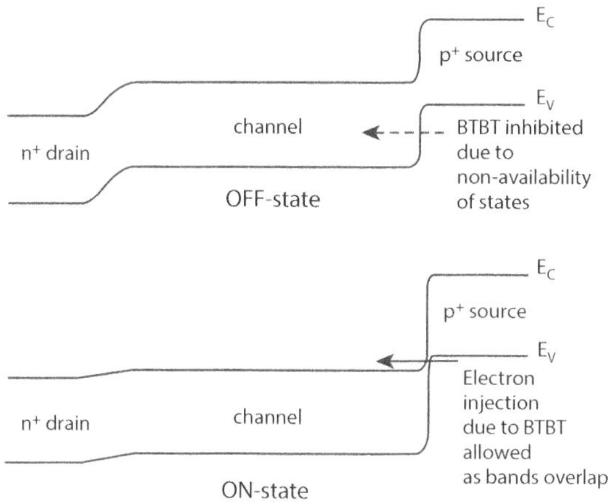

Figure 5.3 (a) OFF-state and (b) ON-state of n-type TFET [3].

changes the carrier density. Figure 5.3 illustrates how band bending occurs at the gate when a sufficiently high voltage is applied, aligning the conduction band in the channel with the valence band in the source. The positive bias of the drain strikes the electrons that tunnel into the channel to the drain terminal. This is the fundamental operation of n-type TFET.

5.2.3 Transfer Characteristics of TFET

The TFET transfer characteristics can be broadly classified into three regions as shown in Figure 5.4 [3]:

- OFF state ($0 < V_{GS} < V_{OFF}$): BTBT is reduced and the TFET has a very low current in this area. The gate voltage, or V_{OFF}, is the point where the drain current begins to flow.

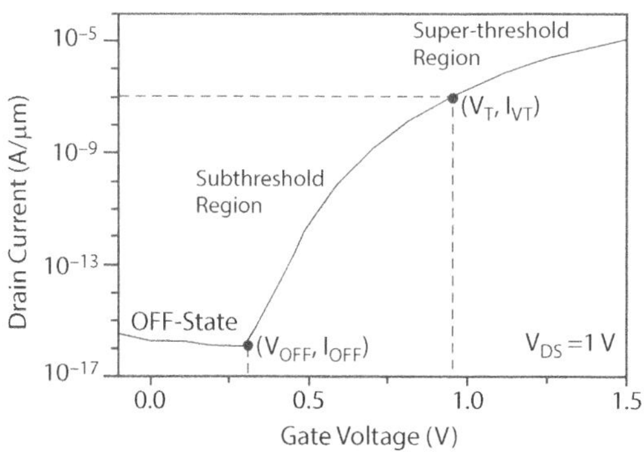

Figure 5.4 Simulated transfer characteristics of the TFET [3].

- Subthreshold area ($V_{OFF} < V_{GS} < V_T$): This area shows a sharp increase in the drain current. V_T is the DGTFET's threshold voltage.
- Superthreshold region ($V_{GS} > V_T$): In this area, the drain current increases in parallel with the gate voltage being lowered.

5.2.4 Different Types of TFETs

TFETs are various types, each with unique characteristics and advantages depending on their design, materials, fabrication feasibility, and intended applications.

Different types of tunnel field-effect transistors (TFETs) in various configurations and their advantages and challenges are discussed in this section:

1. Homojunction TFETs [8, 9]
   - Description: These TFETs use the same semiconductor material for both the source and the drain.
   - Example Materials: Silicon, Germanium.
   - Advantages: Easier fabrication process since only one material is used.
   - Challenges: Limited performance improvements due to the inherent material properties.

2. Heterojunction TFETs [10–12]
   - Description: These TFETs use different semiconductor materials for the source and drain regions, creating heterojunctions.
   - Example Materials: Silicon–Germanium (SiGe), III–V compounds (such as InGaAs).
   - Advantages: Improved band alignment and tunneling efficiency, leading to higher on-state currents.
   - Challenges: More complex fabrication and potential material compatibility issues.

3. III–V TFETs [13, 14]
   - Description: These TFETs use III–V compound semiconductors known for their high electron mobility and favorable bandgap properties.

- Example Materials: InAs, GaSb, InGaAs.
- Advantages: Superior performance due to better tunneling efficiency and a higher on-state current.
- Challenges: Higher manufacturing costs and integration challenges with existing silicon technology.

4. Vertical TFETs [15, 16]
   - Description: These TFETs have a vertical structure wherein the source and drain are stacked on top of each other.
   - Advantages: Better electrostatic control and potentially higher drive currents.
   - Challenges: More complex fabrication processes and difficulty in scaling down to smaller dimensions.

5. Graphene Nanoribbon TFETs (GNR-TFETs) [17, 18]
   - Description: These TFETs use graphene nanoribbons, which are narrow strips of graphene, as channel materials.
   - Advantages: High carrier mobility and excellent electrostatic control.
   - Challenges: Bandgap engineering in graphene is challenging, and the fabrication processes are still under development.

5.2.5 Point Versus Line Tunneling

Two forms of tunneling can be specified in a TFET: Point tunneling and line tunneling, depending on how the electric field is formed in the tunnel paths by the applied gate potential. A typical flow of point and line tunneling is depicted in Figure 5.5, where the point tunneling concept is introduced, with the gate covering the channel. The concept of "point tunneling" is based on the curvature tunneling pathway that lies above the tunnel junction and forms the interface between the gate metal and gate dielectric [19].

A significant overlap of the gate and the source may lead to line tunneling, as shown in Figure 5.5(b) [19]. In contrast to point tunneling, in the

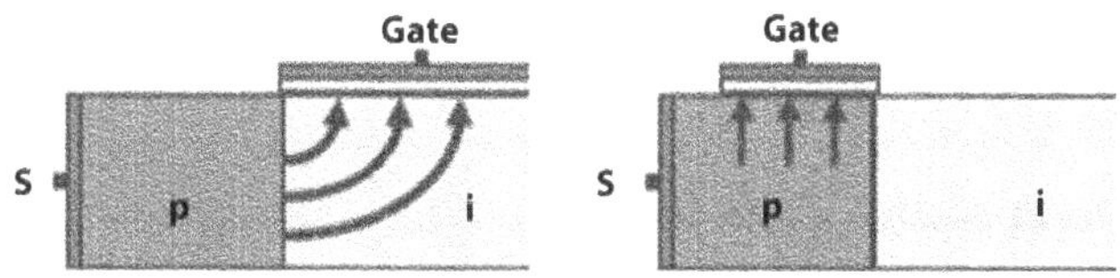

Figure 5.5 (a) Point and (b) line tunneling TFET arrangement [19].

ideal case, where there are no parasitic routes, the tunnel paths lie vertical to the gate dielectric [20]. Researchers are continuously focusing on TFET technology to increase the current efficiency of electronic devices.

Different device architectures such as vertical, source pocket engineered, and L-shaped/U-shaped have been explored for improving the I_{ON}/I_{OFF} ratio [21]. In addition, when the device is turned on, the lateral gate pocket overlap increases the tunneling area and electric field at the top of the tunneling junction [19].

5.3 Applications of TFETs

TFETs have been explored for use in a variety of applications, including memory, sensors, analog circuits, and traditional logic gates. Several theoretical studies have shown that TFET-based circuits can operate better than traditional CMOS circuits at low supply voltages. These findings have prompted scientists to investigate circuit designs that exploit the TFETs' unique characteristics to save energy and power.

5.3.1 Electrical Characteristics

Differentiating TFET from traditional MOSFET involves specific electrical features. The following characteristics are crucial for employing TFETs in circuit implementation.

Subthreshold Swing: At normal temperatures, TFETs can display subthreshold swings of less than 60 mV/decade compared to conventional MOSFET. The TFETs' low subthreshold swing allows for voltage scaling without affecting their functionality.

Unidirectional Conduction: Because of its asymmetric p-i-n structure, a TFET flows current in a single direction compared with a standard MOSFET. In an n-type TFET, a significant current flows only when V_{DS} is positive or the p-i-n structure is reverse-biased for a given gate voltage [22–25]. The unidirectional current flow property significantly affects on the application of the TFETs. By preventing reverse leakage pathways, the TFET's unidirectional characteristics can be used to increase the efficiency of the RF rectifier circuits.

Gate-to-Drain Miller Capacitance: In TFET, the gate-to-drain capacitance (C_{GD}) dominates the total gate capacitance (C_{GG}) under all bias conditions. For the OFF-state TFET, the channel-to-drain resistance was significantly lower than the channel-to-source resistance, and the C_{GD} was much larger than the C_{GS}.

There is a nominal potential drop between the channel and the drain when the TFET is in the ON state, which leads to a high C_{GD} [26]. This is in contrast to a conventional MOSFET, in which the saturation area is dominated by the gate-to-source capacitance (C_{GS}), whereas the gate-to-drain capacitance (C_{GD}) contributes nearly equally in the linear region [26]. The TFET's large C_{GD} results in a larger coupling between its output pin, drain, input pin, or gate in the circuits. This coupling has a significant effect on the behavior of the circuit because of the Miller effect.

Identical Characteristics of N- and P-Type TFETs: It is preferable for the properties of an N-TFET and a P-TFET to be identical to use TFETs in CMOS circuit design. The drive parameters of the NMOS and PMOS transistors were matched for a standard MOSFET by correctly adjusting the transistor width (W) and length (L) ratios. However, there are additional variations in the electrical properties between an N-TFET and a P-TFET, such as the subthreshold swing and the mismatch in the driving strengths [27].

5.3.2 Digital Circuits and Memories

The delay properties, namely propagation delay, power dissipation, rise/fall times, circuit area, and noise resilience, are the key matrices adjusted to standard digital circuits [28]. To evaluate the reasonable advantage of TFETs over MOSFETs, these properties were generated for TFET-based circuits and compared with standard CMOS circuits. Researchers are looking more closely at the features of inverters realized using TFETs because they are the most basic type of digital circuit. In digital circuits other than inverters, TFETs are being investigated for possible application [29, 30]. It is possible to create pass transistor logic (PTL) with TFETs [32]. Because of their excellent subthreshold swing and incredibly low leakage current, TFETs have been thoroughly studied for the purpose of creating SRAMs [31, 32].

5.3.3 Analog Circuits

TFETs have mostly been studied for digital use and as logic application substitutes for traditional MOSFETs. Still, in recent times, scientists have assessed the benefits of utilizing TFETs in various types of analogue circuits [33–37]. These investigations have demonstrated that, owing to their excellent saturation in the output characteristics and minimal subthreshold swing, TFETs are a favorable option for low-power analog circuits. TFETs can be used in amplifier circuits that are similar to circuits that use a traditional MOSFET. A recent comparison between the Si TFET application and CMOS implementation of an identical circuit was developed [36]. When the Si TFET-based amplifier is compared with a comparable CMOS amplifier, it is shown that the Si TFET's high output resistance leads to a higher voltage gain. The charge and discharge times of the hold capacitor in the Track and Hold (T/H) circuit varied significantly as a result of the unidirectional current flow in the TFET [38]. A Schmitt-trigger inverter can be implemented by taking advantage of TFETs to reduce positive drain bias [38].

5.3.4 Future Perception of TFETs in Circuits

Most research has shown that TFETs provide noticeable benefits over traditional MOSFETs in terms of power conservation at low to medium frequencies. The switching rates of TFETs are lower when they function below the quantum capacitance limit [39]. In addition, TFETs are promising for ultralow analog circuits and SRAM applications. However, many investigations have shown that a TFET's greater energy efficiency over CMOS is only possible at low supply voltages.

5.4 Literature Review

It should be mentioned that VTFETs are easier to develop, denser, and less immune to trapping issues than conventional transistors [40–42]. Using the gate-stacking technique, Gupta [43] investigated a Vertical TFET device with an n + SiGe-doped layer using a TCAD simulation platform. Seema [44] suggested using a hetero-dielectric buried oxide (HDB) and a double-gate vertical TFET linearity study for high-frequency applications. Vanlalawmpuia investigated a Ge-source vertical TFET with a Delta-Doped Layer [15]. Here, the tunneling direction and the gate electric field are parallel. An investigation of low-frequency flicker noise on dual dielectric pockets with three distinct gate-source underlap thicknesses and lengths was reported by Das *et al.* The proposed device has a high current ratio and a significantly steep slope [45]. Using low bandgap material reduces the tunneling width and increases the drain current. However, there is a concession between performance and power consumption with staggered heterojunction transistors (TFETs) based on III–V materials, which makes them appropriate for low-voltage operation, as reported by Tomas *et al.* [46].

5.5 Simulation of Dual Metal Double Gate Hetero Pocket V-TFET

5.5.1 Device Structure

The band-to-band-tunneling (BTBT) nonlocal model was used in the TCAD simulation to incorporate the shift of the energy band. The other models used in this simulation are bandgap-narrowing, Shockley–Read–Hall generation-recombination, field-dependent mobility, and Auger recombination, and Fermi–Dirac. To increase the drain current and reliability of the fabrication, a high-k (= 9.3) Al_2O_3 material was used in the gate oxide.

Here, drain current is plotted against gate voltage changing it from −0.2 V to 1 V. The DMDG-hetero-V-TFET consists of different work functions along the channel length. The gate near the source side is termed the tunneling gate, and that near the drain side is called the auxiliary gate. The tunnel gate has a lower work function that controls the ON current (I_{ON}), and the auxiliary gate has a higher work function, which is helpful for controlling the OFF current (I_{OFF}). We generated a sample output file after simulating a Hetero-Pocket-V-TFET using a 2-D ATLAS device simulator (version 5.26.1.R). Figure 5.6 shows the 2-D cross-sectional view of the DG-hetero-pocket-V-TFET. The device parameters for the DG-hetero-pocket-V-TFET are listed in Table 5.1.

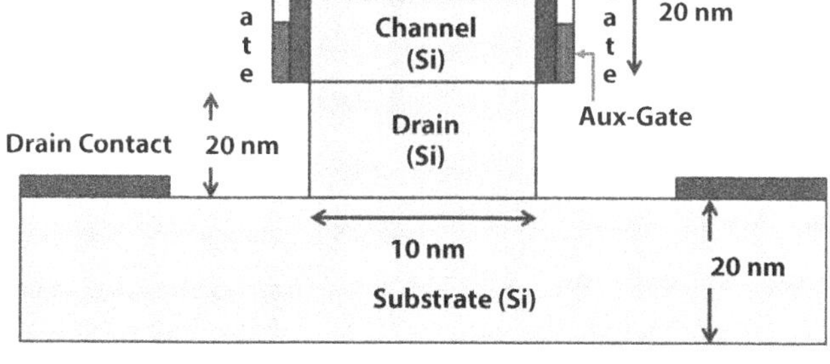

Figure 5.6 Schematic cross-sectional view of the dual metal double gate hetero pocket V-TFET.

Table 5.1 Device parameters for the DG-hetero-pocket-V-TFET.

Parameters	Values (Hetero-V-TFET)
Pocket length (L_p)	3 nm
Al_2O_3 thickness (t_{Al2O3})	2 nm
Substrate length (L_{Si})	20 nm
Channel doping (N_{CH})	1×10^{15} cm^{-3}
Pocket doping (N_D)	6×10^{18} cm^{-3}
Drain doping (n-type)(N_D)	5×10^{19} cm^{-3}
Source doping (p-type)(N_S)	1×10^{20} cm^{-3}
Gate work function Tunnel/Aux	4.1/4.8

5.5.2 Sample Output File

Figure 5.7a shows the drain current characteristics of the suggested pocket hetero-VTFET. Here, drain current is plotted against gate voltage keeping drain voltage (V_{DS}) fixed at 0.9 V and V_{gs} is varying from −0.2 V to 1 V. The present work provides novel structures for Hetero-V-TFETs. After the simulation, the different parameters were extracted shown in Table 5.2. Recalibrating and adjusting other tool-specific parameters may be necessary for the tunneling model, subject to the type of tool used, accuracy level, and the physical model.

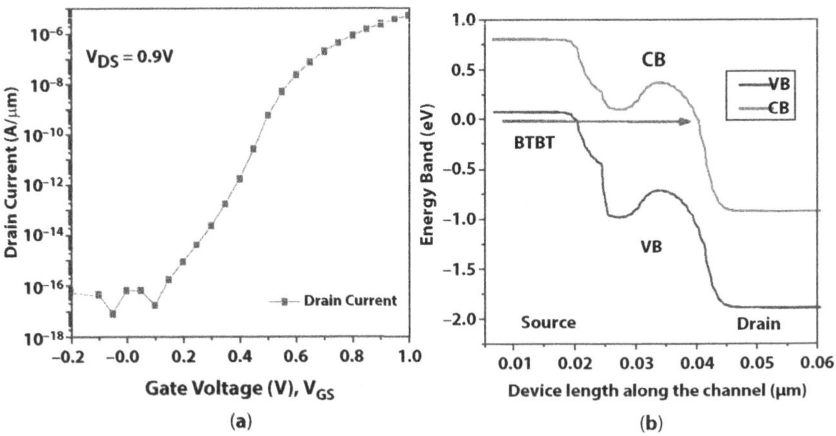

Figure 5.7 (a) I_D-V_{gs} characteristics and (b) energy band diagram of DG-Hetero-V-TFET.

Table 5.2 Different types parameters of Hetero-V-TFET.

Parameters	Values (Hetero-V-TFET)
Subthreshold-Swing (SS)	30 mV/decade
Threshold Voltage (V_{th})	0.38 V
OFF-State Current (I_{OFF})	1.21×10^{-17}
ON-State Current (I_{ON})	5.62×10^{-6}
Ratio (I_{ON}/I_{OFF})	4.63×10^{11}

5.6 Conclusion

This chapter evaluates the present status of TFET research. We examined the features of modern TFETs and their benefits over MOSFETs in terms of their competitiveness. Economically TFETs are more beneficial than conventional MOSFETs for analog applications are being researched, as is the case in many types of analog circuits. We will go through a few of the applications using TFETs in recently proposed analog circuits and the significance of the advantages and disadvantages of TFETs in analog applications. As we have seen, there are three main methods for increasing I_{ON} in TFETs: creating a better gate structure, modifying the tunneling junction, and incorporating new material systems. Owing to technological advancements, TFETs have become the most efficient alternative for low-power commercial applications.

References

1. Das, S., Kumar, B.B., Bundela, P., Singh, K., Performance assessment of Si based dual metal double gate vertical TFET biosensor. *Micro Nanostruct.*, 191, 207864, 2024. https://doi.org/10.1016/J.MICRNA.2024.207864.
2. Han, Z. F., Ru, G. P., Ruan, G., A simulation study of vertical tunnel field effect transistors. In *2011 9th IEEE International Conference on ASIC*, IEEE, pp. 665–668, 2011, October.
3. Saurabh, S. and Kumar, M.J., *Fundamentals of tunnel field-effect transistors*. CRC press, 2016.
4. Boucart, K. and Ionescu, A.M., Double-gate tunnel FET with high-κ gate dielectric. *IEEE Trans. Electron Devices*, 54, 1725–1733, 2007. https://doi.org/10.1109/TED.2007.899389.

5. Toh, E.H., *et al.*, Device physics and design of double-gate tunneling field-effect transistor by silicon film thickness optimization. *Appl. Phys. Lett.*, 90, 26, 2007.
6. Toh, E.H., *et al.*, Device physics and guiding principles for the design of double-gate tunneling field effect transistor with silicon-germanium source heterojunction. *Appl. Phys. Lett.*, 91, 24, 2007.
7. Verhulst, A.S., *et al.*, Groeseneken, G., Tunnel field-effect transistor without gate-drain overlap. *Appl. Phys. Lett.*, 91, 5, 2007.
8. Alian, A., Franco, J., Vandooren, A., Mols, Y., Verhulst, A., El Kazzi, S., Rooyackers, R., Verreck, D., Smets, Q., Mocuta, A., Collaert, N., Lin, D., Thean, A., Record performance InGaAs homo-junction TFET with superior SS reliability over MOSFET. *Tech. Dig. - Int. Electron Devices Meet. IEDM*, 2016, 31.7.1–31.7.4, February (2015). https://doi.org/10.1109/IEDM.2015.7409811.
9. Vandooren, A., Leonelli, D., Rooyackers, R., Hikavyy, A., Devriendt, K., Demand, M., Loo, R., Groeseneken, G., Huyghebaert, C., Analysis of trap-assisted tunneling in vertical Si homo-junction and SiGe hetero-junction Tunnel-FETs. *Solid State Electron.*, 83, 50–55, 2013. https://doi.org/10.1016/J.SSE.2013.01.026.
10. Knoch, J. and Appenzeller, J., Modeling of high-performance p-type IIIV heterojunction tunnel FETs. *IEEE Electron Device Lett.*, 31, 305–307, 2010. https://doi.org/10.1109/LED.2010.2041180.
11. Avci, U.E. and Young, I.A., Heterojunction TFET scaling and resonant-TFET for steep subthreshold slope at sub-9nm gate-length, in: *2013 IEEE International Electron Devices Meeting*, IEEE, 2013.
12. Han, T., Liu, H., Wang, S., Chen, S., Li, W., Yang, X., Cai, M., Yang, K., Design and Investigation of the Junction-Less TFET with Ge/Si0.3Ge0.7/Si Heterojunction and Heterogeneous Gate Dielectric. *Electron*, 8, 476, 2019. https://doi.org/10.3390/ELECTRONICS8050476.
13. Moselund, K.E., Cutaia, D., Schmid, H., Riel, H., Sant, S., Schenk, A., Complementary III-V heterostructure tunnel FETs. *Eur. Solid-State Device Res. Conf.*, 2016, 403–407, October (2016). https://doi.org/10.1109/ESSDERC.2016.7599671.
14. Convertino, C., Zota, C.B., Schmid, H., Ionescu, A.M., Moselund, K.E., III–V heterostructure tunnel field-effect transistor. *J. Phys. Condens. Matter*, 30, 264005, 2018. https://doi.org/10.1088/1361-648X/AAC5B4.
15. Vanlalawpuia, K. and Bhowmick, B., Investigation of a Ge-Source Vertical TFET with Delta-Doped Layer. *IEEE Trans. Electron Devices*, 66, 4439–4445, 2019. https://doi.org/10.1109/TED.2019.2933313.
16. Chen, F., Ilatikhameneh, H., Tan, Y., Klimeck, G., Rahman, R., Switching Mechanism and the Scalability of Vertical-TFETs. *IEEE Trans. Electron Devices*, 65, 3065–3068, 2018. https://doi.org/10.1109/TED.2018.2831688.

17. Zhang, Q., Fang, T., Xing, H., Seabaugh, A., Jena, D., Graphene nanoribbon tunnel transistors. *IEEE Electron Device Lett.*, 29, 1344–1346, 2008. https://doi.org/10.1109/LED.2008.2005650.
18. Aghanejad Ahmadchally, A. and Gholipour, M., Investigation of 6-armchair graphene nanoribbon tunnel FETs. *J. Comput. Electron.*, 20, 1114–1124, 2021. https://doi.org/10.1007/S10825-021-01709-4/FIGURES/8.
19. Verreck, D. and G., G., The tunnel field-effect transistor, in: *Wiley Encycl. Electr. Electron. Eng*, Kuleuven, L., Verreck, BD., Groeseneken, G., Verhulst, A. (Eds.), 2016, W.E. of, undefined, https://doi.org/10.1002/047134608X.W8333.
20. Li, W., Liu, H., Wang, S., Chen, S., Yang, Z., Design of High Performance Si/SiGe Heterojunction Tunneling FETs with a T-Shaped Gate. *Nanoscale Res. Lett.*, 12, 1–8, 2017. https://doi.org/10.1186/S11671-017-1958-3/FIGURES/11.
21. Tripathy, M.R., Singh, A.K., Samad, A., Chander, S., Baral, K., Singh, P.K., Jit, S., Device and Circuit-Level Assessment of GaSb/Si Heterojunction Vertical Tunnel-FET for Low-Power Applications. *IEEE Trans. Electron Devices*, 67, 1285–1292, 2020. https://doi.org/10.1109/TED.2020.2964428.
22. Yang, X. and Mohanram, K., Robust 6T Si tunneling transistor SRAM design, in: *2011 Design, Automation & Test in Europe*, IEEE, 2011.
23. Solomon, P.M., Lauer, I., Majumdar, A., Teherani, J.T., Luisier, M., Cai, J., Koester, S.J., Effect of uniaxial strain on the drain current of a heterojunction tunneling field-effect transistor. *IEEE Electron Device Lett.*, 32, 464–466, 2011. https://doi.org/10.1109/LED.2011.2108993.
24. Mukundrajan, R., Cotter, M., Saripalli, V., Irwin, M.J., Datta, S., Narayanan, V., Ultra low power circuit design using tunnel FETs. *Proc. - 2012 IEEE Comput. Soc. Annu. Symp. VLSI ISVLSI*, 2012, 153–158, 2012. https://doi.org/10.1109/ISVLSI.2012.70.
25. Lee, Y., Kim, D., Cai, J., Lauer, I., Chang, L., Koester, S.J., Blaauw, D., Sylvester, D., Low-power circuit analysis and design based on heterojunction tunneling transistors (HETTs). *IEEE Trans. Very Large Scale Integr. Syst.*, 21, 1632–1643, 2013. https://doi.org/10.1109/TVLSI.2012.2213103.
26. Mookerjea, S., Krishnan, R., Datta, S., Narayanan, V., On enhanced miller capacitance effect in interband tunnel transistors. *IEEE Electron Device Lett.*, 30, 1102–1104, 2009. https://doi.org/10.1109/LED.2009.2028907.
27. Wu, C., Huang, R., Huang, Q., Wang, J., Wang, Y., Design Guideline for Complementary Heterostructure Tunnel FETs with Steep Slope and Improved Output Behavior. *IEEE Electron Device Lett.*, 37, 20–23, 2016. https://doi.org/10.1109/LED.2015.2499183.
28. Wang, S., Pan, A., Chui, C.O., Gupta, P., PROCEED: A Pareto Optimization-Based Circuit-Level Evaluator for Emerging Devices. *IEEE Trans. Very Large Scale Integr. Syst.*, 24, 192–205, 2016. https://doi.org/10.1109/TVLSI.2015.2393852.

29. Zhao, Q.T., Richter, S., Schulte-Braucks, C., Knoll, L., Blaeser, S., Luong, G.V., Trellenkamp, S., Schafer, A., Tiedemann, A., Hartmann, J.M., Bourdelle, K., Mantl, S., Strained Si and SiGe nanowire tunnel FETs for logic and analog applications. *IEEE J. Electron Devices Soc.*, 3, 103–114, 2015. https://doi.org/10.1109/JEDS.2015.2400371.
30. Morris, D.H., Avci, U.E., Rios, R., Young, I.A., Design of low voltage tunneling-FET logic circuits considering asymmetric conduction characteristics. *IEEE J. Emerg. Sel. Top. Circuits Syst.*, 4, 380–388, 2014. https://doi.org/10.1109/JETCAS.2014.2361054.
31. Singh, J., *et al.*, A novel Si-tunnel FET based SRAM design for ultra-low-power 0.3 V VDD applications, in: *2010 15th Asia and South Pacific Design Automation Conference (ASP-DAC)*, IEEE, 2010.
32. Saripalli, V., Mohata, D.K., Mookerjea, S., Datta, S., Narayanan, V., Low power Loadless 4T SRAM cell based on degenerately doped source (DDS) In0.53Ga0.47As Tunnel FETs, in: *Device Res. Conf. - Conf. Dig. DRC*, pp. 101–102, 2010, https://doi.org/10.1109/DRC.2010.5551859.
33. Fulde, M., Heigl, A., Wachutka, G., Knoblinger, G., Schmitt-Landsiedel, D., Complementary multi-gate tunnelling FETs: Fabrication, optimisation and application aspects. *Int. J. Nanotechnol.*, 6, 628–639, 2009. https://doi.org/10.1504/IJNT.2009.025301.
34. Guo, P.F., *et al.*, Temperature independent current biasing employing TFET. *Electron. Lett.*, 46, 11, 786–787, 2010.
35. Senale-Rodriguez, B., *et al.*, Perspectives of TFETs for low power analog ICs, in: *2012 IEEE Subthreshold Microelectronics Conference (SubVT)*, IEEE, 2012.
36. Mallik, A. and Chattopadhyay, A., Tunnel field-effect transistors for analog/mixed-signal system-on-chip applications. *IEEE Trans. Electron Devices*, 59, 888–894, 2012. https://doi.org/10.1109/TED.2011.2181178.
37. Trivedi, A.R., Carlo, S., Mukhopadhyay, S., Exploring tunnel-FET for ultra low power analog applications: A case study on operational transconductance amplifier, in: *Proc. - Des. Autom. Conf*, 2013, https://doi.org/10.1145/2463209.2488868.
38. Sedighi, B., Hu, X.S., Liu, H., Nahas, J.J., Niemier, M., Analog circuit design using tunnel-FETs. *IEEE Trans. Circuits Syst. I Regul. Pap.*, 62, 39–48, 2015. https://doi.org/10.1109/TCSI.2014.2342371.
39. Koswatta, S.O., Lundstrom, M.S., Nikonov, D.E., Performance comparison between p-i-n tunneling transistors and conventional MOSFETs. *IEEE Trans. Electron Devices*, 56, 456–465, 2009. https://doi.org/10.1109/TED.2008.2011934.
40. Chander, S., Baishya, S., Sinha, S.K., Kumar, S., Singh, P.K., Baral, K., Tripathy, M.R., Singh, A.K., Jit, S., Two-dimensional analytical modeling for electrical characteristics of Ge/Si SOI-tunnel FinFETs. *Superlattices Microstruct.*, 131, 30–39, 2019. https://doi.org/10.1016/J.SPMI.2019.05.037.

41. Neves, F.S., Agopian, P.G.D., Martino, J.A., Cretu, B., Rooyackers, R., Vandooren, A., Simoen, E., Thean, A.V.Y., Claeys, C., Low-Frequency Noise Analysis and Modeling in Vertical Tunnel FETs with Ge Source. *IEEE Trans. Electron Devices*, 63, 1658–1665, 2016. https://doi.org/10.1109/TED.2016.2533360.
42. Singh, S. and Raj, B., Two-dimensional analytical modeling of the surface potential and drain current of a double-gate vertical t-shaped tunnel field-effect transistor. *J. Comput. Electron.*, 19, 1154–1163, 2020. https://doi.org/10.1007/S10825-020-01496-4/FIGURES/9.
43. Gupta, S., Wairya, S., Singh, S., Design and Analysis of Triple Metal Vertical TFET Gate Stacked with N-Type SiGe Delta-Doped Layer. *Silicon*, 14, 4217–4225, 2022. https://doi.org/10.1007/S12633-021-01211-3/METRICS.
44. Seema, S.S., Chauhan, Linearity Performance Analysis of Double Gate (DG) VTFET Using HDB for RF Applications. *Silicon*, 13, 1121–1125, 2021. https://doi.org/10.1007/S12633-020-00499-X/METRICS.
45. Das, D. and Chakraborty, U., A Study on Dual Dielectric Pocket Heterojunction SOI Tunnel FET Performance and Flicker Noise Analysis in Presence of Interface Traps. *Silicon*, 13, 787–798, 2021. https://doi.org/10.1007/S12633-020-00488-0/METRICS.
46. Gonzalez, T., Introduction to special issue on noise modeling. *J. Comput. Electron.*, 14, 1–3, 2015. https://doi.org/10.1007/S10825-015-0674-4.

6

Novel Gate All Around FET with Enhanced Performance and Improved Process Sensitivity

Mandeep Singh Narula*, Archana Pandey and Ajay Kumar

Jaypee Institute of Information Technology, Noida, India

Abstract

MOSFETs have been used in the integrated circuits for a long time. Device dimensions are shrinking continuously due to scaling. A reduction in transistor dimensions is essential for high chip density, low power, increased speed of operation, reduced chip cost, and overall enhanced performance. However, the transistor's smaller size results in numerous undesirable short channel effects (SCE), including hot carrier effect, velocity saturation, drain induced barrier lowering (DIBL), and reduced gate control over the channel. In order to overcome these issues, MOSFETs were replaced by FinFET's in the year 2011. But for highly scaled devices, the performance of FinFETs is compromised due to various short channel effects which start degrading the device performance. A promising alternative to FinFETs in the near future is the Gate All Around FET (GAA FET), which has excellent short channel performance. GAA FETs have gates on all four sides, in contrast to FinFETs, which have gates around the channel on three sides only. Because of this, the GAA FET has better gate control, has higher drive current, higher I_{ON}/I_{OFF}, and is less susceptible to process fluctuations. Also, the short channel performance of GAA FET is better than as compared to FinFET.

Multi-gate transistors, such as FinFET and GAA FET, are extremely complex to fabricate. For instance, it is challenging to accurately adjust the fin's height, width, and inclination angle. In the same way, it is challenging to etch the gate oxide uniformly. Another problem is the oxide and channel contact surface roughness, which can lead to scattering. Another significant problem that might deteriorate the electrical properties of the device is the random positioning of dopants. The very limited number of dopants for narrow gate lengths makes them difficult to

**Corresponding author*: mandeep.narula@jiit.ac.in

Ekta Goel, Archana Pandey, Shiromani Balmukund Rahi and Arun Samuel (eds.) Handbook of Advanced Semiconductor Field Effect Transistors, (107–118) © 2026 Scrivener Publishing LLC

manage during construction. The performance of the device may be deteriorated by variations in the threshold voltage brought on by variations in the dopant concentration over various areas of the channel. Because of this, the threshold voltage varies greatly at higher channel doping values and less at lower channel doping values. Dopant concentrations can range from very low in the center to quite high on the edges. This is because the average dopant concentration might change for very tiny gate length values due to the discrete nature of dopant atoms. Therefore, it is crucial to study the impact of process variations due to modifications in device dimensions during the fabrication process. The impact of process variations on device performance is covered in this chapter. A detailed discussion has been done regarding the impact of varying the radius, gate oxide thickness, gate length, and channel doping of GAA FETs. For each of the aforementioned variations, the percentage change in the drive current and threshold voltage has been examined. Changes in oxide thickness, nanowire radius, gate length, channel doping, and nanowire thickness do not significantly affect the drive current or threshold voltage. The sensitivity of oxide thickness and gate length is significantly lower than that of the GAA FET's channel and radius doping. The proposed device structure is therefore appropriate for uses where a high degree of immunity to process fluctuations is needed.

Keywords: FinFET, gate all around, process variation, scaling, drive current

6.1 Introduction

Over the past few decades, chip sizes have been steadily reduced due to scaling. In 1965, Moore noticed that silicon transistors were constantly decreasing. The transistor count of an integrated circuit doubles every two years, as predicted by Moore [1]. The transistor minimum feature size has dropped from 10 µm to sub-10 nm since its measurement, and it will continue to drop. Scaling entails reducing the doping concentration, channel length and width, and oxide thickness of the MOSFETs. Some of the current physical characteristics of transistors are already approaching the upper threshold. In MOSFETs, the gate oxide thickness drops below 1 nm. As a result, the oxide capacitance was higher, which contributed to the high operating current. However, this also leads to an increase in the gate-tunneling current above the threshold, which is undesirable. The leakage current may increase sharply if the oxide layer's thickness is further reduced. Furthermore, it may cause the hot electron effect and trap charge carriers in the oxide, impairing device performance and preventing achievement of the intended electrical properties [2–5]. Other problems can degrade the device, including increased power dissipation and the risk of rupture of the gate oxide. To solve this problem, a unique architecture or technique is required [6–9].

Further improvement was achieved in 2007 using a metal gate with a high dielectric constant. These issues can be avoided by mixing SiO_2 with a dielectric material that has a high dielectric constant. A high-dielectric-constant insulator can have greater thickness, but the effective oxide thickness (EOT) remains the same. Therefore, a SiO_2 layer is necessary in addition to an insulator to provide an outstanding interface quality [10]. The best material for stacking oxide layers is HfO_2, as hafnium oxide has a high bandgap, a very low tunneling probability, and the ability to withstand high temperatures. Researchers have studied several minerals, including SiO_2, HfO_2 (K = 25), Al_2O_3 (k = 10), La_2O_3 (k = 27), and TiO_2 (k = 50). TiO_2 has proven to be a very good material for suppressing leakage current, but the performance of devices is negatively impacted by its high dielectric constant and large capacitance [11, 12].

FinFET was launched in 2011. This marked the first instance of a 2D transistor being swapped out for a 3D transistor. A higher drive current, simpler multigate transistor construction, increased speed, low leakage current, improved resistance to short-channel effects, decreased fluctuations caused by dopants, and scalability are some of the benefits of FinFETs [5, 13, 14].

FinFETs and Gate-All-Around FET (GAA FET) work similarly. Unlike the FinFET, which has a gate that encircles the channel on three sides, the GAA FET has a gate that entirely encircles the channel. Unlike FinFETs, where the gate surrounds the channel from three sides only, GAA FETs have a gate that surrounds the channel from all four sides. However, MOSFET have a planar structure. This is illustrated in Figures 6.1 and 6.2 [15].

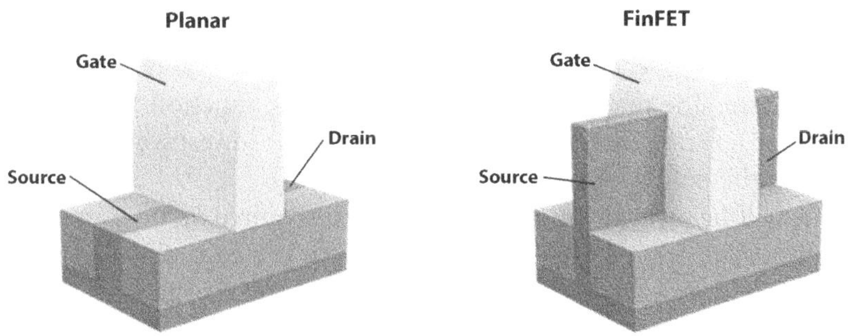

Figure 6.1 MOSFET vs FinFET.

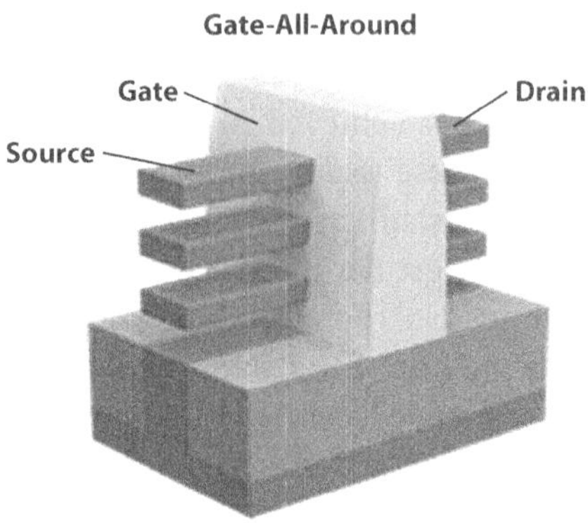

Figure 6.2 Gate all around FET.

In the near future, gate all around (GAA) FET are expected to replace FinFETs in technology nodes because they have a gate that completely wraps the channel, giving it better gate control over the channel and therefore better immunity to short-channel effects than FinFETs [16–18].

6.2 Proposed Structure

The Sentaurus TCAD Suite was used in this work to implement all GAA-FET-based device architectures. Various models have been employed, such as the Drift Diffusion Model, Mobility Model, and Shockley–Read–Hall (SRH) Generation Recombination Model. The calibration of these models was performed at the 10 nm technology node to match the experimental results [19]. The simulated and experimental I_d–V_{gs} properties of the GAA FET correlate well with each other. This is shown in Figure 6.3.

Figure 6.4 displays the proposed structure, which is the triple-material double-gate cylindrical gate all-around (TMDG CGAA) FET. Three materials were chosen, each with a distinct work function value, as illustrated in the figure. The material with the highest work function was closest to the source terminal, whereas the material with the lowest work function is closest to the drain terminal.

This will help set the drain current to the required value by providing an appropriate drain voltage. Furthermore, this increases the electric field, which accelerates and increases the velocity of the charge carriers at the source side. As a result, the device's I_{ON} increases. Titanium is used as M3, gold is used as M1, molybdenum is used as M2, and titanium is used as M3.

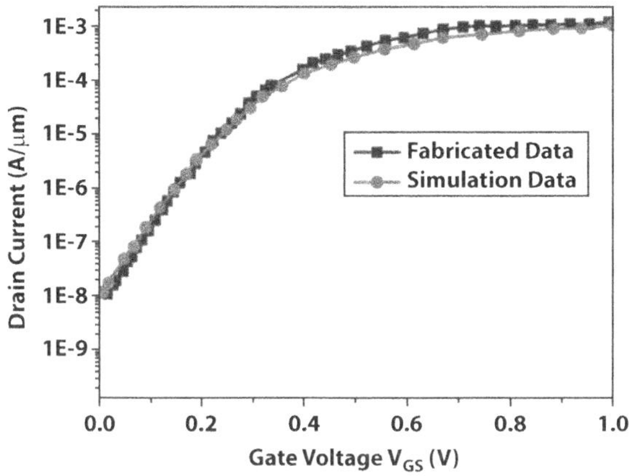

Figure 6.3 Calibration of simulation setup.

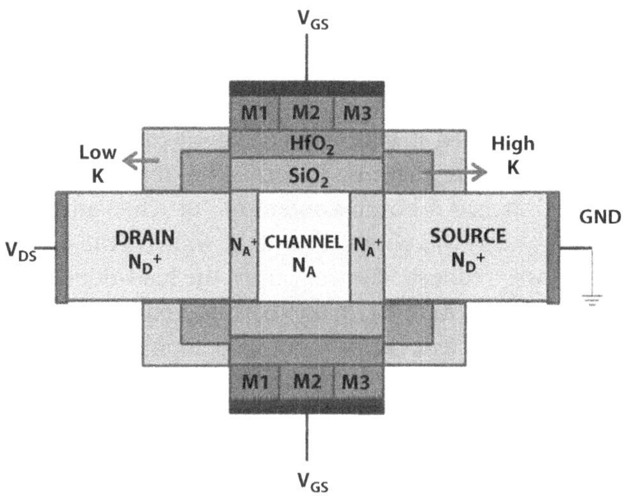

Figure 6.4 Proposed device—TMDG CGAA.

Gold had a work function of 4.8 eV. The lowest work function metal is positioned close to the drain area, and the highest work function metal is located near the source region. The use of three distinct materials with varying work functions contributes to the creation of a stronger electric field. Because the charge carriers are forced to accelerate by a high electric field, the proposed device has a high I_{ON} and transconductance. This also results in enhanced channel-area screening against variations in the drain potential in the TMDG CGAA device.

In brief, a few advantages of gate engineering are a low leakage current, higher drive current, higher transconductance, higher charge carrier velocity, and suppressed short-channel effects. In addition, the use of gate oxide stacking lowers the leakage current. The SiO_2 was stacked with hafnium oxide. Utilizing a layer of hafnium oxide in conjunction with SiO_2 produced a high-quality interface. In additionally, there were fewer faults at the interface and a reduced leakage current.

A coaxial inner gate is another feature of the TMDG CGAA device construction. This was the only distinction between the two devices. The other dimensions of the TMDG CGAA device were the same. The coaxial inner gate is an additional gate that contributes to channel inversion. Consequently, more charge carriers participated in the conduction process. Hence, the TMDG CGAA gadget had greater I_{ON}.

Little research has been conducted on halo doping for all-around gate configurations. To create a unique structure, the advantages of employing multiple gate electrodes, gate oxide stacking, and halo doping have been combined for the first time in triple-material gate electrode-based GAA device construction. This unique structure provides greater resistance to SCEs than traditional GAA. The channel region near the source and drain junctions in a halo-doped structure has a slightly higher doping level than the other sections of the channel. Boron is the dopant and silicon is the channel material. It had a boron content of 10^{17} cm^{-3} after doping. As a result, a larger I_{ON} value was obtained, and the width of the depletion region at the junctions was reduced. Consequently, the halo-doped structure has high immunity to SCE, high I_{ON}, high switching ratio, and high transconductance. Compared to a device without halo doping, the device with halo doping has a somewhat better I_{ON}. Thus, channel engineering may enhance the electrical properties of the devices to some extent. Furthermore, halo doping also suppresses short-channel effects.

High-k spacers provide numerous advantages regarding the DC and RF properties of the apparatus. High-k spacers have the advantages of a high I_{ON}/I_{OFF} ratio, excellent DC characteristics, reduced leakage current, high immunity to short-channel effects, and reasonable RF performance. Table 6.1 displays many parameters of the suggested structure.

The radius was 5 nm and the channel length was 10 nm. The lengths of the source and drain were both 10 nm. In the source and drain regions, the doping was 2×10^{20}, whereas in the channel region, it was 1×10^{17}. Compared to the channel region, the source and drain regions were heavily doped. A high I_{ON} is ensured by the high doping in the source and drain areas. The three gate materials had work function values of 4.8 eV, 4.6 eV, and 4.4 eV. At 0.7 V, the supply voltage V_{DD} is fixed. Only 0.7 V is used for

Table 6.1 Device parameters.

Parameter	Symbol	Value
Supply Voltage	V_{DD} (V)	0.7
Length of Gate	L_G (nm)	10
Length of Source/Drain	$L_{S/D}$ (nm)	10
Radius	R (nm)	5
Effective Oxide Thickness	E_{OT} (nm)	1
Doping of Source/Drain	$N_{S/D}$ (cm^{-3})	2×10^{20}
Channel Doping	N_{CH} (cm^{-3})	1×10^{17}
Work function of M1	Φ_{M1} (eV)	4.4
Work function of M2	Φ_{M2} (eV)	4.6
Work function of M3	Φ_{M3} (eV)	4.8

gate and drain bias in each simulation. Hence, this gadget is appropriate for low-power applications.

6.3 Device Performance

The transfer properties of the proposed device with various spacer types are shown in Figure 6.5(a). Figure 6.5(b) displays the logarithmic-scale transfer characteristics. The TMDG CGAA with a full nitride spacer and the TMDG CGAA with a corner spacer have almost similar ions, but the dual spacer has the lowest I_{ON} of any spacer design. Nonetheless, the I_{OFF} of the dual spacer was the lowest and poorest compared to that of the air spacer. The switching ratio of the TMDG CGAA with the corner spacer was slightly lower than that of the TMDG CGAA with the Si_3N_4 spacer, as shown in Figure 6.5(c). However, there was little switching between SiO_2 and air spacers. As shown in Figure 6.5(d), the subthreshold swing of every spacer design was nearly the same. Therefore, it can be concluded that improved device performance and SCE immunity can be achieved by utilizing triple-metal gates with varying work function values.

A comparison of the suggested structures (TMDG CGAA) with various spacers is presented in Table 6.2. The maximum switching ratio and

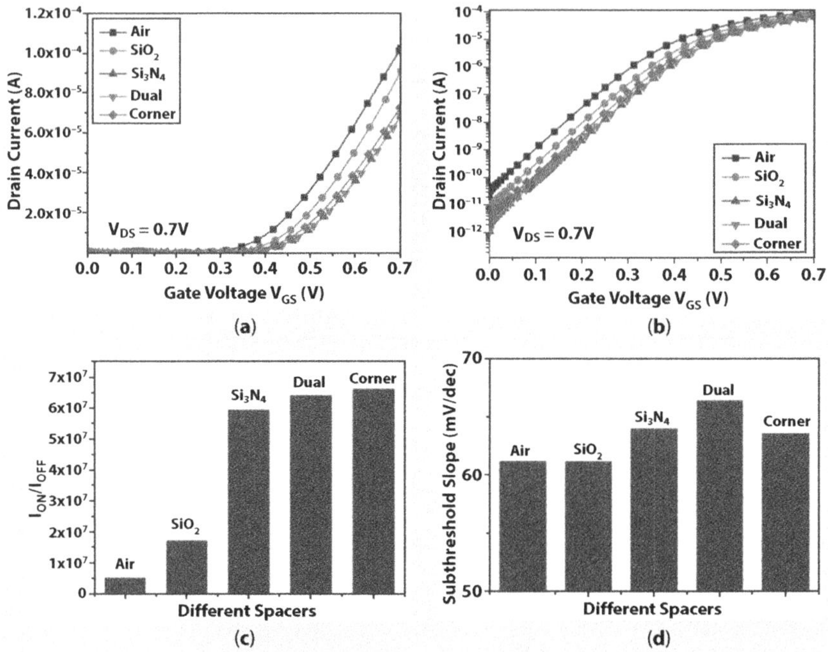

Figure 6.5 (a) Transfer characteristics, (b) transfer characteristics in log scale, (c) switching ratio, and (d) subthreshold slope of TMDG SiNW FET with different spacers.

Table 6.2 Performance of proposed structure.

Device structure	I_{ON} (A)	I_{OFF} (A)	$I_{ON/}I_{OFF}$	C_{gg} (fF)	f_T (GHz)	f_{MAX} (GHz)
TMDG CGAA with Air Spacer	1.02×10^{-4}	1.94×10^{-11}	0.52×10^{7}	0.019	10,200	13,500
TMDG CGAA with SiO_2 Spacer	9.07×10^{-5}	5.27×10^{-11}	1.72×10^{6}	0.21	9,800	13,100
TMDG CGAA with Si_3N_4 Spacer	6.84×10^{-5}	1.05×10^{-12}	6.51×10^{7}	0.31	8,500	10,550
TMDG CGAA with Dual Spacer	6.80×10^{-5}	1.06×10^{-12}	6.41×10^{7}	0.28	9,100	12,200
TMDG CGAA with Corner Spacer	7.29×10^{-5}	1.1×10^{-12}	6.62×10^{7}	0.25	9,400	12,800

minimum leakage current were achieved using the TMDG CGAA and the corner spacer. Therefore, the proposed device was a TMDG CGAA with a corner spacer.

The AC characteristics of the proposed device are listed in Table 6.3. Higher transconductance, suitable gate capacitance, and higher values of the cutoff frequency, maximum oscillation frequency, TFP, and TGP indicate that the RF performance is good.

6.4 Process Sensitivity

Given their superior DC and AC properties, gate all-around FETs may soon replace FinFETs as device architectures. Therefore, research into the effect of process variation on GAA FET is essential. Multigate transistors, such as FinFETs and GAA FET, are extremely complex to fabricate. For instance, it is challenging to adjust the fin height, width, and inclination angle accurately. Similarly, it is challenging to uniformly etch the gate oxide. Another problem that might lead to scattering is the surface roughness of the oxide and channel interface [19]. I_{ON} and leakage current levels are heavily dependent on device physical characteristics, such as radius, gate length, and oxide thickness [20]. The radius of the nanowire is the most sensitive parameter, and even a small fluctuation has the biggest impact on device characteristics. The oxide thickness also affects the device characteristics. If the oxide thickness is increased, the drain current will be reduced because more gate voltage is required to invert the channel. This also increased the threshold voltage of the device. Gate length is another parameter that affects the device performance. Therefore, it is crucial to precisely control the gate length during fabrication. The device's I_{ON} and leakage current were both affected by channel doping. This is a crucial parameter that needs to be checked to provide the device with the desired electrical properties.

Another significant problem that might deteriorate the electrical properties of the device is the random positioning of dopants. The limited number of dopants for narrow gate lengths makes them difficult to manage during construction. The performance of the device may deteriorate by variations in the threshold voltage caused by variations in the dopant concentration over various areas of the channel.

The analysis of process sensitivity is listed in Table 6.4.

Multi-gate transistors, such as FinFET and GAA FET, are extremely complex to fabricate. For instance, it is challenging to accurately adjust the fin's height, width, and inclination angle. In the same way, it is challenging

Table 6.3 AC characteristics.

Parameter	Value
g_m(S)	9.5×10^{-5}
C_{gg} (fF)	0.25
C_{gd} (fF)	0.00141
f_T (THz)	9.4
f_{max} (THz)	12.8
TFP (THz/V)	12.75
TGF (V^{-1})	1.35
GBP (THz)	10
Intrinsic Delay (ps)	0.45

Table 6.4 Process sensitivity analysis.

Parameter	%Variation per nm
Variation in I_{ON} with oxide thickness	0.54%
Variation in V_{th} with oxide thickness	4.8%
Variation in I_{ON} with gate Length	0.4%
Variation in V_{th} with gate Length	4%
Variation in I_{ON} with channel doping	2.81%
Variation in V_{th} with channel doping	9.5%
Variation in I_{ON} with radius	1.6%
Variation in V_{th} with radius	3.7%

to etch the gate oxide uniformly. Another problem is the oxide and channel contact surface roughness, which can lead to scattering. Thus, it is imperative to examine the effect of process variations on GAA FET. Changes in the oxide thickness, nanowire radius, gate length, channel doping, and nanowire thickness did not significantly affect the drive current or threshold voltage. Thus, the suggested gadget performs exceptionally well in both DC and RF.

6.5 Conclusion

We have suggested a novel TMDG CGAA FET device that meets the requirements of a high ION, low leakage current, high transconductance, high cutoff frequency f_T, and high maximum frequency of oscillation f_{MAX}. Excellent DC and AC characteristics were demonstrated by the suggested device using triple metal gates, gate oxide stacking, halo doping, and corner spacers. The optimized device parameters employed were Lg = 10 nm, R = 5 nm, E_{OT} = 1 nm, L_{COR} = 2.5 nm, TCOR = 2.5 nm, M1 = 4.8 eV, M2 = 4.6 eV, and M3 = 4.4 eV in order to achieve better ION, sub threshold swing, and transconductance. The proposed device has a high subthreshold slope of 62.3 mV/dec, high fT of 9,400 GHz, high f_{MAX} of 12,800 GHz, and high ON state current of 14.58 mA/μm. It has been demonstrated that the suggested device is the optimum option for enhancing DC performance and serving analog/RF applications. Consequently, it may be said that the TMDG CGAA FET is a contender to replace the FinFET in the near future.

References

1. Schaller, R.R., Moore's law: past, present and future. *IEEE Spectr.*, 34, 6, 52–59, June 1997.
2. Kumar, A., Saini, S., Gupta, A., Gupta, N., Tripathi, M.M., Chaujar, R., Sub-10 nm high-k dielectric SOI-FinFET for High Performance low power applications. *International Conference on Signal Processing and Communication*, pp. 310–314, 2020.
3. Narula, M.S. and Pandey, A., Dual-Gate Silicon Nanowire FET with a Corner Spacer for High Performance & High Frequency Applications. *J. Electron. Mater.*, 52, 10, 6708–6718, July 2023.
4. Munjal, S., Prakash, N.R., Kaur, J., Evolution of junctionless field effect transistors in semiconductor industry: a review. *Int. J. Innov. Sci. Eng. Technol.*, 8, 8, 94–103, 2021.
5. Vallabhuni, R.R., Yamini, G., Vinitha, T., Performance analysis: D-Latch modules designed using 18nm FinFET Technology. *International Conference on Smart Electronics and Communication*, pp. 1169–1174, 2020.
6. Frank, D.J., *et al.*, Device scaling limits of Si MOSFETs and their application dependencies. *IEEE Proc.*, 8, 259–288, Feb. 2001.
7. Brews, J.R., Fichtne, W., Sze, S.M., Generalized guide for MOSFET miniaturization. *IEEE Electron Device Meeting*, 1980.
8. Narula, M.S. and Pandey, A., A Comprehensive Review on FinFET, Gate All Around, Tunnel FET : Concept, Performance and Challenges. *IEEE International Conference on Signal Processing & Communication*, Dec 2022.

9. Dennard, R.H., Gaensslen, F.H., Yu, H.N., Rideout, E., Leblanc, A.R., Design of ion-implanted MOSFET's with very small physical dimensions. *IEEE J. Solid-State Circuits*, 9, 256–268, Oct 1974.
10. Narula, M.S. and Pandey, A., Gate Engineered Silicon Nanowire FET with Coaxial Inner Gate for Enhanced Performance. *Silicon*, 15, 10, 4217–4227, Feb 2023.
11. Rajveer, K. and Balwinder, S., Comparative Study of Single and Double gate ALL Around Cylindrical FET Structures for High –K Dielectric Materials. *Trans. Electr. Electron. Mater.*, 22, 4, 509–514, 2020.
12. Narula, M.S. and Pandey, A., Performance Evaluation of Stacked Gate Oxide/High K Spacers Based Gate All Around Device Architectures at 10 nm Technology Node. *Silicon*, 14, 5, 2397–2407, April 2022.
13. Kajal, S., Design and Simulation of FinFET Circuits at Different Technologies. *International Conference on Inventive Computational Technologies*, pp. 1–6, 2021.
14. Zhang, S., Review of Modern Field Effect Transistor Technologies for Scaling. *2nd International Conference on Electronic Engineering and Informatics*, 2020.
15. Razavieh, A., Zeitzoff, P., Nowak, E.J., Challenges and limitations of CMOS scaling for FinFET and beyond architectures. *IEEE Trans. Nanotechnol.*, 18, 999–1004, 2019.
16. Liu, T.Y. and Pan, F.M., Characteristics of gate-all-around Junctionless poly-silicon nanowire transistors with twin 20-nm gates. *Int. J. Electron Devices*, 3, 5, 405–409, 2015.
17. Zhuge, J., Wang, R., Huang, R., Tian, Y., Zhang, L., Kim, D.-W., Park, D., Wang, Y.Y., High-performance silicon nanowire gate-all-around nMOS-FETs fabricated on a bulk substrate using CMOS compatible process. *IEEE Electron Device Lett.*, 31, 12, 57–60, 2009.
18. Huang, R., Wang, R., Zhuge, J., Liu, C., Yu, T., Zhang, L., Huang, X., Zou, J., Liu, Y., Fan, J., Wang, Y., Characterization and Analysis of Gate-All-Around Si Nanowire Transistors for Extreme Scaling. *IEEE Custom Integrated Circuit's Conference*, pp. 1–8, 2011.
19. Bangsaruntip, S., *et al.*, Density scaling with gate-all-around silicon nanowire MOSFETs for the 10 nm node and beyond. *IEEE International Electron Devices Meeting*, Washington, DC, USA, 2013.
20. Gupta, N. and Chaujar, R., Optimization of high-k and gate metal workfunction for improved analog and intermodulation performance of Gate Stack (GS)-GEWE-SiNW MOSFET. *Superlattices Microstruct.*, 97, 630–641, 2016.

7

Rise of Tunnel FETs as a Revolutionary MOSFET Alternative

G. Munirathnam[1]* and Y. Murali Mohan Babu[2]

[1]JNTUACEK, Kalikiri, Annamayya District, AP, India
[2]NBKRIT, Vidyanagar, Nellore District, AP, India

Abstract

To meet the growing need for high-performance, energy-efficient electronics, this study investigates Tunnel Field-Effect Transistors (TFETs) as a possible substitute for conventional MOSFETs. TFETs use band-to-band tunneling to obtain lower operating voltages and improved performance in contrast to conventional MOSFETs, which have drawbacks such as short-channel effects and excessive power consumption. The advantages of TFETs in terms of power efficiency and scalability are highlighted in this study's examination of their historical developments, operating principles, and device topologies. It also addresses the difficulties in fabricating TFETs, such as material compatibility and technology integration. This research emphasizes how TFETs can revolutionize semiconductor technology by assessing their performance parameters and investigating diverse applications, especially in low-power and high-speed devices. Creative solutions are crucial to overcome fabrication obstacles and enhance the overall device performance in contemporary electronic systems. This study concludes with proposals for future paths in TFET research.

***Keywords*:** Band-to-band tunneling, low-power electronics, MOSFET alternative, short-channel effects, device scalability, fabrication challenges, high-performance semiconductors

**Corresponding author*: munirathnam.ece@jntua.ac.in

Ekta Goel, Archana Pandey, Shiromani Balmukund Rahi and Arun Samuel (eds.) Handbook of Advanced Semiconductor Field Effect Transistors, (119–168) © 2026 Scrivener Publishing LLC

7.1 Introduction to Tunnel FETs

7.1.1 Overview of Traditional FETs and Their Limitations

Field-Effect Transistors (FETs) are fundamental components of modern electronic devices, and play a crucial role in amplifying and switching electronic signals. Traditional FETs, such as metal-oxide-semiconductor FETs (MOSFETs), are widely used because of their efficiency and scalability. Despite their success, traditional FETs face significant limitations as technology advances, particularly in terms of power consumption, scaling challenges, and short-channel effects. This overview explores the structure, operation, and limitations of traditional FETs, providing a comprehensive understanding of their role in electronics and the challenges that they face.

Structure and Operation of Traditional FETs: Conventional FETs, particularly MOSFETs, comprise three primary areas: the source, drain, and gate. These regions are isolated from the semiconductor channel by a thin insulating layer, which is typically made of silicon dioxide. A MOSFET's main job is to regulate the current flow between its source and drain by placing a voltage on its gate [1]. Applying a voltage to the gate causes an electric field that changes the semiconductor channel's conductivity, enabling or preventing current flow.

There are two types of MOSFETs: n-channel MOSFETs (NMOSs) and p-channel MOSFETs (PMOSs). In NMOS devices, electrons are the charge carriers, whereas in PMOS devices, holes are the charge carriers. Complementary MOS (CMOS) technology, which combines both NMOS and PMOS transistors, is the foundation of modern digital integrated circuits because of its low power consumption and high noise immunity.

Key Parameters and Characteristics:

1. Threshold Voltage (V_{th}): The minimum gate voltage required to create a conductive channel between the source and drain.
2. Transconductance (g_m): The rate at which the drain current changes with respect to the gate voltage, indicating the amplification capability of the transistor.
3. Subthreshold Slope (SS): The rate at which the current increases in the subthreshold region (below the threshold voltage). A lower SS indicates better switching-off characteristics.

Limitations of Traditional FETs:
Despite their widespread use, traditional FETs face several limitations, particularly as the technology nodes shrink below 10 nm. These limitations include the following:

1. Short-Channel Effects (SCEs):
Drain-Induced Barrier Lowering (DIBL): As the channel length decreases, the electric field from the drain affects the barrier at the source, reducing the threshold voltage and increasing the leakage current [2].
Velocity Saturation: At high electric fields, the carrier velocity saturates, limiting the current drive capability and reducing the transistor performance.
Channel Length Modulation: Variations in channel length due to manufacturing imperfections can lead to inconsistencies in transistor behavior, affecting circuit reliability and performance

2. Leakage Currents:
Gate Leakage: Thinning the gate oxide to improve control over the channel also increases the tunneling currents through the gate, leading to power loss and potential reliability issues. Subthreshold Leakage: As devices scale down, maintaining a low threshold voltage becomes challenging, leading to an increased subthreshold leakage current [3] and higher static power consumption.

3. Power Consumption:
Dynamic Power: Although dynamic power consumption (proportional to the switching activity and capacitive load) can be managed through techniques such as clock gating and dynamic voltage scaling, it still poses a significant challenge in high-performance applications.
Static Power: Leakage currents contribute to static power consumption, which is a critical issue in battery-powered and energy-efficient applications.

4. Variability and Reliability:
Process Variability: As dimensions shrink, variations in manufacturing processes lead to inconsistencies in device characteristics, affecting yield and reliability.
Aging Effects: Mechanisms such as Bias Temperature Instability (BTI) and Hot Carrier Injection (HCI) degrade the performance of transistors over time, impacting the long-term reliability of electronic circuits [4].

5. Quantum Effects:
Quantum Tunneling: At very small dimensions, quantum mechanical effects become significant, leading to increased leakage currents through the thin gate dielectric.
Quantum Confinement: The discrete energy levels in ultrathin channels can the alter carrier transport properties [5], affecting device performance.

Potential Solutions and Emerging Technologies: To address the limitations of traditional FETs, researchers have explored various innovative approaches and emerging technologies.

1. FinFETs:
Fin Field-Effect Transistors (FinFETs) have a three-dimensional structure with a fin-shaped channel that provides better control over the channel, reducing short-channel effects and leakage current. They are widely used in advanced technology nodes to enhance performance and energy efficiency.

2. Gate-All-Around (GAA) FETs:
GAA FETs further improve gate control by surrounding the channel with the gate on all sides, providing superior electrostatic control, and reducing leakage currents. Nanowire and nanosheet GAA FETs are being developed for future technological nodes.

3. Tunnel FETs (TFETs):
TFETs leverage quantum tunneling for carrier transport, enabling lower subthreshold slopes and reducing the power consumption. TFETs are promising for ultralow-power applications owing to their steep subthreshold swing.

4. 2D Materials:
Transition-metal dichalcogenides (TMDs) and other 2D materials offer unique electronic properties and scalability benefits. These materials can be used to fabricate transistors with high mobility and low leakage current, potentially overcoming the limitations of traditional FETs.

5. High-k/Metal Gate (HKMG) Stacks:
Replacing the traditional silicon dioxide gate dielectric with high-k materials reduces gate leakage, whereas metal gates improve performance and scalability. HKMG technology has already been used in advanced process nodes.

6. Strain Engineering:
Applying strain to the semiconductor channel can enhance the carrier mobility and improve the performance of the transistors. Strain engineering is a practical technique for boosting the performance of traditional FETs without significant changes to the device structure [6].

7. Carbon-Based Electronics:
Carbon nanotubes (CNTs) and graphene have exceptional electrical properties that can be leveraged to create transistors with high performance and low power consumption. Although still in the research phase, carbon-based electronics holds promise for future applications.

7.1.2 Introduction to Tunnel FETs as a Promising Alternative Transistor Technology

Tunnel field-effect transistors (TFETs) are a viable substitute for conventional MOSFETs and show promise in addressing a number of issues that modern semiconductor devices are confronting. TFETs offer an innovative way to oversome the drawbacks of conventional FETs, particularly with regard to power consumption and scaling, as the technology nodes become smaller and power efficiency becomes more important.

The Need for Alternative Transistor Technologies: Traditional MOSFETs have driven the rapid advancement of electronic technologies for decades. However, as device dimensions continue to scale down, traditional MOSFETs encounter significant issues such as short-channel effects, increased leakage currents, and power consumption challenges. These issues become more pronounced at sub-10 nm technology nodes, where the physical limitations of MOSFETs hinder further performance improvements and energy efficiency.

One of the critical limitations of traditional MOSFETs is the subthreshold slope, which defines how sharply a transistor can transition from off (nonconducting) to on (conducting). The theoretical limit for the subthreshold slope in MOSFETs is 60 mV/decade at room temperature. This limits the ability to reduce the supply voltage (V_{dd}) without incurring significant leakage currents, thereby limiting power savings [7].

Principles of Tunnel FET Operation: TFETs operate on a fundamentally different principle than traditional MOSFETs. Instead of relying on thermionic emission over a barrier, TFETs leverage quantum-mechanical tunneling for carrier transport. This tunneling mechanism allows the carriers to move from the source to the channel through a thin potential barrier when a gate voltage is applied (see Figure 7.1).

In TFET, the source and drain are typically made of different types of semiconductor materials (e.g., p-type and n-type), creating a p-i-n or n-i-p structure. When a gate voltage is applied, it modulates the width of the tunneling barrier at the source-channel junction. If the barrier is sufficiently thin, carriers can tunnel through it, leading to a current flow. This tunneling process enables TFETs to achieve a subthreshold slope that is steeper than the theoretical limit of MOSFETs, potentially reaching values below 60 mV/decade.

Advantages of Tunnel FETs:

1. Lower Subthreshold Slope: The primary advantage of TFETs is their ability to achieve a subthreshold slope below 60 mV/decade, allowing for lower threshold voltages and reduced power consumption without increasing leakage currents.

2. Reduced Power Consumption: TFETs can operate at lower supply voltages due to their steep subthreshold slope, significantly reducing both the dynamic and static power consumption, making them ideal for low-power and energy-efficient applications.

3. Scalability: TFETs are more scalable than traditional MOSFETs because they are less affected by the short-channel effects. This makes them suitable for advanced technology [6] nodes, where the MOSFET performance degrades.

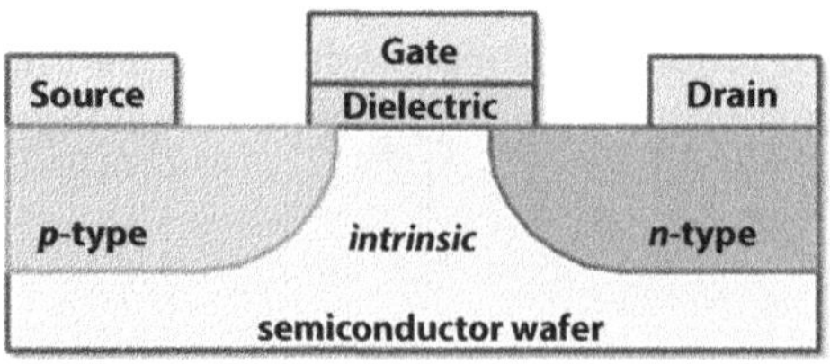

Figure 7.1 Structure of TFET [6].

Challenges and Research Directions: Despite their advantages, TFETs face several challenges that must be addressed for widespread adoption.

1. Material Selection: Identifying appropriate materials for the source, channel, and drain regions is crucial for achieving efficient tunneling and optimal device performance. Research is ongoing to explore various semiconductor materials, including III–V compounds and 2D materials.

2. Fabrication Complexity: The fabrication of TFETs can be more complex than that of traditional MOSFETs because of the need for precise control over doping and material interfaces.

3. Current-Drive Capability: While TFETs excel in low-power applications, their on-current (drive current) is generally lower than that of MOSFETs. Enhancing the current driving capability without compromising the subthreshold slope is an active area of research.

7.1.3 Historical Development and Milestones in TFET Research

Particularly for low-power applications, tunnel field-effect transistors (TFETs) have attracted considerable interest as a potential replacement for conventional MOSFETs. The need to overcome the power efficiency and scalability constraints of conventional FETs has fueled the development of TFETs. This summary traces the evolution of and significant turning points in the history of TFET research.

Early Concepts and Initial Proposals: The concept of utilizing quantum tunneling for transistor operations dates back to the 1970s. Initial theoretical work explored the idea of tunneling-based devices as a means of achieving steeper subthreshold slopes than those possible with thermionic emission in MOSFETs. However, practical implementation remains challenging owing to material and fabrication limitations.

1990s: Emergence of TFET Research

The 1990s saw renewed interest in TFETs driven by advancements in semiconductor fabrication technology and a better understanding of quantum mechanics. Researchers have proposed specific TFET designs and explored their potential benefits. One of the earliest significant works was the proposal of the tunnel FET, which laid the groundwork for subsequent experimental studies.

Early 2000s: Demonstration of Working Prototypes
In the early 2000s, several research groups successfully demonstrated their first experimental TFET prototype. These initial devices were primarily based on silicon and silicon-germanium (SiGe) materials. Notably, in 2004, the first vertical TFET demonstrated the practical feasibility of TFETs, and sparking increased interest in the field.

Mid-2000s: Advances in Material Science
The mid-2000s marked significant progress in TFET research, particularly in the exploration of new materials. Researchers has begun investigating III–V compound semiconductors such as indium arsenide (InAs) and gallium arsenide (GaAs), which offer superior tunneling properties compared to silicon. In 2007, an InAs/Si heterojunction TFET was demonstrated by researchers at the University of Notre Dame, highlighting the potential of heterojunctions to enhance the TFET performance.

Late 2000s: Improved Performance and Scalability
By the late 2000s, TFETs had achieved significant performance improvements. The introduction of high-k dielectrics and advanced doping techniques has helped reduce the leakage currents and improve the on-current. In 2009, a breakthrough was achieved when researchers at the University of California, Berkeley demonstrated a TFET with a subthreshold slope of less than 60 mV/decade, proving the theoretical advantage of TFETs over traditional MOSFETs.

2010s: Integration and Application
The 2010s saw efforts to integrate TFETs into practical applications and commercial technology. Researchers have focused on optimizing TFET designs for specific applications, such as low-power digital circuits and sensors. In 2011, IBM demonstrated a TFET-based SRAM cell, showcasing the potential of TFETs in memory applications. Additionally, research has expanded into two-dimensional (2D) materials, such as transition metal dichalcogenides (TMDs), which offer further opportunities for scaling and performance enhancements.

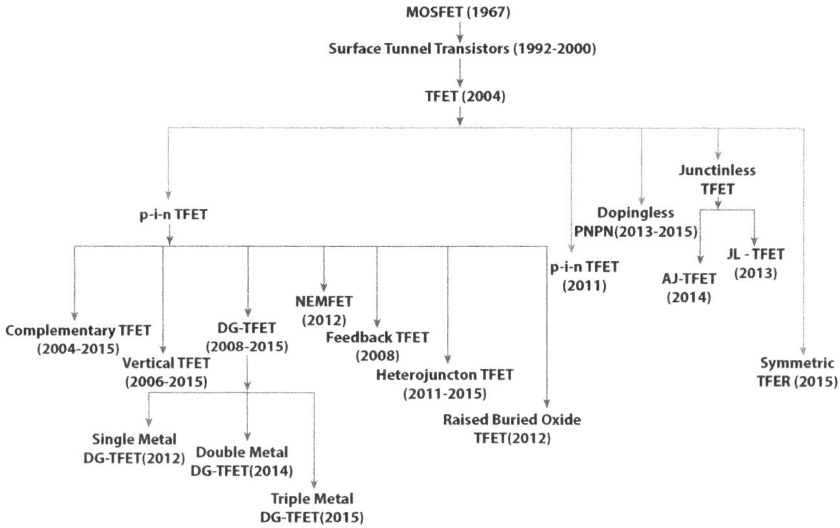

Figure 7.2 Evaluation of tunnel field effect transistor.

Recent Developments and Future Directions:
In recent years, TFET research has continued to advance with significant milestones in both material science and device architecture. Researchers are exploring new material systems, such as black phosphorus and transition metal dichalcogenides, to achieve better tunneling efficiency and on-currents. In 2018, a breakthrough was reported by a team at ETH Zurich, who demonstrated a molybdenum disulfide (MoS2) TFET with excellent subthreshold characteristics.

Current research focuses on addressing the remaining challenges such as improving the drive current and ensuring manufacturability. Novel device architectures, including vertical TFETs (see Figure 7.2) and nanowire TFETs, are being explored to further enhance performance and integration density.

7.2 Working Principles of Tunnel FETs

7.2.1 Basic Principles of TFET Operation: Band-to-Band Tunneling Mechanism

The band-to-band tunneling (BTBT) process is utilized by Tunnel Field-Effect Transistors (TFETs) to facilitate carrier transport, setting them apart from conventional MOSFETs, which depend on thermionic emissions. Owing to this process, TFETs are appealing for low-power electronic

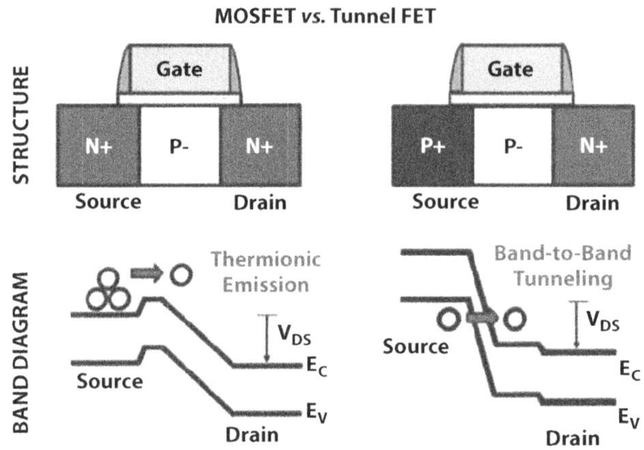

Figure 7.3 Energy band diagram [3].

applications because they can produce greater subthreshold slopes and lower power consumption. To fully comprehend TFETs and their benefits, one must have a solid understanding of the BTBT process [8].

Band-to-Band Tunneling (BTBT) Mechanism: Carriers tunnel straight from the valence band of the source region to the conduction band of the channel region in a process known as band-to-band tunneling. In contrast to thermionic emission, which necessitates that carriers overcome a potential barrier, BTBT permits carriers, even those without sufficient thermal energy to overcome the barrier, to flow through a thin barrier in the presence of a strong electric field (see Figure 7.3).

The source and drain of a TFET are usually made of two distinct types of semiconductor materials, an n-type drain and a highly doped p-type source, which combine to form a p-i-n or n-i-p structure. The width of the tunneling barrier at the source-channel junction was modulated by the application of a gate voltage.

Key Steps in TFET Operation

1. Off-State (No Gate Voltage): The energy bands of the source and channel are mismatched in the off state, and no discernible gate voltage is applied. There is no tunneling current because the channel's conduction band is above the source's valence band.

2. On-State (Gate Voltage Applied): The channel's energy bands are drawn downward in relation to the source by the electric field produced when a gate voltage was applied. A tunneling path is formed when the conduction band of the channel and the source's valence band line are up. In n-type TFET, electrons tunnel from the channel's conduction band to the p-type source's valence band. The current in the on-state is produced by this tunneling mechanism.

3. Tunneling Junction: The key to efficient TFET operation is the tunneling junction, where the electric field is the strongest and band alignment occurs. The thinner and more abrupt the junction, the more efficient is the tunneling process [9].

Advantages of BTBT in TFETs:

1. Steep Subthreshold Slope
TFETs can achieve a subthreshold slope steeper than the 60 mV/decade limit of MOSFETs, because BTBT is not thermally activated. This allows for significantly sharper transitions between the off and on states, enabling lower threshold voltages and reduced power consumption.

2. Lower Power Consumption
Owing to the steep subthreshold slope, TFETs can operate at lower supply voltages (V_{dd}), significantly reducing both the dynamic and static power consumption. This makes TFETs ideal for ultra-low-power applications, such as wearable electronics and Internet of Things (IoT) devices.

3. Enhanced Scalability
TFETs are less susceptible to short-channel effects compared to MOSFETs. The reliance on BTBT rather than thermal emission allows TFETs to maintain performance as the device dimensions scale down, addressing the challenges faced by traditional transistors at the nanometer scale.

Challenges and Research Directions: While the BTBT mechanism provides significant advantages, several challenges need to be addressed to fully realize the potential of TFETs.

1. Material Selection
The BTBT efficiency depends heavily on the choice of materials for the source and channel. Researchers are currently exploring materials with

narrow bandgaps and high tunneling probabilities, such as III–V compounds (e.g., InAs) and 2D materials (e.g., MoS2).

2. Fabrication Techniques
Achieving the precise control required for tunneling junctions and ensuring high-quality interfaces between different materials is challenging. To address these sissues, advanced fabrication techniques are being developed.

3. Current Drive Capability
Although TFETs excel in low-power operations, their on-current is typically lower than that of MOSFETs. Enhancing the current-drive capability [10] while maintaining low power consumption is a critical area of ongoing research.

7.2.2 Comparison of TFET Operation with Conventional FETs (MOSFETs)

Tunnel Field-Effect Transistors (TFETs) and metal-oxide-semiconductor field-effect transistors (MOSFETs) represent two different paradigms in transistor technology. Although both serve as essential components in modern electronics, their operating principles, performance characteristics, and applications vary significantly. This comparison highlights the key differences between the TFETs and conventional MOSFETs.

Operating Principles:
MOSFETs
MOSFETs operate on the basis of the modulation of a conductive channel using an electric field. When a gate voltage (Vg) is applied, it influences the charge carriers in the channel, allowing the current (Id) to flow between the source and drain terminals. The current in a MOSFET can be described by the equation for the linear (ohmic) region has written in Equation 7.1

$$I_D = \mu_n C_{ox} \frac{W}{L}(V_G - V_{th})V_{DS} - \frac{V_{DS}^2}{2} \tag{7.1}$$

and for the saturation region has written in Equation 7.2 and the scientific variable used in Equation 7.1 and Equation 7.2 have following meaning

$$I_D = \frac{1}{2}\mu_n C_{ox} \frac{W}{L}(V_G - V_{th})^2 \tag{7.2}$$

Where:
μ_n is the carrier mobility
C_{OX} is the oxide capacitance per unit area.
W and L are the width and length of the channel, respectively
V_{th} is the threshold voltage.
V_{DS} is the drain–source voltage.

TFETs
Band-to-band tunneling (BTBT) is the basis of TFET operation. The width of the tunneling barrier at the source-channel junction was modulated by the application of a gate voltage (Vg). Electrons (or holes) can tunnel from the source valence band to the channel conduction band if the barrier is sufficiently thin [11]. The current in a TFET can be described by the BTBT current equation 7.3:

$$I_{BTBT} \propto AE^2 exp\left(-\frac{B}{E_{perpenducular}}\right) \tag{7.3}$$

In Equation 7.3, where A and B are constants that depend on the material and tunneling mechanism. The $E_{perpenducular}$ is the electric field perpendicular to the tunneling junction.

Subthreshold Slope:
MOSFETs
The subthreshold slope (S) in MOSFETs is limited by thermal voltage and typically has a minimum value of 60 mV/decade at room temperature follow Equations 7.4 and 7.5

$$S = 1 \Big/ \left(\frac{dI_D}{dV_g}\right) \tag{7.4}$$

TFETs
TFETs can achieve a subthreshold slope steeper than 60 mV/decade because the tunneling mechanism is not limited by thermal voltage. This allows for sharper transitions between the off and on states.

$$S < 60\ mV/decade \tag{7.5}$$

Power Consumption:
MOSFETs
MOSFETs require a higher threshold voltage (V_{th}) and operating voltage (V_{DD}) to maintain low off-state leakage currents, which leads to higher power consumption.
TFETs
Owing to their steep subthreshold slope, TFETs can operate at lower (V_{th}) and (V_{DD}) resulting in significantly lower power consumption. This makes the TFETs particularly suitable for low-power applications.

Leakage Current:
MOSFETs
The leakage current in MOSFETs increases exponentially with the reduction in (V_{th}), posing a significant challenge at sub-10 nm nodes.
TFETs
TFETs exhibit lower leakage currents owing to their sharp switching characteristics, which help maintain low off-state leakage, even at reduced (V_{th}).

Scalability:
MOSFETs
The MOSFET performance degrades at smaller scales owing to short-channel effects and increased leakage currents.
TFETs
TFETs are scalable because their operation relies on tunneling, which is less affected by short-channel effects. This makes TFETs better suited for advanced technology nodes.

Current Drive Capability:
MOSFETs
MOSFETs generally have a higher on-current (I_{ON}) than TFETs, making them suitable for high-performance applications where speed is critical.
TFETs
TFETs typically have lower (I_{ON}), which can limit their performance in high-speed applications. Research is ongoing to improve the current-drive capability of TFETs.

7.2.3 Impact of Tunneling on Device Characteristics and Performance

The tunneling mechanism in Tunnel Field-Effect Transistors (TFETs) fundamentally alters their device characteristics and performance compared with conventional MOSFETs. Understanding this impact is crucial for leveraging TFETs for low-power and high-performance applications. Here, we explore the influence of tunneling on key aspects, such as the subthreshold slope, on-current, leakage current, and scalability.

Subthreshold Slope: One of the most significant impacts of tunneling is on the subthreshold slope (S), which measures how sharply a device can transition from the off to the on state [12]. For MOSFETs, the subthreshold slope is limited by the thermal voltage (kT/q), resulting in a minimum value of approximately 60 mV/decade at room temperature followed Equation 7.6 and Equation 7.7 (see Figure 7.4).

$$S_{MOSFET} = 60\ mV/decade \tag{7.6}$$

In contrast, TFETs utilize band-to-band tunneling (BTBT), allowing for subthreshold slopes steeper than the 60 mV/decade limit. This is because the BTBT process was not thermally activated.

$$S_{TFET} < 60\ mV/decade \tag{7.7}$$

The ability to achieve a steeper subthreshold slope allows TFETs to operate at lower threshold voltages (V_{th}) and supply voltages (V_{DD}), thereby reducing the overall power consumption.

On-Current (I_{ON}): The on-current (I_{ON}) is a crucial performance metric for any transistor, which determines its ability to drive loads and switch rapidly. In MOSFETs, the current in the saturation region followed by Equation 7.8:

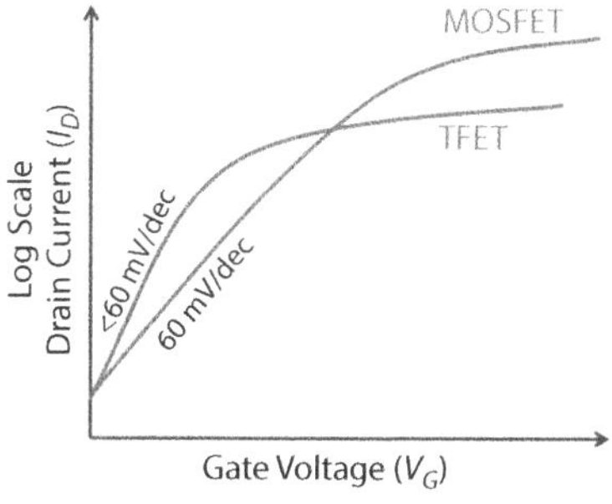

Figure 7.4 Drain current and gate voltage relationship [6].

While in TFETs, the on-current is governed by the tunneling probability T(E), which depends on the electric field ($E_{Perpendicular}$) at the tunneling junction inside device. The band to band tunneling (BTBT) current can be expressed as Equation 7.8:

$$I_{BTBT} \propto AE^2 exp\left(-\frac{B}{E_{perpenducular}}\right) \tag{7.8}$$

In Equation 7.8, A and B are material-dependent constants. $E_{Perpendicular}$ denotes the perpendicular electric field inside TFET device structure.

Although TFETs can achieve lower off-state currents owing to the steep subthreshold slope, their on-current is generally lower than that of MOSFETs. This is a result of the inherently lower tunneling probability [13] compared with the drift-diffusion current in MOSFETs.

Leakage Current (I_{off}): The leakage current, or off-current, is a major concern in modern transistor design, particularly as devices scale down. The MOSFETs suffer from increased leakage currents due to short-channel effects and a reduced V_{th}. The leakage current in MOSFETs is given by 7.9. In Equation V_{th} is threshold voltage of conventional MOSFET, k_B is Boltzmann's constant, q is electronic charge with temperature T in kelvin.

$$I_{off-MOSFET} \propto exp\left(q_{Vth}/k_B T\right) \tag{7.9}$$

In conventional TFET devices, the steep subthreshold slope significantly reduces due to sharp transition from off state current followed the relation 7.10

$$I_{off,-TFET} \ll I_{off,\,MOSFET} \tag{7.10}$$

This reduction in leakage current makes TFETs highly attractive for ultralow-power applications, where minimizing standby power is critical.

Scalability:
Scalability is another area in which tunneling has a profound impact. MOSFETs experience severe short-channel effects as the device dimensions shrink, leading to degraded performance and increased leakage. However, TFETs are less affected due to short channel effects (SCEs) because to tunneling of charge carriers from band to band tunneling(BTBT). The unique

property of lesser SCEs allows more scaling than conventional MOSFET. This can be expressed in Equation 7.11 as

$$L_{min,\,TFET} < L_{min,\,MOSFET} \tag{7.11}$$

The ability to maintain performance at reduced dimensions makes TFETs more scalable and suitable for future technological nodes.

Power Consumption: The overall power consumption of digital circuits is a combination of dynamic and static power. The dynamic power expressed in Equation 7.12 as:

$$P_{dynamic} = C_{load} V_{dd}^{2} f \tag{7.12}$$

Where:
C_{load} denotes the load capacitance.
V_{DD} denotes the supply voltage.
f denotes the operating frequency.

TFETs can operate at a lower (V_{DD}) due to their steep subthreshold slope, which significantly reduces dynamic power consumption. A lower leakage current also minimizes static power consumption.

7.3 TFET Device Structure and Fabrication

7.3.1 Overview of TFET Device Structure: Source, Drain, Channel, and Gate

Tunnel Field-Effect Transistors (TFETs) are a class of transistors that utilize band-to-band tunneling (BTBT) for carrier injection, offering potential advantages in terms of power efficiency and scalability over traditional MOSFETs. The TFET structure comprises several key components [14]: source, drain, channel, and gate. Each component plays a critical role in the device operation and performance.

Source: The source region in a TFET is typically heavily doped with one type of carrier (p-type or n-type), creating a high concentration of either holes or electrons. This heavy doping is crucial for establishing the steep potential gradient necessary for efficient tunneling. In n-type TFET, the source is p-doped, whereas in p-type TFET, the source is n-doped.

The primary function of the source is to provide a high concentration of carriers that can tunnel into the channel region when a sufficient electric field is applied. The abrupt doping profile at the source junction enhances the electric field, which is essential for initiating tunneling.

Drain: The drain region in a TFET, similar to the source, is also heavily doped, but with the opposite type of carrier. In n-type TFET, the drain is n-doped, whereas in p-type TFET, the drain is p-doped. Heavy doping ensures that the drain can efficiently collect carriers that have tunneled through the channel.

The drain collects the carriers that have tunneled through the channel from the source, completing the circuit and allowing current to flow when the device is in the on-state. The electric potential difference between the drain and source helps to control the direction and magnitude of the current [15].

Channel: The channel in a TFET is typically an intrinsic or lightly doped semiconductor material that lies between the source and the drain. Unlike MOSFETs, where carriers flow through the channel by drift and diffusion, in TFETs, carriers move through the channel primarily *via* tunneling.

The channel length [16] and doping profile are crucial for determining the tunneling efficiency and overall device performance. A shorter channel length reduces the tunneling barrier width, enhancing the tunneling probability and increasing the on-current. The material choice for the channel also affects device performance, with materials that have lower effective masses and appropriate bandgap alignments being preferred.

Gate: The gate in a TFET is similar to that in a MOSFET and is used to control the electric field across the channel. The gate is typically made of a conductive material, such as polysilicon or metal, and is separated from the channel by a thin insulating layer, usually silicon dioxide (SiO_2) or high-k dielectric material.

The channel's energy bands in relation to the source are modulated when a voltage is applied to the gate. An n-type TFET with a positive gate voltage allows tunneling by lowering the channel's conduction band edge and lining it up with the source's valence band edge. In contrast, a negative gate voltage in a p-type TFET lines up the channel's valence band edge with the source's conduction band edge.

The gate's effectiveness in controlling the channel's electric field directly influences the device's subthreshold slope, threshold voltage, and overall switching behavior. Advanced gate materials and high-k dielectrics are often employed to improve gate control and reduce leakage current.

7.3.2 Materials and Fabrication Techniques for TFETs

Tunnel Field-Effect Transistors (TFETs) offer promising advantages for low-power high-performance applications. The materials and fabrication techniques used in TFETs are crucial for determining their performance. This section outlines the key materials [17] and fabrication methods employed in TFET development, with a focus on enhancing tunneling efficiency and device scalability.

Materials for TFETs: The choice of materials for TFETs is critical for optimizing their tunneling properties and overall performance (see Table 7.1).

Table 7.1 Materials used for different TFET components.

Component	Common materials	Properties/advantages
Source/Drain	Silicon (Si), Germanium (Ge), III–V semiconductors (InAs, GaSb)	High doping concentration, low bandgap for efficient tunneling
Channel	Silicon (Si), Germanium (Ge), III–V semiconductors (InAs, GaSb)	Low effective mass, suitable band alignment for tunneling
Gate	Polysilicon, Metal (TiN, Mo)	Good control over channel, compatibility with high-k dielectrics
Gate Dielectric	Silicon Dioxide (SiO_2), High-k dielectrics (HfO_2, Al_2O_3)	Improved gate control, reduced leakage current

Source/Drain Materials
Silicon (Si): Widely used due to its compatibility with the existing CMOS technology. However, Si has a higher bandgap, which can limit the tunneling efficiency.
Germanium (Ge): Offers a lower bandgap than Si, resulting in higher tunneling currents.
III–V Semiconductors (e.g., InAs, GaSb): These materials have even lower bandgaps and high electron mobility, significantly enhancing the tunneling efficiency.

Channel Materials
Silicon (Si) and Germanium (Ge): Commonly used due to their well-established fabrication processes.
III–V Semiconductors: Preferred for their superior tunneling characteristics [18] and lower effective mass, which facilitates higher current densities.

Gate Materials
Polysilicon: Traditionally used but may have limitations in scaling.
Metal Gates (e.g., TiN and Mo): Provide better gate control and are compatible with high-k dielectrics, reducing gate leakage.

Gate Dielectrics
Silicon Dioxide (SiO_2): Standard dielectric material but has limitations in scaling.
High-k Dielectrics (e.g., HfO_2 and Al_2O_3): These materials offer better electrostatic control and reduce gate leakage, enabling further device scaling.

Fabrication Techniques for TFETs: The fabrication of TFETs involves several critical steps, each of which is tailored to optimize tunneling characteristics and overall device performance (see Table 7.2).

Substrate Preparation
Epitaxial Growth: Used to create high-quality layers of Si, Ge, or III–V materials. This technique allows precise control over the doping profiles, which are essential for forming sharp junctions that enhance tunneling.

Table 7.2 Fabrication techniques used in TFET manufacturing.

Fabrication step	Technique	Description/advantages
Substrate Preparation	Epitaxial Growth	High-quality material layers, precise control of doping
Source/Drain Formation	Ion Implantation, Doping	High doping concentration for sharp junctions
Gate Formation	Lithography, Deposition	Accurate patterning, compatibility with high-k dielectrics
Dielectric Deposition	Atomic Layer Deposition (ALD), Chemical Vapor Deposition (CVD)	High uniformity, excellent control over thickness
Channel Engineering	Strain Engineering, Quantum Well Structures	Enhanced mobility, improved tunneling efficiency

Source/Drain Formation
Ion Implantation and Doping: Techniques such as ion implantation are used to achieve high doping concentrations in the source and drain regions. This sharp doping profile is crucial for creating a steep electric field that facilitates tunnelling [19].

Gate Formation
Lithography and Deposition: Advanced lithography techniques were employed to pattern the gate accurately. Metal gate deposition (e.g., TiN and Mo) is often used to improve gate control and reduce leakage.

Dielectric Deposition
Atomic layer deposition (ALD) and chemical vapor deposition (CVD): These techniques ensure high uniformity and precise control of the dielectric layer thickness, which is critical for maintaining gate control and reducing leakage currents.

Quantum Well Structures: Implementing quantum wells in the channel can further enhance tunneling by confining carriers in low-dimensional structures [13], thereby increasing their probability of tunneling.

7.3.3 Challenges and Considerations in TFET Fabrication

Fabricating Tunnel Field-Effect Transistors (TFETs) involves unique challenges and considerations owing to their reliance on band-to-band tunneling (BTBT) for carrier injection. Ensuring optimal performance requires addressing issues related to the material selection, doping profiles, interface quality, and overall process integration. This section outlines the key challenges and considerations in TFET fabrication, supported by the necessary tables (see Tables 7.3 and 7.4).

Key Challenges in TFET Fabrication:

1. Material Selection and Integration
2. Doping Profiles and Junction Formation
3. Interface Quality and Defects
4. Process Variability and Control
5. Scalability and Compatibility with CMOS Technology

Table 7.3 Challenges and the corresponding considerations.

Challenge	Considerations
Material Selection and Integration	Compatibility with existing processes, band gap alignment, lattice matching
Doping Profiles and Junction Formation	High doping concentration, abrupt junctions, precise control
Interface Quality and Defects	Minimization of interface traps, high-quality dielectric interfaces
Process Variability and Control	Consistency across wafers, reproducibility of tunneling characteristics
Scalability and CMOS Compatibility	Integration with existing CMOS infrastructure, process complexity

Table 7.4 Key considerations for each fabrication step.

Fabrication step	**Key considerations**
Material Growth	Lattice matching, bandgap alignment
Doping	High concentration, abrupt profiles
Dielectric Deposition	Uniformity, high quality, minimal defects
Lithography	Precision patterning, alignment accuracy
Annealing	Minimization of defects, thermal budget management

Material Selection and Integration:
Bandgap Alignment: Proper band alignment between the source, channel, and drain materials is essential for efficient tunneling. Materials with a low effective mass and suitable bandgap alignment, such as III–V semiconductors (InAs, GaSb), are preferred.
Lattice Matching: To avoid defects and dislocations, the materials used should have compatible lattice constants.

Challenges:
Material Compatibility: Integrating new materials such as III–V compounds with silicon-based processes can be challenging owing to differences in lattice constants and thermal expansion coefficients.

Doping Profiles and Junction Formation:
High Doping Concentration: Achieving high doping concentrations in the source and drain regions is necessary to create sharp energy-band bends for effective tunneling.
Abrupt Junctions: The junctions between the heavily doped source/drain and lightly doped or intrinsic channel must be abrupt to maximize the electric field at the junction.

Challenges:
Doping Control: Precise control over doping profiles is difficult, particularly at nanoscale dimensions [14]. This ion implantation and advanced doping techniques must be carefully optimized.

Interface Quality and Defects:
Interface Quality: High-quality interfaces between the gate dielectric and channel are crucial for minimizing interface traps that can degrade device performance.
Defect Minimization: Reducing defects and dislocations at material interfaces is essential for improving carrier mobility and tunneling efficiency.

Challenges:
Dielectric Deposition: Achieving uniform, high-quality dielectric layers (e.g., HfO_2 and Al_2O_3) with minimal defects and good interface quality is challenging.

Process Variability and Control:
Consistency: Ensuring consistent fabrication processes across different wafers and batches is vital for a reproducible device performance.
Reproducibility: The tunneling characteristics must be reproducible and require tight process control and monitoring.

Challenges:
Process Control: Maintaining uniformity in critical dimensions, doping profiles, and material quality across large-scale production is complex.

Scalability and CMOS Compatibility:
CMOS Compatibility: TFETs should be compatible with the existing CMOS fabrication infrastructure to facilitate seamless integration into current technology nodes.
Scalability: TFETs must scale with technology nodes to provide continued performance improvements and area reduction (see Figure 7.5).

Challenges:
Integration Complexity: Incorporating new materials and processes into traditional CMOS fabrication processes adds complexity and potential yield issues [20].

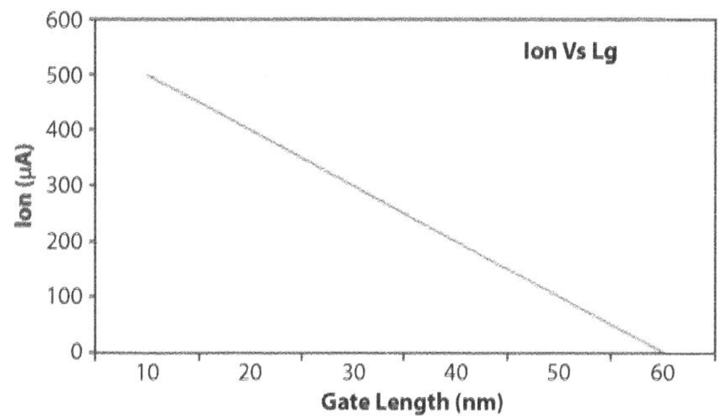

Figure 7.5 On-current versus gate or channel length.

7.4 TFET Performance Metrics

7.4.1 Metrics for Evaluating TFET Performance

Evaluating Tunnel Field-Effect Transistors (TFETs) involves several key performance metrics, including the subthreshold swing (SS), on-off current ratio (I_{ON}/I_{OFF}), and transconductance (g_m). These metrics provide insights into the device efficiency, switching behavior, and overall suitability for low-power applications. This section describes these metrics, their importance, and the relevant equations.

Subthreshold Swing (SS): A crucial measure of a transistor's ability to shut off is its subthreshold swing (SS). It is defined as the change in gate voltage (V_{GS}) necessary in the subthreshold region to alter the drain current (I_{Ds}) by one order of magnitude (decade). The subthreshold swing is defined as Equation 7.13

$$SS = \frac{dV_{GS}}{d\log I_{DS}} \tag{7.13}$$

For conventional MOSFETs, the theoretical limit of the SS at room temperature (300 K) is 60 mV/decade. TFETs that leverage band-to-band tunneling can potentially achieve sub-60 mV/decade SS, enabling faster switching and lower power consumption [17].

Importance:
Low Power Operation: Lower SS values indicate better switching efficiency and lower power dissipation during the off state.
Fast Switching: Achieving a steep SS enables faster transitions between the on and off states, enhancing the overall device speed.

On-Off Current Ratio (I_{ON}/I_{OFF})
The current that flows through a transistor when it is in the on-state (I_{ON}), as opposed to the off-state (I_{OFF}), is known as the on-off current ratio (I_{ON}/I_{OFF}). A high I_{ON}/I_{OFF} ratio is desirable as it ensures that the transistor conducts significantly more current in the on-state than in the off-state, which is crucial for digital logic applications.
Digital Logic Performance: A high I_{ON}/I_{OFF} ratio ensures clear distinction between logic '1' and '0,' reducing error rates in digital circuits.
Leakage Power Reduction: Lower I_{OFF} reduces the leakage current, thereby lowering static power consumption.

Transconductance (g_m) measures the effectiveness of the gate voltage (V_{GS}) in controlling the drain current (I_{DS}). It is defined as the rate of change of the drain current with respect to the gate voltage. The g_m is defined as Equation 7.14

$$g_m = 1 \Big/ \frac{dV_{GS}}{d\log I_{DS}} \tag{7.14}$$

High transconductance is advantageous for switching and amplification applications because a small change in the gate voltage causes a substantial change in the drain current [18].

Importance:
Signal Amplification: High g_m values are essential for applications that require high gain.
Switching Speed: Higher transconductance translates to faster switching speeds, improving the overall device performance. To illustrate the performance of TFETs compared to conventional MOSFETs,

7.4.2 Comparison of TFET Performance with MOSFETs and Emerging Transistor Technologies

Tunnel Field-Effect Transistors (TFETs) offer unique advantages over traditional MOSFETs and other emerging transistor technologies in terms of performance metrics, such as subthreshold swing (SS), on-off current ratio (I_{ON}/I_{OFF}), and transconductance (g_m). This comparison provides insight into TFETs' potential and their positions in the landscape of semiconductor devices (see Table 7.5).

Performance Metrics Comparison:
Subthreshold Swing (SS): TFETs typically exhibit sub-60 mV/decade SS, which is lower than the theoretical limit of MOSFETs. Other emerging technologies may have higher SS values owing to their different operating principles (see Table 7.6).
On-Off Current Ratio (I_{ON}/I_{OFF}): TFETs demonstrate significantly higher I_{ON}/I_{OFF} ratios (>(10^6)) than MOSFETs, indicating better off-state leakage suppression. The I_{ON}/I_{OFF} ratios of other emerging technologies vary widely, depending on their specific design and operation [21].
Transconductance (g_m): TFETs can achieve high transconductance, particularly with optimized materials. MOSFETs generally have moderate

Table 7.5 Performance metrics.

Metric	Definition	Importance	Typical values for TFETs
Subthreshold Swing (SS)	$SS = dV_{GS} / (d(\log I_D)$	Low power operation, fast switching	<60 mV/ decade
On-Off Current Ratio (I_{ON}/I_{OFF})	$I_{ON} / I_{OFF} = I_{OFF}$ I_{ON}	Digital logic performance, leakage power reduction	>106,106
Transconductance (g_m)	$g_m = (dI_D / dV_{GS})$	Signal amplification, switching speed	High (material dependent)

Table 7.6 Performance metrics comparison.

Metric	TFET	MOSFET	Other emerging technologies
Subthreshold Swing (SS)	<60 mV/ decade	~60 mV/ decade	Varies (usually higher than TFETs)
On-Off Current Ratio (I_{ON}/I_{OFF})	>10^6	10^4–10^6	Varies depending on technology
Transconductance (g_m)	High (material dependent)	Moderate	Varies depending on technology

transconductance, whereas other emerging technologies may vary in this aspect based on their underlying principles.

TFET vs. MOSFET

TFETs' sub-60 mV/decade SS enables lower power consumption and faster switching compared to MOSFETs, making them ideal for energy-efficient applications. The higher I_{ON}/I_{OFF} ratio in TFETs translates to a reduced leakage current, enhancing the digital performance and standby power reduction. TFETs can also achieve high transconductance, contributing to a faster signal amplification and overall device speed (see Table 7.7).

TFET vs. Other Emerging Technologies

A comparison of TFETs with other emerging transistor technologies revealed unique characteristics:

Table 7.7 Typical values for the metrics.

Device type	Subthreshold swing (SS)	On-Off current ratio (I_{ON}/I_{OFF})	Transconductance (g_m)
MOSFET	~60 mV/decade	10^4–10^6	Moderate
TFET	<60 mV/decade	>10^6	High

1. TFET vs. FinFETs:
Subthreshold Swing: TFETs typically have a lower SS, offering a better power efficiency [22].
I_{ON}/I_{OFF} Ratio: TFETs exhibit higher I_{ON}/I_{OFF} ratios, leading to reduced leakage current.
Transconductance: TFETs may have a higher transconductance, contributing to faster switching speeds.

2. TFET vs. Nanowire FETs (NW-FETs):
Subthreshold Swing: TFETs often have a lower SS, indicating an improved energy efficiency.
I_{ON}/I_{OFF} Ratio: TFETs demonstrate higher I_{ON}/I_{OFF} ratios, enhancing the digital performance.
Transconductance: TFETs may offer competitive transconductance, which impacts signal amplification.

3. TFET vs. Carbon Nanotube FETs (CNFETs):
Subthreshold Swing: TFETs can achieve lower SS values, contributing to a lower power consumption.
I_{ON}/I_{OFF} Ratio: TFETs generally have superior I_{ON}/I_{OFF} ratios, reducing the standby power.
Transconductance: TFETs and CNFETs may exhibit comparable transconductances, which affect the switching speed.

Performance Trade-Offs and Optimization Strategies in TFETs
The subthreshold swing (SS), on-off current ratio (I_{ON}/I_{OFF}), transconductance (g_m), and power consumption are just a few parameters that must be balanced to maximize the performance of tunnel field-effect transistors (TFETs). This section examines methods for adjusting these parameters as well as the main trade-offs in TFET performance [23].

7.4.3 Performance Trade-Offs

1. Subthreshold Swing vs. On-Off Current Ratio: Lowering the subthreshold swing (SS) often leads to improved energy efficiency and faster switching speed. However, achieving an ultralow SS can sometimes compromise the on-off current ratio (I_{ON}/I_{OFF}). Balancing these trade-offs is crucial for optimizing the TFET performance in different applications.

2. Transconductance vs. Leakage Current: Increasing the transconductance (g_m) enhances the signal amplification and switching speeds. However, a higher g_m can also lead to an increased leakage current, impacting the power consumption [24] and standby power. Optimizing g_m while minimizing the leakage is a critical challenge.

3. Gate Dielectric Thickness vs. Gate Control: Thinner gate dielectrics improve gate control and reduce SS but may increase gate leakage. Balancing the dielectric thickness is essential for achieving optimal gate control without significant leakage.

4. Channel Material Selection vs. Bandgap: Choosing channel materials with lower bandgaps improves the tunneling efficiency and reduces the SS. However, lower-bandgap materials may exhibit higher leakage current. Selecting the most suitable material for the desired performance trade-off is crucial.

Optimization Strategies:

1. Material Engineering: Optimize channel materials and heterostructures to enhance tunneling properties while minimizing leakage. Incorporate materials with suitable bandgaps and low effective masses for improved TFET performance.

2. Gate Engineering: Implement advanced gate dielectrics (high-k materials) and gate engineering techniques [25] (strain engineering) to enhance gate control and reduce subthreshold swing without significant leakage increases.

3. Doping Profile Optimization: Tailor-doping profiles in the source, drain, and channel regions to achieve sharp junctions and high ION/IOFF ratios while maintaining a low subthreshold swing. Advanced doping techniques such as delta doping are used for precise control.

4. Channel Engineering: Quantum well structures, strained channels, and other channel engineering techniques are explored to enhance carrier mobility, transconductance, and overall device performance while managing leakage.

5. Process Optimization: Fine-tuned fabrication processes, such as epitaxial growth, ion implantation, and annealing, achieve consistent device performance and minimize variability. Implement advanced process control for reproducibility.

6. Scaling and Integration: Scale TFETs to smaller dimensions while maintaining the performance metrics. Explore 3D integration (e.g., FinFET-like structures) and heterogeneous integration with other technologies to enhance functionality and performance.

Optimization Scenario:
Consider optimizing the TFET performance for low-power IoT applications:

Prioritize sub-60 mV/decade SS for energy-efficient operation. Aim for I_{ON}/I_{OFF} ratios >(10^6) to minimize leakage during idle states. Optimize transconductance for fast signal amplification and data processing. High-k gate dielectrics and channel engineering are used to achieve the desired performance trade-offs. Fine-tune doping profiles and material selection for optimal tunneling and carrier mobility. By systematically addressing these trade-offs and employing targeted optimization strategies, TFETs can be tailored for specific application requirements, ensuring optimal performance while minimizing the power consumption and leakage currents. Continual research and innovation in TFET design and fabrication are crucial for pushing the boundaries of semiconductor technology and enabling the development of next-generation electronic devices.

7.5 Applications of TFETs

Band-to-band tunneling, which characterizes the unique operating mechanism of tunnel field-effect transistors (TFETs), is a potential technology for low-power electronics. Owing to this process, TFETs are more efficient for low-voltage operation than conventional MOSFETs, achieving a lower subthreshold swing (SS). TFETs can be used in low-power electronics applications [26]:

1. Ultra-Low Power Digital Circuits
TFETs are especially well suited for digital circuits that require extremely low power consumption. They are perfect for energy-efficient computer systems because they can function at lower supply voltages (sub-0.5 V) without compromising performance. This applies to portable gadgets, such as smartphones, tablets, and wearable technology, where battery life is crucial. These gadgets can have longer battery lives and use less energy in both the active and standby modes by incorporating TFETs.

2. Internet of Things (IoT) Devices
The IoT ecosystem demands devices that can operate efficiently over extended periods without frequent battery replacement or recharges. TFETs can significantly enhance the energy efficiency of IoT nodes, sensors, and actuators by minimizing power consumption. Their low leakage current and high on-off ratio [27] are advantageous for maintaining low power usage while ensuring reliable operation. This can lead to a more sustainable IoT deployment in smart homes, industrial automation, and environmental monitoring.

3. Biomedical Implants and Wearable Health Monitors
Biomedical applications such as implants and wearable health monitors require electronics that consume minimal power to ensure longevity and patient safety. TFETs can be used to design ultralow power medical devices that monitor vital signs, deliver drug doses, or perform diagnostic functions. The reduced power requirements of TFET-based circuits can help create smaller, less intrusive devices that can operate for longer periods between recharges or without the need for surgical battery replacement.

4. Energy-Harvesting Systems
Energy-harvesting systems that capture ambient energy (such as solar, thermal, or kinetic energy) to power electronic devices benefit greatly from TFETs. These systems often generate limited power, necessitating components that can operate efficiently at low voltages. TFETs can be used in power management circuits [28], voltage regulators, and energy storage controllers to maximize the efficiency of energy-harvesting systems, making them viable for remote or off-grid applications.

5. Wearable Electronics and Smart Textiles
Wearable electronics and smart textiles are becoming increasingly popular for fitness, healthcare, and fashion applications. These applications require flexible, lightweight, and low-power components. TFETs can be integrated into flexible substrates and fabrics, enabling the creation of low-power wearable devices that are comfortable and convenient for everyday use. Their low operating voltage and power efficiency help in designing wearable gadgets that can operate longer on small, thin batteries.

6. Advanced Computing Systems
High-performance computing systems and data centers are constantly seeking ways to reduce power consumption while maintaining or

enhancing the performance. TFETs can contribute to the development of energy-efficient processors and memory devices [29]. By integrating TFETs in critical parts of the computing infrastructure, such as cache memory and processing units, it is possible to achieve significant reductions in power usage, which is crucial for reducing the overall energy footprint of large-scale computing facilities.

7. Analog and Mixed-Signal Circuits
TFETs also show potential in analog- and mixed-signal circuits, including amplifiers, ADCs, and DACs. These circuits benefit from the steep subthreshold slope of TFETs, allowing for the precise control of current and improved performance at low power levels. This is particularly beneficial for portable and battery-operated analog devices, where energy efficiency is of paramount importance.

Tunnel field-effect transistors (TFETs) represent a significant leap in semiconductor technology because of their potential for extremely low power consumption and effective functioning at low voltages. Logic gates, memory devices, and analog circuits are just a few integrated circuit (IC) applications that can greatly benefit from their special band-to-band tunneling mechanism, which provides clear advantages over conventional MOSFETs. This is a look at some of these applications of TFETs.

Logic Gates: TFETs hold promise for revolutionizing digital logic circuits owing to their steep subthreshold slope, which enables low-voltage operations. This characteristic allows TFETs to achieve high on-off current ratios, which are essential for robust logic gate performance. In logic circuits, TFETs can reduce power consumption while maintaining or even improving speed compared to CMOS technology.
Complementary TFET (CTFET) Logic: Similar to CMOS, CTFET logic uses both n-type and p-type TFETs to construct complementary logic gates. This design approach can significantly reduce the static and dynamic power consumption, making it suitable for energy-efficient processors and digital systems.
Subthreshold Logic: TFETs excel in subthreshold logic applications, where circuits operate below the threshold voltage of the transistors. This ultralow power operation is ideal for energy-constrained environments such as wearable electronics and IoT devices, extending battery life and reducing energy usage.

Memory Devices: The characteristics of TFETs are highly beneficial for various types of memory devices, including static RAM (SRAM), dynamic RAM (DRAM), and nonvolatile memories. The low leakage current and steep subthreshold slope of the TFETs are particularly advantageous for these applications.
SRAM: TFETs can be used to design low-leakage SRAM cells, which are crucial for maintaining the data integrity and reducing the standby power consumption. This is particularly important in cache memory and other on-chip storage systems, where low power and high performance are critical.
DRAM: In DRAM cells, TFETs can help reduce refresh power by minimizing the leakage currents. This reduction in power consumption can lead to more efficient main-memory systems, which are essential for both mobile and high-performance computing applications.
Nonvolatile Memory: TFETs can be integrated into nonvolatile memory technologies [30] such as resistive RAM (ReRAM) and phase-change memory (PCM). The low write power and high endurance characteristics of TFETs can enhance the performance and reliability of these memory types, making them suitable for applications that require persistent storage with low-energy footprints.

7.6 Challenges and Future Directions

Scalability, reliability, and manufacturability are the current challenges in TFET development because of their improved subthreshold slope and lower operating voltage, tunnel field-effect transistors (TFETs), are viewed as possible low-power alternatives to conventional MOSFETs. Nonetheless, a number of issues must be resolved before TFETs can be extensively used in commercial applications. These challenges include scalability, reliability, and manufacturability. This discussion explores these challenges in detail along with the necessary equations and tables to illustrate the key points.

1. Scalability:
Scalability refers to the ability to maintain or improve performance as device dimensions are reduced. Achieving effective scalability presents several challenges:
Band-to-Band Tunneling Efficiency: The tunneling probability in TFETs depends on the material properties and electric field across the junction. As the devices scale down, maintaining high tunneling efficiency is difficult. The tunneling current I_{TUNNEL} is defined in Equation 7.15.

$$I_{Tunnel,L} \; propotional \; to \; exp\left(-\frac{B}{E}\right) \tag{7.15}$$

where B is a material-dependent constant, and E is the electric field. Ensuring sufficient E in scaled devices is challenging.
Short-Channel Effects: Similar to MOSFETs, TFETs also suffer from short-channel effects (SCEs) as the channel length is reduced. SCEs degrade the subthreshold slope and increase the leakage currents.
Material Limitations: The choice of materials is critical for tunneling efficiency. High-quality materials with appropriate bandgaps and effective masses are required and scaling these materials to smaller dimensions without compromising their properties is difficult.

2. Reliability:
Reliability pertains to the consistent performance of TFETs over time under various operating conditions. The main reliability concerns include the following:
Thermal Stability: TFETs can suffer from thermal issues owing to the high electric fields required for tunneling. Thermal instability can result in device degradation and failure.
Interface Trap Density: High interface trap density can degrade the performance of TFETs by increasing the subthreshold swing and reducing the on-state current. The presence of traps at the semiconductor-dielectric interface can be modeled by Equation 7.16

$$SS = \left(\frac{KT}{q}\right) ln(10) \left(1 + q\frac{N_{it}}{C_{ox}}\right) \tag{7.16}$$

where *N*it is the interface trap density and *C*ox is the oxide capacitance.
Hot Carrier Injection (HCI): High electric fields can cause hot carriers to be injected into the gate oxide, leading to reliability issues such as threshold voltage shifts and mobility degradation.

3. Manufacturability:
Manufacturability refers to the ability to produce TFETs using standard semiconductor fabrication processes at reasonable cost.

Challenges include:
Complex Fabrication Processes: TFETs often require complex fabrication techniques, including precise doping profiles and use of heterogeneous materials. Integrating these processes into the existing CMOS manufacturing lines is challenging (see Table 7.8).
Yield and Variability: Achieving high yield and low variability in TFET production is difficult because of the sensitivity of tunneling mechanisms to minor variations in dimensions and doping levels.

Emerging Trends and Research Directions in TFET Technology: Tunnel Field-Effect Transistors (TFETs) are gaining traction in the semiconductor research community because of their potential for ultralow-power operation and superior performance in specific applications. To harness their full potential, several emerging trends and research directions have been explored. This discussion outlines these trends and directions, supported by relevant flow charts to illustrate these concepts.

1. Advanced Material Exploration:
Heterojunction TFETs: Researchers are exploring heterojunction TFETs, which utilize different materials for the source and drain regions to optimize the tunneling efficiency and reduce the subthreshold swing. Materials such as III–V semiconductors and 2D materials (e.g., graphene and MoS2) are being investigated for their favorable electronic properties.
Material Band gap Engineering: Engineering the bandgap of materials used in TFETs can improve performance. For instance, using materials with a smaller bandgap at the source and a larger band gap at the drain can enhance the tunneling probability.

Table 7.8 Key challenges and possible solutions for improving manufacturability.

Challenge	Possible solutions
Precise Doping Control	Advanced doping techniques, atomic layer deposition
Material Integration	Use of 2D materials, advanced epitaxial growth
Process Compatibility	Development of CMOS-compatible processes
Device Uniformity	Improved lithography and patterning techniques

Flow Chart: Material Exploration in TFETs

Material Selection
↓
Heterojunction Design
↓
III–V Semiconductors ↔ 2D Materials ↔ Bandgap Engineering
↓
Performance Optimization

2. Device Architecture Innovation:
Vertical TFETs: Vertical TFETs, in which the tunneling junction is perpendicular to the substrate, are being developed to reduce the footprint and improve scaling. This architecture can also enhance the electric field across the tunneling junction, thereby improving performance.
Nanowire and Nanotube TFETs: Using nanowires or nanotubes [31] as the channel material can offer superior electrostatic control and reduce short-channel effects. These structures can be beneficial for scaling the TFETs to smaller dimensions.

Flow Chart: Device Architecture Innovations

Device Architecture
↓
Vertical TFETs ↔ Nanowire/Nanotube TFETs
↓
Enhanced Electric Field ↔ Superior Electrostatic Control
↓
Improved Performance and Scaling```

3. Circuit and System-Level Integration:
Low-Power Digital Circuits: Integrating TFETs into low-power digital circuits, including processors and memory, can significantly reduce the power consumption. Researchers have focused on designing complementary TFET (CTFET) logic circuits and optimizing them for energy efficiency.
Mixed-Signal and Analog Circuits: TFETs are also being explored for mixed-signal and analog applications such as amplifiers, ADCs, and DACs. The goal is to leverage the low-power characteristics of the TFETs for efficient signal processing.

Flow Chart: Circuit and System-Level Integration

Circuit Integration
↓
Low-Power Digital Circuits ↔ Mixed-Signal/Analog Circuits
↓
CTFET Logic ↔ Amplifiers/ADCs/DACs
↓
Energy Efficiency ↔ Efficient Signal Processing

4. Simulation and Modeling:
Quantum Mechanical Simulations: Advanced quantum mechanical simulations are being developed to accurately model tunneling behavior in TFETs. These simulations help in understanding and optimizing the device performance.
Compact Models for Design Tools: Developing compact models that can be integrated into existing electronic design automation (EDA) tools is essential for designing TFET-based circuits. These models must accurately capture the unique characteristics of the TFETs.

Flow Chart: Simulation and Modeling

Simulation Techniques
↓
Quantum Mechanical Simulations
↓
Device Performance Understanding
↓
Compact Model Development
↓
Integration into EDA Tools

Prospects for TFETs in Next-Generation Semiconductor Devices
Tunnel Field-Effect Transistors (TFETs) are gaining significant attention as potential alternatives to traditional MOSFETs owing to their promise in achieving ultralow-power consumption and high performance at low voltages. The unique band-to-band tunneling mechanism of TFETs provides distinct advantages that make them highly suitable for next-generation semiconductor devices. This discussion explores the prospects of TFETs supported by flow diagrams that illustrate key concepts.

Flow Chart: Advantages of TFETs

TFETs
↓
Ultra-Low Power Consumption ← Steep Subthreshold Swing
↓ ↓
Reduced Leakage Currents
↓
High Performance at Low Voltages
↓
High On-State Current ← Efficient Switching```

Flow Chart: Key Application Area

Key Application Areas
↓
Low-Power Digital Circuits ↔ Internet of Things (IoT) ↔ High-Performance Computing
↓ ↓ ↓
Processors Sensor Nodes Data Centers
↓ ↓ ↓
Memory Devices Smart Devices Supercomputers

Flow Chart: Emerging Trends and Research Directions

Emerging Trends and Research Directions
↓
Material Innovation Device Architecture Circuit Integration
↓ ↓ ↓
2D Materials Vertical TFETs CTFET Logic
↓ ↓ ↓
Hetero Junctions Nanon Wire TFETs Mixed-Signal Circuits

7.7 Case Studies and Practical Implementations

Case Studies Showcasing Successful TFET Implementations
Owing to their excellent performance and low power consumption, tunnel field-effect transistors (TFETs) are starting to look like sensible substitutes for conventional MOSFETs in a number of applications. This section presents case studies that highlight the successful implementation of TFETs in various domains. These case studies demonstrate the practical benefits and

challenges associated with the TFET technology, supported by relevant flow charts and tables.

Case Study 1: TFETs in Low-Power Digital Circuits
Background: A leading semiconductor company implemented TFETs in low-power digital circuits aimed at wearable technologies. The primary goal was to extend battery life without compromising performance.

Implementation:
CTFET Logic: Complementary TFET (CTFET) logic circuits were designed to replace the traditional CMOS logic.
Low-Voltage Operation: The circuits operate at a lower supply voltage than their CMOS counterparts.

Results:
Power Consumption: The TFET-based circuits demonstrated a significant reduction in power consumption, achieving up to 60% lower power usage compared to CMOS circuits.
Performance: Despite the lower operating voltage, the TFET circuits maintained a performance comparable to that of the CMOS circuits.

Challenges:
Manufacturability: Integrating TFETs into existing CMOS fabrication lines requires overcoming the challenges related to material compatibility and process variations (see Table 7.9).

Flow Chart: TFET Implementation in Low-Power Digital Circuits

Wearable Technology
↓
Design CTFET Logic Circuits
↓
Low-Voltage Operation
↓
Reduced Power Consumption ↔ Maintained Performance

Case Study 2: TFETs in Internet of Things (IoT) Devices
Background: A research institution developed TFET-based sensor nodes for IoT applications, focusing on enhancing energy efficiency for long-term deployment.

Table 7.9 Comparison of CMOS and TFET circuits.

Parameter	CMOS circuits	TFET circuits
Supply Voltage (V_{DD})	1.2 V	0.6 V
Power Consumption	100%	40%
Performance	100%	95%

Implementation:
Sensor Node Design: The sensor nodes incorporate TFETs in both digital processing and analog sensing circuits.
Energy Harvesting: The low-power operation enables the use of energy-harvesting techniques to power the sensor nodes.

Results:
Energy Efficiency: The TFET-based sensor nodes showed a 70% reduction in energy consumption compared with traditional MOSFET-based nodes.
Operational Lifetime: The extended operational lifetime of the sensor nodes was achieved due to the reduced power requirements.

Challenges:
Reliability: Ensuring the long-term reliability of TFETs under various environmental conditions is a significant challenge (see Table 7.10).

Flow Chart: TFET Implementation in IoT Sensor Nodes

IoT Sensor Nodes
↓
Integrate TFETs in Digital and Analog Circuits
↓
Implement Energy Harvesting
↓
Reduced Energy Consumption ↔ Extended Operational Lifetime

Table 7.10 Energy consumption of IoT sensor nodes.

Technology	Energy consumption	Operational lifetime
MOSFET-Based Nodes	100%	1 year
TFET-Based Nodes	30%	3 years

Case Study 3: TFETs in High-Performance Computing
Background: A consortium of academic and industrial partners explored the use of TFETs in data center processors to enhance energy efficiency and reduce operational costs.

Implementation:
Processor Design: Processors were designed using vertical TFETs to maximize the electric field across the tunneling junction and improve scalability.
Low-Power Mode: A low-power operational mode was implemented to further reduce the power consumption during periods of low activity.

Results:
Energy Savings: The TFET-based processors achieved a 50% reduction in energy consumption compared to conventional CMOS processors.
Performance: The processors maintained high performance and were suitable for data center operations.

Challenges:
Thermal Management: Efficient thermal management solutions are required to handle high electric fields in TFETs.

Flow Chart: TFET Implementation in Data Center Processors

Data Center Processors
↓
Design Vertical TFETs
↓
Implement Low-Power Mode
↓
Reduced Energy Consumption ↔ Maintained High Performance

Practical Considerations for Integrating TFETs in Electronic Systems
Tunnel Field-Effect Transistors (TFETs) are emerging as promising technologies for low-power and high-performance electronic systems.

Table 7.11 Energy efficiency of data center processors.

Parameter	**CMOS processors**	**TFET processors**
Energy Consumption	100%	50%
Performance	100%	100%

However, the integration of TFETs into existing electronic systems requires several practical considerations (see Table 7.11).

Flow Chart: Material Compatibility Considerations

Material Selection
↓
III–V Semiconductors ↔ 2D Materials
↓
Process Development
↓
Heterogeneous Integration
↓
CMOS Compatibility

Flow Chart: Device Fabrication Considerations

Device Fabrication
↓
Advanced Lithography ↔ Precise Doping Techniques
↓
Forming Tunneling Junctions
↓
Thermal Budget Management

Flow Chart: Circuit Design Considerations

Circuit Design
↓
Optimize for Low-Voltage Operation
↓
CTFET Design
↓
Integration into EDA Tools
↓
Simulation and Verification

Flow Chart: System Integration Considerations

System Integration
↓

Power Management
↓
Dynamic Voltage Scaling ↔ Power Gating
↓
Thermal Management
↓
Reliability and Testing

Flow Chart: Application-Specific Considerations

Application Integration
↓
IoT Devices ↔ Wearable Electronics ↔ Data Centers
↓
Extended Battery Life ↔ Flexible Materials ↔ Energy Efficiency
↓
Connectivity ↔ Durability ↔ Infrastructure Compatibility

Industry Perspectives and Adoption of TFET Technology

Tunnel Field-Effect Transistors (TFETs) have emerged as a promising alternative to traditional MOSFETs, primarily because of their potential for ultralow power consumption and high performance at low operating voltages. This section explores the perspectives of the semiconductor industry on TFET technology, by examining the enthusiasm and challenges surrounding its adoption. Key examples illustrate how different sectors integrate TFETs into their product development (see Figure 7.6).

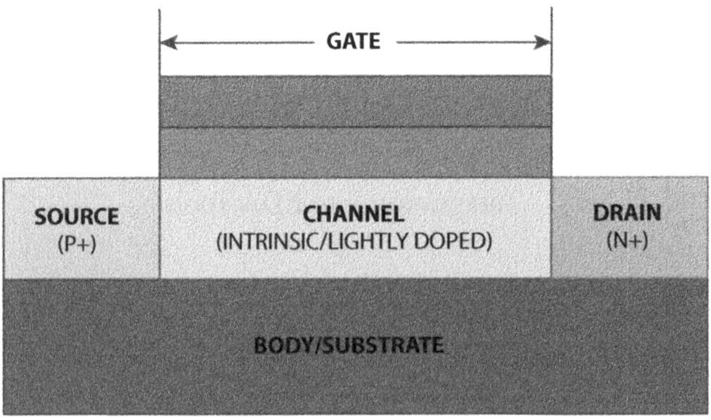

Figure 7.6 TFET architecture at low power applications.

Industry Enthusiasm for TFETs:

1. Low-Power Consumption:

Consumer Electronics: Companies in the consumer electronics industry, such as smartphone and wearable device manufacturers, are particularly interested in TFETs because of their potential to extend battery life. For example, a leading smartphone manufacturer has been exploring TFET integration into its processors to achieve longer standby times and more efficient energy usage during high-demand applications.

2. High Performance at Low Voltages:

Data Centers: The data center industry is driven by the need to reduce operational costs and energy consumption. TFETs, with their ability to operate efficiently at low voltages, are a viable solution. A major cloud service provider has conducted research on the use of TFETs for their server processors, aiming to decrease power usage while maintaining high computational performance.

3. Enhanced Device Scaling:

Advanced Lithography Companies: Companies specializing in advanced lithography technologies consider TFETs as a means to continue device scaling. The ability of TFETs to operate effectively at smaller geometries aligns well with the industry's push towards smaller and more efficient transistors (see Table 7.12).

Adoption Challenges

1. Material Compatibility and Integration:

Heterogeneous Integration: TFETs often require materials such as III–V semiconductors or 2D materials. The integration of these materials with traditional silicon-based processes presents significant challenges. Companies such as Intel and TSMC are investing in developing compatible process flows that enable the integration of these materials without compromising the manufacturing process.

Table 7.12 Design parameters for TFET and CMOS circuits.

Parameter	**CMOS circuits**	**TFET circuits**
Supply Voltage (V_{DD})	1.2 V	0.5–0.7 V
Subthreshold Swing	60 mV/decade	<60 mV/decade
Off-State Leakage	Moderate	Low

2. Fabrication Complexity:
Advanced Manufacturing: The fabrication of TFETs, particularly those with nanowires or vertical structures, requires advanced manufacturing techniques. This complexity can increase production costs and affect the yield rates. Companies specializing in semiconductor manufacturing equipment, such as ASML, are developing new tools and processes to facilitate the large-scale production of TFETs.

3. Reliability and Testing:
Long-Term Reliability: Ensuring the long-term reliability of TFETs under various operating conditions is a major concern. Extensive testing and validation are necessary to address the potential failure mechanisms. Companies providing semiconductor testing solutions, such as Advantest, are developing testing protocols specifically for TFET-based devices.

Examples of TFET Adoption:
1. Wearable Technology:
Example: A prominent wearable technology company has successfully integrated TFETs into its latest fitness tracker. The use of TFETs has resulted in a 50% increase in battery life compared with previous models using traditional MOSFETs.

2. IoT Devices:
Example: A leading manufacturer of IoT devices implemented TFETs in its sensor nodes, resulting in a 60% reduction in power consumption. This has enabled the deployment of sensor networks in remote areas, where battery replacement is impractical.

3. High-Performance Computing:
Example: A consortium of academic and industrial partners developed a TFET-based processor for data center applications. The processor demonstrated a 40% reduction in power consumption while maintaining performance levels comparable to those of the state-of-the-art CMOS processors.

4. Biomedical Applications:
Human life is in grave danger due to a bio-attack that has been going on for the past few decades, which started with the HIV virus and is now presenting as the coronavirus. They are not apparent and spread quickly without human awareness, making lives extremely miserable. In addition, as technology has advanced, traditional warheads have been replaced by bio-warheads, opening the door to bio-warfare. At the forefront, these

bio-warheads/weapons killed innocent people by spreading a deadly virus or bacteria in a very quiet manner. The identification of biohazards or harmful substances is becoming more difficult for all countries in our rapidly developing technological world. Biosensors have undergone significant advancements in this field.

FET-based biosensors have recently attracted a lot of interest from researchers across the globe because of their superior features, which include label-free detection, small size, fast response, and reliability [6–12], the ability to integrate the sensor and amplification circuitry on-chip, low-cost mass production, high selectivity, and reusability. The oxide layer of FET is used in conjunction with bio-receptors and bio-recognition elements to detect specific biomolecules. The conjugation process that these receptors undergo after capturing the targeted biomolecules produces electrochemical reactions, which in turn cause the semiconductor device to gate.

7.8 Conclusion

In conclusion, TFETs represent a revolutionary advancement in transistor technology, offering significant potential for low-power electronics. Their unique tunneling mechanisms and favorable performance metrics make them key candidates for next-generation semiconductor devices. Addressing the challenges related to scalability, reliability, and manufacturability through sustained research and development is crucial for unlocking TFETs' full potential. As advancements continue, TFETs are likely to play a pivotal role in the semiconductor industry's evolution, enabling innovative applications and improved performance in electronic systems.

References

1. Lu, B., Cui, Y., Guo, A., Wang, D., Lv, Z., Zhou, J., Miao, Y., Characteristics of InAs/GaSb Line-Tunneling FETs With Buried Drain Technique. *IEEE Trans. Electron Devices*, 2021.
2. Gupta, N., Makosiej, A., Shrimali, H., Amara, A., Vladimirescu, A., Anghel, C., Tunnel FET Negative-Differential-Resistance based 1T1C Refresh-Free-DRAM, 2T1C SRAM and 3T1C CAM. *IEEE Trans. Nanotechnol.*, 20, 270–278, 2021.
3. Lin, Z., Li, L., Wu, X., Peng, C., Lu, W., Zhao, Q., Hybrid MOSFET-TFET 11T SRAM Cell with High Write Speed and Free Half-Selected Disturbance. *Microelectron. J.*, 106, Article 106498, 2024.

4. Kim, H.W. and Kwon, D., Low-Power Vertical Tunnel Field-Effect Transistor Ternary Inverter.
5. Lyu, Z., Lv, H., Zhang, Y., Zhang, Y., Zhu, Y., Sun, J., Li, M., Lu, B., A Novel High-Performance Planar InAs/GaSb Face-Tunneling FET With Implanted Drain for Leakage Current Reduction. *IEEE Trans. Electron Devices*, 68, 3, 1313–1317, MARCH 2021.
6. Kumar, P. and Sivani, K., A Tunnel field effect transistor is a substitute for ultra-low power applications. IEEE, 2016, 978-1-4673-8810-8/16/$31.00.
7. Hu, C., Chou, D., Patel, P., Bowonder, A., Green transistor-A VDD scaling path for future low power ICs, in: *VLSI Symp. Tech. Dig.*, pp. 14–15, 2008.
8. Nagavarapu, V., Jhaveri, R., Woo, J.C.S., The tunnel source (PNPN)n-MOSFET: A novel high performance transistor. *IEEE Trans. Electron Devices*, 55, 4, 1013–1019, Apr. 2008.
9. Kim, S.H., Kam, H., Hu, C., Liu, T.K., Germanium-source tunnel field effect transistors with record high ION/IOFF, in: *VLSI Symp. Tech.Dig.*, pp. 178–179, 2009.
10. Bowonder, A., Patel, P., Jeon, K., Oh, J., Majhi, P., Tseng, H.H., Hu, C., Low-voltage green transistor using ultra shallow junction and hetero-tunneling, in: *Proc. IWJT*, pp. 93–96, 2008.
11. Krishnamohan, T., Kim, D., Raghunathan, S., Saraswat, K.C., Doublegate strained-Ge heterostructure tunneling FET (TFET) with record highdrive current and <60 mV/dec subthreshold slope, in: *IEDM Tech. Dig.*, pp. 947–949, 2008.
12. Seabaugh, A., Quantum tunneling is a limitation in today's transistors, but could be the key to future devices, IEEE spectrum, 30th September, 2013.
13. Aswathy, M., Biju, N.M., Komaragiri, R., Comparison of a 30nm Tunnels Field Effect Transistor and CMOS Inverter Characteristics. *3rd International Conference on Advances in Computing and Communications*, Cochin, pp. 149–152, Aug. 2013.
14. Dennard, R., Gaensslen, F., Yu, H.-N., Rideout, V., Bassous, E., LeBlanc, A., Design of Ion-Implanted MOSFETs with Very Small Physical Dimensions. *IEEE J. Solid-State Circuits*, 9, 5, 256–268, October 1974.
15. Deepak Kumar's M. TECH thesis, Tunnel FETs and Its Application to Digital Circuits, Indian Institute of Technology, Kanpur, June 2015.
16. Packan, P., Short Course. *International Electron Devices Meeting short courses (IEDM)*, 2007.
17. Nilsson, P., Arithmetic Reduction of the Static Power Consumption in Nanoscale CMOS. *13th IEEE International Conference on Electronics, Circuits and Systems*, Nice, 10–13, pp. 656–659, December 2006.
18. Ionescu, A.M. and Riel, H., Tunnel field effect transistors as energy efficient electronic switches. *Nature*, 479, 329–337, Nov. 2011.
19. Thapa, N., Maurya, L., Mehra, R., Performance advancement of high-k dielectric MOSFET. *Int. J. Innov. Adv. Comput. Sci.*, 3, 3, 98–103, May 2014.

20. Deepak Kumar's M. TECH thesis, Tunnel FETs and Its Application to Digital Circuits, Indian Institute of Technology, Kanpur, pp. 1–4, June 2015.
21. Calhoun, B.H. and Chandrakasan, A., Characterizing and Modelling Minimum Energy Operation for Subthreshold Circuits. *International Symposium Low Power Electronics and Design*, pp. 90–95, August 2004.
22. Nirschl, T., Henzler, S., Pacha, C., Wang, P.-F., Hansch, W., Georgakos, G., Schmitt-Landsiedel, D., The Tunneling Field Effect Transistor (TFET) used in a Single-EventUpset (SEU) insensitive 6 transistor SRAM cell in ultra-low voltage applications. *4th IEEE Conference on Nanotechnology*, pp. 402–405, 2004.
23. Boucart, K. and Ionescu, A.M., Double-Gate Tunnel FET With High-k Gate Dielectric. *IEEE Trans. Electron Devices*, 54, 7, 1725–1733, Jul. 2007.
24. Jhaveri, R., Nagavarapu, V., Woo, J.C.S., Effect of Pocket Doping and Annealing Schemes on the Source-Pocket Tunnel Field-Effect Transistor. *IEEE Trans. Electron Devices*, 58, 1, 80–86, Jan. 2011.
25. Liu, H. and Dutta, S., III-V Tunnel FET Model Manual, Network for Computational Nanotechnology (NCN), West Lafayette, Indiana, USA, 2015, [online] Available: https://nanohub.org/publications/12/2.
26. Lattanzio, L., Michielis, D.L., Ionescu, A.M., Electron-hole bilayer tunnel FET for steep subthreshold swing and improved ON current. *IEEE Solid-State Dev. Research Conference (ESSDERC)*, pp. 259–262, Sept. 2011.
27. Asra, R., Shrivastava, M., VRM Murali, K., Pandey, R.K., Gossner, H., Rao, V.R., A Tunnel FET for VDD Scaling Below 0.6 V With a CMOS - Comparable Performance. *IEEE Trans. Electron Devices*, 58, 7, 1855–1863, Jul. 2011.
28. Jhaveri, R., Nagavarapu, V., Woo, J.C.S., AsymmetricSchottkyTunneling Source SOI MOSFET Design for Mixed-Mode Applications. *IEEE Trans. Electron Devices*, 56, 1, 93–99, Jan. 2009.
29. Mishra, A., Jha, K.K., Pattanaik, M., Parameter variation aware hybrid TFET-CMOS based power gating technique with a temperature variation tolerant sleep mode. *Microelectron. J.*, 45, 11, 1515–1521, Nov. 2014.
30. Zhou, G., Yeqing, L., Rui, L., Qin, Z., Wan, S.H., Qingmin, L., Tim, V., *et al.*, Vertical InGaAs/InP tunnel FETs with tunneling normal to the gate. *IEEE Electron Device Lett.*, 32, 11, 1516–1518, Nov. 2011.
31. Sant, S. and Schenk, A., Methods to enhance the Performance of InGaAs/InP Heterojunction Tunnel FET's. *IEEE Trans. Electron Devices*, 63, 5, 2169–2175, May 2016.

8

Tunnel Field Effect Transistors: Harnessing Light Sensitivity for Optical Sensing

Jagritee Talukdar[1]*, Malvika[2], Basab Das[3] and Kavicharan Mummaneni[2]

[1]*National Institute of Technology Meghalaya, Saitsohpen Sohra, East Khasi Hills, Meghalaya, India*
[2]*Department of Electronics and Communication Engineering, National Institute of Silchar, Silchar, Cachar, India*
[3]*Department of Electronics and Communication Engineering, Girijananda Chowdhury University, Guwahati, Assam, India*

Abstract

Tunnel field effect transistors (TFET), characterized by gated p-i-n structures, are renowned for their rapid switching capabilities owing to their band-to-band tunneling mechanism, which makes them capable of achieving a low subthreshold swing (SS). Extensively researched as potential substitutes for MOSFETs, particularly for low-voltage and low-power applications. The numerous benefits of TFETs extend their utility across sensing applications including biosensing, image sensors, and temperature sensing. This study primarily focuses on exploring TFETs' optical sensitivity. Optical sensing has wide-ranging applications in medical diagnostics, chemical-mixing configurations, military equipment, and energy harvesting. Recently, optical sensors based on FETs have garnered significant attention owing to their compatibility with integrated circuits in low-power applications.

In this study, we explore an extended source double-gate TFET (ES-DGTFET) photosensor. The device facilitates lateral tunneling as well as vertical band-to-band tunneling owing to the extended source. As a result, the device was capable of achieving a high ON current and low SS. The upper gate of the ES-DGTFET was replaced by an optical gate that was responsible for the generation of photovoltage upon excitation by light of a specific wavelength and intensity. To investigate the performance of the TFET photosensor, different parameters, such as photocurrent, response, and quantum efficiency, were observed for the proposed device. The photosensor is capable of achieving a current of 5.08×10^{-6}A/μm, responsivity

**Corresponding author*: jagriteetalukdar5@gmail.com

Ekta Goel, Archana Pandey, Shiromani Balmukund Rahi and Arun Samuel (eds.) Handbook of Advanced Semiconductor Field Effect Transistors, (169–182) © 2026 Scrivener Publishing LLC

of approximately 10^5, and quantum efficiency of more than 50% for an incident light of wavelength 450 nm and intensity of 0.8 W/cm^2.

***Keywords*:** Tunnel FET, photosensor, MOSFET, responsivity, quantum efficiency

8.1 Introduction

Over the past several decades, dramatic advancements in computing power and information processing have transformed nearly every aspect of work and life. Computational progress has been made by the constant scaling and growing integration of metal-oxide-semiconductor field-effect transistors (MOSFETs), which follow Moore's law. Moore projected in 1965 that transistor sizes would be cut by half every two years, resulting in a two-fold increase in the number of transistors per unit area every two years [1, 2]. From that moment until a few years ago, at each new technological node, the geometric scaling of these transistors enabled the fabrication of devices with smaller sizes, improved performance, and improved switching energy. However, as the feature size approached values in the range of micrometers to nanometers, it became evident that maintaining effective electrostatic control of Short-Channel Effects (SCEs), heat, and power dissipation was no longer feasible [3, 4]. This is primarily because with aggressive scaling, the reduction in supply voltage cannot keep up with the shrinking feature size. On the other hand, the switching power of MOSFET is given by the product of switching frequency, output capacitance and power supply voltage. Hence, to reduce the switching power consumption, the supply voltage should be reduced, which demands a significant reduction in the threshold voltage (voltage needed to turn the MOSFET ON). However, the reduction in the threshold voltage will come at the cost of increased OFF-stat current as the subthreshold swing (SS) of the MOSFET is fixed [5, 6]. Hence, in light of these challenges, various new transistor concepts, such as tunneling field-effect transistors (TFET), have been devised to mitigate the I_{OFF} and I_{ON} trade-off by reducing the SS below Boltzmann's limitation. The TFET represents a novel transistor structure engineered to achieve an SS lower than the conventional MOSFET limit of 60 mV/decade. They accomplished this by leveraging quantum mechanical tunneling within the source and channel from the valence band of the former to the conduction band of the latter. This process is called band-to-band tunneling (BTBT), which is the main working principle behind TFET, rather than relying on thermionic carrier injection. The fundamental structure of TFETs resembles that of conventional MOSFETs, which comprise a source drain as the contact region and an intrinsic channel region coated by an oxide layer atop the gate contact. However, unlike MOSFETs, the source regions of TFETs feature doping with opposite polarity, resulting in the creation of a p–i–n profile in the case of n-TFETs. After undergoing considerable research over the last decade, TFETs have become one of the most promising technologies for ultralow power applications [7–9].

In an integrated optoelectronic platform, a field-effect transistor (FET)-based photodetector offers numerous advantages compared to conventional photodetectors, including compatibility with CMOS technology, scalability, minimal power consumption, high switching ratio, higher stability and sensitivity, higher noise immunity, and lower dark current. Previous photodetectors, such as PIN photodiodes, avalanche photodiodes, and PN junction photosensors, face many issues regarding a lower photon absorption rate, chance of impact ionization, higher power dissipation, and internal noise [10, 11]. Optical sensing has numerous applications, such as environmental monitoring, medical diagnostics, and industrial automation [12–14]. The use of FETs in photodetectors is an active area of research with a focus on enhancing the photoresponse through advancements in device physics, architecture, and materials. It has been noted that the SS is the reason for determining the photosensitivity of FET-based photosensors. In this context, TFET-based photosensors are considered advantageous compared with MOSFET-based photosensors because of their very low SS [15].

There are two established methods through which incident light can access the region in the FET-based photosensor. First, the gap between the gate source and gate drain is used for light exposure, through which light can reach the substrate [16]. In the second approach, light is allowed to fall on the gate, and in this case, the gate is made of semimetal. The thickness of the metal gate is maintained such that light can be transmitted to the underlying semiconductor from the gate [17]. Several architectures were defined by changing the gate material pr architecture, etc., to improve light absorption. These techniques, along with the ongoing research on FET-based photosensors, have spurred the development of photogating techniques, which represent a highly active area of recent research.

A brief overview of this study on TFET-based photosensors is provided in *Section 8.2*. In *Section 8.3*, the working principle and conventional geometry of TFETs as photosensor sensors using the photogating technique are discussed. *Section 8.4* discusses the technology computer-aided design (TCAD) tool for various physics-based models that need to be considered when simulating a TFET photosensor. *Section 8.5* defines various parameters that determine the quality of the photosensor. In *Section 8.6*, an extended source-based TFET photosensor is discussed through TCAD simulation, and the chapter is concluded in *Section 8.7* and provides insights into future prospects.

8.2 A Brief Overview

Traditionally, semiconductor photodetectors work by converting incoming photons into free carriers, resulting in an output photocurrent. These photodetectors primarily operate based on the photogating effect, which is a technique in which the device's gate is exposed to light. This illumination induces a gate field or voltage that modulates the device's electrical properties. The photoactivated gate is separated from the channel by a dielectric layer, which facilitates the accumulation of carriers at the interface [18]. Tiwari *et al.* proposed and compared a Z-shaped source and hetero-stacked TFET-based photosensor. In their work, it was shown that in the case of signal-to-noise ratio and photosensitivity, the hetero-stacked TFET performed better than the Z-shaped source TFET-based photosensor [19]. The same group of authors also demonstrated a Si-based, Ge-source, heterogate TFET photosensor for the visible spectrum using a TCAD simulator [20]. Joshi *et al.* demonstrated a split-gate FET-based photosensor by using a TMD channel. In addition, the positions of these split gates are varied to comment on their effect on the photosensor performance [21]. Wangkheirakpam *et al.* reported a dual-MOSCAP vertical TFET designed for near-infrared light detection [22]. Ghosh *et al.* proposed an extended dual-source double-gate structure for a tunnel FET to detect distinctly separated spectral lines of near-infrared light and conducted a comparative analysis of its spectral response at different wavelengths [23]. Panda *et al.* reported the effectiveness of a single-gate SiGe/Si heterostructure tunnel field-effect transistor as a photosensor in the visible spectrum. They utilized a transparent zinc oxide (ZnO) layer as the optical region over the channel to sense incident light [24]. Tamersit *et al.* demonstrated an asymmetric dual-gate junctionless graphene nanoribbon tunneling field-effect phototransistor using a quantum-simulation approach. Their work focused on improving photosensitivity electrostatically by introducing asymmetry in the coupling capacitances [25]. Kim *et al.* reported broadband detection with high sensitivity and a fast photoresponse in an IR proximity sensor by integrating graphene or MoS_2 with NiO/Ni via van der Waals forces. This integration introduces tunneling, where graphene or MoS_2 acts as the IR absorber, and NiO/Ni serves as the hole-selective transport layer electrode [26]. Ashok *et al.* reported that the optical performance can be improved by an elevated source TFET with an extended back-gate-based photosensor. This design improves the detection of incident light with closely spaced wavelengths (100 nm) and a low luminous intensity in the near-infrared range. By using both the elevated top gate and extended back gate, the line tunneling at the source-channel interface region can be increased, which leads to an improved performance of the photosensor, consequently enhancing the photosensor performance [27]. Dagtekin *et al.* presented a partially gated pin structure that featuring an extended channel coated with transparent material. This study investigated the interaction between band-to-band tunneling and diverse illumination conditions for photosensing applications [28]. Manikanta *et al.* investigated the optical characteristics of a tunnel-enabled double-gate reverse T-shaped channel TFET. They examined how the choice of source material and variations in illumination specifications on the photosensitive area affect device performance [29].

8.3 Photo Sensing in TFETs: Principle of Operation and Geometry

In contrast to MOSFETs, where thermionic emission governs transport, a conventional TFET operates as a gated reverse-biased p-i-n structure. Tunneling between bands is the primary mode of transport in TFETs. In an n-TFET, as the positive gate bias increases, the energy bands are suppressed. Consequently, the separation between the valence band of the p+ source and conduction bands of the intrinsic channel diminished. This width reduction facilitated the tunneling of electrons from the valence band to the conduction band. Consequently, there was a contribution to the drain current. This process highlights the significance of gate bias in controlling the flow of electrons through the device, which is crucial for its operation as a transistor [30]. The source-channel junction of the TFET features a triangular barrier model representing the tunnel barrier. Using the Wentzel–Kramers–Brillouin (WKB) approximation, the computation of tunneling probability through this triangular barrier is conducted in [31] as:

$$T_E = \exp\left(\frac{-4\mathrm{L}_{\mathrm{Tunnel}}\sqrt{2\mathrm{m}_e\mathrm{E}_g{}^{1.5}}}{3q\hbar\left(\Delta\mathrm{E}+\mathrm{E}_g\right)}\right) \tag{8.1}$$

In the given equation, m_e, E_g represents the effective mass, energy band gap, ΔE denotes the convergence of energy bands at the tunnel junction, L_{Tunnel} represents the tunneling length through which tunneling takes place, q and ħ represents the electronic charge and reduced Planck's constant. The tunneling length (L_{Tunnel}) is defined as [31]:

$$L_{Tunnel} = \sqrt{\xi_{SC} T_{SC} T_{OX} / \xi_{OX}} \tag{8.2}$$

where ξ_{SC} and ξ_{OX} are the semiconductor and oxide dielectric constants, T_{SC} and T_{OX} and are their thicknesses, respectively.

The tunneling length is directly affected by the applied gate voltage; however, when the TFET is used as a photosensor, the gate voltage V_{gs} is modified because of the addition of the voltage. A pictorial representation of the conventional structure of a TFET-based photosensor is shown in Figure 8.1, where the gate is extended in length for use as a photogate. Once light falls on the gate, extra charge carriers are generated in that region; this happens efficiently only when the wavelength of the light energy exceeds the bandgap of the material used for the gate. When a positive gate voltage is applied in the case of the n-TFET, the holes move near the gate electrode, and electrons repel toward the gate-oxide interface. This event separates the electron and hole quasi-Fermi levels in the gate region. Owing to this separation of carriers, an extra voltage develops in the gate region, which is termed photovoltage [21].

A bend diagram illustrating this process is shown in Figure 8.1(b). This generated photovoltage, along with the applied gate voltage, increased the effective voltage in the device. This increase in the effective gate voltage increases the tunneling in the source channel junction.

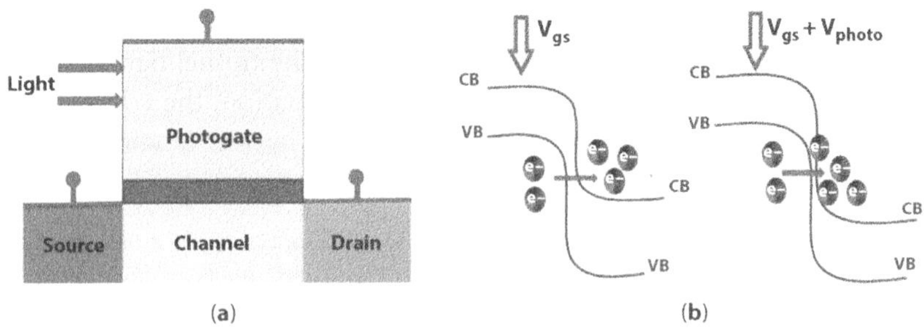

Figure 8.1 (a) Schematic of a conventional TFET based photosensor, (b) band diagram illustrating the band bending in presence and absence of photovoltage.

8.4 Simulation Strategy for TFET-Based Photosensor

The most efficient method for analyzing a photosensor is to use a computational platform, in which physics-based models help assess its performance. Many industrial simulators are equipped with features to define geometries and run iterative simulations using various models from their libraries. This work uses Sentaurus TCAD as a simulator, and models based on it are explained in detail [32].

In a TFET, the source and drain are highly doped because during simulation, it is necessary to activate the Fermi distribution model so that large bend banding can be achieved during equilibrium, and field-dependent mobility and recombination models can also be included. Next, to establish tunneling from the source to drain in the TFET BTBT model, it is necessary to consider the physics file of the simulator. To fabricate a photogate-based TFET, the gate material must be capable of efficiently absorbing light to facilitate optical generation in the gate. In the physics section of the TCAD, the light wavelength, intensity, and angle of incidence are specified under the complex refractive index model to activate optical illumination in the gate region. The gate where the light falls should be defined in a 2D coordinate system along the length of the illumination in the physics section. The light intensity, wavelength, and polarization angle on this region are defined in the physics section. The polarization angle was set to 0° and 90° for the illuminated and dark states, respectively. When the angle of incidence is 0°, light strikes the photo gate perpendicularly; otherwise, for 90°, light passes parallel to the photo gate. To investigate the sensitivity of the photosensor, various optical parameters, such as the absorbed photon density, optical generation rate, photocurrent, and complex refractive index must be extracted after being specified in the plot section. Once these parameters are extracted, the different sensitivity metrics for the photosensor can be calculated. The details of the sensitivity parameters are discussed in the following section.

8.5 Sensitivity Parameters of Photosensor

The effectiveness of a photosensor, whether it is a FET, diode, or another type, depends on the efficiency of light absorption. The factors that define the sensitivity are photosensitivity, responsivity, quantum efficiency, gain, and detectivity. The signal-to-noise ratio (SNR) is another important optical factor that has a considerable influence on the performance of the photosensor and can be considered as a performance metric for a photosensor.

Photosensitivity: The photosensitivity of a photosensor depends on the current in the illuminated and dark states and is given by $(I_{photo} - I_{dark})/I_{dark}$, where I_{photo} represents the current in the illuminated state and I_{dark} is the current when no light falls on the gate, i.e., the dark state. The sensitivity of the photosensor mainly depends on the difference between the photocurrents in the illuminated and dark states.

Responsivity: Responsivity describes a sensor's reaction to the incidence of light illumination. It is influenced by the incident wavelength, light modulation frequency, applied bias, and the gain of the photosensor. It can be given by the ratio of photocurrent to incident optical power, as given below [33]:

$$Responsivity = \frac{Photocurrent}{Optical\ Intensity} = (QE)\frac{q\lambda G}{\hbar c}\frac{1}{\sqrt{1+(2\pi f_T)^2}} \quad (8.3)$$

where, (λ) is the wavelength, c is the velocity of light and ħ is the Plank's constant, and f is the frequency.

Quantum efficiency (QE): This is the conversion of the light flux to the electric flux. Furthermore, QE considers various external losses, such as reflection and scattering, along with any additional losses at the interface. The quantum efficiency can be articulated in two ways: internal quantum efficiency and external quantum efficiency. The second includes all external losses, along with the conversion efficiency. It is proportional to the responsivity and can be given by photons absorbed by the incident photons.

$$Responsivity \propto QE \quad (8.4)$$

Gain (G): Another figure of merit for a well-designed photosensor is the gain. In typical scenarios, one type of carrier is trapped, whereas the other freely traverses the channel. If the trapping time (lifetime) is shorter than the transit time of the free carriers, a photoconductive gain occurs, which may lead to high responsivity. The gain-bandwidth trade-off highlights that a higher gain leads to a slower response time; hence, while designing a photosensor, this trade-off must be taken into consideration.

Detectivity: Another important parameter of a photosensor is detectivity. These parameters depend on the wavelength, frequency, temperature, and applied bias. This can be defined using the following equation:

$$Directivity = Area.Bandwidth \, / \, Noisepower \tag{8.5}$$

Signal-to-noise ratio (SNR): SNR can also serve as a performance metric for photosensors. After the illumination of light in the photosensitive area during the conversion of light flux to electric flux, noise may be generated. The measurement can be given by the SNR, as defined in Eq. (8.6). The higher the SNR, the higher the detectivity of the light signal.

$$SNR = 20log\left(\mathrm{I_{photo}} \, / \, \mathrm{I_{dark}}\right) \tag{8.6}$$

where, I_{photo} and I_{dark} are current due to presence and absence of light, respectively.

8.6 An Extended Source TFET-Based Photosensor

This section presents an extended source TFET-based photosensor design and discusses some related results. As the name suggests, the base of the TFET design is an extended source that has been used to prevent vertical tunneling and lateral tunneling, as discussed in [34]. The device mainly uses silicon for the source channel, drain, and gate. where source is highly p-type, the channel is intrinsic, and the drain and gate are doped with an n-type dopant. The overall length of the device is 100 nm, with the channel accounting for 40 nm. The design was modified to function as a photosensor by making the gate a photosensitive component, as illustrated in Figure 8.2a. In this configuration, the gate width was maintained at 50 nm to allow for light exposure. In the gate region, we allowed light of wavelength 450 nm and intensity of 0.8 W/cm^2 to fall, which led to the generation of electron and hole pairs in the gate region. Next, when a gate voltage (V_{gs}) of 1.5 V was applied to the gate, the electron moved toward the gate electrode, and the hole moved to the SiO_2–gate interface. This distance generates a voltage referred to as the photovoltage, resulting from the separation of quasi-electron and hole Fermi levels, as explained in *Section 8.3*.

Figure 8.2b shows the drain current of the photosensor at a wavelength of 450 nm and an intensity of 0.8 W/cm^2. Figure 8.2a shows that because

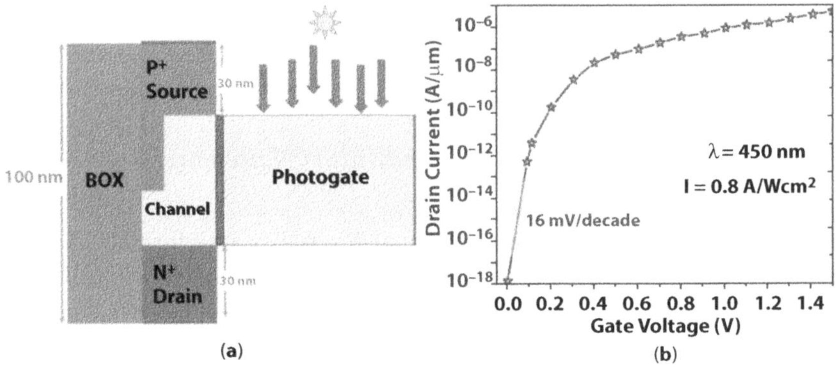

Figure 8.2 (a) Schematic of proposed extended source TFET based photosensor and (b) transfer characteristics of (a) for a wavelength of 450 nm and intensity of 0.8 W/cm^2.

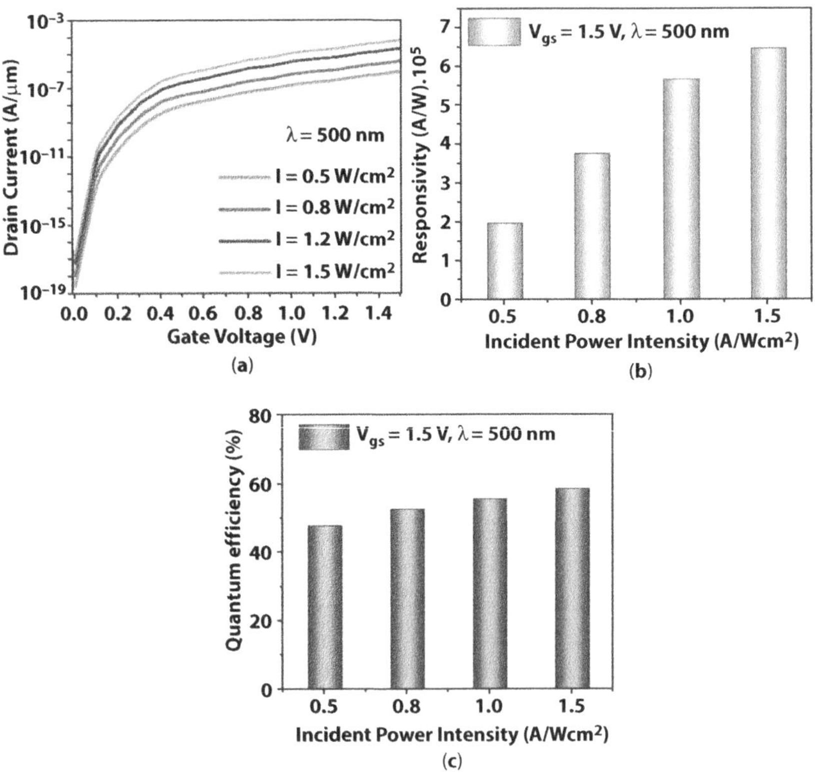

Figure 8.3 (a) Transfer characteristics, (b) responsivity, and (c) quantum efficiency of the extended source-based TFET photosensor for varying intensity from 0.5 W/cm^2 to 1.5 W/cm^2 at a wavelength of 500 nm.

of the higher tunneling rate, the ON current obtained is 5.08×10^{-6}A/μm and the OFF current is 1.36×10^{-18}A/μm, leading to an ON-OFF current ratio of approximately 10^{12}. Next, the steepness of the device is also quite impressive, which is 16 mV/decade marking the effectiveness of the proposed extended source based TFET photosensor.

To further analyze the proposed photosensor's performance, we varied the intensity of incoming light from 0.5 W/cm^2 to 1.5 W/cm^2 at a wavelength of 500 nm, as shown in Figure 8.3. It can be observed that, an increase in optical intensity results in a higher photocurrent in the device. This is because a greater optical intensity enhances the optical generation, leading to more electron-hole pairs. As a result, the photovoltage increases, which in turn increases the effective gate voltage and promotes tunneling. Figures 8.3a,b also illustrate that both the responsivity and quantum efficiency (QE) increase with the optical intensity.

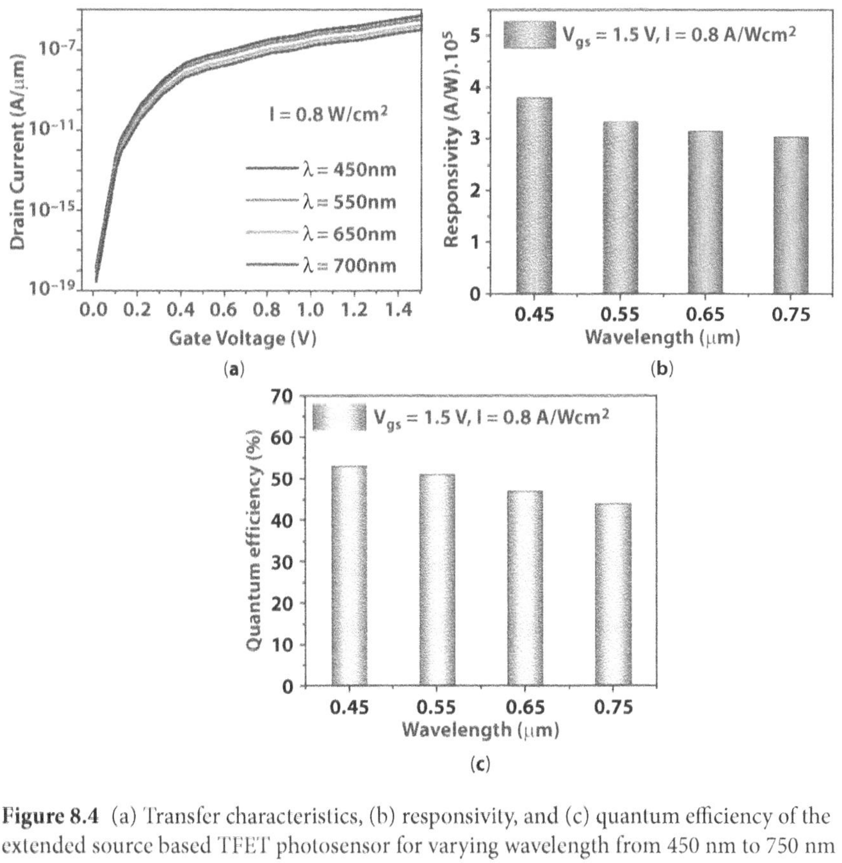

Figure 8.4 (a) Transfer characteristics, (b) responsivity, and (c) quantum efficiency of the extended source based TFET photosensor for varying wavelength from 450 nm to 750 nm at an intensity 0.8 W/cm^2.

Figure 8.4 depicts the variation in the drain current as a function of wavelength; as the wavelength increases from 450 nm to 750 nm, the current decreases because of the decreased optical energy with increased wavelength. Again, in the case of responsivity, a decreasing trend can be observed with increasing wavelength, which is attributed to an exponential decrease in the absorption coefficient. Finally, a higher QE was obtained at lower frequencies, where the energy of light substantially exceeded the bandgap energy of the material.

8.7 Conclusion

This chapter presents a summary of the tunnel field-effect transistors (TFETs) as photosensors. TFETs have become highly significant in the field of low-power applications because of their immunity towards scaling and their capability to attain extremely low subthreshold swing (SS). With the growing interest in TFET-based photosensors, this chapter also provides a brief overview, highlighting recent research on TFETs as photosensors. Furthermore, a detailed photosensor design and its operation are presented along with the simulation techniques. A detailed description of the parameters necessary to investigate the photosensor performance was also included in the study. Finally, a new design based on an extended-source TFET-based photosensor is presented. The photocurrent responsivity and quantum efficiency of the proposed design were investigated by varying the incident wavelength and intensity of the light. The proposed device delivered an impressive performance parameter value in the visible wavelength spectrum. Proper simulation and modeling can predict various effects of such photosensors under visible and infrared spectra, and physics-based models are crucial for validating these results with the TCAD tool.

References

1. Schaller, R.R., Moore's law: past, present and future. *IEEE Spectr.*, 34, 6, 52–59, 1997.
2. Khan, H.N., Hounshell, D.A., Fuchs, E.RH., Science and research policy at the end of moore's law. *Nat. Electron.*, 1, 1, 14–21, 2018.
3. Khanna, V.K. and Khanna, V.K., Short-channel effects in mosfets, in: *Integrated Nanoelectronics: Nanoscale CMOS, Post-CMOS and Allied Nanotechnologies*, pp. 73–93, 2016.

4. Hiblot, G., Rafhay, Q., Boeuf, F., Ghibaudo, G., Impact of short- channel effects on velocity overshoot in mosfet, in: *2015 IEEE 13th International New Circuits and Systems Conference (NEWCAS)*, IEEE, pp. 1–4, 2015.
5. Vandamme, E.P., Jansen, P., Deferm, L., Modeling the subthreshold swing in mosfet's. *IEEE Electron Device Lett.*, 18, 8, 369–371, 1997.
6. Zhang, Q., Zhao, W., Seabaugh, A., Low-subthreshold-swing tunnel transistors. *IEEE Electron Device Lett.*, 27, 4, 297–300, 2006.
7. Avci, U.E., Morris, D.H., Young, I.A., Tunnel field-effect transistors: Prospects and challenges. *IEEE J. Electron Devices Soc.*, 3, 3, 88–95, 2015.
8. Boucart, K. and Ionescu, A.M., Double-gate tunnel fet with high-k gate dielectric. *IEEE Trans. Electron Devices*, 54, 7, 1725–1733, 2007.
9. Talukdar, J. and Mummaneni, K., A non-uniform silicon tfet design with dual- material source and compressed drain. *Appl. Phys. A*, 126, 1, 81, 2020.
10. Fahs, B., Chowdhury, A.J., Zhang, Y., Ghasemi, J., Hitchcock, C., Zarkesh-Ha, P., Hella, M.M., Design and modeling of blue-enhanced and bandwidth-extended pn photodiode in standard cmos technology. *IEEE Trans. Electron Devices*, 64, 7, 2859–2866, 2017.
11. Moszynski, M., Energy resolution of scintillation detectors, in: *SPIE Optical Engineering + Applications 2023*, vol. 12696, pp. 16–29, SPIE, 25 October 2023.
12. Han, H., Cho, H.C., Jang, S.M., Choi, C., Monolithic 3d integration with photosensor and cmos circuits using ion-cut layer transfer. *IEEE Electron Device Lett.*, 43, 3, 430–433, 2022.
13. Mennel, L., Molina-Mendoza, A.J., Paur, M., Polyushkin, D.K., Kwak, D., Giparakis, M., Beiser, M., Andrews, A.M., Mueller, T., A photosensor employing data-driven binning for ultrafast image recognition. *Sci. Rep.*, 12, 1, 14441, 2022.
14. Gupta, P.S., Rahaman, H., Sinha, K., Chattopadhyay, S., An optoelectronic band-to-band tunnel transistor for near-infrared sensing applications: Device physics, modeling, and simulation. *J. Appl. Phys.*, 120, 084510, 2016. https://doi.org/10.1063/1.4961426.
15. Choi, W.Y., Park, B.-G., Lee, J.D., King Liu, T.-J., Tunneling field- effect transistors (tfets) with subthreshold swing (ss) less than 60 mv/dec. *IEEE Electron Device Lett.*, 28, 8, 743–745, 2007.
16. Seeds, A.J. and De Salles, A.A.A., Optical control of microwave semiconductor devices. *IEEE Trans. Microwave Theory Tech.*, 38, 5, 577–585, 1990.
17. Singh, V.K., Chattopadhyay, S.N., Pal, B.B., Optically controlled characteristics in an ion- implanted silicon mesfet. *Solid State Electron.*, 29, 7, 707–711, 1986.
18. Fang, H. and Hu, W., Photogating in low dimensional photodetectors. *Adv. Sci.*, 4, 12, 1700323, 2017.
19. Tiwari, S. and Saha, R., Optical performance of split-source z-shaped horizontal-pocket and hetero-stacked tfet-based photosensors. *J. Electron. Mater.*, 52, 3, 1888–1899, 2023.

20. Tiwari, S. and Saha, R., Improved optical performance in near visible light detection photosensor based on tfet. *Microelectron. J.*, 129, 105554, 2022.
21. Joshi, S., Dubey, P.K., Kaushik, B.K., Photosensor based on split gate tmd tfet using photogating effect for visible light detection. *IEEE Sens. J.*, 20, 12, 6346–6353, 2020.
22. Wangkheirakpam, V.D., Bhowmick, B., Pukhrambam, P.D., Near infra-red photosensor using optically gated d-mos vertical tfet, in: *2021 International Conference on Numerical Simulation of Optoelectronic Devices (NUSOD)*, IEEE, pp. 47–48, 2021.
23. Ghosh, P., Tripathi, S., Devi, W.V., An extended dual source double-gate tfet-based optical sensor for near-infrared-sensing applications. *Appl. Phys. A*, 129, 11, 781, 2023.
24. Panda, S. and Dash, S., Implementing a single gate heterostructure tunnel fet as a low-power photosensor with improved sensitivity. *ECS J. Solid State Sci. Technol.*, 11, 12, 123004, 2022.
25. Tamersit, K., Asymmetric dual-gate junctionless gnr tunnel fet as a high-performance photosensor with an electrostatically improved photosensitivity: A quantum simulation study. *IEEE Sens. J.*, 23, 22, 27431–27437, 15 Nov. 15, 2023, doi: 10.1109/JSEN.2023.3316913.
26. Kim, Y.R., Phan, T.L., Cho, K.W., Kang, W.T., Kim, K., Lee, Y.H., Yu, W.J., Infrared proximity sensors based on photo-induced tunneling in van der waals integration. *Adv. Funct. Mater.*, 31, 31, 2100966, 2021.
27. Ashok, T. and Pandey, C.K., Performance investigation of elevated source ebg tfet based photosensor for near-infrared light sensing applications. *Micro Nanostruct.*, 188, 207803, 2024.
28. Dagtekin, N. and Ionescu, A.M., Investigation of partially gated si tunnel fets for low power integrated optical sensing, in: *2014 44th European Solid State Device Research Conference (ESSDERC)*, IEEE, pp. 190–193, 2014.
29. Manikanta, K. and Nanda, U., Optical performance of the double gate reverse t-shaped channel tfet in near visible light photodetector. *Trans. Electr. Electron. Mater.*, 25, 2, 160–172, 2024.
30. Le Royer, C. and Mayer, F., Exhaustive experimental study of tunnel field effect transistors (tfets): From materials to architecture, in: *2009 10th International Conference on Ultimate Integration of Silicon*, IEEE, pp. 53–56, 2009.
31. Knoch, J. and Appenzeller, J., A novel concept for field-effect transistors-the tunneling carbon nanotube fet, in: *63rd Device Research Conference Digest, 2005*. DRC'05, vol. 1, IEEE, pp. 153–156, 2005.
32. *TCAD Sentaurus Device Manual*, Property of Synopsys, CA, USA, 2013.
33. Kufer, D. and Konstantatos, G., Photo-fets: Phototransistors enabled by 2d and 0d nanomaterials. *ACS Photonics*, 3, 12, 2197–2210, 2016.
34. Talukdar, J., Rawat, G., Mummaneni, K., A novel extended source tfet with delp+- sige layer. *Silicon*, 12, 1–9, 10 2020.

9

2D Material Based FET Sensors

Archana Pandey[1]*, Jyoti Pant[2], Medha Joshi[3], Nitanshu Chauhan[4] and Mandeep Singh[1]

[1]Department of ECE, Jaypee Institute of Information Technology, Noida, India
[2]BGN Technologies Private Limited, Ghaziabad, UP, India
[3]Department of ECE, Indian Institute of Technology, Roorkee, Uttarakhand, India
[4]Department of ECE, Indian Institute of Technology, Bhilai, India

Abstract
The development of field-effect transistor (FET) sensors has undergone significant progress in recent years, owing in large part to the emergence of two-dimensional (2D) materials. 2D materials, such as graphene, transition metal dichalcogenides (TMDs), and black phosphorus, have unique electrical, mechanical, and chemical properties that make them ideal candidates for sensing applications. This chapter provides an overview of the ongoing progress in FET sensors using 2D materials and explains their major standards, creation procedures, and different applications. This chapter begins with an introduction to 2D material-based FET sensors and their working principles. In next section interesting properties and applications of 2D material will be explained, for example, graphene, transition metal dichalcogenides (TMDs), and dark phosphorus, which make them reasonable for sensor applications. It further investigates the procedure utilized for the fabrication of FET sensors. An overview of the sensing mechanisms used will be explained further. This chapter highlights the challenges and future scope of 2D material-based FET sensors.

Keywords: 2D material, field effect transistor (FET), sensor

9.1 Introduction

A Field-Effect Transistor (FET) sensor, often referred to as an FET-based sensor, is an electronic device used to detect and measure various physical,

**Corresponding author*: archana.pandey@jiit.ac.in

Ekta Goel, Archana Pandey, Shiromani Balmukund Rahi and Arun Samuel (eds.) Handbook of Advanced Semiconductor Field Effect Transistors, (183–198) © 2026 Scrivener Publishing LLC

chemical, and biological properties [1, 2] by monitoring changes in the electrical conductance or capacitance of the transistor channel. FET sensors have gained significant importance in diverse fields, including electronics, environmental monitoring, medical diagnostics, and chemical analysis, due to their high sensitivity, rapid response, and ability to be integrated into microscale and nanoscale devices [1–4].

Using a 2D material as a sensing element in a Field-Effect Transistor (FET) is a powerful approach for developing highly sensitive and selective sensors for various applications [5]. In 2D material-based sensors, the 2D material serves as the active sensing layer in the FET, and its electrical properties change in response to the presence of specific analytes or environmental conditions.

A 2D material (e.g., graphene, TMDs, and black phosphorus) is integrated as a thin film or monolayer onto the FET structure. This 2D material serves as the sensing element, which is called as Active Sensing Layer. The FET includes a gate electrode, a source electrode, and a drain electrode. A 2D material is typically positioned between the source and drain electrodes. Modification of the 2D material's surface enhances its sensitivity to the target analyte (e.g., through chemical functionalization). When the analyte or target molecule of interest interacts with the 2D material's surface, it can induce changes in the material's electrical properties. These changes can occur through various mechanisms such as charge transfer, chemical adsorption, and doping. Interaction with analytes may change the carrier concentration, mobility, or band structure of 2D materials. As a result, the electrical conductivity of the 2D material is modulated [6].

This chapter focuses on the latest advancements in 2D material-based FET sensors, highlighting their key attributes, applications, and challenges. Field-effect transistor (FET) sensor using two-dimensional (2D) materials have arisen as a promising candidate in sensing technology.

9.1.1 FET Operation

A gate electrode was used to apply voltage to the 2D material. The changes in the electrical properties of the 2D material, caused by the presence of analytes, result in a change in the FET current (IDS) when a fixed source-drain voltage (VDS) is applied.

9.1.2 Sensing Response

The output current (IDS) is monitored, and any changes in the current were correlated with the concentration or presence of the target analyte.

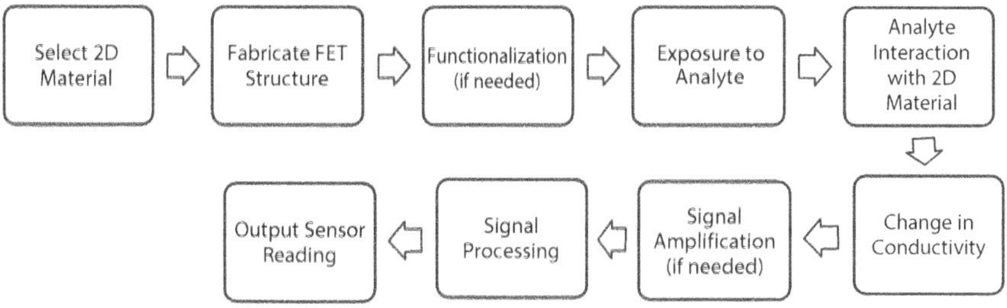

Figure 9.1 Sensing mechanism process flow of 2D material-based FET sensor.

This change in the current serves as a quantitative or qualitative indication of the analyte concentration or the environmental conditions being sensed.

9.1.3 Selectivity and Sensitivity

The choice of 2D materials and functionalization techniques can provide high selectivity and sensitivity for specific analytes. Functional groups or receptors on the 2D material's surface can enhance their specificity for particular molecules. The working of the 2D material-based FET sensor is summarized in Figure 9.1.

The development of field-effect transistor (FET) sensors has made significant progress in recent years, owing in large part to the emergence of two-dimensional (2D) materials. These 2D materials, have unique electrical, mechanical, and chemical properties that make them ideal candidates for sensing applications. This review focuses on the latest advancements in 2D material-based FET sensors, highlighting their key attributes, applications, and challenges.

9.2 Properties and Applications of 2D Materials

2D materials possess remarkable properties that are advantageous for FET sensor applications [7]. These properties include high carrier mobility, large surface area, and tunable electronic properties. Graphene, for example, offers exceptional electrical conductivity, whereas TMDs exhibit a bandgap suitable for various sensing tasks [8, 9]. 2D material-based FET sensors are likely to play a crucial role in environmental monitoring, including the detection of pollutants, gases, and pathogens [10, 11].

The applications of various 2D materials, including graphene and transition metal dichalcogenides, in FET-based chemical and biological sensors were reviewed in [5]. The use of 2D materials, such as graphene and

MoS2, in FET sensors for environmental gas monitoring, highlighting their sensitivity and selectivity, was explored in [6]. [7] provided insights into the use of 2D materials beyond graphene in FET sensors for a wide range of sensing applications. This study focuses on graphene-based FET sensors for gas detection and pH sensing, demonstrating their potential in environmental and healthcare applications [12–14]. [15, 16] discussed the use of transition metal dichalcogenides (TMDs) in FET sensors for gas detection, emphasizing their potential as highly selective sensors.

The use of black phosphorus in FET sensors provides insights into its properties and applications in various sensing tasks [17–20]. [21–23] covers the use of 2D materials in FET-based biosensors and highlights their potential for healthcare and biotechnological applications.

In addition, the atomically thin nature of 2D materials enables efficient charge transfer and excellent sensitivity [8, 9]. Two-dimensional (2D) materials have gained significant attention in the field of field-effect transistors (FETs) owing to their unique properties, which make them suitable for various applications. The key properties and applications of commonly used 2D materials in FETs are listed in Tables 9.1 to Table 9.6.

Graphene is an ideal 2D material for FET biosensors. It has unique properties, such as high electrical conductivity, large surface area, mechanical strength, and high carrier mobility. These properties enhance the sensor performance. Therefore, graphene-based FET sensors are particularly useful in advanced applications, such as medical diagnostics, environmental monitoring, and food safety. Table 9.1 summarizes the key properties of graphene materials that make them suitable for the given applications.

Table 9.1 Graphene [5, 12–14].

Properties	Applications
Single layer of carbon atoms in hexagonal lattice	High-frequency electronics
Exceptionally high electron mobility	Transparent conductive electrodes
High electrical conductivity	Gas sensors
Transparency and mechanical flexibility	Biological sensors [24]
Broadband optical absorption	Flexible electronics

Table 9.2 summarizes the key properties of TMDs material that make it suitable for the given applications.

Table 9.3 summarizes the key properties of black phosphorus material that make it suitable for the given applications.

Table 9.2 Transition metal dichalcogenides (TMDs) (e.g., MoS2, WS2) [5, 10, 15, 16].

Properties	Applications
Layered structures with strong in-plane covalent bonding and weak interlayer van der Waals interactions [25].	Digital logic transistors
Possess a sizable bandgap, suitable for digital logic apps [26].	Optoelectronic devices and Photodetectors [27]
Thickness-dependent electronic properties.	Gas sensors
Moderate to high carrier mobility.	Photovoltaics

Table 9.3 Black phosphorus (phosphorene) [17–20, 28].

Properties	Applications
Layered structure with a direct bandgap that varies with the number of layers.	FETs for analog and digital electronics.
High charge carrier mobility.	Photodetectors.
Anisotropic electrical properties.	Thermoelectric devices.
Excellent thermal conductivity.	Battery electrodes.

Table 9.4 Hexagonal boron nitride (hBN) [29].

Properties	Applications
Insulating 2D material, Good electrical insulator.	Dielectric layers in 2D material-based devices.
Excellent thermal conductivity.	Encapsulation layers to protect other 2D materials.
High mechanical strength.	Quantum devices.

Table 9.5 Phosphorene oxides [30].

Properties	**Applications**
Oxidized form of black phosphorus.	Stable FETs and sensors.
Improved stability in air and humidity compared to black phosphorus.	Photodetectors.
Tunable electronic properties.	Photovoltaics.

Table 9.6 MXenes (e.g., Ti3C2Tx) [31].

Properties	**Applications**
Transition metal carbides/nitrides with layered structures.	Energy storage (batteries and supercapacitors).
High electrical conductivity.	Electromagnetic interference shielding.
Good mechanical strength.	Gas sensors.
Hydrophilic surface properties.	Flexible electronics.

9.3 Sensing Mechanisms

Various sensing mechanisms have been employed in 2D material-based FETs, including charge transfer, electrostatic gating, and chemical functionalization. These mechanisms allow the detection of a wide range of analytes, including gases, biomolecules, and ions, with high sensitivity and selectivity.

9.3.1 Gas Sensing

Gas sensing mechanism is explained with the help of the following flowchart in Figure 9.2.

2D material-based FET sensors have demonstrated exceptional promise for gas-sensing applications. Graphene-based FETs exhibit high sensitivity to gases, such as NO2, NH3, and CO2. TMD-based FETs offer tunable selectivity for specific gas molecules, making them suitable for environmental monitoring and gas detection in industrial settings [10]. Figure 9.3 shows a schematic of the MoS_2 transistor-based gas sensing device.

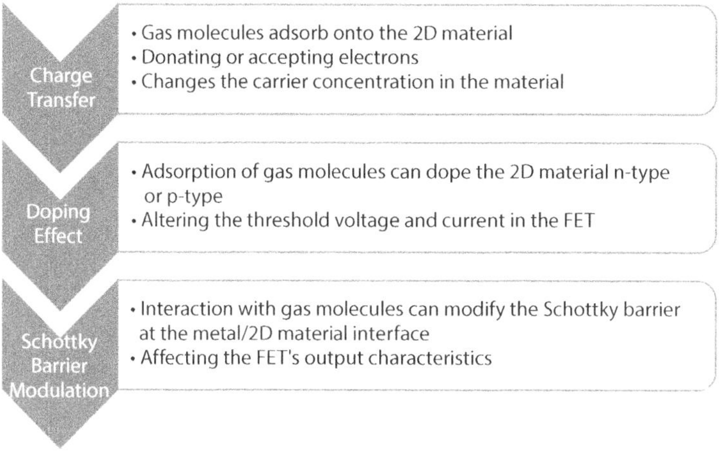

Figure 9.2 Flow chart of the gas sensing mechanism.

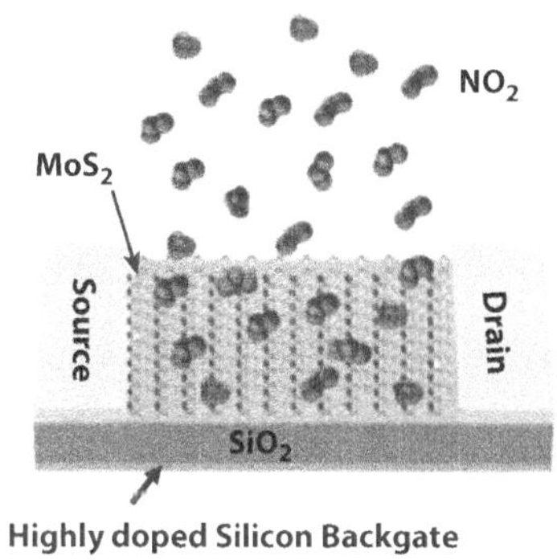

Figure 9.3 Schematic of MoS_2 transistor-based gas sensing device [32].

9.3.2 Biosensing

The biocompatibility and high surface-to-volume ratio of 2D materials make them well suited for biosensing applications. Functionalization with biomolecules or aptamers enables the detection of specific proteins, DNA, and other biomarkers. These sensors have applications in medical diagnostics, disease monitoring, and point-of-care testing [21, 22]. Figure 9.4 shows a schematic of the MoS2 biosensor with HfO2 dielectric coating, bioreceptors, and biomolecules.

Various biosensing mechanisms are summarized below in Figure 9.5.

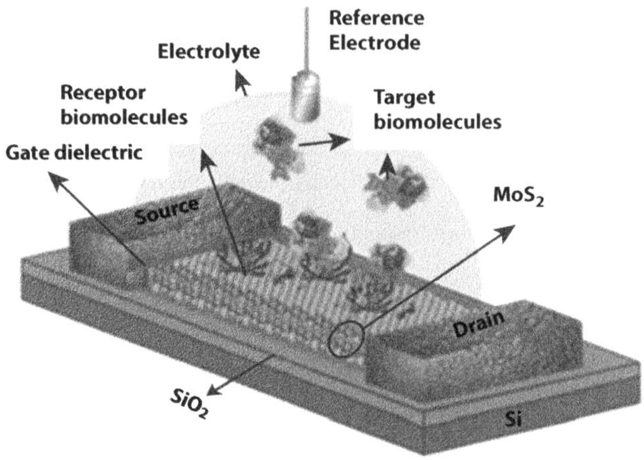

Figure 9.4 Schematic of MoS2 biosensor with HfO2 dielectric coating, bioreceptor, and biomolecules [23].

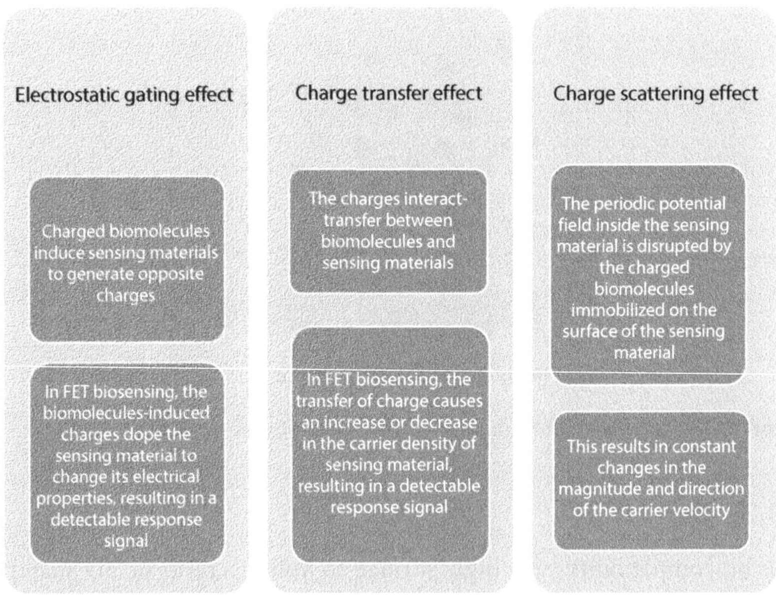

Figure 9.5 Biosensing mechanisms.

9.3.3 Environmental and Chemical Sensing

2D material-based FET sensors have been employed to detect a wide range of chemical analytes, including heavy metals, pollutants, and pH levels

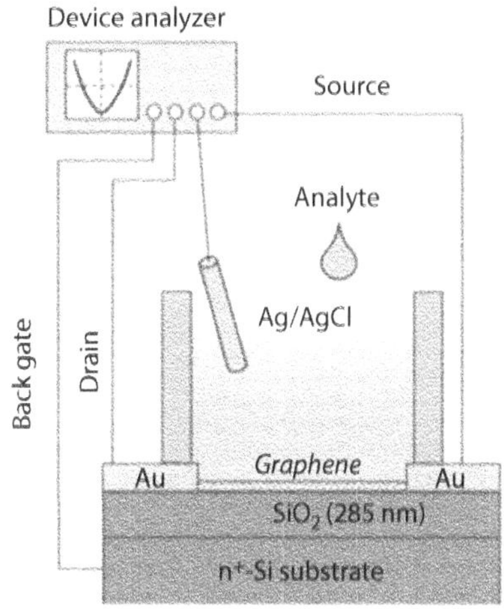

Figure 9.6 A schematic illustration of the experimental setup of a graphene field effect transistor for chemical sensing [34].

[23, 33]. Their sensitivity, coupled with their ability to functionalize the sensor surface, makes them valuable tools for environmental monitoring and water quality assessment. Figure 9.6 shows a schematic illustration of the experimental setup of a graphene field-effect transistor for chemical sensing.

9.4 Challenges and Future Directions

Despite their many advantages, 2D material-based FET sensors face certain challenges. These include reproducibility issues, scalability of production, and long-term stability [35, 36]. Future research efforts should focus on addressing these challenges and exploring novel 2D materials and hybrid structures to further enhance the sensor performance.

9.4.1 Challenges

9.4.1.1 Sensitivity and Specificity

Achieving high sensitivity and specificity for target analytes remains a challenge. Enhancing the selectivity of 2D material-based FET sensors to differentiate between similar molecules is crucial, especially in medical and environmental monitoring applications [37, 38].

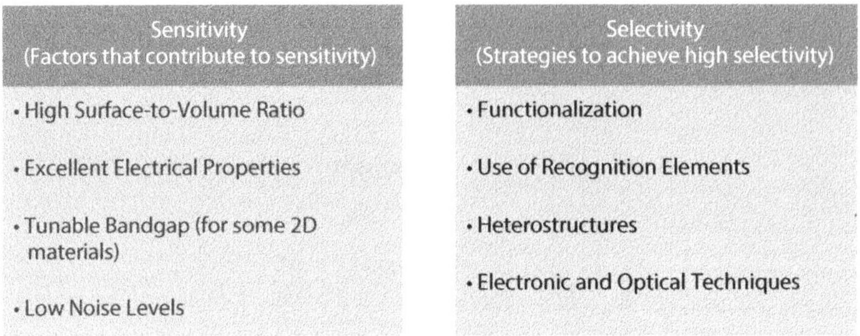

Figure 9.7 Factors that contribute to high sensitivity of 2D material FET sensor and strategies to achieve high selectivity.

2D materials exhibit a high surface-to-volume ratio, exposing a significant portion of their atoms to enhance interactions with analytes, thereby improving sensitivity. Their excellent electrical properties, such as high electron mobility and conductivity in materials such as graphene, enable the detection of minute changes in the charge distribution due to analyte adsorption. Additionally, some 2D materials, such as transition metal dichalcogenides (TMDs), possess a tunable bandgap that can be optimized for specific analytes, further enhancing sensitivity. The atomic thinness of these materials also contributes to lower noise levels in electronic measurements, thereby enhancing the signal-to-noise ratio. Figure 9.7 summarizes the factors that contribute to the high sensitivity of 2D material FET sensors and strategies to achieve high selectivity.

Enhancing the selectivity of 2D material FET sensors can be achieved using several methods. Functionalizing the surface with specific molecules or polymers allows the attachment of functional groups with high affinity for particular analytes, thereby improving selectivity. Incorporating biological recognition elements can further enhance selectivity by enabling specific binding to target molecules. Heterostructures created by stacking different 2D materials can offer selective pathways for different analytes.

Advanced electronic techniques (e.g., back gating and dual gating) and optical methods (e.g., Raman spectroscopy) can differentiate signals based on unique electronic or optical signatures. For instance, graphene-based FET sensors functionalized with metal nanoparticles or polymers show increased selectivity for gases such as NO_2, NH_3, and CO_2. MoS_2 and other TMD-based FET sensors are used for detecting biomolecules, such as glucose, DNA, and proteins, with improved selectivity achieved through aptamers or antibodies. Black phosphorus FET sensors can detect chemicals, such as heavy metals or organic solvents, with selectivity tailored by modifying the surface chemistry.

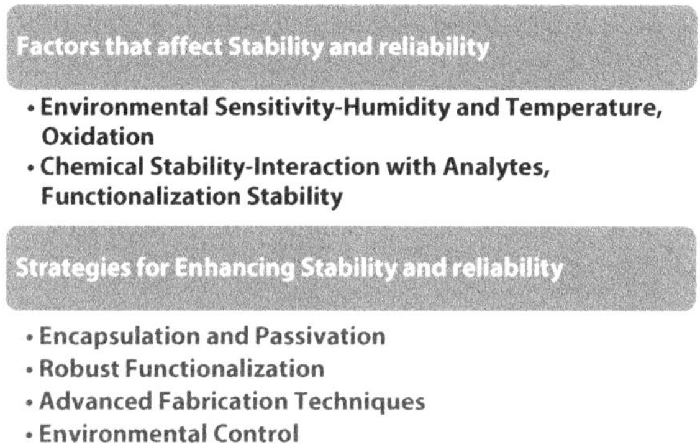

Figure 9.8 Factors that impact stability and reliability of strategies for enhancing them.

9.4.1.2 Stability and Reliability

Ensuring the long-term stability and reliability of 2D material-based sensors, particularly in harsh or changing environments, is a significant challenge. Materials may degrade over time, affecting the sensor performance [39, 40].

Several strategies can be employed to enhance the stability and reliability of 2D material-based FET sensors. Encapsulation with stable layers such as hexagonal boron nitride (h-BN) or aluminum oxide (Al_2O_3) protects 2D materials from environmental degradation and improves their stability. Robust functionalization techniques are essential to ensure that attached functional groups or molecules remain effective over extended periods. Utilizing advanced and standardized fabrication techniques helps minimize variability and enhance the reproducibility of these sensors. Additionally, implementing environmental control measures, such as maintaining a controlled atmosphere around the sensor, can mitigate the impact of external factors on sensor performance. Figure 9.8 summarizes the factors that impact the stability and reliability of strategies for enhancing them.

9.4.1.3 Scalability and Miniaturization and Integration

Scalable production of high-quality 2D materials is challenging. Developing cost-effective methods for large-scale synthesis while maintaining material quality is essential for commercial applications [41, 42]. Achieving miniaturization and portability of 2D material-based FET sensors is essential for applications such as wearable devices and point-of-care diagnostics [43].

Integrating 2D material-based FET sensors into existing systems and technologies can be complex. Compatibility with electronics, signal processing, and data analysis tools is necessary for practical application.

9.4.1.4 Biocompatibility

In biomedical applications, ensuring the biocompatibility of 2D materials and sensor interfaces is crucial for avoiding adverse effects on biological systems. The choice of 2D materials is critical. Although some 2D materials, such as graphene, are generally considered biocompatible, others may require surface modifications to enhance biocompatibility. For example, transition metal dichalcogenides (TMDs) and black phosphorus may require surface functionalization to reduce potential toxicity. Surface functionalization involves modifying the 2D material's surface with biocompatible molecules or polymers. This can enhance biocompatibility, reduce cytotoxicity, and enable specific interactions with biomolecules [44].

9.4.2 Future Scope

The discovery and development of new 2D materials beyond graphene, such as transition metal dichalcogenides (TMDs) and black phosphorus, offers opportunities for novel sensor designs and improved performance [22]. Tailoring the surface properties of 2D materials by functionalization with specific molecules or groups can enhance their selectivity for target analytes [45, 46].

Future sensors may be capable of detecting multiple analytes simultaneously, enabling more comprehensive and real-time monitoring in diverse applications [47, 48]. Advancements in flexible and stretchable electronics will enable the creation of conformal 2D material-based FET sensors for wearable and healthcare applications [49]. Integration with the Internet of Things (IoT) for remote monitoring and data sharing can expand the scope of 2D material-based FET sensors in smart cities and industries. Machine learning algorithms and artificial intelligence (AI) can be used to analyze complex sensor data patterns and improve the accuracy and reliability of 2D material-based FET sensors [50].

There has been rapid advancement in the field of novel semiconductor FET device design [51–55]. Development of low-power sensor designs is essential for battery-powered and energy-efficient applications [22, 56]. The use of 2D material-based FET sensors in healthcare, including in disease diagnosis and drug screening, is a promising area for future research. Research efforts should focus on sustainable synthesis methods for 2D materials and the use of environmentally friendly materials for sensor fabrication. As research and development in 2D material-based FET sensors continues to advance, addressing these challenges and exploring their future prospects will pave the way for innovative sensor technologies with a wide range of applications.

9.5 Conclusion

The integration of 2D materials into FET sensors has led to significant advancements in sensing applications. These sensors offer high sensitivity, selectivity, and versatility, making them invaluable tools for gas-sensing, biosensing, and environmental monitoring. As researchers continue to explore and improve upon the unique properties of 2D materials, we anticipate even more exciting developments in the field of 2D material-based FET sensors, with the potential to revolutionize a wide range of industries and applications.

References

1. Cui, Y., Wei, Q., Park, H., Lieber, C.M., Nanowire nanosensors for highly sensitive and selective detection of biological and chemical species. *Science*, 293, 5533, 1289–1292, 2001.
2. Zhou, G., *et al.*, Ultrasensitive mercury ion detection using DNA-functionalized molybdenum disulfide nanosheet/gold nanoparticle hybrid field-effect transistor device. *ACS Sens.*, 1, 3, 295–302, 2016.
3. Sorgenfrei, S., Chiu, C.Y., Gonzalez Jr., R.L., Yu, Y.J., Kim, P., Nuckolls, C., Shepard, K.L., Label-free single-molecule detection of DNA-hybridization kinetics with a carbon nanotube field-effect transistor. *Nat. Nanotechnol.*, 6, 2, 126–132, 2011.
4. Li, Z., Xiao, M., Jin, C., Zhang, Z., Toward the Commercialization of Carbon Nanotube Field Effect Transistor Biosensors. *Biosensors*, 13, 3, 326, 2023.
5. Choi, W., Choudhary, N., Han, G.H., Park, J., Akinwande, D., Lee, Y.H., Recent development of two-dimensional transition metal dichalcogenides and their applications. *Mater. Today*, 20, 3, 116–130, 2017.
6. Noyce, S.G., *et al.*, Understanding and mapping sensitivity in MoS2 field-effect-transistor-based sensors. *ACS Nano*, 14, 9, 11637–11647, 2020.
7. Liu, C., *et al.*, Two-dimensional materials for next-generation computing technologies. *Nat. Nanotechnol.*, 15, 7, 545–557, 2020.
8. Geim, A.K. and Novoselov, K.S., The rise of graphene. *Nat. Mater.*, 6, 3, 183–191, 2007.

9. Mak, K.F., Lee, C., Hone, J., Shan, J., Heinz, T.F., Atomically thin MoS2: A new direct-gap semiconductor. *Phys. Rev. Lett.*, 105, 13, 136805, 2010.
10. Buckley, D.J., Black, N.C., Castanon, E.G., Melios, C., Hardman, M., Kazakova, O., Frontiers of graphene and 2D material-based gas sensors for environmental monitoring. *2D Mater.*, 7, 3, 032002, 2020.
11. Bhimanapati, G.R., *et al.*, Recent advances in two-dimensional materials beyond graphene. *ACS Nano*, 9, 12, 11509–11539, 2015.
12. He, Q., Wu, S., Yin, Z., Zhang, H., Graphene-based electronic sensors. *Chem. Sci.*, 3, 6, 1764–1772, 2012.
13. Huang, H., Su, S., Wu, N., Wan, H., Wan, S., Bi, H., Sun, L., Graphene-based sensors for human health monitoring. *Front. Chem.*, 7, 399, 2019.
14. Danielson, E., *et al.*, Graphene based field-effect transistor biosensors functionalized using gas-phase synthesized gold nanoparticles. *Sens. Actuators B Chem.*, 320, 128432, 2020.
15. Lam, C.Y.K., Zhang, Q., Yin, B., Huang, Y., Wang, H., Yang, M., Wong, S.H.D., Recent advances in two-dimensional transition metal dichalcogenide nanocomposites biosensors for virus detection before and during COVID-19 outbreak. *J. Compos. Sci.*, 5, 7, 190, 2020.
16. Pi, L., Li, L., Liu, K., Zhang, Q., Li, H., Zhai, T., Recent progress on 2D noble-transition-metal dichalcogenides. *Adv. Funct. Mater.*, 29, 51, 1904932, 2019.
17. Liu, H., Hu, K., Yan, D., Chen, R., Zou, Y., Liu, H., Wang, S., Recent advances on black phosphorus for energy storage, catalysis, and sensor applications. *Adv. Mater.*, 30, 32, 1800295, 2018.
18. Eswaraiah, V., Zeng, Q., Long, Y., Liu, Z., Black phosphorus nanosheets: synthesis, characterization and applications. *Small*, 12, 26, 3480–3502, 2016.
19. Li, S., Ma, L., Zhou, M., Li, Y., Xia, Y., Fan, X., Luo, H., New opportunities for emerging 2D materials in bioelectronics and biosensors. *Curr. Opin. Biomed. Eng.*, 13, 32–41, 2020.
20. Xu, Q., Zong, B., Yang, Y., Li, Q., Mao, S., Black phosphorus quantum dots modified monolayer Ti3C2Tx nanosheet for field-effect transistor gas sensor. *Sens. Actuators B: Chem.*, 373, 132696, 2022.
21. Shinde, P.V., Kumar, A., Late, D.J., Rout, C.S., Recent advances in 2D black phosphorus based materials for gas sensing applications. *J. Mater. Chem. C*, 9, 11, 3773-3794, 2021.
22. Zhu, C., Du, D., Lin, Y., Graphene and graphene-like 2D materials for optical biosensing and bioimaging: A review. *2D Mater.*, 2, 3, 032004, 2015.
23. Masurkar, N., Varma, S., Mohana Reddy Arava, L., Supported and suspended 2D material-based FET biosensors. *Electrochem*, 1, 3, 260–277, 2020.
24. Ohno, Y., Maehashi, K., Matsumoto, K., Chemical and biological sensing applications based on graphene field-effect transistors. *Biosens. Bioelectron.*, 26, 4, 1727–1730, 2010.
25. Das, S., Chen, H.-Y., Penumatcha, A.V., Appenzeller, J., High performance multilayer MoS2 transistors with scandium contacts. *Nano Lett.*, 13, 1, 100–105, 2012.

26. Zhang, Y., Chang, T.-R., Zhou, B., Cui, Y.-T., Yan, H., Liu, Z., Schmitt, F., Lee, J., Moore, R., Chen, Y., Lin, H., Jeng, H.-T., Mo, S.-K., Hussain, Z., Bansil, A., Shen, Z.-X., Direct observation of the transition from indirect to direct bandgap in atomically thin epitaxial MoSe2. *Nat. Nanotechnol.*, 9, 2, 111–115, 2014.
27. Wang, Q.H., Kalantar-Zadeh, K., Kis, A., Coleman, J.N., Strano, M.S., Electronics and optoelectronics of two-dimensional transition metal dichalcogenides. *Nat. Nanotechnol.*, 7, 11, 699–712, 2012.
28. Xia, F., Wang, H., Jia, Y., Rediscovering black phosphorus as an anisotropic layered material for optoelectronics and electronics. *Nat. Commun.*, 5, 4458, 2014.
29. Zhang, K., *et al.*, Two dimensional hexagonal boron nitride (2D-hBN): synthesis, properties and applications. *J. Mater. Chem. C*, 5, 46, 11992–12022, 2017.
30. Wang, G., Pandey, R., Karna, S.P., Phosphorene oxide: stability and electronic properties of a novel two-dimensional material. *Nanoscale*, 7, 2, 524–531, 2015.
31. Zhu, J., *et al.*, Recent advance in MXenes: a promising 2D material for catalysis, sensor and chemical adsorption. *Coord. Chem. Rev.*, 352, 306–327, 2017.
32. Kim, T.H., *et al.*, Two dimensional transition metal disulfides for chemoresistive gas sensing: perspective and challenges. *Chemosensors*, 5, 2, 15, 2017.
33. Liu, Y., Duan, X., Gao, H.J., Graphene and graphene oxide: Two ideal choices for the fabrication of a nanodevice for the detection of DNA and pH. *Chem. Commun.*, 46, 7, 1276–1278, 2010.
34. Lee, C.W., Suh, J.M., Jang, H.W., Chemical sensors based on two-dimensional (2D) materials for selective detection of ions and molecules in liquid. *Front. Chem.*, 7, 708, 2019.
35. Ren, H., *et al.*, Field-Effect Transistor-Based Biosensor for pH Sensing and Mapping. *Adv. Sens. Res.*, 2, 8, 2200098, 2023.
36. Chen, F., *et al.*, Structures, properties, and challenges of emerging 2D materials in bioelectronics and biosensors. *InfoMat*, 4, 5, e12299, 2022.
37. Fan, Q., *et al.*, Solution-gated transistors of two-dimensional materials for chemical and biological sensors: status and challenges. *Nanoscale*, 12, 21, 11364–11394, 2020.
38. Sedki, M., *et al.*, Non-carbon 2D materials-based field-effect transistor biosensors: recent advances, challenges, and future perspectives. *Sensors*, 20, 17, 4811, 2020.
39. Das, S., *et al.*, Transistors based on two-dimensional materials for future integrated circuits. *Nat. Electron.*, 4, 11, 786–799, 2021.
40. Shrivastava, M. and Ramgopal Rao, V., A roadmap for disruptive applications and heterogeneous integration using two-dimensional materials: State-of-the-art and technological challenges. *Nano Lett.*, 21, 15, 6359–6381, 2021.
41. Gablech, I., *et al.*, Monoelemental 2D materials-based field effect transistors for sensing and biosensing: phosphorene, antimonene, arsenene, silicene,

and germanene go beyond graphene. *TrAC, Trends Anal. Chem.*, 105, 251–262, 2018.
42. Miao, J., *et al.*, Black phosphorus electronic and optoelectronic devices. *2D Mater.*, 6, 3, 032003, 2019.
43. Byakodi, M., *et al.*, Emerging 0D, 1D, 2D, and 3D nanostructures for efficient point-of-care biosensing. *Biosens. Bioelectron.: X*, 12, 100284, 2022.
44. Liu, G., *et al.*, Biocompatible Material-Based Flexible Biosensors: From Materials Design to Wearable/Implantable Devices and Integrated Sensing Systems. *Small*, 19, 27, 2207879, 2023.
45. Zhang, J., *et al.*, A review on two-dimensional materials for chemiresistive-and FET-type gas sensors. *Phys. Chem. Chem. Phys.*, 23, 29, 15420–15439, 2021.
46. Viswanathan, S., *et al.*, Graphene–protein field effect biosensors: glucose sensing. *Mater. Today*, 18, 9, 513–522, 2015.
47. Soikkeli, M., *et al.*, Wafer-scale graphene field-effect transistor biosensor arrays with monolithic CMOS readout. *ACS Appl. Electron. Mater.*, 5, 9, 4925–4932, 2023.
48. Dai, C., *et al.*, Graphene Transistors for *In Vitro* Detection of Health Biomarkers. *Adv. Funct. Mater.*, 33, 31, 2301948, 2023.
49. Kim, D.C., *et al.*, Material-based approaches for the fabrication of stretchable electronics. *Adv. Mater.*, 32, 15, 1902743, 2020.
50. Cao, G., *et al.*, 2D material based synaptic devices for neuromorphic computing. *Adv. Funct. Mater.*, 31, 4, 2005443, 2021.
51. Pandey, A., Recent Trends in Novel Semiconductor Devices. *Silicon*, 14, 9211–9222, 2022, https://doi.org/10.1007/s12633-022-01694-8.
52. Pandey, A., Raycha, S., Maheshwaram, S., Manhas, S.K., Dasgupta, S., Saxena, A.K., Anand, B., Effect of load capacitance and input transition time on FinFET inverter capacitances. *IEEE Trans. Electron Devices*, 61, 1, 30–36, 2014.
53. Narula, M.S. and Pandey, A., Performance Evaluation of Stacked Gate Oxide/High K Spacers based Gate All Around Device Architectures at 10nm Technology Node. *Silicon*, 14, 5, 2397–2407, April 2022, doi: doi.org/10.1007/s12633-022-01685-9.
54. Narula, M.S. and Pandey, A., A Comprehensive Review on FinFET, Gate All Around, Tunnel FET: Concept, Performance and Challenges. *8th International Conference on Signal Processing and Communication (ICSC)*, Noida, India, pp. 554–559, 2022, doi: 10.1109/ICSC56524.2022.10009504.
55. Singh, A. and Pandey, A., Gate-All-Around Nanosheet FET Device Simulation Methodology Using a Sentaurus TCAD, in: *Advanced Field-Effect Transistors*, pp. 232–259, CRC Press, Boca Raton, 2023, ISBN- 9781003393542.
56. Zhang, L., *et al.*, Recent advances in emerging 2D material-based gas sensors: potential in disease diagnosis. *Adv. Mater. Interfaces*, 6, 22, 1901329, 2019.

10

2D Material-Based FETs for Next Generation Integrated Circuits

Aruru Sai Kumar[1]*, V. Bharath Sreenivasulu[2], K. Sarangam[3†], P. Ravi Sankar[4] and K. Nishanth Rao[5]

[1]School of Electronics Engineering, VIT-AP University, Amaravati, Andhra Pradesh, India
[2]Department of ECE, Manipal Institute of Technology Bengaluru, Manipal Academy of Higher Education, Manipal, Karnataka, India
[3]Department of Electronics and Communication Engineering, National Institute of Technology, Warangal, India
[4]Department of CSE, SAHE University, Vijayawada, Andhra Pradesh, India
[5]Department of ECE, MLR Institute of Technology, Hyderabad, India

Abstract

In the last few years, a significant interest has been shown in 2D field effect transistor (2D-FETs) as an appropriate candidate for advanced electronics devices. This study of a two-dimensional FET (2D-FET) works with a channel made of MoS_2, $MoSe_2$, and $MoTe_2$ and different dielectric materials. Materials such as molybdenum disulfide (MoS_2), molybdenum diselenide ($MoSe_2$), and molybdenum ditelluride ($MoTe_2$) belong to the transition metal dichalcogenide (TMD) family and are known for their fascinating electronic properties. Transition metal dichalcogenides like MoS_2 have emerged as promising candidates, boasting satisfactory bandgaps, thermal stability, high carrier mobility, and compatibility with silicon CMOS processes. Their ability to reduce scattering events, electrical insulating qualities, and suitability for 2D semiconductor channels make them excellent choices for low-power setups. The ideal dielectric, however, must also consider the particular needs of the device, including its operating environment, targeted performance indicators, and 2D semiconductor channel material. These materials have showcased immense potential in device applications, offering superior electrostatics and resilience against short-channel effects compared to traditional

**Corresponding author*: asaikumar.nitw@gmail.com
†*Corresponding author*: sarangam_7@nitw.ac.in

Ekta Goel, Archana Pandey, Shiromani Balmukund Rahi and Arun Samuel (eds.) Handbook of Advanced Semiconductor Field Effect Transistors, (199–216) © 2026 Scrivener Publishing LLC

silicon channels. From the classic Metal-Oxide-Semiconductor FET (MOSFET) to the evolution of dual-gate FETs, tri-gate devices, and the current pinnacle, Gate-All-Around Field-Effect Transistors (GAA FETs), each technological era has presented unique opportunities and challenges. These materials have different band gaps, carrier mobilities, and intrinsic properties that make them ideal candidates for 2D transistor channels. The I_{ON}/I_{OFF} graph is a critical metric for evaluating the performance of Field-Effect Transistors (FETs), providing insights into the on-state (I_{ON}) and off-state (I_{OFF}) current characteristics. In scenarios involving a HfO_2 spacer, MoS_2 consistently demonstrates superior performance compared to alternative 2D materials such as $MoTe_2$ and $MoSe_2$. The I_{ON}/I_{OFF} graph for MoS_2 FETs typically exhibits a more favorable ratio, emphasizing a higher on-state current coupled with a lower off-state current. This advantageous combination is attributed to MoS_2's unique electronic properties, including its favorable band structure and superior carrier mobility. The HfO_2 spacer enhances the gate control over the channel, contributing to efficient modulation of the FET's conductivity. In contrast, $MoTe_2$ and $MoSe_2$ may exhibit comparatively higher off-state currents or reduced on-state currents due to factors like lower carrier mobility or less effective gating. The I_{ON}/I_{OFF} graph underscores MoS_2's prowess as a promising material for FET applications with HfO_2 spacers, highlighting its potential for high-performance electronic devices. In this study, we investigate the influence of different dielectric materials on the performance of 2D FETs with MoS_2, $MoSe_2$, and $MoTe_2$ as channel materials, with a strong focus on high-speed operation. Our study analyzes performance metrics like on-current (I_{ON}), off-current (I_{OFF}), and I_{ON}/I_{OFF} ratio which will be more when compared with Si-based FETs. Notably, the choice of dielectric significantly influences these metrics, with high-k dielectrics and HfO_2 often surpassing SiO_2.

***Keywords*:** 2D-FET, MoS_2 FET, scattering transport model, high-k dielectrics, low power and transition metal dichalcogenides (TMDs)

10.1 Introduction

The semiconductor industry has always been working toward the goal of developing new technological developments to conform to the requirements set forth by the ITRS. To comply with the law established by Moore, which specifies that the number of transistors on a chip must be increased by a factor of two every 18 months, the primary motive for all of the developments is to comply with this legislation. One of the most effective ways to increase the number of transistors is to decrease the size of each transistor [1]. Channel length reduction, which was initially the most successful approach to improve technology nodes, is a method used to reduce the size of transistors during the process. During the first age of the FET family, the conventional MOSFET, which was constructed in 1959, was used in a variety of application domains. Until SCEs were prominent, MOSFETs were the dominant technology. Consequently, the performance started to decrease [2].

In recent years, the exploration of novel materials and device architectures has been pivotal for advancing the field of nanoelectronics. Among these materials, Transition Metal Dichalcogenides (TMDs) have emerged as promising candidates for channel materials in Field-Effect Transistors (FETs) owing to their unique electronic properties and two-dimensional nature. In parallel, the development of Nanosheet FETs has garnered significant attention owing to their potential to overcome the limitations associated with conventional FET structures. Nanosheet FETs offer several advantages over the traditional planar and FinFET architectures. Their thin layered structure enables superior electrostatic control and reduces short-channel effects, making them highly suitable for scaled-down devices. Furthermore, the two-dimensional geometry of the nanosheets facilitates efficient gate control over the channel region, leading to enhanced device performance and reduced power consumption [1, 3, 4].

Nanosheets represent the next evolutionary stage in transistor technology and are poised to supersede FinFET devices for technology nodes below 7 nm and beyond. As a member of the multigate transistor family, the Nanosheet Gate-All-Around (GAA) FET embodies a specialized transistor configuration, distinguished by its nanoscale channel thickness. This design refinement enables Nanosheet Field-Effect Transistors (NSFETs) to exert exceptional electrostatic control over the channel area, thereby mitigating the short-channel effect (SCE) characteristic of conventional transistors. NSFETs also exhibit a remarkable capacity to drive substantial capacitive loads, further enhancing their utility in advanced electronic systems [5–10].

Research has shown that individual single-layer nanosheets outperform both FinFETs and stacked nanowires significantly, underscoring their superior performance characteristics. Despite the common practice of vertically stacking nanosheets to enhance current-driving capability, this approach introduces the potential drawback of increased parasitic capacitance owing to the heightened device structure. While stacking nanosheets facilitates driving substantial currents, the resulting increase in device height may exacerbate the parasitic capacitance issues. NSFETs, with their inherent advantages, are poised to supplement traditional MOSFETs [11–15]. By combining the benefits of the GAA Nanosheet FETs with vertical stacking, the overall device performance was further enhanced. This integration affords robust leakage control, increased effective width, improved electrostatic performance over the channel, and amplification of driver current. However, it is noteworthy that the number of vertically stacked nanosheets also influences the device performance, potentially exacerbating parasitic capacitance issues. Consequently, this study imposes a limitation on the stacking of nanosheets into two layers to mitigate these concerns and optimize the device performance [16].

The miniaturization of transistors has led to problems, such as leakage currents and short-channel effects. To overcome these limitations and maintain high performance with low power consumption, researchers are exploring 2D materials as replacements for conventional silicon. Graphene, the first discovered 2D material, initially attracted attention due to its conductivity, but its lack of a bandgap hinders its use in transistors. This has shifted the focus to transition metal dichalcogenides (TMDs) such as MoS_2 and WS_2, which possess the desired optical and electrical properties and a bandgap. The ultrathin structure of 2D materials such as MoS_2 effectively reduces short-channel effects in scaled devices owing to its superior electrostatics. However, these materials exhibit limitations in on-current (I_{on}) owing to their low saturation velocity, even at minimal channel lengths. Additionally, their low saturation velocity makes the increasing mobility less impactful on the on-current [17]. Owing to their two-dimensional structure, they can overcome the limits imposed by Moore's law, which has long prevented the further performance development of MOSFETs. This enables them to demonstrate great potential for downsized and low-power transistors. Second-dimensional nanomaterials have the potential to be more sensitive in sensing applications because their two-dimensional nanostructures have a high specific surface area, which allows for the dense alteration of binding sites [18].

There are two main carrier transport mechanisms: scattering and ballistic transport. These factors determine how carriers move within a material, thereby affecting device performance. The specific transport mode of 2D FETs depends on the device's configuration and characteristics. Ballistic transport, which is ideal for high-speed FETs, minimizes the carrier collisions within the channel. In this study, we focused on evaluating the performance of nanosheet FETs with TMD materials as channel materials. Specifically, we investigated the influence of varying the gate length and dielectric materials on the device characteristics. The choice of gate length and dielectric plays a crucial role in determining key metrics, such as the on-current (I_{on}), off-current (I_{off}), and I_{on}/I_{off} ratio, which are pivotal indicators of device performance [19].

The on-current (I_{on}) represents the maximum current flowing through the device when the transistor is in the on-state, reflecting its capability to effectively conduct charge carriers. Conversely, the off current (I_{off}) signifies the leakage current when the transistor is in the off state, which directly affects the power consumption and overall efficiency of the device. The I_{on}/I_{off} ratio serves as a measure of the transistor's switching efficiency, indicating how effectively it can switch between on and off states. Understanding the influence of gate-length scaling and dielectric selection on these performance metrics is crucial for optimizing the operation of nanosheet FETs with TMD channels. By elucidating these relationships, we aimed to provide valuable insights into the design and engineering of next-generation nanoelectronic devices tailored for advanced applications in computing, communication, and beyond.

The remainder of this research study is structured as follows: *Section 10.2* provides a description of the literature survey, and *Section 10.3* presents the proposed technique that describes the monolayer nanosheet field-effect transistor (NSFET). An analysis of the results is presented in *Section 10.4*, and the conclusions of this study are discussed in *Section 10.5*.

10.2 Literature Survey

In light of the proliferation of research on TMDC FETs, our objective is to provide a relevant overview of TMDC-based FET sensors specifically designed for environmental investigations in a variety of media. First, the fundamentals of TMDC and FET sensors are presented in a theoretical context. Subsequently, the most recent developments in TMDC FET sensors for the detection of pollutants in gaseous and aqueous systems are discussed. Finally, a discussion of the future prospects and difficulties associated with the practical application and commercialization of the TMDC FET sensors is presented. Chen *et al.* [19] provided an overview of TMDC sensors for a wide variety of analytes, with a particular emphasis on the growing demand for improved sensing technologies in environmental monitoring.

Ilatikhameneh *et al.* [20] examined the efficiency of tunnel field-effect transistors (TFETs) using atomistic quantum transport simulations focus on 2-D transition metal dichalcogenide (TMD) materials. A significant obstacle faced by TFETs is their low ON currents. 2-D material-based Tunnel Field-Effect Transistors (TFETs) exhibit precise gate control and generate strong electric fields at the tunnel junction. Consequently, they have the potential to produce significant ON currents while maintaining a subthreshold swing (SS) below 60 mV/decade. Our simulations demonstrate that achieving high-performance TMD TFETs requires both effective gate control and the selection of an appropriate channel material with an optimal bandgap, effective mass, and source/drain doping level. In contrast to earlier studies, this study employs a comprehensive atomistic tight-binding method that incorporates a full-band approach. The method is self-consistently coupled with a three-dimensional Poisson equation to accurately simulate ballistic quantum transport in these devices. The impact of the selection of TMD material on the device performance and its transfer properties was analyzed. In addition, the criteria for high ON currents were elucidated using a straightforward analytical model, which demonstrated the associated fundamental elements. Ultimately, the subthreshold swing (SS) and energy delay of these Tunnel Field-Effect Transistors (TFETs) are contrasted with those of ordinary complementary metal-oxide-semiconductor (CMOS) devices.

Advancements in the field of atomically thin two-dimensional transition metal dichalcogenides (2D TMDs) have resulted in the development of several promising technologies such as nanoelectronics, photonics, sensing, energy storage, and optoelectronics. Choi *et al.* [21] provided an overview of the latest advancements in 2D materials that go beyond graphene. It primarily focuses on transition metal dichalcogenides (TMDs) such as MoS_2, WS_2, $MoSe_2$, and WSe_2. These materials are used in specific and specialized applications for advanced electronics and optoelectronic devices that depend on extremely thin atomic layers. Despite encountering several obstacles in the process of creating scalable and flawless TMDs on specific substrates, novel growth methods compatible with both standard and atypical substrates have been devised to fulfill the growing need for superior quality and precise control in real-world applications. The focus is on creating new 2D transition metal dichalcogenides (TMDs) with unique properties and chemistry. Finally, alongside electronics, a significant amount of effort is dedicated to thoroughly discussing the utilization of these materials for energy and sensing applications.

Recently, there has been increasing interest in the use of Monolayer MoS_2 (ML-MoS_2) with different polymorphic phases for device applications. Zong *et al.* [22] introduced a gas sensor using a monolayer of MoS_2 with a heterophase consisting of a 1T metallic phase and 2H semiconducting phase. The nanosheets of MoS_2 exfoliated with lithium possess a monolayer structure that contains several active sites suitable for gas adsorption. By subjecting the lithium-exfoliated 1T-phase MoS_2 to thermal annealing in the temperature range of 50–300 °C, a progressive transformation occurred where the initial 1T-phase MoS_2 changed into the 2H phase. This transformation allows for the modulation of the heterophases 1T and 2H. Heterophase MoS_2, with a 1T to 2H ratio, exhibits p-type semiconducting characteristics and a notable ability to adsorb NO_2 molecules. The highest level of reaction is reported for MoS_2 annealed at 100 °C, with a composition of 40% 1T phase and 60% 2H phase. This material has a sensitivity of up to 25% towards 2 ppm NO_2 at room temperature within a relatively short time frame of 10 s. In addition, it has a lower limit of detection (as low as 25 ppb). This study shows that the gas detection capacity of ML-MoS_2 can be enhanced through heterophase fabrication, providing novel perspectives for gas sensors made of transition-metal dichalcogenides.

10.3 Proposed Methodology

The proposed work aims to comprehensively evaluate the performance of Nanosheet Field-Effect Transistors (FETs) utilizing Transition Metal Dichalcogenide (TMD) materials (MoS_2, $MoSe_2$, and $MoTe_2$) as channel materials, with a focus on varying gate lengths and dielectric materials (HfO2, Si_3N_4, and SiO_2). By leveraging the simulation capabilities of Nanohub.org, the study will calculate the on-current (I_{on}), off-current (I_{off}), and I_{on}/I_{off} ratio for each combination of TMD material, gate length, and dielectric material facilitate a comprehensive analysis of device performance. Optimization strategies will be explored to identify optimal combinations of parameters, considering potential trade-offs between gate-length scaling, dielectric selection, and TMD material choice. Through comparative analysis and discussions on future directions, this study aims to advance our understanding of the field and provide insights for the design of high-performance nanosheet FETs tailored for advanced nano-electronic applications.

The investigation's core goal is to determine the suggested device's thermal stability using temperature-dependent investigations. Changes in spacer thickness are anticipated to impact carrier transport qualities by altering the gate's electrostatic control over the channel. A variety of theoretical models and simulations have been performed, along with considerations of scattering processes, thermal effects, and quantum mechanical effects. Figure 10.1 shows the monolayer nanosheet field-effect transistor (NSFET). This circuit allows for the exploration of different 2D channel materials and dielectrics to better understand their influence on FET performance, which is crucial for designing transistors with specific threshold voltages. The inherent semiconductor properties of TMDs, combined

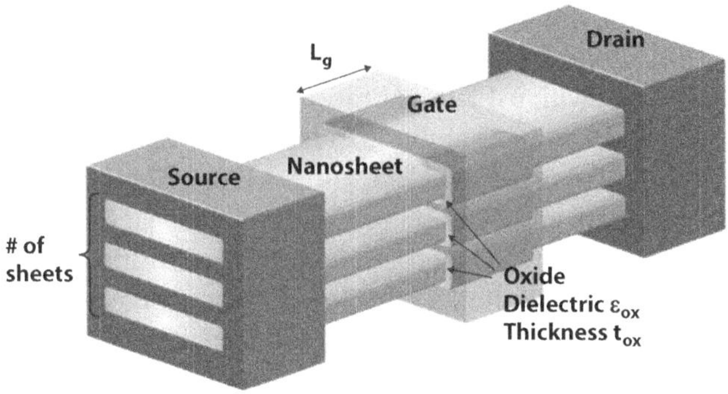

Figure 10.1 Monolayer nanosheet field-effect transistor (NSFET) [24].

with their atomically thin structures, result in a low off-state current (I_{off}). This property is important for minimizing the power consumption of electronic devices in the idle or standby mode. Owing to their unique qualities, TMDs are frequently employed as channel materials in field-effect transistors (FETs). TMDs' atomically thin structure allows superior electrostatic channel control, which enhances gate control and lowers leakage currents. Additionally, TMDs are compatible with high-k dielectrics in transistor design [23]. This compatibility improves gate control, further reduces power consumption, and renders TMD-based transistors promising candidates for energy-efficient applications.

10.4 Result Analysis

Using TMDs as channel materials and a variety of dielectrics, we have conducted an analysis of the performance of nanosheet FET applications. After calculating the transfer characteristics (I_D versus V_{GS}) and switching ratio (I_{on} versus I_{off}), it was determined that the performance of a 2DFET with TMDs as channel materials and HfO_2 as a dielectric is superior to that of a 2DFET with Air and SiO_2 as dielectric materials. An open online simulation tool known as nanohub.org was used to do simulations of nanosheet field-effect transistors (FETs). This tool was used to change the channel materials to simulate the various dielectrics.

10.4.1 I_D vs V_{GS}

The graph depicts the transfer characteristics (I_D vs V_{GS}) for different gate lengths and different spacer materials in Figure 10.2. The drain current

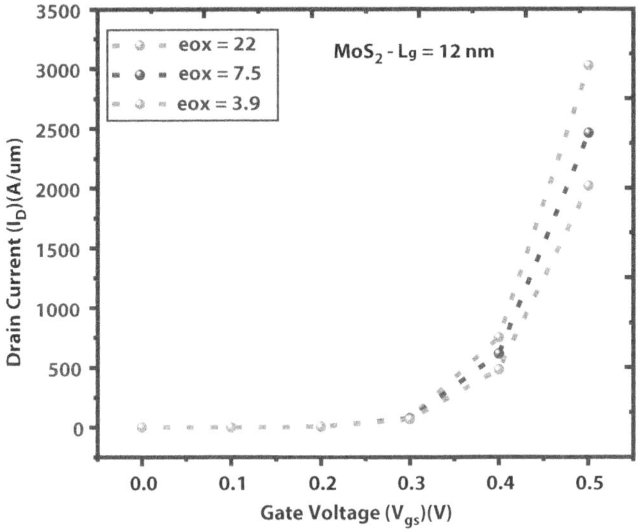

Figure 10.2 I_D vs V_{GS} for various dielectric materials while the channel materials is MoS_2.

(I_D) is high for a gate length of 10 nm compared to the others. The graph that shows the relationship between the drain current (I_D) and gate-source voltage (V_{GS}) for the field effect transistor (FET) serves as a fundamental characterization of the device's behavior. This curve illustrates the device response to varying gate-source voltages while maintaining a constant drain-source voltage. This graph typically exhibits distinct regions, such as the cutoff, saturation, and linear regions, each representing different operational states of the transistor.

In the cutoff region, where the gate-source voltage is below the threshold voltage (V_{th}), the transistor operates in the OFF state, and the drain current remains negligible. As the gate-source voltage increases beyond the threshold voltage, the transistor enters the linear region, where the drain current increases linearly with Vgs. This region demonstrates the device's ability to amplify input signals with a linear response. Beyond a certain gate-source voltage, the FET enters the saturation region, where further increases in V_{GS} do not result in significant changes in Id. In this region, the FET operates as an active switch with the drain current reaching its maximum saturation value. The transition from the linear region to the saturation region highlights the device's ability to provide full conduction with a minimal additional input voltage.

The five materials shown in Figure 10.3 are: MoS_2, $MoSe_2$, $MoTe_2$, WS_2, and WSe_2. The legend indicates that all materials have a channel length (L)

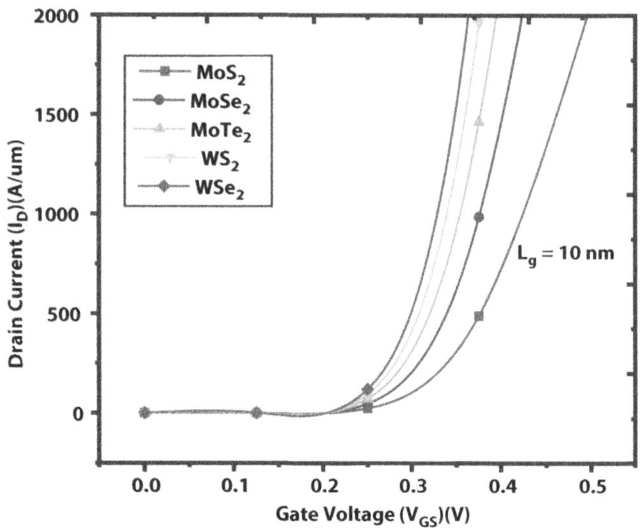

Figure 10.3 I_D vs V_{GS} for various channel material and L_g = 10 nm.

of 10 nm. In all TMDCs, the drain current increased as the gate voltage increased. This suggests that these materials are n-type semiconductors, meaning that the majority carriers are electrons. The rate of increase of the drain current with gate voltage is different for different TMDCs. MoS_2 had the highest rate of increase, whereas $MoTe_2$ had the lowest rate of increase. This suggests that MoS_2 may be a better conductor than $MoTe_2$ is.

10.4.2 I_D vs V_{DS}

Figure 10.4 shows the output traits (I_D vs V_{DS}) for exceptional dielectric materials. The drain current vs. drain-source voltage (I_D vs V_{DS}) is referred to as the "output characteristic" of the field effect transistor (FET), because it essentially depicts the output behavior of the FET in response to versions inside the implemented drain-source voltage. It shows how the transistor responds to external load and voltage changes, presenting complete knowledge of its overall performance characteristics under distinct operating conditions. This indicates how the drain current (I_D), which acts because of the output of the transistor, modifies the applied voltage throughout its terminals. This graph illustrates the relationship between the drain current (I_D) flowing through the transistor and the voltage across the drain and supply terminals (V_{DS}) while maintaining a steady gate-supply voltage (V_{GS}). It is calculated using various drain-supply voltages, even when

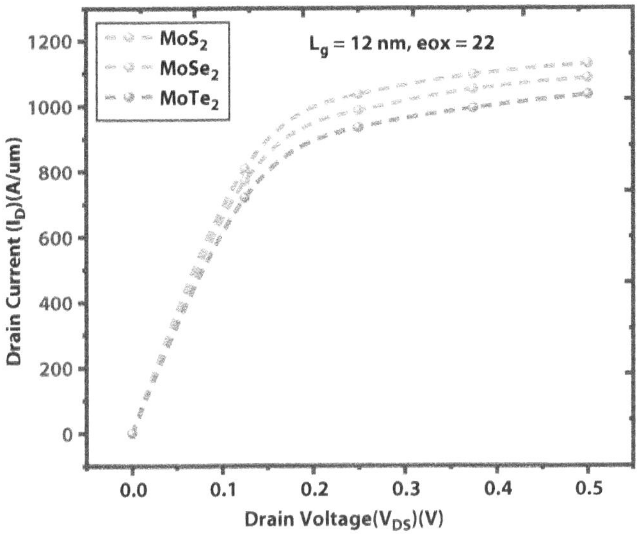

Figure 10.4 I_D vs V_{DS} for various channel materials while the dielectric is HfO_2(eox = 22) and gate length is 12 nm.

measuring the corresponding drain current at a fixed gate-supply voltage. The output characteristics provide important insights into the behavior of the FET, specifically concerning its saturation, linear, and breakdown regions.

10.4.3 I_{ON}/I_{OFF}

The Ion/Ioff ratio, often termed the "switching ratio," is a critical parameter in field-effect transistors (FETs) that quantifies the difference between the ON current (I_{ON}) and OFF current (I_{OFF}). It is calculated by dividing the ON current by the OFF current, providing a measure of the transistor's ability to effectively switch between the ON and OFF states. This ratio is essential for evaluating the performance of FETs because it directly reflects the device's switching efficiency and power consumption characteristics. A higher switching ratio signifies a more significant distinction between the ON and OFF states, indicating better control over the current flow and reduced leakage when the transistor is in the OFF state. The switching ratio serves as a crucial metric for assessing the overall performance and reliability of FET-based devices, as shown in Figure 10.5. By accurately calculating and optimizing this ratio, circuits can be designed with improved energy efficiency, faster response times, and enhanced signal integrity, thereby ensuring optimal performance across various electronic systems.

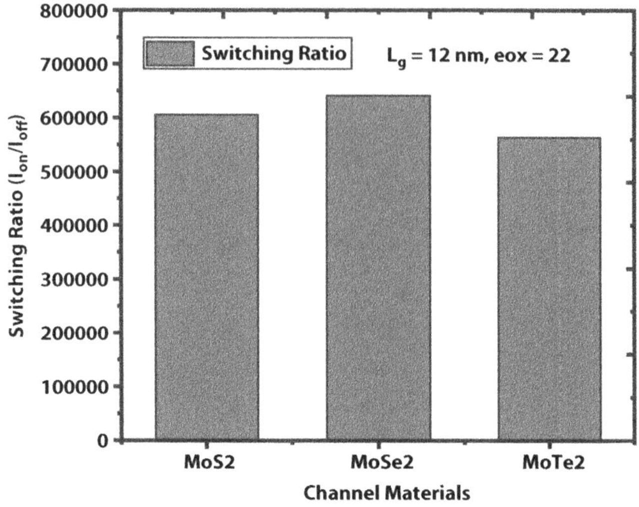

Figure 10.5 I_{ON}/I_{OFF} (switching ratio) for various channel materials while the dielectric is HfO_2(eox = 22), and gate length is 12 nm.

10.4.4 Electron Density per Sheet

Figure 10.6 illustrates the energy band profile that exists within a nanosheet field-effect transistor (NSFET) when subjected to different gate voltages (Vg). There is a separate Vg level represented by each line. When Vg increased, the energy bands moved upward in the spectrum. This illustrates the control that Vg exerts on the conductivity of the channel by changing the energy landscape of the nanosheet. However, in contrast to bulk transistors, the electron density in nanosheet field-effect transistors

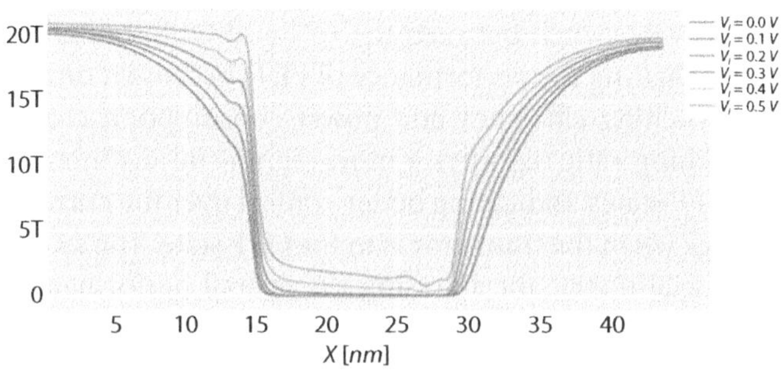

Figure 10.6 The energy band profile within a nanosheet field-effect transistor (NSFET) under varying gate voltages (Vg).

(FETs) is more complex. Electrons are not spread evenly across the channel because of their sheet-like shape; rather, they tend to concentrate close to the channel's innermost point. A single density measurement is not the only factor that engineers rely on to understand the quantity of electrons and where they are located within the nanosheet. They investigated the manner in which electrons were distributed over the width and thickness of the channel.

10.4.5 Current Spectrum

Figure 10.7 shows the current spectrum graph of a field-effect transistor (FET) composed of nanosheets. The channel of this particular form of transistor is achieved through the utilization of a nanosheet, which is a very thin layer of material that is only a few atoms thick. When the gate voltage is applied to a gate electrode, which is a conductor that regulates the flow of current between the source and drain electrodes, a gate voltage is applied. When the gate voltage was increased to a higher positive value, the graph demonstrated that the current flowing through the nanosheet increased. This is due to the fact that a positive gate voltage draws electrons to the channel, which in turn enhances the conductivity of the channel. Approximately −0.4 eV is the point at which the current begins to significantly increase, and it finally achieves its maximum value at approximately 0.2 eV. It may be deduced from this that the threshold voltage for this specific nanosheet FET is somewhere in the vicinity of −0.4 eV. The gate voltage for the channel is the threshold voltage, which is the voltage at which the channel begins to conduct a current. The shape of the curve in the graph can be influenced by a variety of parameters, including the material

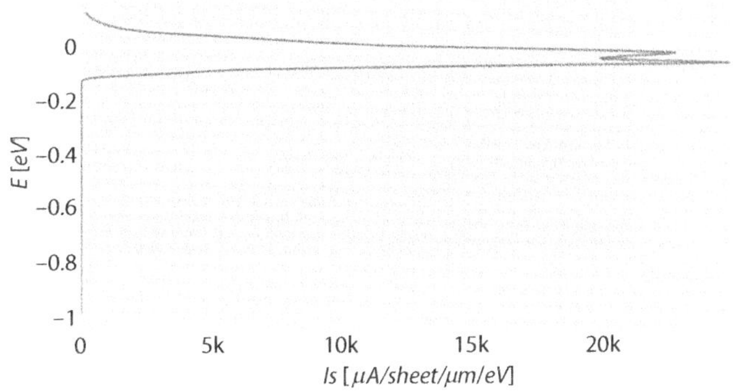

Figure 10.7 Current spectrum graph of a nanosheet field-effect transistor (FET).

properties of the nanosheet, material that serves as the gate dielectric, and geometry of the device. The current spectrum graph of a nanosheet field-effect transistor (FET) is an efficient means of characterizing the electrical properties of a device. The graph illustrates how the current flowing through the nanosheet changes as the gate voltage increases or decreases. For a wide range of applications, nanosheet field-effect transistors (FETs) can be designed and optimized using this information.

10.4.6 Density Spectrum

A nanosheet field-effect transistor (FET) has a density spectrum that shows the energy levels of electrons within the material, as shown in Figure 10.8. These energy levels are displayed on the x-axis and range from the valence band to the conduction band. The peaks and troughs that appear along the y-axis represent the density of states, which displays the number of possible energy states per unit volume. It is possible to recognize characteristics such as the valence and conduction bands, Dirac peaks (for graphene-based nanosheets), and doping effects. These distinguishing characteristics provide insights into the structure of electronic bands, the concentration of carriers, and optimization of device performance. It is helpful for researchers to understand these features because they allow them to design nanosheet FETs with improved functionality and efficiency.

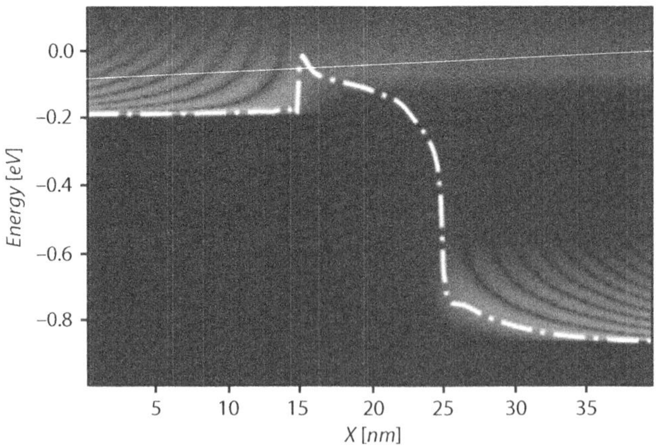

Figure 10.8 A density spectrum in a nanosheet FET.

10.5 Conclusion

In conclusion, our simulation-based study on Nanosheet Field-Effect Transistors (FETs) with Transition Metal Dichalcogenide (TMD) channel materials has provided valuable insights into their performance characteristics. By varying the gate length and dielectric materials, we comprehensively evaluated the key metrics, including the on-current (I_{ON}), off-current (I_{OFF}), and I_{ON}/I_{OFF} ratio. Our findings underscore the significant impact of TMD materials in enhancing the performance of nanosheet FETs. TMDs exhibit favorable electronic properties that contribute to improved carrier mobility and reduced scattering effects, thereby enhancing device operation. Additionally, the unique two-dimensional structure of the nanosheets offers inherent advantages over traditional FET architectures, such as superior electrostatic control and reduced short-channel effects, leading to enhanced Ion/Ioff ratios. Furthermore, our investigation of dielectric materials highlights their crucial role in determining the overall performance of nanosheet FETs. Dielectrics such as HfO_2, Si_3N_4, and SiO_2 influence the device's gate capacitance, leakage current, and subthreshold behavior, thereby directly impacting I_{ON}, I_{OFF}, and I_{ON}/I_{OFF} ratios.

The I_{ON} metric reflects the device's ability to conduct charge carriers effectively in the on state, which is crucial for high-performance transistor operation. Conversely, I_{OFF} represents the leakage current when the transistor is in the off state, which directly influences power consumption and device efficiency. The I_{ON}/I_{OFF} ratio serves as a key indicator of the transistor's switching efficiency, with higher ratios indicating superior performance in terms of power consumption and speed. Overall, our study underscores the importance of optimizing the gate length, dielectric selection, and channel materials to maximize the performance of Nanosheet FETs. By understanding and manipulating these factors, we can pave the way for the development of next-generation nanoelectronic devices tailored for advanced applications in computing, communication, and other fields.

Acknowledgments

We sincerely acknowledge nanohub.org for the experimental tool support. We also acknowledge Jing Guo, Ning Yang, Qimao Yang, University of Florida for the simulation tool support.

References

1. Sreenivasulu, V.B. and Narendar, V., Junctionless SOI FinFET with advanced spacer techniques for sub-3 nm technology nodes. *AEU Int. J. Electron. Commun.*, 145, 154069, August, 2022, doi: 10.1016/j.aeue.2021.154069.
2. Sai Kumar, A., Deekshana, M., Bharath Sreenivasulu, V., Prasad Somineni, R., Kanthi Sudha, D., Characterization for Sub-5nm Technology Nodes of Junctionless Gate-All-Around Nanowire FETs. *2022 13 th Int. Conf. Comput. Commun. Netw. Technol. ICCCNT 2022*, 2022, doi: 10.1109/ICCCNT54827.2022.9984269.
3. Kumari, N.A. and Prithvi, P., Device and circuit-level performance comparison of GAA nanosheet FET with varied geometrical parameters. *Microelectron. J.*, 125, 105432, 2022, doi: 10.1016/j.mejo.2022.105432.
4. Kumari, N.A., Sreenivasulu, V.B., Prithvi, P., Impact of Scaling on Nanosheet FET and CMOS Circuit Applications. *ECS J. Solid State Sci. Technol.*, 12, 3, 033001, 2023, doi: 10.1149/2162-8777/acbcf2.
5. Kumari, N.A., *et al.*, Performance Comparison of Nanosheet FET, CombFET, and TreeFET: Device and Circuit Perspective. *IEEE Access*, 12, 9563–9571, 2024.
6. Panigrahy, A.K., *et al.*, Spacer Dielectric Analysis of Multi-Channel Nanosheet FET for Nanoscale Applications. *IEEE Access*, 12, 73160–73168, 2024.
7. Sreenivasulu, V.B., Neelam, A.K., Kola, S.R., Singh, J., Li, Y., Exploring the Performance of 3-D Nanosheet FET in Inversion and Junctionless Modes: Device and Circuit-Level Analysis and Comparison. *IEEE Access*, 11, 90421–90429, 2023.
8. Sreenivasulu, V.B., Neelam, A.K., Panigrahy, A.K., Vakkalakula, L., Singh, J., Singh, S.G., Benchmarking of Multi-Bridge-Channel FETs Toward Analog and Mixed-Mode Circuit Applications. *IEEE Access*, 12, 7531–7539, 2024.
9. Sreenivasulu, V.B. and Narendar, V., Design Insights of Nanosheet FET and CMOS Circuit Applications at 5-nm Technology Node. *IEEE Trans. Electron Devices*, 69, 8, 4115–4122, 2022.
10. Bharath Sreenivasulu, V., Bhandari, S., Prasad, M., Mani, P., Subba Reddy, C., Durga Prakash, M., Spacer engineering on multichannel FinFET for advanced wireless applications. *AEU Int. J. Electron. Commun.*, 178, 155298, 2024.
11. Kumar, A.S., *et al.*, Nanosheet FET for Future Technology Scaling, in: *Integrated Devices for Artificial Intelligence and VLSI*, pp. 25–47, 2024.
12. Kumar, A.S., Rao, K.N., Sujith, A., Dhanuja, T., Vinay, M.V.S., Design and Implementation of 1KB SRAM array in 45 nm Technology for Low-Power Applications. *2023 3rd International Conference on Advances in Computing, Communication, Embedded and Secure Systems (ACCESS)*, India, pp. 245–250, 2023.

13. Kumar, A.S., Deekshana, M., Sreenivasulu, V.B., Kumari, N.A., Shanthi, G., Device Analysis of Vertically Stacked GAA Nanosheet FET at Advanced Technology Node. *2023 3rd International Conference on Advances in Computing, Communication, Embedded and Secure Systems (ACCESS)*, India, pp. 274–279, 2023.
14. Kumar, A.S., *et al.*, Nanosheet Field Effect Transistor Device and Circuit Aspects for Future Technology Nodes. *ECS J. Solid State Sci. Technol.*, 12, 8, 083009, 2023.
15. Sreenivasulu, V.B. and Narendar, V., Circuit Analysis and Optimization of GAA Nanowire FET Towards Low Power and High Switching. *Silicon*, 14, 10401–10411, 2022.
16. Goel, E. and Pandey, A. (Eds.), *Nanoscale Field Effect Transistors: Emerging Applications*, Bentham Science Publishers, UAE, 2023.
17. Goel, E., Impact of high-K gate stack on subthreshold performance of double-gate (DG) MOSFETs. *Silicon*, 14, 17, 11539–11544, 2022.
18. Babu, K.M.C. and Goel, E., Circuit Level Implementation of Negative Capacitance Source Pocket Double Gate Tunnel FET for Low Power Applications. *ECS J. Solid State Sci. Technol.*, 13, 5, 053011, 2024.
19. Chen, X., Liu, C., Mao, S., Environmental analysis with 2D transition-metal dichalcogenide-based field-effect transistors. *Nano Micro Lett.*, 12, 1–24, 2020.
20. Ilatikhameneh, H., Tan, Y., Novakovic, B., Klimeck, G., Rahman, R., Appenzeller, J., Tunnel field-effect transistors in 2-D transition metal dichalcogenide materials. *IEEE J. Explor. Solid-State Comput. Devices Circuits*, 1, 12–18, 2015.
21. Choi, W., Choudhary, N., Han, G.H., Park, J., Akinwande, D., Lee, Y.H., Recent development of two-dimensional transition metal dichalcogenides and their applications. *Mater. Today*, 20, 3, 116–130, 2017.
22. Zong, B., Li, Q., Chen, X., Liu, C., Li, L., Ruan, J., Mao, S., Highly enhanced gas sensing performance using a 1T/2H heterophase MoS2 field-effect transistor at room temperature. *ACS Appl. Mater. Interfaces*, 12, 45, 50610–50618, 2020.
23. Guo, J., Yang, N., Yang, Q., Nanosheet FET, 2023, https://nanohub.org/resources/nanosheetfet, (DOI: 10.21981/1R27-EN52).
24. Liu, L., Lu, Y., Guo, J., Monolayer MoS2 Field-Effect Transistors at the Scaling Limit. *IEEE Trans. Electron Devices*, 60, 4133, 2013.

11

MOSHEMT—Device Background, Materials, and Structures for Different Applications

Ananya Dastidar[1*], Tapas Kumar Patra[1] and Sushanta Kumar Mohapatra[2]

[1]School of Electronic Sciences, Odisha University of Technology and Research, Bhubaneswar, India
[2]School of Electronics Engineering, KIIT University, Bhubaneswar, India

Abstract

Scaling down the metal oxide semiconductor field-effect transistors (MOSFETs) to achieve exhibiting superior behavior and integration density is accompanied by various challenges. Leakage current with a reduction in gate dielectric and tunneling effects compromises device reliability. Integrating MOSFETs with other semiconductor technologies poses challenges in terms of compatibility and performance optimization. Therefore, addressing these issues is essential for advancing the MOSFET technology and ensuring its viability in future electronic systems. One non-classical device, the high electron mobility transistors (HEMTs), has several advantages over MOSFETs because it exhibits significantly higher electron mobility, enabling faster electron transport and improved high-frequency performance with higher switching speeds. HEMTs excel in high-power applications because of their high current-carrying capacity, making them ideal for power amplifiers in RF and microwave circuits. Additionally, HEMTs offer low-noise figures, especially in the microwave and mm-wave frequency domains, making them suitable for low-noise amplifiers in communication systems and sensitive measurements. The monolithic integration capability of HEMTs allows them to be easily unified with other semiconductor devices, thereby enabling the design of complex and high-performance integrated circuits. Furthermore, HEMTs based on materials such as Gallium Nitride (GaN) exhibit good radiation tolerance, making them suitable for space- and high-energy applications. Metal-oxide-semiconductor high electron mobility transistors (MOSHEMTs) combine the advantages of

Corresponding author: mananyad@yahoo.com

Ekta Goel, Archana Pandey, Shiromani Balmukund Rahi and Arun Samuel (eds.) Handbook of Advanced Semiconductor Field Effect Transistors, (217–254) © 2026 Scrivener Publishing LLC

both MOSFETs and HEMTs by offering improved gate control, allowing precise modulation of the channel and better control over device characteristics such as threshold voltage and transconductance. They also reduce gate leakage, enhancing the power efficiency by utilizing the gate oxide layer as a barrier against current leakage. Additionally, the gate oxide layer in MOSHEMTs enhances device reliability by protecting against hot-carrier effects and prolonging the device lifespan. MOSHEMTs can be seamlessly integrated with Complementary MOS (CMOS) technology, enabling complex systems that leverage the strengths of both MOSFETs and HEMTs, such as power amplifiers with digital control circuits. Furthermore, MOSHEMTs exhibit enhanced performance in terms of power handling capability, speed, and linearity, which is attributed to the enhanced channel modulation enabled by the MOS structure. This chapter presents the MOSHEMT structure, working principle, and its use in different application domains, particularly for biosensing.

Keywords: MOSFET, HEMT/MOSHEMT, dielectric modulation, biosensor, biomarker, sensitivity

11.1 Classical MOSFETs and their Issues

MOSFETs are warhorses in the field of Very Large-Scale Integration (VLSI), and the sole cause behind this is that MOSFETs are scalable units based on the basic concept of VLSI, i.e., integrating millions of transistors and electrical components into a compact silicon area. MOSFETs [1] have evolved from single-gate bulk structures to multi-gate structures. These devices have issues related to subthreshold leakage, DIBL, oxide breakdown, Fowler–Nordheim Tunneling (FNT), band-to-band-tunneling (BTBT) and threshold voltage shift [1, 2]. A MOSFET is a device in which the metal forms the gate terminal, the oxide layer is formed by an insulator (oxide), and a semiconductor material or compound makes up the substrate. With the rapid increase in the small geometry effects owing to the decrease in device dimensions, newer structures evolving out of conventional MOSFETs have been explored. Reported structures include the Multi-Gate (MUG) MOSFETs [3] classification, including the Single-Gate (SG) Silicon-on-Insulator (SOI) MOSFETs [4], Double-Gate (DG) MOSFETs [5], Fin-Field Effect Transistor (FinFET) [6] and the Cylindrical Surrounding-Gate (CSG) MOSFETs) [7], which have been able to, arrest the issues related to Short Channel Effects (SCE) and provide improved current drive capacity. The primary reason for this is the channel length, which is instrumental in determining the threshold voltage variation in MOSFETs. One of the structures that can reduce the SCE of the MOSFET is a heterostructure known as a high electron mobility transistor (HEMT).

11.2 HEMT and Its Challenges

Forty years have passed since the announcement of the first HEMT by Fujitsu Electronics developed by Mimura *et al.* [8, 9]. Conventional HEMTs were developed using Gallium Arsenide (GaAs) [10], but research has now focused on other binary compounds such as the high-frequency Indium Phosphide (InP) [11] and Gallium Nitride (GaN) [12] for efficient low-power designs. The HEMT is a multilayered stack of dissimilar semiconductors that forms a heterojunction based on the physics of electrons confined in ultrathin conductive layers. Unlike the classical PN junction known as a homojunction, the heterojunction can be composed between of two p- or two n-type semiconductors, of which one is a wide bandgap (WBG) material and the other is a narrow bandgap (NBG) material. This type of HEMT is known as the isotype. On the other hand, when the heterojunction is formed using a WBG n-type semiconductor and an NBG p-type semiconductor, or vice versa, we obtain an aniso-type HEMT. To develop this junction, one of the important parameters that must be considered is lattice match. If the lattice match between the materials that make up the junction is not maintained, cracking may occur. Even a lattice constant difference of 0.5 between the two materials, will cause atoms of the junction materials to be adjusted to obtain a single crystal structure, thereby increasing the number of defects at the junction. This issue is more pronounced when the layers are thinner (0.5 μm or fewer). The ultimate effect of a defective heterojunction is reduced mobility, which, if increased by doping, will lead to stress. Despite their many qualities, HEMT fabrication faces many challenges [13]. Defects were generated at high voltages at the barrier and near the gate border, resulting in a reduced gate current. The amount of defect generation is supplemented by the presence of mechanical stress, which leads to an increase in leakage currents, thereby affecting the reliability of the HEMT [14]. Therefore, to develop a good heterojunction with enhanced mobility, proper lattice matching is necessary.

HEMTs are heterostructures of different materials and band gap energies, where buffered layers are employed to reduce lattice mismatch. However, defects and traps in the structure make it challenging to mitigate gate leakage current, minimize noise, and enhance drain current, thereby impacting commercial use and reproducibility. AlGaN/GaN HEMTs encounter self-heating within the conducting channel and require a minimum gate current to remain ON. The gate leakage current increases with the device temperature owing to surface traps [15]. At high frequencies, dual-gate AlGaN/GaN HEMTs offer immunity to short-channel effects, and multiple 2DEG channel regions enhance transport characteristics and linearity. Double-gate (DG) HEMTs offer improved drain characteristics in comparison to single-gate (SG) HEMTs, including improved transconductance, capacitance, cutoff frequency, subthreshold slope, on-resistance, and sensitivity. The low-noise characteristics with the reliable low-temperature operation of HEMTs are currently a driving technology in optoelectronics applications, including optical communication, radio communication, Radio Frequency (RF) communications [16, 17], millimeter-wave infrastructure design [18], and navigation systems. HEMT are also used in power devices [19, 20] for amplifier design [21]. With the ever-increasing speed of communication, device engineers have directed their interest in the design of higher-frequency and faster devices. The suitability of the HEMT as a biosensor arises from its intrinsic properties of enhanced electron mobility, 2DEG, chemical inertness, and improved speed [22, 23]. However, the SCE at low frequency and leakage are in the order of $\sim mA$ [24]. Therefore, it is necessary to consider device optimization to address these issues in the HEMT.

11.3 MOSHEMT

MOSHEMTs were developed by combining the advantages of both MOSFETs and HEMTs to address the issues faced by these devices independently. Increasing the efficiency of the device is possible as it is possible to scale down the MOSHEMT structure. Achieving a higher frequency of operation up to 640 GHz has been reported in [8]. The use of different insulators and high-dielectric (high-κ materials helps to overcome gate leakage. MOSHEMTs offer high-speed capabilities, making them valuable for applications in high-speed communication systems, radars, sensing, and optoelectronics. The MOSHEMT structure typically consists of lattice-matched (LM) wideband semiconductor ternary/binary compound layers (barriers) such as AlGaN/AlGaAs/InAlN/AlN and narrowband binary semiconductor compound layers (channels) such as GaN/GaAs/InP. The gate oxide in MOSHEMTs is made of high-κ oxides, such as titanium oxide (Ti_2O_3)/aluminum oxide (Al_2O_3)/hafnium oxide (HfO_2). These devices are typically fabricated on silicon (Si), silicon carbide (SiC), or sapphire substrate. The following subsections discuss the working principle, different layers, and structural engineering for different application domains of MOSHEMT.

11.3.1 Working Principle

The heterointerface formed between a wide- and a narrow-bandgap material is the key to the operation of the MOSHEMT as seen in Figure 11.1. At this interface, 2DEG was formed which is characterized by high electron mobility, which allows efficient electron transport. The structure of a MOSHEMT consists of a gate electrode, gate dielectric layer, and the source–drain (S/D) junctions. Applying a positive bias to the gate generates an electric field across the oxide layer, thereby modulating the 2DEG electron density. Varying the gate voltage controls the electron concentration within the channel, allowing precise control of device characteristics. In AlGaN/GaN MOSHEMTs, the junctions create a conductive channel from the electrons in the 2DEG, thereby allowing current flow. The combination of high electron mobility and the wide bandgap of the channel material makes MOSHEMT an efficient and robust device that is appropriate for a diverse array of applications.

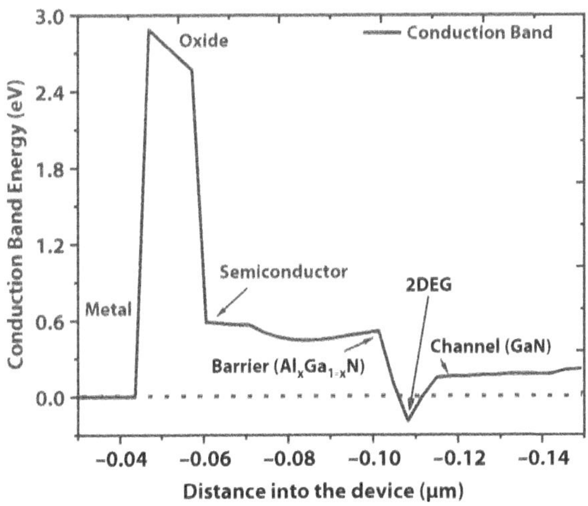

Figure 11.1 The 2DEG formation in MOSHEMT heterointerface.

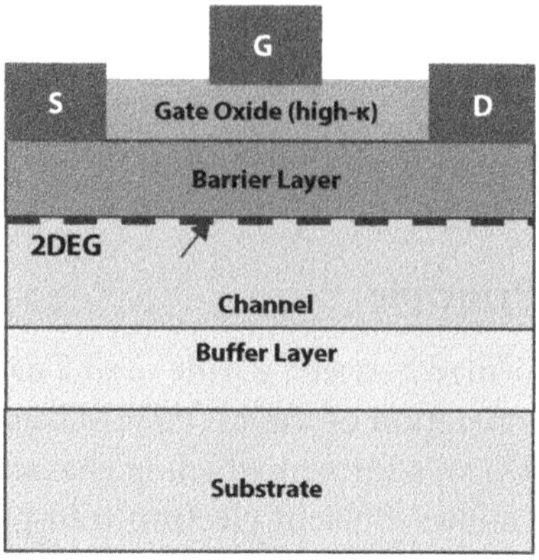

Figure 11.2 General structure of a MOSHEMT.

11.3.2 MOSHEMT Structure

The MOSHEMT structure is comprised of a heterostructure grown on a compatible substrate. It integrates a gate electrode, gate oxide layer, and source–drain (S/D) region.

The gate electrode, which is often constructed from metal, is isolated from the heterostructure by a thin insulating layer called gate oxide. The S/D regions were established by the substantial doping of the GaN layer on both sides of the heterostructure. The general structure of MOSHEMT is shown in Figure 11.2.

11.3.3 Substrate

The substrate provides the foundation upon which the MOSHEMT is built, and the choice of substrate material significantly influences the device's performance based on specific application needs. A widely used substrate material is silicon [25] owing to its favorable properties, such as enhanced resistivity, low cost, abundant availability in nature, low current collapse, and very high breakdown voltage (~1,528 V), but with a ~μA/mm leakage current [26]. SiC [27] is another material that has superior device properties, exhibited lower leakage currents owing to the reduced dislocation density, and exhibited a high on/off ratio ($\sim > 10^{10}$), but its high cost becomes a bottleneck for its choice [28].

Sapphire, which is somewhat lesser expensive, has poor thermal conductivity [29]. In InP, the enhanced electron mobility is caused by the 2DEG [30], which is used for its high selectivity over InAlAs. The impact of the use of InP as a substrate material was presented in [31], where a heterostructure based on InP displayed reduced noise, lower power dissipation, high frequency, and superior gain characteristics. This makes InP-based devices suitable for a plethora of domains such as sub-mm, THz, cryogenic, sensors, biomedical, Low Noise Amplifier (LNA), and digital logic applications [31]. Noise in semiconductor (SC) devices occurs at the SC/insulator interface owing to fluctuations in carrier density resulting from channel/defects charge exchange. The performance of heterostructures on GaAs substrates shows higher values for very small gate lengths, with higher transconductance, but InP substrates showed lower values for similar gate lengths.

11.3.4 Gate Dielectric

The gate insulator or gate oxide, acting as an insulator between the gate junction and the 2DEG, is a critical component whose selection significantly affects the MOSHEMT performance. Compared with HEMTs, MOSHEMTs benefit from the presence of an oxide layer, which reduces the leakage current. Using high-κ oxides like Al_2O_3 [32], HfO_2 [33, 34], ZrO_2 [35, 36], TiO_2 [37], HfSiOx [38], and ZnO [39] as gate dielectrics, along with metal gates, enhances the gate field effect. In addition, thicker dielectric layers help reduce gate leakage. Researchers have explored various structural engineering approaches, such as modifying the gate [40], adjusting the dimensions of different layers (oxide layer [41], barrier layer [5, 42], spacer layer [43]), channel engineering [38], and threshold voltage engineering [44], to improve or understand their impact on the MOSHEMT performance for various application domains.

The introduction of an AlGaN Barrier results in a drop in the leakage current (I_{leak}) between the junctions, as demonstrated by Lee *et al.* in InAlN/GaN MOSHEMTs with a reported $I_{leak} \sim 10^{-10}$ A/mm by employing SiO_2 as the gate oxide material [45]. The use of Al_2O_3 was used to facilitate the mitigation of device degradation by minimizing the impact of current-collapse effects in MOSHEMTs [46]. There are reports of a reduction in the leakage current by seven times, reduced traps, and improvement of its electrical characteristics with a slightly higher gate-to-drain current, thereby making MOSHEMT a potential candidate for power electronic applications [47]. An investigation based on the 2D Technology Computer-Aided Design (TCAD) tool has shown that introducing a thin layer of insulator under the gate is not a fool-proof method for avoiding the collection of holes at the gate; however, this issue is greatly improved by the introduction of a gate oxide. Hole collection has been observed at the gate even in an off-device with HfO_2 gate dielectric but has been reduced significantly with the use of Al_2O_3 owing to its wider bandgap [48]. Freedsman *et al.* presented E-mode Al_2O_3/AlGaN/GaN MOSHEMTs-on-Si. The authors claimed that the presence of negative charges in the oxide interface aids in the enhanced operation of the device. The reported device exhibited an elevated drain current along with reduced leakage and heightened breakdown voltage, making it suitable for high-power applications [49].

The use of Titanium Dioxide (TiO_2) films on GaN layers shows that the current density of the MOSHEMT is superior to that of HEMTs with a wider gate voltage swing. The transconductance and threshold voltage of both structures remained similar, while the results indicated a successful reduction of both the leakage current density and current collapse for MOSHEMTs [50]. The gate current was influenced by the oxide/barrier material, as reported by Rawat *et al.*, who demonstrated gate leakage reduction in fabricated TiO_2/AlGaN/GaN MOS-HEMTs. The TiO_2/AlGaN-based MOS-HEMT showed improved DC and RF characteristics compared to the conventional HEMT. Lower subthreshold leakage and a larger I_{ON}/I_{OFF} ratio were found in the drain characteristics of MOS-HEMTs [51]. They also show improvements in the breakdown voltage, drain current, and I_{ON}/I_{OFF} ratio, proving their suitability in high-power RF circuits [52]. The use of TiO_2 as a gate insulator resulting in a low-leakage device was reported in [37]. There are very few interface traps in TiO_2/GaN MOS capacitors. Meng *et al.* proposed an AlGaN/GaN MOSHEMT with a TiO_2/NiO dielectric that resulted in a lower gate leakage current, higher maximum current, and drain breakdown voltage than a similar geometry HEMT. Although the maximum transconductance of the devices is equal, owing to the high-k gate oxide layer, it resulted in a minor decrease in the current gain and cut-off frequency in the MOSHEMT than in the HEMT structure. This makes MOSHEMT promising for microwave amplifier design in communication and radar systems [53, 54].

Gadolinium Oxide (Gd_2O_3) is an inorganic rare earth material that acts as a gate oxide [55]. Gao *et al.* used Gd_2O_3 as a gate dielectric that is capable of withstanding electrical degradation to the gate thereby, make it a good choice for robust designs in radio-active environment use [55]. The thermal stability test of an AlGaN/GaN MOSHEMT with Gd_2O_3 as its gate insulator [56] exhibited lower leakage and better thermal stability than traditional HEMT and traditional diodes, making them suitable for high-temperature applications [56]. Another material selected as a gate dielectric is high-k material Zirconium Dioxide (ZrO_2), which has been found to exhibit low hysteresis and high breakdown voltage, effective suppression of current collapse [36], and reduced parasitic capacitances [35] in different MOSHEMT structures. The use of Magnesium Calcium Oxide

(MgCaO) as a gate resulted in a significant improvement in the drain current and on/off ratio with a lower subthreshold. Kumar *et al.* demonstrated that tantalum pentoxide (Ta_2O_5) as a high-κ oxide exhibited gate leakage in nA/mm [57]. MOSHEMTs help suppress leakage currents, which is a bottleneck in HEMT structures. The use of Scandium Oxide (Sc_2O_3) as a gate dielectric and passivation layer in MOSHEMT was reported by Mehandru *et al.* [58]. Sc_2O_3 helps to maintain low surface states and aids in the effective elimination of current collapse in comparison to its HEMT counterpart. Lee *et al.* used ZnO as a gate dielectric to show superior performance to conventional HEMTs [59]. The presence of traps at the oxide/semiconductor edge adversely affects device performance owing to the capability to trap charge carriers. Therefore, the reduction of its effects using different high-κ dielectrics has been reported in many studies.

A comparison of the HEMT and MOSHEMT with HfO_2 as a gate insulator at room temperature under various thermal and stress conditions for electrical stability and stress analysis showed that the device exhibited improved electrical and thermal performance with low leakage and higher drain current because the HfO_2 layer can prevent the oxidation of GaN, thereby preventing structural deformation [60]. A ternary compound, $HfLaO_x$, comprising HfO_2 and La_2O_3, was used as the gate dielectric in [61], where the device was found to display a high breakdown voltage of 840 V. The $HfSiO_x$ layer has been reported to have a significantly improved breakdown lifetime compared to its counterpart with an undoped gate dielectric layer of HfO_2 [33]. The reduced surface density is the outcome of a high-quality oxide/barrier interface that has been obtained by chlorine surface treatment, as reported by Chiou *et al.* [39]. Improvement of the interface layer performance by reducing the trap density was presented by Liao *et al.* using a thermally oxidized HfO_2/Y_2O_3 gate dielectric layer. The results exhibited excellent drain-suppressed current collapse [62]. The effect of the oxide/barrier interface states (N_{ox}) on the threshold voltage was investigated in [63], where the developed model was validated for different insulator materials and N_{ox}. The change in the threshold voltage is more pronounced for N_{ox} at higher insulator thicknesses, whereas it is negligible otherwise. A normally off action was realized by modulating the amount of N_{ox} in the device with Al_2O_3 or HfO_2 as gate insulators [63]. Liu *et al.* reported the use of H_2O_2 oxidation for the growth of aluminum oxide (AlO_x) on an AlGaN barrier. The fabricated device has been reported to show enhanced performance in terms of drain current, leakage current, transconductance, and breakdown voltage, making it suitable for low-noise and high-power applications [64].

11.3.5 Barrier

Another layer whose material selection affects device performance is the barrier layer. This layer helps confine the electrons within the device. The barrier comparison of lattice-matched InAlN/GaN and the AlGaN/GaN HEMT structures showed that owing to a larger bandgap offset, the InAlN/GaN showed a higher 2DEG, and because of its thinner barrier, it had lesser short-channel effects. However, one issue faced by the InAlN/GaN structure is the high gate leakage and interface trap owing to its thin barrier, which in turn degrades the device's off-state performance, resulting in a decreased off-state breakdown voltage [65]. Analysis of DC reliability parameters for different MOS-based FET/HBTs by Putcha *et al.* showed that MOSHEMT/MISHEMT exhibited robust architecture owing to the barrier layer, which lessens the impact of degradation mechanisms [66].

The introduction of a barrier is an important advantage of the MOSHEMT structure compared to the MOSFET structure. The results obtained showed an improved f_T of 200 GHz for a 100 nm gate length of the PM HEMT, which is 20 GHz higher than that of the LM MOSHEMT [67]. Tiwari *et al.* presented a simulation of a wideband $Al_xGa_{1-x}N/GaN$ MOSHEMT, where with a higher thickness of AlGaN, the device displayed better RF performance than with a lower thickness, owing to the elimination of SCE. In the case of higher thicknesses, the maximum f_T has been reported to be high, and the DC performance of the device with a thicker AlGaN layer shows better performance than that with a thinner AlGaN layer.

So, it can be said that short channel effects are more pronounced for thicker AlGaN layers in comparison to the thinner ones [68]. Yadav *et al.* demonstrated the effects of a barrier layer, delta doping, and mobility on a fabricated device [69]. Mim *et al.* presented that barrier thickness variation governs the gate. The increase in capacitance stems from the carrier population near the upper edge of the barrier region [70]. Ternary compound material InAlN has been found to show the highest drain current, while a device with an InGaP barrier showed the lowest drain current. AlGaN, which is a widely used material in GaN/AlGaN heterostructures, exhibits a consistent performance. The variation in gate capacitance with barrier thickness in the nanoscale regime for a GaN-based MOSHEMT in an AlN/GaN MOSHEMT showed a remarkable reduction in gate capacitance when the thickness of the barrier increased in comparison to the MOSHEMT, which showed an enhancement in the RF characteristics of the device while reducing the device propagation delay [71]. An InGaN back barrier of a MOSHEMT was showcased in prior studies of [72] using a TCAD tool. The back barrier resulted in improved DEG confinement and improved SCE, analog, and RF performance of the device, thereby sealing their candidature for upcoming analog and RF applications [72].

11.3.6 Other Layers

In many MOSHEMT structures, a buffer layer alleviates the effects of lattice mismatch between the substrate and subsequent layers. This helps to improve the quality of the subsequent layers and reduce defects. This layer isolates the active device from any material defect present near the substrate interface. The most widely used material as a buffer is GaN [73], followed by binary compounds like AlGaN [25], InP [10], and ternary compound InAlAs A channel layer, is needed for providing a conductive path for electron flow. In some designs, a spacer layer is used to further augment the confinement of electrons and improve the device performance.

Cap layers are used to reduce the dispersion effects near the barrier interface by negative polarization charge compensation near the cap/barrier interface. Nitrides and oxides have been widely used as capping materials, such as GaN [74], SiN [25], and Al_2O_3 [75]. The other layer is the nucleation or interlayer, which helps to reduce the interface roughness and improve the 2DEG mobility, and mostly InGaAs [10], GaN [46], and AlN [54] have been used for this layer. The thickness of this layer helps to provide the necessary stress, thereby suppressing the cracking of the subsequent layers.

S/D are heavily doped regions formed within the channel layer on either side of the 2DEG. These regions facilitated the injection and extraction of electrons into and from the channel, respectively. S/D regions are typically formed using techniques such as ion implantation or selective-area epitaxy.

The semiconductor industry is currently shifting its focus from pure silicon-based devices to semiconductor compounds, such as GaAs [76], GaN [77], and InP [78]-based devices. GaN-based MOSHEMTs are gaining attention for multiple target applications owing to their superior performance compared with MOSFETs or HEMTs based devices. This shift opens up new possibilities for the design, manufacture, and production of advanced devices. Future high-power GaN MOSHEMT devices will enable faster and more reliable data transmissions across radio channels. Moreover, GaN MOSHEMTs have applications in imaging radar systems [79], thereby enhancing the accuracy of autonomous systems. Additionally, these devices can serve as high-resolution and precision biosensor systems [80], augmenting medical diagnosis and contributing to advancements in biomedical research and development.

11.4 MOSHEMT Structural Engineering

11.4.1 Single Gate and Gate Stack MOSHEMT for RF Applications

A Single Gate (SG) MOSHEMT has a simple HEMT-like structure with a single gate. Studies using these structures are usually based on comparisons with conventional HEMT as well as models for different applications. The DC characteristics of a conventional HEMT with a high-κ oxide layer (HfO_2) at the gate/barrier layer were enhanced to obtain an improved MOSHEMT structure. A comparison of the DC and RF parameters between AlN/GaN HEMTs and MOSHEMTs with varying oxide and barrier layer dimensions revealed that the MOSHEMT exhibits near-zero current flow between the G/S terminals. This characteristic renders the device well suited for low-power applications [81]. Investigations of the critical thickness of the InAlN barrier layer in an ultrathin normally off LM In0.17Al0.83N/GaN MOSHEMT with a 10 nm oxide layer and a 3 nm barrier device to obtain a threshold voltage of 1 V show the feasibility of the structure for power electronic applications [82]. The conduction band level increases and confinement issues close to the channel are eliminated by using an AlInN back barrier, which is modeled using numerical analysis by employing a Cubic Spline Interpolation (CSI) technique [42].

A gate stack structure includes a stack of metals for the fabrication of the gate. A gate stack MOSHEMT with different gate dielectrics, such as SiO_2 and HfO_2 with a thin AlGaN barrier is insufficient to form an adequate 2DEG at the heterointerface at zero gate bias. Therefore, the dielectric arrangement of the HfO_2/SiO_2/AlGaN/GaN MOSHEMT structure is used to achieve a better transconductance and a positive threshold voltage, making it suitable for power electronic applications, while the SiO_2/HfO_2 AlGaN/GaN structure has been shown to have a cut-off frequency of 16 Ghz, thereby making it suitable for RF applications [83].

A stacked gate dielectric-based AlGaAs/InGaAs MOSHEMT was fabricated using sputtering and ozone water oxidation, which demonstrated excellent switching characteristics, such as improved SS and I_{ds}. In comparison to its HEMT counterpart, MOSHEMT exhibited a positive shift in I_{ds} and g_m owing to passivation effects [84].

11.4.2 Dual/Triple Gate and Recess Gate MOSHEMT for Power Electronic Applications

Dual-gate MOSHEMT structure [85] and a triple-gate nanowire E-mode InAlN/GaN MOSHEMT on silicon reveals its huge potential in areas of power device applications [86]. The integration of p-GaN, triple-gate, and a MOS exhibited positive threshold voltage with lower R_{ON}. p-GaN uses band engineering to minimize the density of carriers in the triple-gate structure, which is used for achieving gate control, thereby aiding E-mode operation. The MOS section adds flavors with excellent switching characteristics. The tri-gate device exhibited superior channel control, reduced leakage, higher $\frac{I_{ON}}{I_{OFF}}$, and low SS as compared to other reported planar p-GaN devices, thereby making these devices prospective candidates for power electronics applications [87].

When a section of the barrier is etched to grow the gate, we obtain a recess gate structure. A recessed-gate InAlA/InGaA-MOSHEMT employing a thin oxide layer produced through the ozone process resulted in enhanced transconductance characteristics [30]. Lattice Matched (LM) MOSHEMTs with a recess gate self-alignment fabricated by thermal oxidation (dry) showed a current and density of approximately 2,400 mA/mm and 6,000 mW/mm, respectively, with a f_{max} of 112 GHz. Such devices show promise for reliable operation at high frequencies [73]. Surface damage causes threshold voltage instability and can be eliminated by KOH passivation, as reported for GaN-on-Si MOSHEMTs with gate recess [19]. Fabrication of normally off recess gate structures can be achieved by employing LPCVD Si_3N_4 passivation techniques [88], and the fabricated structure exhibited a high BDV of 1,400 V. Parallel oxidation followed by dissolving treatment in a hybrid alkaline solution consisting of H_2O_2 and KOH during the fabrication process, followed by wet etching was employed to show the normally off operation of the Al_2O_3/GaN MOSHEMT. The results showed a high threshold voltage and breakdown voltage. The gate recess method is claimed to be appropriate for low damage and inexpensive fabrication [89]. Electrochemical wet etching and oxidation methods are being used for the fabrication of a recess gate MOSHEMT [90]. A recessed gate AlGaN/GaN MOSHEMT with Al_2O_3 as the dielectric fabricated by a low damage process and owing to the high-quality Al_2O_3 and the passivation techniques employed, it is possible to reduce the DC-RF dispersion [91].

11.4.3 Asymmetric Gate MOSHEMT for High Frequency Applications

An AlN/GaN heterostructure with a HfO_2 gate insulator and a T-gate exhibited a gain of 4.6 dB at 17% power-added efficiency, which makes the device capable of use in mm-wave applications [92]. Another T-gate MMOSHEMT exhibited improved performance owing to a reduction in the length and resistance of the gate [93]. A simulated T-structured gate E-mode MOSHEMT with a TiO_2 dielectric on a SiC substrate reported a high f_T and f_{MAX} of 524 GHz and 758 GHz, respectively [94].

A quad gate and quad T-gate structure in fin-nanochannel array MOSHEMTs [95] in comparison to single-gate structures showed improved drain current, improved maximum transconductance, increased cut-off and maximum oscillation frequency, and reduced parasitic resistance. An InAlN/AlN/GaN on SiC MOSHEMTs with a T-gate using a TCAD tool showed that the use of an AlN back barrier aids the device breakdown voltage and suppresses gate leakage towards an enhanced device with small and large signal performance [96]. The I_{ds} of the device was found to be 2,700 mA/mm, gm of 920 mS/mm, and with a SS of 35 mV/dec and break down voltage of 81 V with leakage current $\sim 10^{-11}$ A/mm and further it was found that f_T and f_{MAX} were, respectively, 426 and 366 GHz, respectively [96]. The AlGaN/GaN MOSHEMTs using lattice-matched magnesium calcium oxide as gate insulators, where the gate had a T-structure [97], showed decreased leakage currents exhibiting improved RF performance.

The Γ-gate MOSHEMT fabrication based on AlGaAs/InGaAs was reported in [98] using techniques such as ozone water oxidation and shifted exposure. A comparison between a Schottky gate, normal MOSHEMT, and a Γ-gate MOSHEMT showed that the Γ-gate MOSHEMT outperformed the other two on almost all device parameters such as maximum drain current, transconductance, breakdown voltage, and small-signal parameters [98]. An asymmetric π-Gate MOSHEMT was demonstrated in [99] using a TCAD tool where the structure exhibited better RF frequency characteristics than its HEMT counterpart.

11.4.4 Double-Gate and Double-Channel MOSHEMT for Low Leakage Performance

The use of structures with two gates, i.e., a double-gate (DG) structure, has been reported to exploit the benefits of two 2DEG, more gate control, and low leakage. A DG InP/InGaA structure has been presented in [100], where TCAD simulations have been conducted for the gate and barrier dimension variations. The proposed device exhibited improved device parameters except for *SS* and $\frac{I_{ON}}{I_{OFF}}$. A double-delta-doped DG-MOSHEMT with a multi-cap layer, with a thin InAlA barrier, and HfO_2 as a dielectric material was used in the high-frequency domain. The obtained parameters of this structure when compared with an SG-HEMT structure, maximum transconductance, and higher cut-off frequency were reported in the former structure [101]. Simulation of an underlap double-gate (U-DG) MOSHEMT with HfO_2 as the gate oxide indicated improved analog performance over a Schottky HEMT, thereby making it suitable for high-power enhanced frequency devices [102]. Double gate structures are better at suppressing SCE than SG structures double metal double gate (DMDG) MOSFET devices but scaling the device affects device performance [103].

A symmetric U-DG MOS-HEMT with a doped AlGaN barrier and spacer of varying width and its effect on the different analog metrics showed that scaling down the doped region width leads to improved gate control as well as a high ON current [41]. A DG n-AlGaN/GaN MOSHEMT, where the doping concentration of the AlGaN was manipulated, and the device performance was compared with a conventional HEMT as well as a MOSFET device [104]. While higher doping showed higher drain current and improved transconductance, lower doping showed better RF performance, showcasing the device's potential for low-power communication [104]. A DG underlap AlInN/GaN MOSHEMT for Analog and RF performance measurements using a TCAD tool shows that through the gate and underlap dimension scaling, there is an increase in the device parameter values such as drain current, transconductance, and frequency, but it caused a drop in the on-resistance of the simulated device. These findings suggest that the device is viable for high-power and high-frequency applications [105]. An increase in the effective barrier thickness resulted in an increase in I_{ds} and SCE, similar to DIBL and SS. Moreover, a reduction in delay and a negative threshold voltage shift were observed. A decrease in the barrier thickness resulted in an increase in the gate capacitance, output resistance, transconductance, and improved cut-off and maximum oscillation frequency [5].

In addition to gate structure engineering, channel engineering for fabricating more than one channel to produce a double-channel (DC) MOSHEMT has also been reported. A normally off GaN DC-MOSHEMT with an AlN insertion layer under the barrier/GaN interface creates an additional channel at the AlN-insertion layer/GaN interface [106], where the top and bottom channels are a MOS channel and a hetero-channel, respectively. This double-channel MOSHEMT exhibits a high breakdown voltage and a steep subthreshold swing. Multiple-channel GaN MOSHEMTs for sensing applications consider on-current, transconductance, and leakage currents as the sensing performance parameters [107]. Incorporating an MOS field plate as an electric field shield at the drain of a double-channel reverse-biased AlGaN/GaN MOSHEMT on a Si substrate, combined with a metal-2DEG Schottky contact towards the drain, aids in enhancing the device's blocking capacity during its off-state. This improvement is achieved while maintaining low leakage, along with reduced turn-on voltage and on-resistance [108]. The use of double-channel structures with gate insulators has proved to have better performance as E-mode transistors. Using a DC structure does not ensure a low on-resistance if the electrical coupling between the two channels is given proper consideration because the channel-to-channel coupling reduces the on-resistance owing to current convergence between the two channels. For two devices with channel-to-channel distances of 7.5 nm and 11.5 nm, respectively, a lower R_{ON} has been reported for the former, while a higher sheet resistance has been reported for the latter. Thus, the lower the thickness, the greater the electrical coupling and the less the R_{ON} [109]. The inclusion of a thin AIN interlayer under the channel layer supplements the channel formed by the creation of a second channel, thereby impacting the gate control on the channel. Moreover, the effect of the recess gate on the Vth distribution was negligible if the recess depth did not cross the upper channel boundary.

Different gate oxide stacks such as HfO_2, SiO_2, and HfO_2-SiO_2 stacks have been presented [41], where the HfO_2-SiO_2 gate stack structure showed performance intermediate to only SiO_2 and only HfO_2 gate oxide structure because of the increased overall capacitance, which improves the drain current and transconductance performance. HfO_2, as an oxide material with a high dielectric constant ($k \sim 25$), provides the advantage of reduced leakage [110]. The performance evaluation of a 100 to 300 nm channel length MOS-HEMT with an S/D underlap for RF applications is presented.

A double-gate device with a double heterointerface helps to achieve a higher on-current and better device linearity [111], where the authors have used Al_2O_3 as the gate dielectric, which provides a greater conduction band offset to GaN (~1.96 eV), although it possesses a comparably lower dielectric constant ($k \sim 9$). Scaling the L_G. from 0.8 μm to 100 nm enhances the performance over SG-MOSHEMT. A preference for high-k materials is necessary for channel controllability, where a substantial conduction band offset is essential to mitigate the gate leakage current.

11.4.5 Novel MOSHEMT Structures

While single-gate and recess-gate structures have ruled for years, trends are now towards multi-gate and fin-gated structures. The recent Fin MOSHEMT is a structure whose impact on device performance has been demonstrated by using AlGaN/GaN fin-MOSHEMTs on silicon for different fin widths using MgCaO as the gate dielectric. Apart from the minimal I_D-V_{GS} hysteresis and current collapse, fin-MOSHEMT is a promising contender for power applications [112]. A detailed fabrication process of the fin-MOSHEMTs has also been presented [113]. A triple-gate nanowire MOSHEMT with Al_2O_3 as the gate insulator exhibited superior electrical performance, steep SS, and large BDV with lower leakage [114]. The effect of the fin width on the triple-gate AlGaN/GaN MOSHEMT's parameters, such as the threshold voltage, ON-resistance, and current capability [115], has also been observed. GaN-based HEMTs feature properties such as a wide bandgap, higher electron mobility, and a high breakdown field, making them suitable for power electronic applications.

However, owing to the 2DEG density, conventional GaN HEMT inherently operate in a normally on state. On the other hand, small channel widths are sufficient for a conventional tri-gate (FinFET) GaN HEMT to achieve normally off operation [116]. A comparison of the performance of planar, submicron-fin, and nano-channeled fin MOSHEMTs and the effect of scaling the fin channels have also been investigated. It was found that gate control improved the fin-gated channel AlGaN MOSHEMT by decreasing the width of the channel, thereby decreasing the lateral electric field effect in comparison to its planar counterpart. Device electrical parameters as well as the frequency of operation of a narrower channel array showed improvement with a reduction in the channel width [117]. Many new structures have been proposed for different applications, such as the recently reported structure of Circular MOSHEMT (C-MOSHEMT) for pH detection. All the structures reported earlier are linear MOSHEMTs (L-MOSHEMTs), while the gate all-round structure is known as a Circular MOSHEMT(C-MOSHEMT). C-MOSHEMT reported an increase in the device sensitivity by 41% for an increased sensing area (width of the device) [118, 119].

For different applications, structural engineering and gate engineering are important research areas for obtaining optimized and enhanced device behavior towards the application domain.

11.5 MOSHEMT for Biosensing Applications

11.5.1 Biosensors

Biosensors have a rich history, beginning with Clark, who introduced the first biosensor for glucose detection in 1962 [120]. Advancements in biotech, nanotech, and electronics have improved the biosensors' design and performance. Enzyme-based biosensors became prominent in the 1980s and the 1990s, transforming medical diagnostics and environmental monitoring. The 21st century witnessed the rapid development of DNA and immunosensors. Miniaturization and wearable biosensors enable real-time health monitoring. They have applications in medical diagnostics, environmental monitoring, food safety, and biotechnology. The use of semiconductor devices for biomolecule detection has resulted in a paradigm shift in the field.

A biosensor comprises a receptor element (e.g., enzymes or antibodies) that interacts selectively with the biosample and a transducer to convert the biological response into a measurable signal (e.g., current, light intensity, or mass change as presented in Figure 11.3). Semiconductor devices offer unique advantages in biosensing, including high sensitivity, miniaturization, integration, selectivity, label-free detection, and reusability. These features make semiconductor-based biosensors highly promising for a plethora of applications including medicine, environmental monitoring, food health monitoring, and biosecurity. The performance metrics

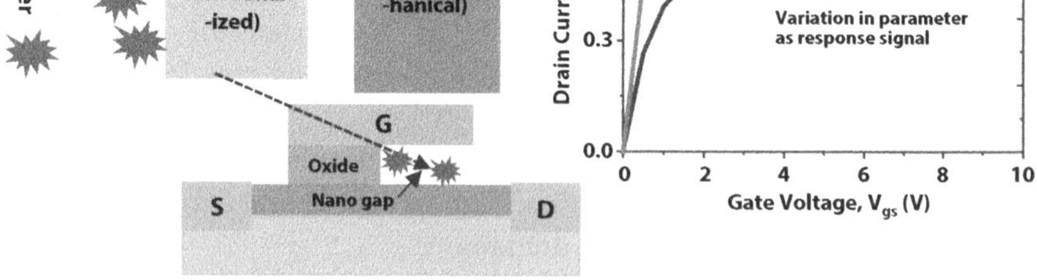

Figure 11.3 A biosensor block schematic [121].

of biosensors are of utmost importance for analyzing their efficiency. Although semiconductor devices are used in a myriad of domains, one trending domain of application is biosensing. MOSHEMTs have distinctive characteristics, such as high electron mobility and exceptional current-voltage characteristics, rendering them highly sensitive to surface charge changes. Based on the literature, we comprehend the use of different biomarkers for biomolecule detection, which is presented in the next section.

11.5.2 Biomarkers

Biomarkers are measurable indicators such as molecules or genes that provide information about various biological processes, conditions, or diseases in living organisms. They are used in medical diagnostics, research, and personalized medicine. They can be either neutral or charged. Neutral biomarkers are associated with permittivity only. Some biomarkers, along with their reported permittivity (κ), are being used for the study of biosensing applications of semiconductor devices, including the following [122].

Uricase ($\kappa = 1.54$) plays a crucial role in uric acid detection, and is used in the medical diagnosis of gout, kidney disease, and uric acid disorders.

Urease ($\kappa = 1.64$) can be used as an agricultural biomarker to assess soil health, nitrogen availability, and environmental monitoring.

Streptavidin ($\kappa = 2.1$) can be used as a biomarker in the food industry to test food quality.

Protein ($\kappa = 2.5$) biomarkers are vital in various fields including medicine, diagnostics, and research.

Biotin ($\kappa = 2.63$) has been used in DNA sequencing to incorporate biotin into DNA strands, which allows for subsequent detection or purification of the labeled DNA.

Cholesterol oxidase ($ChOx$) ($\kappa = 3.28$) catalyzes the oxidation of cholesterol, which is used to measure the total cholesterol levels in blood samples.

Glucose oxidase (GOx) ($\kappa = 3.46$) is used as a tool or reagent for glucose detection as a point-of-care device for diabetes management.

APTES (3-Aminopropyltriethoxysilane) ($\kappa = 3.57$) is used for drug delivery and bioimaging.

Zein ($\kappa = 5$) finds use in the food industry and pharmaceutical applications.

Keratin ($\kappa = 8$) can be used for medical diagnosis related to skin and hair disorders and certain types of cancer.

Charged biomolecules are biomarkers of permittivity and charge.

One such charged biomarker includes DNA ($\kappa = 2-72$), which is widely used in medicine, and research continues to expand, offering new opportunities for improved disease diagnosis, treatment, and patient care.

Another example are the ammino acids ($\kappa = 11-25$) Amino acids can serve as important biomarkers for various medical and research applications. Blood amino acid levels can indicate the nutritional status of an individual. They can be used for liver function, kidney function, cancer diagnosis and prognosis, neurological disorders, such as Alzheimer's disease or epilepsy, and metabolic disorders.

11.5.3 Performance Metrics of Biosensors

Each biosensor has performance metrics that can be used to analyze the performance of the biosensor [123].

Variations in device parameters are used to detect changes in device parameters, such as threshold voltage, on-current, off-current, transconductance, drain current, and output conductance. The threshold voltage variation in the absence and presence of biomarkers is expressed as

$$\Delta V_t = V_t^{Absence\ of\ Biomarker} - V_t^{Presence\ of\ Biomarker} \tag{11.1}$$

Sensitivity, as a performance metric in a biosensor, is essential in devices such as field-effect transistors (FETs) or MOSHEMTs, which can detect changes in electrical properties when interacting with biological molecules. This sensitivity allows for detection of minute concentrations of biomarkers or analytes, enabling highly sensitive biosensing. The threshold voltage sensitivity can be expressed as

$$S_{V_t} = \frac{\Delta V_t}{V_t^{Absence\ of\ Biomarker}} \tag{11.2}$$

Reproducibility refers to the sensor's capability to yield consistent and reliable responses when repeatedly exposed to similar test sample concentrations, operating under comparable conditions across subsequent cycles.

Responsivity serves as a metric for evaluating and comparing sensor actions in various sensing areas. This metric enables design engineers to make informed decisions when selecting the most suitable sensor for specific applications.

The limit of detection (LoD) of a biosensor is defined as the smallest quantity of analyte that can be detected reliably.

Label-free detection is possible in semiconductor-device-based biosensors, eliminating the need for cumbersome and time-consuming labeling processes. This simplifies the biosensing workflow, reduces costs, and allows for real-time monitoring.

Reusability indicates the number of times the biosensor can reliably sense the biomarker by regenerating the sensing surface. This reusability reduces cost and enhances the practicality of biosensors for routine use.

Optimum performance metrics can be extracted using an optimized biosensor design.

11.5.4 Key Considerations for Biosensor Design

Various design considerations have been applied to obtain desirable performance metrics for biosensors [124].

The first is *material selection*, which offers desirable electrical and biocompatibility properties. *Surface functionalization*, where the surface of the device needs to be selected to enhance its sensitivity and selectivity towards the target biomolecules, can be achieved through techniques such as chemical functionalization, self-assembled monolayers (SAMs), fabrication techniques, or specific receptor immobilization methods.

An appropriate *gate electrode design* was used to ensure efficient coupling with the biomolecular interactions occurring on the surface. A larger gate area can provide higher sensitivity by increasing the interaction surface area, thereby improving the detection capabilities of the biosensor. Selection of optimized *channel dimensions* to achieve a suitable balance between sensitivity and device performance. Reducing the channel length can improve the sensitivity by enhancing the electric field effect, whereas wider channels can increase the overall device current and transconductance. Choice of a *suitable gate dielectric* with high capacitance to enhance gate controllability of the channel. Alumina (Al_2O_3) and Hafnium Oxide (HfO_2) are commonly used dielectric materials for MOSHEMT biosensors. The *contacts* were designed to ensure low contact resistance and efficient charge injection into the channel. Ohmic contacts or metal alloy structures can be used to achieve low-resistance electrical connections. Proper *Device geometry*, including the gate-to-channel distance, gate-junction spacing, and dimensions of the junction regions, can influence the sensitivity and electrical characteristics of the MOSHEMT biosensor.

To achieve the optimal structural engineering of a MOSHEMT biosensor, different structures may be implemented to analyze its performance metrics.

11.5.5 MOSHEMT-Based Biosensors

HEMT-based biosensors are a specific type of semiconductor-based biosensor utilized for biomolecule detection and analysis. HEMTs exhibit high electron mobility and superior electrical properties. HEMT-based biosensors operate based on the principle of charge modulation. The presence of charged biomarkers leads to surface modifications, which in turn affect the conductivity of the HEMT channel, enabling the detection and quantification of the target analytes. HEMT-based biosensors allow for label-free detection, thereby eliminating the need for fluorescent or enzymatic labels that simplify the assay process and enable real-time monitoring.

A MOSHEMT can also be used as a sensor based on detecting modifications in the electrical characteristics of the semiconductor channel owing to the entry of a target analyte, causing a variation in the electrical properties of the semiconductor channel. This change in electrical properties can be detected by measuring the device's electrical characteristics, such as the current–voltage (I–V) curves or transconductance. Gallium nitride (GaN) is a popular choice for MOSHEMTs because of its advantageous properties including high electron mobility, low leakage current, and good chemical stability [125].

MOSHEMT biosensors offer several advantages including high sensitivity, rapid response time, and compatibility with CMOS technology. These characteristics make them suitable for various applications such as point-of-care diagnostics, environmental monitoring, and biomedical research. Literature has shown that HEMT/MOSHEMT-based biosensors can be implemented by using a cavity near the gate or by coating the gate with a bio-analyte to detect specific biomolecules.

A nanocavity near the gate impacts the oxide capacitance of the device, which, in turn, impacts the different devices. Single-gate MOSHEMT with a cavity under the gate where the bio-hybridization affects the 2DEG; hence, the device parameters have been reported [126]. A double layer was created when a biomolecule entered the cavity of the device. MOSHEMT with a sensing area on the gate where the biomarker is placed on gate MOSHEMT has been presented for cancer detection using the concept that the concentration of the DEG and the potential in the channel are responsive to changes in device characteristics caused by the presence of biomarkers [127]. This allows for sensitivity analysis, where variations in these parameters can be used to detect and analyze the impact of biomarkers on the device, making GaN-based MOSHEMTs promising for biosensing applications. Increasing the sensitivity of a dual-cavity device under a single-gate MOSHEMT has been reported [126]. This device has been used for the detection of neutral biomarkers by measuring changes in the device parameters. Many structural changes in MOSHEMTs have been reported for biomolecule detection. To utilize the benefits of double-gate MOSHEMTs, a MOSHEMT with a top and bottom gate with four cavities was proposed in [128].

In label-free biomolecule detection, different device parameters such as gain, cavity and gate dimensions, gate material, etc., are used to achieve higher sensitivity [128] and accuracy in detecting and analyzing biomolecules without the need for additional labeling techniques. The drain current sensitivity ($S_{I_{drain-source}}$) of the device towards biomolecules are calculated using the equation

$$S_{I_{drain-source}} = \frac{\Delta I_{drain-source}}{I_{drain-source}} \tag{11.3}$$

Here, $\Delta I_{drain-source}$ is the shift in the filled to Source Current for an empty cavity-filled cavity. The $I_{drain-source}$ is the current of an empty cavity. During the exploration of barrier layer thickness (d_{AlGaN}) and fill height (h_{cavity}) for different barrier layer thicknesses and filled/half-filled cavities, a constraint arises owing to the correct functioning of the device. Increasing the AlGaN layer thickness too much results in an increase in the surface-channel distance, but excessive reduction can lead to the thickness falling below the critical threshold, preventing the formation of the 2DEG/channel [129]. This limitation poses challenges in the design of nanoregime biosensors, where precise control of geometrical parameters is essential to ensure optimal device performance for biomolecule detection and sensitivity.

While silicon-based biosensors are easily degraded owing to chemical exposure, have poor ruggedness, low chemical stability, low sensitivity, and are non-reusable, MOSHEMTs can be fabricated using biocompatible materials and processes, making them suitable for interfacing with biological systems without adverse effects. Biocompatibility is vital for applications that require direct contact with biological samples or *in vivo* sensing.

Table 11.1 presents the existing MOSHEMT-based biosensor designs and their findings. Increasing the cavity dimensions affects the sensitivity as the interaction area with the biomolecule increases. MOSHEMTs for biosensing applications use the effect of dielectric modulation to analyze the

Table 11.1 Existing MOSHEMT biosensor.

Ref.	Type of sensing area	Analysis method and tools used	Type of biomolecule (BM) and the device parameters used for sensing	FoM investigated	Findings
[130]	Single Cavity	Dielectric Modulation (TCAD)	Neutral BM	Sensitivity	Dielectric Modulation approach to detect modification
[80]	Top of Gate	Charge deduction (Fabrication and (TCAD))	Charged BM ($\frac{\Delta g_d}{I_{ds}}$, $\Delta E(x)$, $\Delta chpot$, Δg_d as Sensing Metric)	Sensitivity	Different metrics for different biomarkers can be used
[131]	Dual Cavity	Dielectric Modulation (TCAD)	Neutral and Charged BM ΔV_{th}	Variation	Improved area for improved sensitivity
[126]	Dual Cavity	Dielectric Modulation (TCAD)	Neutral BM ΔI_{ds} and ΔV_{th}	Variation	Improved sensitivity than GAA structures
[129]	Single Cavity	Dielectric Modulation (TCAD)	Neutral and Charged BM ΔV_{th} and ΔI_{ds}	Variation	Constraint of AlGaN layer thickness
[132]	Single Cavity	Dielectric Modulation (TCAD)	Neutral and Charged BM ΔV_{th}, ΔI_{ds}, and ΔI_{ON}	Variation	The cavity placed towards the source side to improve I_{ON}
[133]	Single Cavity	Dielectric Modulation (TCAD)	Neutral BM SI_{ON}	Sensitivity	Use of ZnO/MnZnO for the heterostructure
[134]	Single Cavity	Dielectric Modulation (TCAD)	Neutral BM ΔI_{ds}	Variation	Virus detection
[135]	Single Cavity	Dielectric Modulation (TCAD)	Neutral BM Protein (I_{ds}, g_m)	Sensitivity	Specific protein detection

effect of neutral biomolecules (NBMs) and charged biomolecules (CBMs) on device behavior. As the biomolecule enters the cavity, the oxide capacitance is influenced, which affects the device metrics, such as the threshold voltage, drain current, and transconductance. Variations in these parameters can be used to analyze the sensitivity of a structure toward specific biomolecules. This sensitivity enables the detection of minute alterations in the biointerface's electrical properties, facilitating the measurement of biological signals. The surface of MOSHEMTs can be functionalized with receptors, such as antibodies, enzymes, or DNA probes. This specific interaction with target analytes, such as biomarkers or pathogens, induces changes in the device's electrical properties, allowing the selective and sensitive detection of specific biological targets. These biosensors allow for label-free detection, thereby eliminating the need for additional chemical or fluorescent labels. The device's electrical properties directly respond to binding events at the surface, enabling the real-time and direct monitoring of biological interactions without the use of extra reagents. Immediate feedback on biomarker–receptor kinetics can be obtained, thereby laying the foundation for the use of MOSHEMT-based biosensors in point-of-care diagnostics.

11.6 Summary

Different applications require the selection of materials for different layers of heterostructures. The substrate material and gate dielectric significantly affected the device behavior. Si is widely used owing to its high resistivity and low cost; however, it has a leakage current issue. Silicon Carbide (SiC) offers superior properties but is costly. Indium Phosphide (InP) exhibits enhanced electron mobility. Different gate dielectric materials, such as SiO_2, Al_2O_3, TiO_2, and high-k dielectrics, improve device performance by reducing leakage current and current collapse effects. The choice of the barrier layer, such as InAlN or AlGaN, affects the device characteristics, considering the 2DEG, gate leakage, and interface traps. MOSHEMT offers advantages such as a low leakage current and reduced current collapse effects, making it suitable for different application domains. Moreover, the selection of a proper material and structure of a MOSHEMT to achieve high sensitivity, robust architecture, and the ability to operate in harsh environments make it a promising choice for biosensing applications. The wide-bandgap material GaN lays the foundation for its enabling operation in high-power and high-temperature environments. Second, a barrier layer between the channel and the oxide layer helps mitigate degradation mechanisms and enhances device reliability. Additionally, the MOSHEMT structure offers improved performance compared with traditional MOSFETs, including higher transconductance and lower on-resistance. MOSHEMT biosensors are promising candidates for the sensitive, label-free, and real-time recognition of biological analytes. They hold significant potential for various applications in healthcare, environmental monitoring, food safety, and biotechnology, contributing to advancements in diagnostics, disease monitoring, and personalized medicine. Thus, as the MOSHEMT device structure moves from a single gate to a fin gate structure, its effect on device performance changes, which opens up possibilities for its use in different applications.

References

1. Borel, J., Advanced MOSFET technologies: A review, in: *ESSCIRC 76: 2nd European Solid State Circuits Conference*, Sep. 1976, vol. 1976-Septe, pp. X–XI, doi: 10.1109/ESSCIRC.1976.5469254.
2. Xie, Q., Xu, J., Taur, Y., Review and critique of analytic models of MOSFET short-channel effects in subthreshold. *IEEE Trans. Electron Devices*, 59, 6, 1569–1579, 2012, doi: 10.1109/TED.2012.2191556.
3. Sekigawa, T. and Hayashi, Y., Calculated threshold-voltage characteristics of an XMOS transistor having an additional bottom gate. *Solid State Electron.*, 27, 8–9, 827–828, 1984, doi: 10.1016/0038-1101(84)90036-4.
4. Ortiz-Conde, A., *et al.*, A review of core compact models for undoped double-gate SOI MOSFETs Serigraphy of Solar Cells View project Harmonic distortion in FETs View project A Review of Core Compact Models for Undoped Double-Gate SOI MOSFETs. *IEEE Trans. Electron Devices*, 54, 1, 2007, doi: 10.1109/TED.2006.887046.
5. Pardeshi, H., Raj, G., Pati, S., Mohankumar, N., Sarkar, C.K., Influence of barrier thickness on AlInN/GaN underlap DG MOSFET device performance. *Superlattices Microstruct.*, 60, 47–59, 2013, doi: 10.1016/j.spmi.2013.04.015.
6. Bhowmick, B., Emerging device: FINFET, tunnel FET and their applications. *CSI Trans. ICT*, 7, 3, 221–225, 2019, doi: 10.1007/s40012-019-00240-z.
7. Smaani, B., Latreche, S., Iñiguez, B., Compact drain-current model for undoped cylindrical surrounding-gate metal-oxide-semiconductor field effect transistors including short channel effects. *J. Appl. Phys.*, 114, 22, 232, Dec. 2013, doi: 10.1063/1.4844395/396602.
8. Mimura, T., Hiyamizu, S., Fujii, T., Nanbu, K., A new field-effect transistor with selectively doped GaAs/n-AlxGa1–xAs heterojunctions. *Jpn. J. Appl. Phys.*, 19, 5, L225–L227, 1980, doi: 10.1143/JJAP.19.L225.
9. Mimura, T., The early history of the high electron mobility transistor (HEMT). *IEEE Trans. Microw. Theory Tech.*, 50, 3, 780–782, Mar. 2002, doi: 10.1109/22.989961.
10. Li, Q., Zhou, X., Tang, C.W., Lau, K.M., High-performance inverted In 0.53Ga 0.47As MOSHEMTs on a GaAs substrate with regrown source/drain by MOCVD. *IEEE Electron Device Lett.*, 33, 9, 1246–1248, 2012, doi: 10.1109/LED.2012.2204431.

11. Menozzi, R., Hot electron effects and degradation of GaAs and InP HEMTs for microwave and millimetre-wave applications. *Semicond. Sci. Technol.*, 13, 10, 1053, Oct. 1998, doi: 10.1088/0268-1242/13/10/001.
12. Xiao, M., Duan, X., Zhang, J., Hao, Y., Enhancement-Mode AlN/GaN HEMTs with High Ion/Ioff and Low-Leakage-Current Fabricated with Low-Power Surface Oxidation Treatment. *IEEE Electron Device Lett.*, 39, 5, 719–722, 2018, doi: 10.1109/LED.2018.2816166.
13. Roccaforte, F., *et al.*, Challenges for energy efficient wide band gap semiconductor power devices. *Phys. Status Solidi (A) Appl. Mater. Sci.*, 211, 9, Wiley-VCH Verlag, 2063–2071, 2014, doi: 10.1002/pssa.201300558.
14. Del Alamo, J.A. and Joh, J., GaN HEMT reliability. *Microelectron. Reliab.*, 49, 1200–1206, 2009, doi: 10.1016/j.microrel.2009.07.003.
15. Haziq, M., Falina, S., Manaf, A.A., Kawarada, H., Syamsul, M., Challenges and Opportunities for High-Power and High-Frequency AlGaN/GaN High-Electron-Mobility Transistor (HEMT) Applications: A Review. *Micromachines*, 13, 12, 2022, doi: 10.3390/mi13122133.
16. Fletcher, A.S.A. and Nirmal, D., A survey of Gallium Nitride HEMT for RF and high power applications. *Superlattices Microstruct.*, 109, 519–537, 2017, doi: 10.1016/j.spmi.2017.05.042.
17. Husna Hamza, K. and Nirmal, D., A review of GaN HEMT broadband power amplifiers. *AEU Int. J. Electron. Commun.*, 116, Elsevier GmbH, 153040, Mar. 01, 2020, doi: 10.1016/j.aeue.2019.153040.
18. Vasallo, B.G., *et al.*, Comparison between the dynamic performance of double- and single-gate AlInAs/InGaAs HEMTs. *IEEE Trans. Electron Devices*, 54, 11, 2815–2822, Nov. 2007, doi: 10.1109/TED.2007.907801.
19. Tsai, T.-H., Yang, M., Chang, L.-C., Wu, C.-H., Reduction of threshold voltage instability in recessed-gate AlGaN/GaN MOSHEMTs by KOH passivation. *2016 Compound Semiconductor Week (CSW) [Includes 28th International Conference on Indium Phosphide & Related Materials (IPRM) & 43rd International Symposium on Compound Semiconductors (ISCS)*, pp. 1–2, Toyama, Japan, 2016, doi: 10.1109/ICIPRM.2016.7528681.
20. Bae, J.H., *et al.*, Characterization of traps and trap-related effects in recessed-gate normally-off AlGaN/GaN-based MOSHEMT. *Tech. Dig. - Int. Electron Devices Meet. IEDM*, pp. 303–306, 2012, doi: 10.1109/IEDM.2012.6479034.
21. Then, H.W., *et al.*, High-performance low-leakage enhancement-mode high-K dielectric GaN MOSHEMTs for energy-efficient, compact voltage regulators and RF power amplifiers for low-power mobile SoCs. *Dig. Tech. Pap. - Symp. VLSI Technol.*, 2015-Augus, T202–T203, 2015, doi: 10.1109/VLSIT.2015.7223674.
22. Regmi, A., *et al.*, High sensitivity cardiac troponin I detection in physiological environment using AlGaN/GaN High Electron Mobility Transistor (HEMT) Biosensors. *Biosens. Bioelectron.*, 100, 282–289, 2017, doi: 10.1016/j.bios.2017.09.018.

23. Sarangadharan, I., *et al.*, Review—High Field Modulated FET Biosensors for Biomedical Applications. *ECS J. Solid State Sci. Technol.*, 7, 7, Q3032–Q3042, 2018, doi: 10.1149/2.0061807jss.
24. Rao, M., Ranjan, R., Kashyap, N., Sarin, R.K., Performance Analysis of Normally-on Dual Gate Algan/Gan Hemt. *Trans. Electr. Electron. Mater.*, 22, 5, 691–699, Oct. 2021, doi: 10.1007/S42341-021-00286-4.
25. Medjdoub, F., Van Hove, M., Cheng, K., Marcon, D., Leys, M., Decoutere, S., Novel E-Mode GaN-on-Si MOSHEMT using a selective thermal oxidation. *IEEE Electron Device Lett.*, 31, 9, 948–950, 2010, doi: 10.1109/LED.2010.2052014.
26. Tao, M., *et al.*, Characterization of 880 v Normally off GaN MOSHEMT on Silicon Substrate Fabricated with a Plasma-Free, Self-Terminated Gate Recess Process. *IEEE Trans. Electron Devices*, 65, 4, 1453–1457, 2018, doi: 10.1109/TED.2018.2808345.
27. Inoue, K., Ikeda, Y., Masato, H., Matsuno, T., Nishii, K., Novel GaN-based MOS HFETs with thermally oxidized gate insulator. *Tech. Dig. Electron Devices Meet.*, pp. 577–580, 2001, doi: 10.1109/IEDM.2001.979572.
28. Zhu, M., *et al.*, Comparing buffer leakage in PolarMOSH on SiC and free-standing GaN substrates. *Proc. 25th Bienn. Lester Eastman Conf. High Perform. Devices, LEC 2016*, pp. 27–30, 2016, doi: 10.1109/LEC.2016.7578926.
29. Medjdoub, F., *et al.*, Barrier layer downscaling of InAlN/GaN HEMTs. *65th DRC Device Res. Conf.*, vol. 45, no. 32, pp. 109–110, 2007, doi: 10.1109/DRC.2007.4373673.
30. Nakamura, K., Paul, N.C., Takebe, M., Iiyama, K., Takamiya, S., Depletion/enhancement mode InAlAs/InGaAs - MOSHEMTs with nm - Thin gate insulating layers formed by oxidation of the InAlAs layer. *Conf. Proc. - Int. Conf. Indium Phosphide Relat. Mater.*, pp. 191–194, 2004, doi: 10.1109/iciprm.2004.1442643.
31. Brown, R., *et al.*, A sub-critical barrier thickness normally-Off AlGaN/GaN MOS-HEMT. *IEEE Electron Device Lett.*, 35, 9, 906–908, 2014, doi: 10.1109/LED.2014.2334394.
32. Lin, S., *et al.*, A GaN HEMT structure allowing self-terminated, plasma-free etching for high-uniformity, high-mobility enhancement-mode devices. *IEEE Electron Device Lett.*, 37, 4, 377–380, 2016, doi: 10.1109/LED.2016.2533422.
33. Li, S., Hu, Q., Wang, X., Li, T., Li, X., Wu, Y., Improved Interface Properties and Dielectric Breakdown in Recessed AlGaN/GaN MOS-HEMTs Using HfSiOx as Gate Dielectric. *IEEE Electron Device Lett.*, 40, 2, 295–298, 2019, doi: 10.1109/LED.2018.2888486.
34. Hfo, M.W., Using, D., Prajoon, P., Impact of AlInN Back-Barrier Over AlGaN /GaN Spline Interpolation Technique. *IEEE Trans. Electron Devices*, 67, 9, 1–6, 2020.
35. Markman, B., Brunelli, S.T.S., Goswami, A., Guidry, M., Rodwell, M.J.W., In.Ga.As/InAs Composite Channel MOS-HEMT Exhibiting 511 GHz fτand

256 GHz fmax. *IEEE J. Electron Devices Soc.*, 8, July, 930–934, 2020, doi: 10.1109/JEDS.2020.3017141.
36. Jiang, H., Liu, C., Ng, K.W., Tang, C.W., Lau, K.M., High-Performance AlGaN/GaN/Si Power MOSHEMTs With ZrO2 Gate Dielectric. *IEEE Trans. Electron Devices*, 65, 12, 5337–5342, 2018, doi: 10.1109/TED.2018.2874075.
37. Lin, Y.S. and Lu, C.C., Improved AlGaN/GaN Metal-Oxide-Semiconductor High-Electron Mobility Transistors with TiO2 Gate Dielectric Annealed in Nitrogen. *IEEE Trans. Electron Devices*, 65, 2, 783–787, 2018, doi: 10.1109/TED.2017.2781141.
38. Hu, Q., Li, S., Li, T., Wang, X., Li, X., Wu, Y., Channel engineering of normally-OFF AlGaN/GaN MOS-HEMTs by atomic layer etching and high-κDielectric. *IEEE Electron Device Lett.*, 39, 9, 1377–1380, 2018, doi: 10.1109/LED.2018.2856934.
39. Chiou, Y.L. and Lee, C.T., Band alignment and performance improvement mechanisms of chlorine-treated ZnO-gate AlGaN/GaN metal-oxide-semiconductor high-electron mobility transistors. *IEEE Trans. Electron Devices*, 58, 11, 3869–3875, 2011, doi: 10.1109/TED.2011.2163721.
40. Hu, Q., *et al.*, Improved Current Collapse in Recessed AlGaN/GaN MOS-HEMTs by Interface and Structure Engineering. *IEEE Trans. Electron Devices*, 66, 11, 4591–4596, 2019, doi: 10.1109/TED.2019.2940749.
41. Mondal, A., Roy, A., Mitra, R., Kundu, A., Comparative Study of Variations in Gate Oxide Material of a Novel Underlap DG MOS-HEMT for Analog/RF and High Power Applications. *Silicon*, 12, 9, 2251–2257, 2020, doi: 10.1007/s12633-019-00316-0.
42. Sandeep, V., Pravin, J.C., Ramesh Babu, A., Prajoon, P., Impact of AlInN Back-Barrier over AlGaN/GaN MOS-HEMT with HfO Dielectric Using Cubic Spline Interpolation Technique. *IEEE Trans. Electron Devices*, 67, 9, 3558–3563, Sep. 2020, doi: 10.1109/TED.2020.3010710.
43. Jena, K. and Lenka, T.R., Effect of AlN spacer thickness on device characteristics of AlInN/AlN/GaN MOSHEMT, in: *IEEE Region 10 Annual International Conference, Proceedings/TENCON*, Feb. 2017, pp. 3253–3256, doi: 10.1109/TENCON.2016.7848652.
44. Hahn, H., *et al.*, Threshold voltage engineering in GaN-based HFETs: A systematic study with the threshold voltage reaching more than 2 V. *IEEE Trans. Electron Devices*, 62, 2, 538–545, 2015, doi: 10.1109/TED.2014.2381292.
45. Lee, H.S., Piedra, D., Sun, M., Gao, X., Guo, S., Palacios, T., 3000-V 4.3-mΩ · cm 2 InAlN/GaN MOSHEMTs with AlGaN back barrier. *IEEE Electron Device Lett.*, 33, 7, 982–984, 2012, doi: 10.1109/LED.2012.2196673.
46. Basu, S., Singh, P.K., Lin, S.K., Sze, P.W., Wang, Y.H., Effects of short-term DC-bias-induced stress on n-GaN/AlGaN/GaN MOSHEMTs with liquid-phase-deposited Al2O3 as a gate dielectric. *IEEE Trans. Electron Devices*, 57, 11, 2978–2987, 2010, doi: 10.1109/TED.2010.2071130.

47. Liu, H.Y., Lee, C.S., Lin, C.W., Chiang, M.H., Hsu, W.C., Gate structure engineering for enhancement-mode AlGaN/GaN MOSHEMT. *Device Res. Conf. - Conf. Dig. DRC*, 100, 4–5, 2017, doi: 10.1109/DRC.2017.7999446.
48. Samsel, I.K., *et al.*, Charge collection mechanisms in AlGaN/GaN MOS high electron mobility transistors. *IEEE Trans. Nucl. Sci.*, 60, 6, 4439–4445, 2013, doi: 10.1109/TNS.2013.2289383.
49. Freedsman, J.J., Kubo, T., Egawa, T., High Drain Current Density E-Mode Al_2O_3/AlGaN/GaN MOS-HEMT on Si With. *IEEE Trans. Devices*, 60, 10, 1–5, 2013.
50. Wu, T.Y., *et al.*, AlGaN/GaN MOSHEMTs with liquid-phase-deposited TiO_2 as gate dielectric. *IEEE Trans. Electron Devices*, 56, 12, 2911–2916, 2009, doi: 10.1109/TED.2009.2032745.
51. Rawat, A., Surana, V.K., Meer, M., Bhardwaj, N., Ganguly, S., Saha, D., Gate Current Reduction and Improved DC/RF Characteristics in GaN-Based MOS-HEMTs Using Thermally Grown TiO2 as a Dielectric. *IEEE Trans. Electron Devices*, 66, 6, 2557–2562, 2019, doi: 10.1109/TED.2019.2910608.
52. Chou, B.Y., *et al.*, TiO2-dielectric AlGaN/GaN/Si metal-oxide-semiconductor high electron mobility transistors by using nonvacuum ultrasonic spray pyrolysis deposition. *IEEE Electron Device Lett.*, 35, 11, 1091–1093, 2014, doi: 10.1109/LED.2014.2354643.
53. Meng, D., *et al.*, Low leakage current and high-cutoff frequency AlGaN/GaN MOSHEMT using submicrometer-footprint thermal oxidized TiO2/NiO as Gate Dielectric. *IEEE Electron Device Lett.*, 34, 6, 738–740, 2013, doi: 10.1109/LED.2013.2256102.
54. Di, M., *et al.*, Characteristics of submicron-footprint TiO2 based AlGaN/GaN MOSHEMT on SiC substrate. *2013 IEEE Int. Conf. Electron Devices Solid-State Circuits, EDSSC 2013*, vol. 2, pp. 2–3, 2013, doi: 10.1109/EDSSC.2013.6628138.
55. Gao, Z., Romero, M.F., Redondo-Cubero, A., Pampillon, M.A., San Andres, E., Calle, F., Effects of Gd2O3 Gate Dielectric on Proton-Irradiated AlGaN/GaN HEMTs. *IEEE Electron Device Lett.*, 38, 5, 611–614, May 2017, doi: 10.1109/LED.2017.2682795.
56. Gao, Z., Romero, M.F., Pampillón, M.A., San Andrés, E., Calle, F., Thermal Assessment of AlGaN/GaN MOS-HEMTs on Si Substrate Using Gd2O3 as Gate Dielectric. *IEEE Trans. Electron Devices*, 63, 7, 2729–2734, 2016, doi: 10.1109/TED.2016.2564301.
57. Kumar, S., *et al.*, Investigation of Ta 2 O 5 as an alternative high-k dielectric for InAlN/GaN MOS-HEMT on Si. *IEEE Trans. Electron Devices*, 66, 3, 1230–1235, 2019, doi: 10.1109/TED.2019.2893288.
58. Mehandru, R., *et al.*, AlGaN/GaN metal-oxide-semiconductor high electron mobility transistors using Sc2O3 as the gate oxide and surface passivation. *Appl. Phys. Lett.*, 82, 15, 2530–2532, Apr. 2003, doi: 10.1063/1.1567051.

59. Lee, C.T., Chiou, Y.L., Sen Lee, C., AlGaN/GaN MOS-HEMTs with gate ZnO dielectric layer. *IEEE Electron Device Lett.*, 31, 11, 1220–1223, 2010, doi: 10.1109/LED.2010.2066543.
60. Gao, Z., Romero, M.F., Calle, F., Thermal and Electrical Stability Assessment of AlGaN/GaN Metal-Oxide-Semiconductor High-Electron Mobility Transistors (MOS-HEMTs) with HfO2 Gate Dielectric. *IEEE Trans. Electron Devices*, 65, 8, 3142–3148, 2018, doi: 10.1109/TED.2018.2842205.
61. Li, S., Hu, Q., Wang, X., Wang, M., Wu, Y., AlGaN/GaN E-mode MOS-HEMT Using Atomic- Layer-Deposited HfLaO x as Gate Dielectric. *2018 1st Work. Wide Bandgap Power Devices Appl. Asia (WiPDA Asia)*, pp. 153–155.
62. Liao, C., *et al.*, Low Interface Trap Densities and Enhanced Performance of AlGaN/GaN MOS High- Electron Mobility Transistors Using Thermal Oxidized Y2O3 Interlayer. *IEEE Electron Device Lett.*, 36, 12, 1284–1286, Dec. 2015, doi: 10.1109/LED.2015.2486818.
63. Swain, R., Jena, K., Lenka, T.R., Oxide interfacial charge engineering towards normally-off AlN/GaN MOSHEMT. *Mater. Sci. Semicond. Process.*, 53, 66–71, Oct. 2016, doi: 10.1016/j.mssp.2016.06.008.
64. Liu, H.Y., Chou, B.Y., Hsu, W.C., Lee, C.S., Ho, C.S., A simple gate-dielectric fabrication process for AlGaN/GaN metal-oxide-semiconductor high-electron-mobility transistors. *IEEE Electron Device Lett.*, 33, 7, 997–999, 2012, doi: 10.1109/LED.2012.2197370.
65. Zhou, H., *et al.*, High-Performance InAlN/GaN MOSHEMTs Enabled by Atomic Layer Epitaxy MgCaO as Gate Dielectric. *IEEE Electron Device Lett.*, 37, 5, 556–559, 2016, doi: 10.1109/LED.2016.2537198.
66. Putcha, V., *et al.*, Exploring the DC reliability metrics for scaled GaN-on-Si devices targeted for RF/5G applications. *IEEE Int. Reliab. Phys. Symp. Proc.*, 2020-April, 0–7, 2020, doi: 10.1109/IRPS45951.2020.9129251.
67. Mo, J.J., *et al.*, Lattice matched and Pseudomorphic InGaAs MOSHEMT with fT of 200GHz. *Conf. Proc. Int. Conf. Indium Phosphide Relat. Mater.*, pp. 44–47, 2012, doi: 10.1109/ICIPRM.2012.6403314.
68. Tiwari, N., Gupta, S.K., Mishra, S.N., Impact of AlGaN barrier layer on DC and AC characteristics of AlxGa1-xN/GaN MOSHEMT. *2017 1st Int. Conf. Electron. Mater. Eng. Nano-Technology, IEMENTech 2017*, pp. 4–7, 2017, doi: 10.1109/IEMENTECH.2017.8076958.
69. Yadav, S., *et al.*, High mobility In0.30Ga0.70As MOSHEMTs on low threading dislocation density 200 mm Si substrates: A technology enabler towards heterogeneous integration of low noise and medium power amplifiers with Si CMOS, in: *Technical Digest-International Electron Devices Meeting, IEDM*, Jan. 2018, pp. 17.4.1–17.4.4, doi: 10.1109/IEDM.2017.8268409.
70. Mim, N.G., Bin Kutub, S., Haque, A., Characterization of top barrier thickness from gate capacitance of high mobility III-V semiconductor MOS-HEMT devices. *Results Phys.*, 6, 233–234, 2016, doi: 10.1016/j.rinp.2016.04.015.

71. Panda, J., Jena, K., Lenka, T.R., Comparative analysis of GaN based MOSHEMT devices for RF applications, Aug. 2015, doi: 10.1109/ICECCT.2015.7226141.
72. Swain, S.K., Adak, S., Pati, S.K., Sarkar, C.K., Impact of InGaN back barrier layer on performance of AlInN/AlN/GaN MOS-HEMTs. *Superlattices Microstruct.*, 97, 258–267, Sep. 2016, doi: 10.1016/j.spmi.2016.06.032.
73. Alomari, M., *et al.*, InAlN/GaN MOSHEMT with self-aligned thermally generated oxide recess. *IEEE Electron Device Lett.*, 30, 11, 1131–1133, 2009, doi: 10.1109/LED.2009.2031659.
74. Kim, Y.S., Ha, M.W., Seok, O., Ahn, W.J., Han, M.K., High breakdown voltage AlGaN/GaN MOS-HEMTs-on-Si with atomic-layer-deposited Al2O3 gate insulator. *Proc. Int. Symp. Power Semicond. Devices ICs*, pp. 207–211, 2013, doi: 10.1109/ISPSD.2013.6694482.
75. Zhou, H., *et al.*, InAlN/GaN MOSHEMTs with high drain current of 2.3 A/mm high on/off ratio of 1012 and low SS of 64 mV/dec enabled by atomic-layer-epitaxial MgCaO as gate dielectric, in: *Device Research Conference - Conference Digest, DRC*, Aug. 2015, vol. 2015-Augus, pp. 57–58, doi: 10.1109/DRC.2015.7175552.
76. Thome, F., Heinz, F., Leuther, A., InGaAs MOSHEMT W-Band LNAs on Silicon and Gallium Arsenide Substrates. *IEEE Microw. Wirel. Compon. Lett.*, 30, 11, 1089–1092, 2020, doi: 10.1109/LMWC.2020.3025674.
77. Takakura, K., *et al.*, Low-Frequency Noise Investigation of GaN/AlGaN Metal-Oxide-Semiconductor High-Electron-Mobility Field-Effect Transistor with Different Gate Length and Orientation. *IEEE Trans. Electron Devices*, 67, 8, 3062–3068, 2020, doi: 10.1109/TED.2020.3002732.
78. Kume, E., *et al.*, InAs MOS-HEMT power detector for 1.0 THz on quartz glass. *2017 IEEE Electron Devices Technol. Manuf. Conf. EDTM 2017 – Proc.*, pp. 196–197, 2017, doi: 10.1109/EDTM.2017.7947562.
79. Leuther, A., Merkle, T., Weber, R., Sommer, R., Tessmann, A., THz Frequency HEMTs: Future Trends and Applications. *2019 Compd. Semicond. Week, CSW 2019 – Proc.*, pp. 1–2, 2019, doi: 10.1109/ICIPRM.2019.8819000.
80. Varghese, A., Periasamy, C., Bhargava, L., Fabrication and charge deduction based sensitivity analysis of gan mos-hemt device for glucose, mig, c-erbb-2, kim-1, and psa detection. *IEEE Trans. Nanotechnol.*, 18, 747–755, 2019, doi: 10.1109/TNANO.2019.2928308.
81. Swain, R., Jena, K., Gaini, A., Lenka, T.R., Comparative study of AlN/GaN HEMT and MOSHEMT structures by varying oxide thickness, in: *2014 IEEE 9th Nanotechnology Materials and Devices Conference, NMDC 2014*, Dec. 2014, pp. 128–131, doi: 10.1109/NMDC.2014.6997439.
82. Swain, R. and Lenka, T.R., Investigation of critical barrier thickness in lattice matched InAlN/GaN MOSHEMT towards normally-off operation, in: *IEEE Region 10 Annual International Conference, Proceedings/TENCON*, Jan. 2016, vol. 2016-Janua, doi: 10.1109/TENCON.2015.7373087.

83. Swain, R. and Lenka, T.R., Normally-off Al0.25Ga0.75N/GaN MOSHEMT with stack gate dielectric structure, in: *Proceedings of the 2015 IEEE International Conference on Electron Devices and Solid-State Circuits, EDSSC 2015*, Sep. 2015, pp. 567–570, doi: 10.1109/EDSSC.2015.7285177.
84. Lee, C.S., Liao, Y.H., Chou, B.Y., Liu, H.Y., Hsu, W.C., Composite HfO2/Al2O3-dielectric AlGaAs/InGaAs MOS-HEMTs by using RF sputtering/ozone water oxidation. *Superlattices Microstruct.*, 72, 194–203, Aug. 2014, doi: 10.1016/j.spmi.2014.04.016.
85. Lee, C.T., Yang, C.L., Tseng, C.Y., Chang, J.H., Horng, R.H., GaN-Based Enhancement-Mode Metal-Oxide-Semiconductor High-Electron Mobility Transistors Using LiNbO3 Ferroelectric Insulator on Gate-Recessed Structure. *IEEE Trans. Electron Devices*, 62, 8, 2481–2487, 2015, doi: 10.1109/TED.2015.2446990.
86. Huang, T., Zhu, X., Lau, K.M., Enhancement-mode AlN/GaN MOSHEMTs fabricated by selective area regrowth of AlGaN barrier layer. *Proc. Int. Symp. Power Semicond. Devices ICs*, pp. 199–202, 2013, doi: 10.1109/ISPSD.2013.6694480.
87. Zhu, M., *et al.*, P-GaN Tri-Gate MOS Structure for Normally-Off GaN Power Transistors. *IEEE Electron Device Lett.*, 42, 1, 82–85, 2021, doi: 10.1109/LED.2020.3037026.
88. Gao, J., *et al.*, Gate-recessed normally off GaN MOSHEMT with high-temperature oxidation/wet etching using LPCVD Si3N4 as the mask. *IEEE Trans. Electron Devices*, 65, 5, 1728–1733, 2018, doi: 10.1109/TED.2018.2812215.
89. Tang, C., Xie, G., Sheng, K., Enhancement-mode GaN-on-Silicon MOS-HEMT using pure wet etch technique. *Proc. Int. Symp. Power Semicond. Devices ICs*, vol. 2015-June, pp. 233–236, 2015, doi: 10.1109/ISPSD.2015.7123432.
90. Chiou, Y.L., Huang, L.H., Lee, C.T., Photoelectrochemical function in gate-recessed AlGaN/GaN metaloxidesemiconductor high-electron-mobility transistors. *IEEE Electron Device Lett.*, 31, 3, 183–185, 2010, doi: 10.1109/LED.2009.2037983.
91. Hao, Y., Ling Y., Ma, X., Ma, J., Cao, M., Pan, C., Wang, C., Zhang, J., Highperformance microwave gate-recessed AlGaN/AlN/GaN MOS-HEMT with 73% power-added efficiency. *IEEE Electron Device Lett.*, 32, 5, 626–628, 2011. doi: 10.1109/LED.2011.2118736.
92. Meyer, D.J., *et al.*, High electron velocity submicrometer AlN/GaN MOS-hemts on freestanding GaN substrates. *IEEE Electron Device Lett.*, 34, 2, 199–201, 2013, doi: 10.1109/LED.2012.2228463.
93. Lee, C.T., Chen, W.S., Lee, H.Y., Quadruple gate-embedded T structured GaN-based metal-oxide-semiconductor high-electron mobility transistors. *IEEE J. Electron Devices Soc.*, 6, 1, 63–67, 2017, doi: 10.1109/JEDS.2017.2769112.
94. Zine-eddine, T., Zahra, H., Zitouni, M., Design and analysis of 10 nm T-gate enhancement-mode MOS-HEMT for high power microwave applications. *J. Sci. Adv. Mater. Devices*, 4, 1, 180–187, 2019, doi: 10.1016/j.jsamd.2019.01.001.

95. Lee, H., Jian, J., Lee, C., Fellow, L., Quadruple Gate-Recessed Array Metal-Oxide-Semiconductor High-Electron Mobility Transistors. in: *IEEE Transactions on Electron Devices*, vol. 68, no. 1, pp. 42–48, Jan. 2021, doi: 10.1109/TED.2020.3037274.
96. Murugapandiyan, P., *et al.*, Performance analysis of HfO2/InAlN/AlN/GaN HEMT with AlN buffer layer for high power microwave applications. *J. Sci. Adv. Mater. Devices*, 5, 2, 192–198, 2020, doi: 10.1016/j.jsamd.2020.04.007.
97. Zhou, H., *et al.*, DC and RF Performance of AlGaN/GaN/SiC MOSHEMTs with Deep Sub-Micron T-Gates and Atomic Layer Epitaxy MgCaO as Gate Dielectric. *IEEE Electron Device Lett.*, 38, 10, 1409–1412, 2017, doi: 10.1109/LED.2017.2746338.
98. Lee, C.S., *et al.*, Investigations of novel Γ-gate MOS-HEMTs by ozone water oxidation and shifted exposure techniques. *IEEE Trans. Electron Devices*, 58, 9, 2981–2989, 2011, doi: 10.1109/TED.2011.2158650.
99. Sehra, K., Kumari, V., Gupta, M., Mishra, M., Rawal, D.S., Saxena, M., An Asymmetric π - Gate MOSHEMT Architecture for High Frequency Applications, in: *Lecture Notes in Networks and Systems*, Dec. 2021, vol. 147, pp. 453–458, doi: 10.1007/978-981-15-8366-7_67.
100. Pati, S.K., Pardeshi, H., Raj, G., Mohan Kumar, N., Sarkar, C.K., Impact of gate length and barrier thickness on performance of InP/InGaAs based Double Gate Metal-Oxide-Semiconductor Heterostructure Field-Effect Transistor (DG MOS-HFET). *Superlattices Microstruct.*, 55, 1, 8–15, Mar. 2013, doi: 10.1016/j.spmi.2012.12.002.
101. Saravana Kumar, R., Poornachandran, R., Baskaran, S., Mohan Kumar, N., Sandhiya, S., Shanmugapriya, K.U., DC and RF Characterization of InAs based Double Delta Doped MOSHEMT Device. *Proc. Int. Conf. 2018 IEEE Electron Device Kolkata Conf. EDKCON 2018*, pp. 509–512, 2018, doi: 10.1109/EDKCON.2018.8770418.
102. Mukherjee, H., Dasgupta, R., Kar, M., Kundu, A., A Comparative Analysis of Analog Performances of Underlapped Dual Gate AlGaN/GaN Based MOS-HEMT and Schottky-HEMT. *2020 IEEE Calcutta Conf. CALCON 2020 - Proc.*, pp. 412–416, 2020, doi: 10.1109/CALCON49167.2020.9106420.
103. Chakrabarti, H., Maity, R., Baishya, S., Maity, N.P., An Accurate Model for Threshold Voltage Analysis of Dual Material Double Gate Metal Oxide Semiconductor Field Effect Transistor. *Silicon*, 13, 6, 1851–1861, Jun. 2021, doi: 10.1007/S12633-020-00553-8.
104. Mitra, R., Roy, A., Kundu, A., Kar, M., Impact of AlGaN Doping Concentration on the Analog/RF Performance of a Double Gate Underlap n-AlGaN/GaN MOSHEMT. *3rd Int. Symp. Devices, Circuits Syst. ISDCS 2020 - Proc.*, pp. 3–6, 2020, doi: 10.1109/ISDCS49393.2020.9263012.
105. Pardeshi, H., Analog/RF performance of AlInN/GaN underlap DG MOS-HEMT. *Superlattices Microstruct.*, 88, 508–517, Dec. 2015, doi: 10.1016/j.spmi.2015.10.009.

106. Wei, J., *et al.*, Low On-Resistance Normally-Off GaN Double-Channel Metal-Oxide-Semiconductor High-Electron-Mobility Transistor. *IEEE Electron Device Lett.*, 36, 12, 1287–1290, 2015, doi: 10.1109/LED.2015.2489228.
107. Ma, J., *et al.*, 1200 v Multi-Channel Power Devices with 2.8 Ωmm ON-Resistance, in: *Technical Digest - International Electron Devices Meeting, IEDM*, Dec. 2019, vol. 2019-Decem, doi: 10.1109/IEDM19573.2019.8993536.
108. Lei, J., *et al.*, Reverse-blocking normally-OFF GaN double-channel MOS-HEMT with low reverse leakage current and low ON-state resistance. *IEEE Electron Device Lett.*, 39, 7, 1003–1006, 2018, doi: 10.1109/LED.2018.2832180.
109. Bhuiyan, M.A., *et al.*, Total-Ionizing-Dose Responses of GaN-Based HEMTs with Different Channel Thicknesses and MOSHEMTs with Epitaxial MgCaO as Gate Dielectric. *IEEE Trans. Nucl. Sci.*, 65, 1, 46–52, 2018, doi: 10.1109/TNS.2017.2774928.
110. Roy, A., Mitra, R., Mondal, A., Kundu, A., Analog/RF and Power Performance Analysis of an Underlap DG AlGaN/GaN Based High-K Dielectric MOS-HEMT. *Silicon*, 14, 5, 2211–2218, 2022, doi: 10.1007/s12633-021-01020-8.
111. Verma, M. and Nandi, A., Design and Analysis of AlGaN/GaN Based DG MOSHEMT for High-Frequency Application. *Trans. Electr. Electron. Mater.*, 21, 4, 427–435, 2020, doi: 10.1007/s42341-020-00196-x.
112. Zhou, H., Lou, X., Kim, S.B., Chabak, K.D., Gordon, R.G., Ye, P.D., Enhancement-Mode AlGaN/GaN Fin-MOSHEMTs on Si Substrate with Atomic Layer Epitaxy MgCaO. *IEEE Electron Device Lett.*, 38, 9, 1294–1297, 2017, doi: 10.1109/LED.2017.2731993.
113. Lee, C.T. and Juo, H.Y., Multiple-Submicron Channel Array Gate-Recessed AlGaN/GaN Fin-MOSHEMTs. *IEEE J. Electron Devices Soc.*, 6, 1, 183–188, 2018, doi: 10.1109/JEDS.2017.2786866.
114. Huang, Y.P., Hsu, W.C., Liu, H.Y., Lee, C.S., Enhancement-Mode Tri-Gate Nanowire InAlN/GaN MOSHEMT for Power Applications. *IEEE Electron Device Lett.*, 40, 6, 929–932, 2019, doi: 10.1109/LED.2019.2911698
115. Ma, J., Santoruvo, G., Nela, L., Wang, T., Matioli, E., Impact of Fin Width on Tri-Gate GaN MOSHEMTs. *IEEE Trans. Electron Devices*, 66, 9, 4068–4074, 2019, doi: 10.1109/TED.2019.2925859.
116. Huang, Y.P., Lee, C.S., Hsu, W.C., Normally-Off InAlN/GaN Fin-MOSHEMT with Fluorine Treatment. *Device Res. Conf. - Conf. Dig. DRC*, 2020-June, 2016, 2019–2020, 2020, doi: 10.1109/DRC50226.2020.9135151.
117. Lee, C.T. and Guo, J.C., Fin-gated nanochannel array gate-recessed AlGaN/GaN metal-oxide-semiconductor high-electron-mobility transistors. *IEEE Trans. Electron Devices*, 67, 5, 1939–1945, May 2020, doi: 10.1109/TED.2020.2981138.
118. Varghese, A., Periasamy, C., Bhargava, L., Bin Dolmanan, S., Tripathy, S., Fabrication and Modeling-Based Performance Analysis of Circular GaN MOSHEMT-Based Electrochemical Sensors. *IEEE Sens. J.*, 21, 4, 4216–4224, 2021, doi: 10.1109/JSEN.2020.3032175.

119. Varghese, A., Periasamy, C., Bhargava, L., Bin Dolmanan, S., Tripathy, S., Linear and Circular AlGaN/AlN/GaN MOS-HEMT-based pH Sensor on Si Substrate: A Comparative Analysis. *IEEE Sens. Lett.*, 3, 4, 2019–2022, 2019, doi: 10.1109/LSENS.2019.2909291.
120. Bollella, P., Katz, E., Biosensors—Recent Advances and Future Challenges. *Sensors*, 20, 6645, 2020, doi: 10.3390/s20226645.
121. Libertino, S., Aiello, V., Scandurra, A., Renis, M., Sinatra, F., Lombardo, S., Feasibility studies on Si-based biosensors. *Sensors*, 9, 5, 3469–3490, Apr. 2009, doi: 10.3390/S90503469.
122. Gudkov, A.G., Agasieva, S.V., Tikhomirov, V.G., Zherdeva, V.V., Klinov, D.V., Shashurin, V.D., Perspective in the development of biosensors based on AlGaN/GaNHEMT. *Biomed. Eng.*, 53, 3, 196–200, Sep. 2019, doi: 10.1007/s10527-019-09908-x.
123. Wadhera, T., Kakkar, D., Wadhwa, G., Raj, B., Recent Advances and Progress in Development of the Field Effect Transistor Biosensor: A Review. *J. Electron. Mater.*, 48, 12, 7635–7646, Dec. 2019, doi: 10.1007/S11664-019-07705-6.
124. Rapp, B.E., Gruhl, F.J., Länge, K., Biosensors with label-free detection designed for diagnostic applications. *Anal. Bioanal. Chem.*, 398, 6, 2403–2412, 2010, doi: 10.1007/s00216-010-3906-2.
125. Stutzmann, M., *et al.*, GaN-based heterostructures for sensor applications. *Diam. Relat. Mater.*, 11, 3–6, 886–891, Mar. 2002, doi: 10.1016/S0925-9635(02)00026-2.
126. Shaveta, Ahmed, H.M.M., Chaujar, R., Rapid detection of biomolecules in a dielectric modulated GaN MOSHEMT. *J. Mater. Sci. - Mater. Electron.*, 31, 19, 16609–16615, 2020, doi: 10.1007/s10854-020-04216-7.
127. Varghese, A., Das, P., Tallur, S., A complete analytical model for MOS-HEMT biosensors: Capturing the effect of Stern layer and charge screening on sensor performance. *IEEE Sens. Lett.*, 1472, AUGUST 2020, 2021, doi: 10.1109/LSENS.2021.3065509.
128. Hemaja, V. and Panda, D.K., A Comprehensive Review on High Electron Mobility Transistor (HEMT) Based Biosensors: Recent Advances and Future Prospects and its Comparison with Si-Based Biosensor. *Silicon*, 14, 5, 1873–1886, 2022, doi: 10.1007/s12633-020-00937-w.
129. Bhat, A.M., Varghese, A., Shafi, N., Periasamy, C., A Dielectrically Modulated GaN/AlN/AlGaN MOSHEMT with a Nanogap Embedded Cavity for Biosensing Applications. *IETE J. Res.*, 69, 3, 1419–1428, 2023, doi: 10.1080/03772063.2020.1869593.
130. Pal, P., Pratap, Y., Gupta, M., Kabra, S., Modeling and Simulation of AlGaN/GaN MOS-HEMT for Biosensor Applications. *IEEE Sens. J.*, 19, 2, 587–593, 2019, doi: 10.1109/JSEN.2018.2878243.
131. Mishra, S.N. and Jena, K., A Dielectric-Modulated Normally-Off AlGaN/GaN MOSHEMT for Bio-Sensing Application: Analytical Modeling Study and Sensitivity Analysis. *J. Korean Phys. Soc.*, 74, 4, 349–357, 2019, doi: 10.3938/jkps.74.349.

132. Poonia, R., Bhat, A.M., Periasamy, C., Sahu, C., Performance Analysis of MOS-HEMT as a Biosensor: A Dielectric Modulation Approach. *Silicon*, 14, 15, 10023–10036, Oct. 2022, doi: 10.1007/S12633-022-01742-3/METRICS.
133. Kiran, G., Krishna, R., Dwivedi, P., Sharma, P., Singh, R., Analytical Modeling of MgZnO/ZnO MOSHEMT Based Biosensor for Biomolecule Detection. *Micro Nanostruct.*, 163, 107130, 2022, doi: 10.1016/J.SPMI.2021.107130.
134. Mouffoki, F., Bouguenna, D., Dahou, F.Z., Beloufa, A., Loan, S.A., Performance evaluation of electrical properties of GaN MOS-HEMTs based biosensors for rapid detection of viruses. *Mater. Today Commun.*, 33, 104726, Dec. 2022, doi: 10.1016/J.MTCOMM.2022.104726.
135. Bouguenna, A., Bouguenna, D., Stambouli, A.B., Loan, S.A., Bhat, A.M., Performance Analysis of AlGaN MOSHEMT Based Biosensors for Detection of Proteins. *Trans. Electr. Electron. Mater.*, 24, 3, 188–193, Jun. 2023, doi: 10.1007/S42341-023-00434-Y/METRICS.

12

Quantum Computing and Digital Twins with Development of Semiconductor Field Effect Transistors

Shiromani Balmukund Rahi[1] **and Young Suh Song**[2]*

[1]University School of Information and Communication Technology, Gautam Buddha University, Uttar Pradesh, India
[2]Department of Computer Science, Korea Military Academy, Seoul, Republic of Korea

Abstract
This chapter explains the general concepts of quantum computing and digital twins using technological terminology. Specifically, considering the development of modern semiconductor technology and quantum computing technology, this chapter explores the interconnected domains of quantum computing, digital twins, and semiconductor development (especially non-memory semiconductor development), analyzing their combined influence on technological progress. This chapter investigates the basics and fundamentals of quantum computing, the progression and utilization of digital twins in diverse industries, and the essential role of emerging semiconductor technologies in making them operational. Through a multidisciplinary approach, this chapter reveals the synergies, obstacles, and potential uses in various applications, illuminating the considerable impact of these technologies across a wide range of fields. Furthermore, we address ethical concerns and outline future research directions and collaborative opportunities, offering informative insights into the technological landscape of the future.

Keywords: Quantum computing, digital twins, technological development, semiconductor industry, semiconductor

**Corresponding author*: songyeongseo84@gmail.com

Ekta Goel, Archana Pandey, Shiromani Balmukund Rahi and Arun Samuel (eds.) Handbook of Advanced Semiconductor Field Effect Transistors, (255–276) © 2026 Scrivener Publishing LLC

12.1 Introduction to Quantum Computing: Concept, History, and Principles

Nowadays, quantum computing has drawn significant attention owing to its ultrafast computation speed and technological merits. It has unlimited potential, and it is expected that quantum computing can be widely utilized in many industries. Quantum computing introduces a groundbreaking and state-of-the-art method of computation by intelligently utilizing the fundamental principles of quantum mechanics to execute operations that are challenging for classical computers [1–4]. Quantum computing can be explained by referring to computing related to quantum bits, or qubits, which can exist in multiple states [5–7]. This unique capability enables quantum computers to explore numerous unlimited potential solutions concurrently, resulting in significantly enhanced performance efficiency in solving specific types of problems [8].

The roots of quantum computing can be understood by tracing back to the early 20th century, particularly the work of pioneers such as Max Planck, Albert Einstein, and Niels Bohr [9–13]. However, the concept of quantum computing as we know it today began to take shape in the 1980s and 1990s with groundbreaking research by physicists such as Richard Feynman, David Deutsch, and Peter Shor [9–13]. Since then, significant progress has been made in both theoretical understanding and experimental implementation, leading to the emergence of several quantum-computing technologies and platforms.

To understand quantum computing, it is crucial to grasp key concepts, such as superposition, entanglement, and interference [14, 15] (Figure 12.1). The double-slit experiment, as shown at the bottom of Figure 12.1, which is a fundamental illustration in quantum mechanics, elegantly integrates the principles of superposition, interference, uncertainty, measurement, and quantization. When light traverses a narrow aperture such as a slender slit, it diffuses. When two slender slits are positioned adjacent to each other, the light passing through one aperture diffuses into the light passing through the other. Observing the light at a distance from the apertures reveals areas of constructive interference, where there is some light, and areas of destructive interference, where there is no light. This phenomenon can be understood as interference between light waves emanating from the left and right apertures. Superposition enables qubits to exist simultaneously in multiple states, significantly increasing the computational ability of quantum systems [1–3]. Entanglement links the states of multiple qubits such that the state of one qubit is closely related to the state

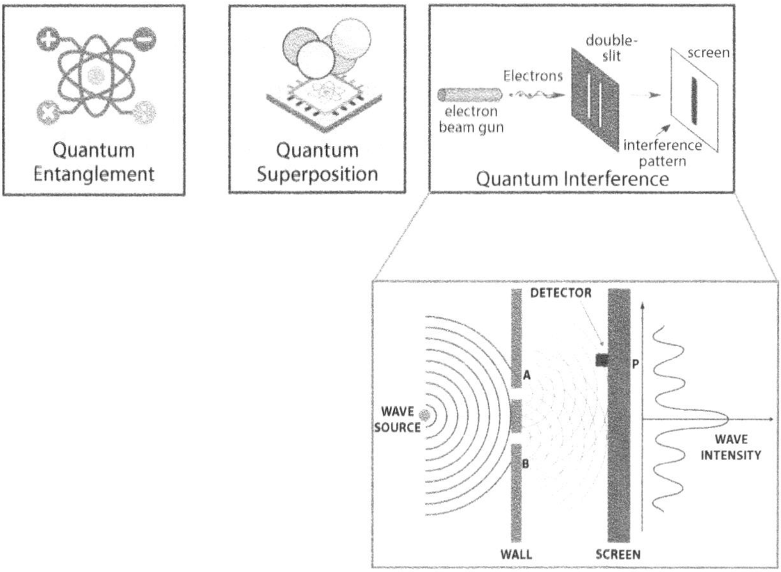

Figure 12.1 Illustration explaining the concept of quantum entanglement, quantum superposition, and quantum interference.

of another, even when separated by large distances. Interference is the phenomenon in which quantum states affect (i.e., interfere with) each other, leading to constructive or destructive interference and influencing the final outcome of computations.

Quantum computing research is experiencing rapid growth and innovation across the academia, industry, and government sectors worldwide. Major technology companies are heavily investing in quantum computing research and development. Significant milestones have been achieved, such as quantum supremacy and advancements in error correction, qubit coherence, and scalability. However, quantum computers are still in the early stages of development and face challenges such as undesirable noise, error rates, and limited qubit connectivity.

The possible uses and applications of quantum computing extend across diverse industries, including cryptography, cybersecurity, drug discovery, material science, optimization, and entertainment. It holds the promise of transforming areas such as machine learning (ML), finance, logistics, and healthcare services by tackling and improving intricate problems that are presently beyond the capabilities of classical computers. As quantum technology advances, it is anticipated to significantly influence society, the economy, and our approach to computational dilemmas in coming years.

12.2 Understanding Digital Twins

12.2.1 Definition and Conceptual Framework

Digital twins represent a sophisticated concept in the realm of technology that encapsulates the idea of creating virtual replicas or simulations of physical entities, systems, or processes. At the heart of digital twins lies the notion of mirroring the behavior, attributes, and interactions of their real-world counterparts within a digital environment. This replication goes beyond mere static representations: digital twins dynamically synchronize with their physical counterparts, constantly updating to reflect real-time changes and conditions.

The conceptual framework underlying digital twins entails a mutually beneficial relationship between the physical and digital realms, facilitated by a range of technologies such as sensors, nanodevices, data analytics, artificial intelligence, and connectivity solutions such as the Internet of Things (IoT). These technologies enable digital twins to establish an ongoing feedback loop between a physical entity and its digital counterpart, enabling real-time monitoring, analysis, and optimization.

Moreover, the conceptual framework of digital twins extends beyond individual entities to encompass entire systems, processes, or ecosystems. In this broader context, digital twins serve as comprehensive models that capture intricate interdependencies and interactions within complex systems. By replicating the behavior of interconnected components, digital twins offer stakeholders valuable insights into system dynamics, performance, and areas that need improvement.

Essentially, the definition and conceptual framework of digital twins revolve around seamlessly integrating physical and digital elements, empowering stakeholders to gain deeper insights, make informed decisions, and optimize processes across various domains, such as manufacturing, healthcare, transportation, and urban planning.

12.2.2 Evolution of Digital Twin Technology

The evolution of digital twin technology has been marked by significant advancements in computing power, data analytics, connectivity, and simulation techniques [16]. Initially conceived in the realm of manufacturing and engineering, digital twins have undergone a remarkable transformation, expanding their applications across diverse industries and domains.

The early stages of digital twin development primarily focused on creating static models of physical assets or systems for simulation and analysis. These models, while are useful for design and prototyping, lack the dynamic capabilities necessary to capture real-time data and respond to changes in the physical environment [17]. However, with the advent of sensor technology and the proliferation of IoT devices, digital twins have begun to evolve into more dynamic and data-driven representations.

The incorporation of sensors into physical assets facilitates real-time collection of data on diverse parameters such as temperature, pressure, vibration, and performance metrics. This data influx laid the groundwork for dynamic digital twins capable of continually monitoring and analyzing the behavior of their physical counterparts. Simultaneously, advancements in data analytics, machine learning, and artificial intelligence have empowered digital twins to derive actionable insights from the vast volumes of data generated by sensors and IoT devices.

Furthermore, the advent of cloud computing and edge computing technologies has accelerated the evolution of digital twins. Cloud platforms offer scalable infrastructure and computational resources for processing and analyzing large datasets. This distributed computing architecture enhances the responsiveness and agility of digital twins, enabling them to function effectively in dynamic and complex environments with minimal latency.

Presently, digital twins have transcended their roots in manufacturing and engineering, finding applications across a broad spectrum of sectors, including healthcare, transportation, energy, and smart cities [18]. From patient-specific digital twins in healthcare for tailoring treatment strategies to digital twins in urban environments to facilitate city planning and optimization, potential applications are unlimited. As digital twin technology continues to advance, fueled by technological progress and innovation, its impact on society, the economy, and our interaction with the physical world is poised to expand exponentially.

12.2.3 Types of Digital Twins

Digital twins can be classified into several types based on their scope, functionality, and level of abstraction. Each type serves a specific purpose and caters to different use cases across various industries and domains. These types are summarized in Table 12.1.

Table 12.1 Type of digital twins.

Type	Description
Product Twins	Product twins replicate individual physical assets, products, or components in a digital environment. These digital replicas are used to monitor the performance, condition, and behavior of specific products throughout their lifecycle. Product twins enable manufacturers to track product usage, detect anomalies, and optimize maintenance schedules. For example, in the automotive industry, product twins can simulate the behavior of vehicles to predict maintenance requirements and optimize performance.
System Twins	System twins represent interconnected systems or ecosystems comprising of multiple components, subsystems, or entities. These digital replicas capture the interactions and dependencies between different elements within a system, thereby providing holistic insights into system behavior, performance, and resilience. System twins enable stakeholders to analyze system dynamics, predict emergent behaviors, and optimize overall system performance. Examples include digital twins of smart cities, energy grids, and healthcare systems, which facilitate urban planning, energy management, and patient care optimization.
Process Twins	Process twins simulate entire and overall workflows, processes, or systems, providing insights into their operation, efficiency, and performance. These digital replicas capture the interactions and dependencies between different components or stages within a process, allowing stakeholders to identify bottlenecks, streamline operations, and improve resource allocations. Process twins find applications in manufacturing, supply chain management, and logistics, where they enable process optimization, inventory management, and predictive analytics.
Asset Twins	Asset twins focus on individual assets or equipment within a large system or facility. These digital replicas provide detailed insights into asset health, performance, and maintenance requirements, thereby enabling proactive maintenance, condition monitoring, and lifecycle management. Asset twins find applications in industries such as manufacturing, utilities, and aerospace, where they help optimize asset utilization, minimize downtime, and extend the asset lifespan.

(*Continued*)

Table 12.1 Type of digital twins. (*Continued*)

Type	Description
Enterprise Twins	Enterprise twins encompass comprehensive digital representations of all organizations, including their assets, processes, and stakeholders. These digital replicas integrate data from across organizations, enabling holistic insights into operations, performance, and decision-making. Enterprise twins can systematically support strategic planning, risk management, and organizational optimization by providing executives and decision makers with a unified view of the organization's resources, capabilities, and challenges.

12.2.4 Challenges and Opportunities in Digital Twin Implementation

The implementation of digital twins entails potential challenges and additional desirable opportunities for organizations across various industries. While digital twins offer transformative potential in terms of optimization, predictive maintenance, and innovation, their successful deployment requires overcoming several hurdles and leveraging opportunities effectively. There are a couple of challenges and opportunities, and they are explained in Tables 12.2 and 12.3.

Table 12.2 Challenges in digital twin implementation.

Type of challenge	Description
Data Integration and Quality	One of the primary challenges in digital twin implementation is the integration of data from diverse sources and ensuring its quality, consistency, and accuracy. Data may be collected from sensors, IoT devices, operational systems, and external sources, leading to issues such as data silos, interoperability, and inconsistency. Ensuring seamless data integration and maintaining data quality are essential for the reliability and effectiveness of digital twins.

(*Continued*)

Table 12.2 Challenges in digital twin implementation. (*Continued*)

Type of challenge	**Description**
Scalability and Complexity	Digital twins in complex systems or large-scale deployments pose scalability and complexity challenges. As the number of assets, processes, or components increases, managing computational resources, storage requirements, and computational complexity becomes increasingly challenging. Scalability issues may arise in terms of processing real-time data, performing simulations, and updating digital twins in a timely manner.
Interoperability and Standards	Interoperability and standards are crucial for enabling seamless communication and integration among different digital twin instances, systems, and platforms. A lack of interoperability standards may hinder data exchange, collaboration, and integration across organizational boundaries or between different technology providers. Establishing common data formats, protocols, and interfaces is essential for overcoming interoperability challenges and promoting interoperability among digital twins.
Security and Privacy	Security and privacy concerns are significant challenges in digital twin implementation, particularly concerning sensitive data, intellectual property, and critical infrastructure. Digital twins may be vulnerable to cyber-attacks, data breaches, and unauthorized access, posing risks to the integrity, confidentiality, and availability of data. Implementing robust security measures, encryption techniques, access controls, and privacy policies is essential to safeguard digital twins and the data they handle.

Table 12.3 Opportunities in digital twin implementation.

Type of opportunities	Description
Predictive Maintenance and Optimization	Digital twins enable predictive maintenance and optimization by leveraging real-time data, simulation models, and predictive analytics. By monitoring asset health, performance trends, and operational parameters, organizations can anticipate maintenance needs, detect anomalies, and optimize asset utilization. Predictive maintenance reduces downtime, extends the asset lifespan, and improves operational efficiency, leading to cost savings and enhanced reliability.
Innovation and Product Development	Digital twins facilitate innovation and product development by enabling virtual prototyping, simulations, and experimentation. Organizations can use digital twins to simulate product designs, test different configurations, and evaluate performance characteristics before physical prototypes are built. Digital twins accelerate the product development cycle, reduce time to market, and enhance product quality and functionality.
Decision Support and Risk Management	Digital twins provide decision support and risk management capabilities by enabling stakeholders to analyze complex scenarios, assess alternative strategies, and proactively mitigate risks. Organizations can use digital twins to simulate what-if scenarios, evaluate the impact of decisions, and optimize resource allocation. Digital twins improve decision-making accuracy, resilience, and agility, enabling organizations to respond effectively to changing conditions and uncertainties.
Enhanced Customer Experience	Digital twins enable organizations to deliver personalized, responsive, and immersive customer experiences by understanding customer preferences, behavioral patterns, and needs. Organizations can tailor offerings, anticipate customer requirements, and deliver customized solutions by creating digital replicas of customer environments, products, and services. Digital twins enhance customer satisfaction, loyalty, retention, business growth, and competitive advantage.

12.3 Semiconductor Development: Past, Present, and Future

The semiconductor industry plays a pivotal role in modern technology and serves as the foundation for electronic devices, wearable devices, networks, computations, and systems. Semiconductors are materials with electrical conductivity between those of conductors and insulators, enabling the control and manipulation of electrical signals. The semiconductor industry includes the design, manufacturing (also known as fabrication), and distribution of semiconductor devices, such as integrated circuits (ICs), transistors, and diodes (Figure 12.2). The key players in the semiconductor industry include semiconductor manufacturers, foundries, design firms, and equipment suppliers. Industrial growth is driven by advancements in computing, telecommunications, consumer electronics, automotive, and industrial applications.

The history of semiconductor development has been marked by several significant milestones that have shaped the industry's evolution. The development of integrated circuits (ICs) by Jack Kilby at Texas Instruments in 1958 further accelerated the miniaturization and integration of electronic components, leading to the emergence of the modern semiconductor industry. Subsequent milestones include the introduction of complementary metal-oxide-semiconductor (CMOS) technology, proliferation of microprocessors, rise of memory devices, and advent of advanced fabrication techniques such as photolithography and semiconductor scaling.

The current state of semiconductor technology is characterized by continuous innovation and advancements in device performance, energy efficiency, and integration density [19]. Semiconductor manufacturers produce chips with increasingly smaller feature sizes, enabling higher transistor densities and improved performance per watt [20]. Key technological trends include the transition to FinFET and gate-all-around (GAA)

Figure 12.2 The example of semiconductor: CPU.

transistor architectures (as shown in Figure 12.3), adoption of advanced packaging techniques such as 3D stacking and heterogeneous integration, and development of novel materials and manufacturing processes [21–25]. Additionally, semiconductor companies are focusing on specialized applications such as artificial intelligence (AI), the Internet of Things (IoT), computer design, high-performance CPU design, and automotive electronics to address emerging market demands [26, 27].

Emerging trends and technologies in semiconductors are shaping the future of the industry and driving innovation across various domains [28–31]. One such trend is the proliferation of edge computing and AI, which require low-power, high-performance semiconductor solutions capable of processing and analyzing data at the edge of the network. Another trend is the growing importance of silicon photonics and quantum computing, which leverage semiconductor materials and fabrication techniques, respectively, to enable faster data transfer and quantum information processing. In addition, the semiconductor industry is witnessing increased investment in sustainability and environmental stewardship, with a focus on reducing energy consumption, minimizing waste, and developing eco-friendly manufacturing processes.

Semiconductor development plays a crucial role in advancing quantum computing and digital twins by enabling the development of the underlying hardware infrastructure and computational capabilities. As the transistor size decreases and the integrity of the CPU improves, it is possible to significantly improve the operation of quantum computing. Quantum computing relies on semiconductor-based qubits to perform quantum operations and manipulate the quantum states. Semiconductor manufacturers are exploring novel materials, device architectures, and fabrication techniques to improve the qubit

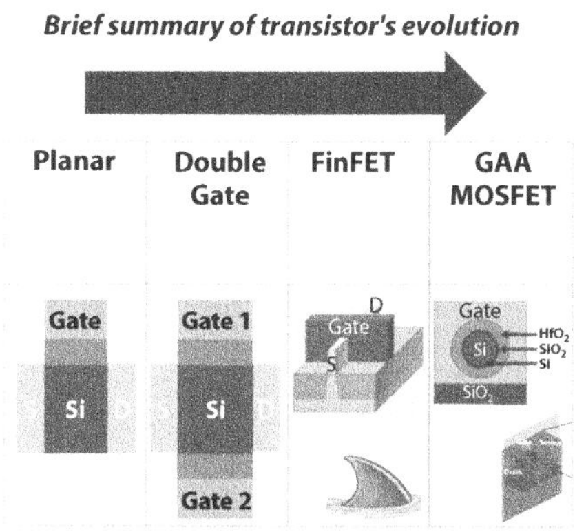

Figure 12.3 Brief summary of the evolution of transistor.

performance, coherence, and scalability. Similarly, digital twins leverage semiconductor technology for data acquisition, processing, and communication, enabling the real-time monitoring, analysis, and optimization of physical systems. Semiconductor companies are developing specialized chips and sensors optimized for digital twin applications, enabling more efficient and intelligent data-driven decision making. Overall, semiconductor development is poised to drive the next wave of innovation in quantum computing, digital twins, and beyond, shaping the future of the technology and transforming industries.

12.4 Integration of Quantum Computing and Digital Twins

12.4.1 Synergies Between Quantum Computing and Digital Twin Technologies

In this section, we discuss the intertwin synergies between two state-of-the-art technologies: quantum computing and digital twin technology. In essence, the integration of quantum computing and digital twin technologies presents exciting synergies that have the potential to revolutionize various industries.

The synergy between quantum computing and digital twins lies in their complementary strengths. Quantum computing can enhance the capabilities of digital twins by providing advanced analytics, optimization algorithms, and simulation techniques. Quantum algorithms can be used to optimize processes, predict system behavior, and identify optimal configurations in real-time. By leveraging quantum computing, digital twins can perform complex calculations and simulations more efficiently, enabling faster decision making and improved performance.

Moreover, digital twins might benefit quantum computing by providing real-world data, scenarios, and feedback loops for algorithm development and validation. Digital twins can generate training data, simulate experimental conditions, and test quantum algorithms in virtual environments before being deployed in the real world. This integration enables quantum computing to address real-world challenges and applications more effectively, thereby accelerating the adoption and impact of quantum technology across industries.

12.4.2 Case Studies and Use Cases

Several case studies and use cases have demonstrated the potential of integrating quantum computing and digital twin technologies in various domains. For example, in manufacturing, digital twins can simulate production processes, predict equipment failures, and optimize supply chain logistics, whereas quantum computing can optimize production schedules, perform advanced material simulations, and optimize inventory management algorithms.

Within the healthcare sector, the integration of digital twins representing patient physiology with quantum algorithms offers opportunities for advancements in drug discovery, personalized treatment strategies, and disease modeling. Quantum computing is capable of analyzing genomic data, stimulating molecular interactions, and identifying optimal drug candidates. Concurrently, digital twins can provide immediate patient data and insights for the validation and refinement of algorithms.

Digital twins in financial markets can be augmented with quantum algorithms for risk management, portfolio optimization, and fraud detection. Quantum computing can analyze market data, simulate economic models, and optimize trading strategies, whereas digital twins can provide real-time market data and scenario analysis for algorithm testing and validation.

12.4.3 Challenges and Solutions in Integration

Despite the promising synergies between quantum computing and digital twins, several challenges exist in their integration. One challenge is the compatibility and interoperability between quantum computing platforms and digital twin environments. Integrating quantum algorithms with the existing digital twin infrastructure and data systems requires standardized interfaces, protocols, and data formats.

Another challenge is the scalability and complexity of quantum algorithms in real-world applications. Many quantum algorithms are still in the research and development stages, with limited scalability and applicability to large-scale problems. Overcoming these challenges requires advances in quantum hardware, software, and algorithm design as well as collaboration between quantum computing researchers and digital twin practitioners.

Furthermore, ensuring the security and privacy of data in integrated quantum-digital twin systems is crucial. Quantum computing has the potential to break traditional encryption schemes, posing risks to the privacy of sensitive data and information. Implementing robust encryption, authentication, and access control mechanisms is essential for safeguarding data in integrated quantum-digital twin environments.

12.4.4 Future Prospects and Possibilities

Despite these challenges, the integration of quantum computing and digital twin technologies holds immense promise for the future. As quantum hardware matures and quantum algorithms improve, the capabilities of integrated quantum-digital twin systems will continue to expand. Industries such as manufacturing, healthcare, finance, and logistics stand to benefit from the enhanced predictive analytics, optimization, and decision-support capabilities enabled by this integration.

Moreover, the emergence of quantum Internet and distributed quantum computing architectures will further accelerate the adoption and impact of integrated quantum-digital twin systems. Quantum networks enable seamless communication and collaboration between distributed digital twin instances, allowing for real-time data sharing, analysis, and decision making across organizational boundaries and geographical locations.

12.5 Applications and Impact Across Industries

Considering the wide variety of industries involved, it seems like there are many possible applications. Among these, the promising application might be 'healthcare,' 'manufacturing,' 'finance and banking,' 'transportation,' 'energy,' and 'entertainment.' A specific explanation of each application can be found in Table 12.4.

Table 12.4 Possible applications related to digital twins and quantum computing.

Categories	Explanation
Healthcare	In the healthcare sector, the integration of advanced technologies such as artificial intelligence (AI), the Internet of Things (IoT), and digital twins is revolutionizing patient care, medical research, and healthcare management. Digital twins in patient physiology enable personalized treatment planning, predictive diagnostics, and virtual surgery simulations, leading to improved patient outcomes and reduced healthcare costs. Additionally, AI algorithms powered by quantum computing can analyze genomic data, identify disease biomarkers, and optimize drug discovery processes, thereby accelerating the development of novel therapies and treatments.

(Continued)

Table 12.4 Possible applications related to digital twins and quantum computing. (*Continued*)

Categories	Explanation
Manufacturing	In manufacturing, digital twins transform traditional production processes by enabling predictive maintenance, real-time monitoring, and process optimization. Digital twins in manufacturing plants simulate production workflows, predict equipment failures, and optimize production schedules, leading to increased operational efficiency and reduced downtime. Moreover, the integration of quantum computing in manufacturing enables advanced material simulations, optimization of supply chain logistics, predictive analytics for demand forecasting, and driving innovation and competitiveness in the industry.
Finance and Banking	The finance and banking sector leverages digital twins and quantum computing to enhance risk management, portfolio optimization, and fraud detection capabilities. Digital twins in financial markets provide real-time insights into market dynamics, enabling traders and investors to make informed decisions and effectively manage risks. Quantum algorithms powered by quantum computing can analyze vast amounts of financial data, simulate economic models, and optimize trading strategies, leading to improved investment returns and financial stability.
Transportation	In transportation, digital twins are revolutionizing vehicle design, traffic management, and logistics optimization. Digital twins of vehicles enable predictive maintenance, performance monitoring, and autonomous driving simulations, thereby improving the safety and efficiency of roads. Quantum computing in transportation facilitates route optimization, traffic prediction, and fleet management, thereby reducing congestion, emissions, and fuel consumption. Additionally, digital twins in smart cities enable urban planners to design sustainable transportation systems, optimize public transit routes, and improve overall mobility for residents.

(*Continued*)

Table 12.4 Possible applications related to digital twins and quantum computing. (*Continued*)

Categories	Explanation
Energy	The energy sector harnesses digital twins and quantum computing to optimize energy production, distribution, and consumption. Digital twins in power plants simulate energy generation processes, predict equipment failures, and optimize energy efficiency, leading to reduced operational costs and environmental impacts. Quantum algorithms powered by quantum computing can optimize energy grid operations, perform renewable energy simulations, and enable the predictive maintenance of energy infrastructure, thereby supporting the transition to a sustainable energy future.
Entertainment and Gaming	In the entertainment and gaming industries, digital twins and quantum computing drive innovation in content creation, immersive experiences, and virtual environments. Digital twins in gaming environments enable realistic simulations, personalized gaming experiences, and dynamic content generation, thereby enhancing player engagement and retention. Quantum computing in entertainment enables advanced graphics rendering, complex physics simulations, and realistic artificial intelligence behavior, pushing the boundaries between interactive entertainment and virtual reality applications.

12.6 Ethical and Societal Implications

As digital twins and quantum computing have become increasingly integrated into various aspects of our lives, concerns about privacy and data security have become paramount. Digital twins generate vast amounts of data, including sensitive personal and operational information, raising concerns regarding data breaches, unauthorized access, and misuse. Quantum computing also poses security risks because quantum algorithms can potentially break conventional encryption methods, compromising data confidentiality and integrity. Addressing these concerns requires robust security measures, encryption techniques, access controls, and privacy policies to safeguard data and ensure trust and transparency in digital twin and quantum computing systems.

The adoption of digital twins and quantum computing technologies may exacerbate the existing disparities in access to technology, resources, and opportunities. Affordability, infrastructure availability, and technical expertise are barriers to technology adoption, particularly in underserved communities and in developing regions. Moreover, biases in data collection, algorithm design, and decision-making processes can perpetuate inequality and discrimination, reinforcing the existing social and economic disparities. Ensuring equitable access to technology and promoting diversity, inclusion, and accessibility in technology development and deployment are essential for addressing equity concerns and fostering equitable and inclusive societies.

The proliferation of digital twins and quantum computing technologies has implications for environmental sustainability, resource consumption, and the carbon footprint. The energy-intensive nature of quantum computing, particularly in the cooling and operation of quantum hardware, raises concerns regarding energy consumption and environmental impact. Similarly, the manufacturing, operation, and disposal of digital twin-enabled devices and infrastructure contributes to electronic waste and resource depletion. Adopting energy-efficient hardware, sustainable manufacturing practices, and recycling initiatives, as well as investing in renewable energy sources, can mitigate the environmental footprint of digital twins and quantum computing technologies, thereby promoting environmental sustainability and responsible stewardship of resources.

12.7 Future Directions

12.7.1 Anticipated Developments in Quantum Computing and Digital Twins

The future of quantum computing and digital twins holds immense promise for further innovation, advancement, and transformative impacts across industries and domains. In quantum computing, anticipated developments include the realization of fault-tolerant quantum computers capable of solving complex optimization and simulation problems with unprecedented speed and efficiency. Quantum algorithms and software frameworks are expected to mature, enabling broader adoption and application in areas such as drug discovery, materials science, cryptography, and financial modeling. Moreover, the development of quantum Internet and distributed quantum computing architectures will expand the reach and capabilities of quantum computing, facilitating seamless collaboration and information sharing across global networks.

In the realm of digital twins, future developments are expected to focus on enhancing realism, interactivity, and intelligence in virtual environments. Advancements in sensor technology, augmented reality (AR), and machine learning (ML) will enable more immersive and interactive digital twin experiences, blurring the boundaries between physical and virtual worlds. Digital twins will evolve from static replicas to dynamic adaptive systems capable of autonomously responding to changes, optimizing processes, and predicting future scenarios. Additionally, the integration of digital twins with emerging technologies such as blockchain, 5G, and edge computing will enable new use cases and applications, further expanding the utility and impact of digital twins across industries and domains.

12.7.2 Opportunities for Further Research and Collaboration

Opportunities for further research and collaboration abound at the intersection of quantum computing and digital twins. Interdisciplinary research efforts are needed to explore the synergies, challenges, and applications of integrating quantum computing and digital twin technologies in diverse domains. Collaborative initiatives between academia, industry, and government can drive innovation, knowledge exchange, and technology transfer in quantum computing and digital twins. Moreover, international collaboration and standardization efforts are essential for advancing the field, fostering cooperation, and addressing ethical, regulatory, and societal issues. Areas for further research include quantum-enhanced digital twin algorithms, quantum-secure data encryption, and quantum-assisted optimization techniques for digital twin applications.

12.8 Conclusion

Overall, the convergence of quantum computing and digital twins represents a paradigm shift in how we model, simulate, and optimize complex systems and processes. These technologies offer unprecedented opportunities for innovation, efficiency, and transformation across various industries and domains. However, realizing the full potential of quantum computing and digital twins requires addressing technical, ethical, and societal challenges, fostering collaboration, and promoting responsible innovation. By embracing these challenges and opportunities, we can harness the transformative power of quantum computing and digital twins to create a brighter, more sustainable future.

Acknowledgment

In this chapter, the knowledge in previous research has been widely utilized, and the source of information is listed as reference format (e.g., [1, 2], ...). The author would like to sincerely thank the authors who have made their sincere efforts to publish many articles and research related to this chapter.

References

1. de Andrade, M.G., Dai, W., Guha, S., Towsley, D., A quantum walk control plane for distributed quantum computing in quantum networks. *2021 IEEE International Conference on Quantum Computing and Engineering (QCE)*, Broomfield, CO, USA, pp. 313–323, 2021, doi: 10.1109/QCE52317.2021.00048.
2. de Andrade, M.G., Dai, W., Guha, S., Towsley, D., Optimal Policies for Distributed Quantum Computing with Quantum Walk Control Plane Protocol. *2021 IEEE International Conference on Quantum Computing and Engineering (QCE)*, Broomfield, CO, USA, pp. 452–453, 2021, doi: 10.1109/QCE52317.2021.00074.
3. Lokes, S., Mahenthar, C.S.J., Kumaran, S.P., Sathyaprakash, P., Jayakumar, V., Implementation of Quantum Deep Reinforcement Learning Using Variational Quantum Circuits. *2022 International Conference on Trends in Quantum Computing and Emerging Business Technologies (TQCEBT)*, Pune, India, pp. 1–4, 2022, doi: 10.1109/TQCEBT54229.2022.10041479.
4. Mykhailova, M., Teaching Quantum Computing Using Microsoft Quantum Development Kit and Azure Quantum. *2023 IEEE International Conference on Quantum Computing and Engineering (QCE)*, Bellevue, WA, USA, pp. 15–20, 2023, doi: 10.1109/QCE57702.2023.20320.
5. Alghadeer, M., Aldawsari, E., Selvarajan, R., Alutaibi, K., Kais, S., Alharbi, F.H., Psitrum and Universal Simulation of Quantum Computers. *2022 IEEE International Conference on Quantum Computing and Engineering (QCE)*, Broomfield, CO, USA, pp. 837–838, 2022, doi: 10.1109/QCE53715.2022.00137.
6. Lohof, F., Götting, N., Gies, C., Leveraging Quantum Dynamics for Physical Computing Applications. *2023 IEEE International Conference on Quantum Computing and Engineering (QCE)*, Bellevue, WA, USA, pp. 320–321, 2023, doi: 10.1109/QCE57702.2023.10262.
7. Arun, G. and Mishra, V., A review on quantum computing and communication. *2014 2nd International Conference on Emerging Technology Trends in Electronics, Communication and Networking*, Surat, India, pp. 1–5, 2014, doi: 10.1109/ET2ECN.2014.7044953.

8. Ferrari, D., Carretta, S., Amoretti, M., A Modular Quantum Compilation Framework for Distributed Quantum Computing. *IEEE Trans. Quantum Eng.*, 4, 1–13, 2023, Art no. 2500213, doi: 10.1109/TQE.2023.3303935.
9. Mykhailova, M., Developing Programming Assignments for Teaching Quantum Computing and Quantum Programming. *2022 IEEE International Conference on Quantum Computing and Engineering (QCE)*, Broomfield, CO, USA, pp. 688–692, 2022, doi: 10.1109/QCE53715.2022.00092.
10. Chen, K.-C., Short-Depth Circuits and Error Mitigation for Large-Scale GHZ-State Preparation, and Benchmarking on IBM's 127-Qubit System. *2023 IEEE International Conference on Quantum Computing and Engineering (QCE)*, Bellevue, WA, USA, pp. 207–210, 2023, doi: 10.1109/QCE57702.2023.10214.
11. Ferrari, D., Cacciapuoti, A.S., Amoretti, M., Caleffi, M., Compiler Design for Distributed Quantum Computing. *IEEE Trans. Quantum Eng.*, 2, 1–20, 2021, Art no. 4100720, doi: 10.1109/TQE.2021.3053921.
12. Parekh, R., Ricciardi, A., Darwish, A., DiAdamo, S., Quantum Algorithms and Simulation for Parallel and Distributed Quantum Computing. *2021 IEEE/ACM Second International Workshop on Quantum Computing Software (QCS)*, St. Louis, MO, USA, pp. 9–19, 2021, doi: 10.1109/QCS54837.2021.00005.
13. Liao, Y.-P., Cheng, Y.-L., Zhang, Y.-T., Wu, H.-X., Lu, R.-C., The interactive system of Bloch sphere for quantum computing education. *2022 IEEE International Conference on Quantum Computing and Engineering (QCE)*, Broomfield, CO, USA, pp. 718–723, 2022, doi: 10.1109/QCE53715.2022.00097.
14. Anantraj, I., Umarani, B., Karpagavalli, C., Usharani, C., Lakshmi, S.J., Quantum Computing's Double-Edged Sword Unravelling the Vulnerabilities in Quantum Key Distribution for Enhanced Network Security. *2023 International Conference on Next Generation Electronics (NEleX)*, Vellore, India, pp. 1–5, 2023, doi: 10.1109/NEleX59773.2023.10420896.
15. Ohkura, Y., Satoh, T., Van Meter, R., Simultaneous Execution of Quantum Circuits on Current and Near-Future NISQ Systems. *IEEE Trans. Quantum Eng.*, 3, 1–10, 2022, Art no. 2500210, doi: 10.1109/TQE.2022.3164716.
16. Talasila, P., Gomes, C., Mikkelsen, P.H., Arboleda, S.G., Kamburjan, E., Larsen, P.G., Digital Twin as a Service (DTaaS): A Platform for Digital Twin Developers and Users. *2023 IEEE Smart World Congress (SWC)*, Portsmouth, United Kingdom, pp. 1–8, 2023, doi: 10.1109/SWC57546.2023.10448890.
17. Zhu, Y., Chen, D., Zhou, C., Lu, L., Duan, X., A knowledge graph based construction method for Digital Twin Network. *2021 IEEE 1st International Conference on Digital Twins and Parallel Intelligence (DTPI)*, Beijing, China, pp. 362–365, 2021, doi: 10.1109/DTPI52967.2021.9540177.
18. Zheng, M. and Tian, L., Knowledge-based Digital Twin Model Evolution Management Method for Mechanical Products. *2021 IEEE 1st International*

Conference on Digital Twins and Parallel Intelligence (DTPI), Beijing, China, pp. 312–315, 2021, doi: 10.1109/DTPI52967.2021.9540181.

19. Wang, S., *et al.*, Wafer Defect Map Similarity Search using Deep Learning in Semiconductor Manufacturing. *2021 China Semiconductor Technology International Conference (CSTIC)*, Shanghai, China, pp. 1–4, 2021, doi: 10.1109/CSTIC52283.2021.9461569.
20. Rathore, S., *et al.*, Self-Heating Aware Threshold Voltage Modulation Conforming to Process and Ambient Temperature Variation for Reliable Nanosheet FET. *2023 IEEE International Reliability Physics Symposium (IRPS)*, Monterey, CA, USA, pp. 1–5, 2023, doi: 10.1109/IRPS48203.2023.10117918.
21. Song, Y.S., Kim, S., Kim, J.H., Kim, G., Lee, J.-H., Choi, W.Y., Enhancement of Thermal Characteristics and On-Current in GAA MOSFET by Utilizing Al2O3-Based Dual-κ Spacer Structure. *IEEE Trans. Electron Devices*, 70, 1, 343–348, Jan. 2023, doi: 10.1109/TED.2022.3223321.
22. Jaisawal, R.K., *et al.*, Self-Heating and Interface Traps Assisted Early Aging Revelation and Reliability Analysis of Negative Capacitance FinFET. *2023 7th IEEE Electron Devices Technology & Manufacturing Conference (EDTM)*, Seoul, Korea, Republic of, pp. 1–3, 2023, doi: 10.1109/EDTM55494.2023.10103127.
23. Song, Y.S., Kim, J.H., Kim, G., Kim, H.-M., Kim, S., Park, B.-G., Improvement in Self-Heating Characteristic by Incorporating Hetero-Gate-Dielectric in Gate-All-Around MOSFETs. *IEEE J. Electron Devices Soc.*, 9, 36–41, 2021, doi: 10.1109/JEDS.2020.3038391.
24. Greenwood, B., *et al.*, Gate oxide yield improvement for 0.18μm power semiconductor devices with deep trenches: DP: Discrete and power devices. *2017 28th Annual SEMI Advanced Semiconductor Manufacturing Conference (ASMC)*, Saratoga Springs, NY, USA, pp. 346–351, 2017, doi: 10.1109/ASMC.2017.7969258.
25. Tayal, S., *et al.*, Incorporating Bottom-Up Approach Into Device/Circuit Co-Design for SRAM-Based Cache Memory Applications. *IEEE Trans. Electron Devices*, 69, 11, 6127–6132, Nov. 2022, doi: 10.1109/TED.2022.3210070.
26. Song, Y.S., Tayal, S., Rahi, S.B., Kim, J.H., Upadhyay, A.K., Park, B.-G., Thermal-Aware IC Chip Design by Combining High Thermal Conductivity Materials and GAA MOSFET. *2022 5th International Conference on Circuits, Systems and Simulation (ICCSS)*, Nanjing, China, pp. 135–140, 2022, doi: 10.1109/ICCSS55260.2022.
27. Kang, S.J., Kim, J.H., Song, Y.S., Go, S., Kim, S., Investigation of Self-Heating Effects in Vertically Stacked GAA MOSFET With Wrap-Around Contact. *IEEE Trans. Electron Devices*, 69, 3, 910–914, March 2022, doi: 10.1109/TED.2022.3140283.
28. Adly, F., Yoo, P.D., Muhaidat, S., Al-Hammadi, Y., Lee, U., Ismail, M., Randomized General Regression Network for Identification of Defect

Patterns in Semiconductor Wafer Maps. *IEEE Trans. Semicond. Manuf.*, 28, 2, 145–152, May 2015, doi: 10.1109/TSM.2015.2405252.

29. Choi, Y., *et al.*, Simulation of the effect of parasitic channel height on characteristics of stacked gate-all-around nanosheet FET. *Solid State Electron.*, 164, 107686, Feb. 2020, doi: 10.1016/j.sse.2019.107686.
30. Kim, J.H., *et al.*, Analysis of current variation with work function variation in l-shaped tunnel-field effect transistor. *Micromachines*, 11, 8, 780, Aug. 2020, doi: 10.3390/mi11080780.
31. Song, Y.S., *et al.*, Reliability improvement of self-heating effect, hot-carrier injection, and on-current variation by electrical/thermal co-design. *Solid State Electron.*, 197, 108436, Nov. 2022, doi: 10.1016/j.sse.2022.108436.

13

Low Voltage Circuit Design with FinFETs

Sarita Yadav[1]* and Nitanshu Chauhan[2]

[1]*Department of Electronics Engineering, National Institute of Technology, Uttarakhand, India*
[2]*Department of Electronics and Communication Engineering, Indian Institute of Technology, Bhilai, India*

Abstract

The continued advancements in the field of CMOS technology have led to aggressive scaling of device dimensions with the goal of boosting the chip integration density. However, it has become increasingly hard to keep pace with Moore's law due to short channel effects that worsens with scaling. To address these challenges, the semiconductor industry has explored various solutions. One of the most promising solutions was adopting alternative device architectures like FinFETs which are associated with double or triple gates. These multiple gates enhanced the control exercised by the gate over the channel thus mitigating the severity of several second order effects referred as short channel effects.

With adoption of FinFET Technology, it became a norm of packing smaller devices with more features on a chip. This has led to a greater focus on ensuring these devices perform efficiently within the limits of their power supply. This becomes critical for areas like mobile computing, IoT devices, biosensors and etc where reliable operation is to be ensured with lower than nominal supply voltages. Hence, there is a lot of emphasis today on low voltage circuit design. FinFETs have exhibited superior subthreshold characteristics when compared to planar devices Moreover, it has been established that the minimum value to which the supply voltage can be scaled in a CMOS inverter realized with FinFET devices is quite small. This voltage is well below 36mV, which is the fundamental limit for planar devices at room temperature. Owing to these, FinFETs present themselves as an excellent candidate for low voltage and low power circuit designs.

However simply replacing the planar devices with FinFETs is not the right approach for optimized performance of circuits at low voltages. There are several

**Corresponding author*: sarita.yadav@nituk.ac.in

Ekta Goel, Archana Pandey, Shiromani Balmukund Rahi and Arun Samuel (eds.) Handbook of Advanced Semiconductor Field Effect Transistors, (277–304) © 2026 Scrivener Publishing LLC

device- circuit interaction issues arising that need to be addressed in a comprehensive manner before finalizing the circuit design. FinFETs have quantized widths due to discrete number of fins and circuits cannot be designed for arbitrary widths like in conventional CMOS technology. By virtue of its structure, FinFETs inherently have source and drain extension (SE and DE) regions. The carrier density in these regions is bias dependent. This results in several interesting device level phenomenon. One critical observation with FinFETs based circuits is the dependence of effective capacitances on circuit parameters like supply voltage of operation, the transition time of input signals and the fanout the circuit. It has also been reported that saturation phenomenon in FinFETs is quite distinct when compared with planar devices. This significantly affects the performance of circuits designed from the analog perspective as well. A systematic methodology for FinFET based low voltage circuit design is required for meeting performance criteria be it a digital circuit or an analog circuit. This chapter first presents an in-depth review of the existing literature on low voltage circuit design using FinFETs. It focuses on first reviewing the critical device-circuit co-design challenges in low voltage domain for FinFETs. After developing this understanding, we review the existing methodologies used to design low voltage circuits using FinFETs in the literature.

***Keywords*:** Analog design, effective capacitance, extension length, FinFETs, logical effort, low voltage, inverter delay, supply voltage

13.1 Introduction

Intel, a leading force in technological innovation, was the first major company to introduce three-dimensional transistors by adopting 22-nm FinFETs as a replacement for traditional planar CMOS devices. This move marked a significant milestone in the semiconductor industry. FinFETs have emerged as an advanced solution, especially at the most cutting-edge technology nodes, where they demonstrate numerous benefits under low-supply voltage conditions. Specifically, FinFETs provide enhanced performance metrics, minimize leakage current, and lower dynamic power consumption while reducing variability in device behavior. These advantages make FinFETs exceptionally well suited for digital designs that require low-voltage operation to achieve efficiency. While nanosheets are now being explored as a potential successor to FinFET devices, the FinFET architecture remains the most extensively studied among various multiple-gate technologies. FinFETs have superior subthreshold characteristics compared with planar devices [1]. Interestingly, studies have shown that the minimum supply voltage achievable in a CMOS inverter using FinFET technology is as low as 25 mV [2], which is a significant reduction from the 36-mV limit inherent to planar

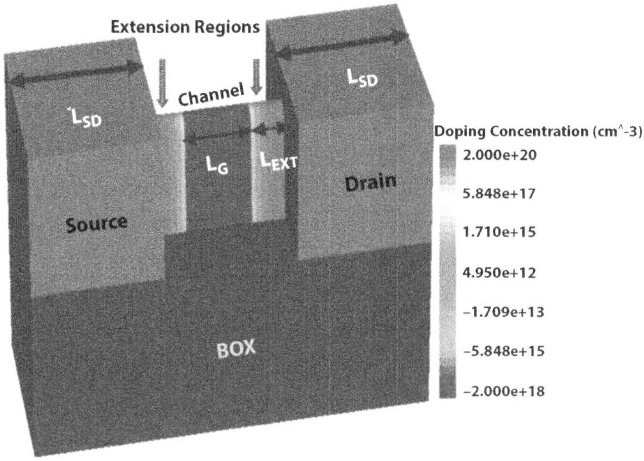

Figure 13.1 The three-dimensional structure of a 14-nm SOI FinFET was simulated using Synopsys TCAD, illustrating a typical doping profile across the source/drain pads, source and drain extension regions, and the fin region.

devices at room temperature [3]. These attributes make FinFETs an ideal choice for designing low-voltage, low-power circuits.

FinFETs exhibit atypical circuit behavior when compared to planar devices, which is a significant point of interest in low-voltage designs [4, 5]. This behavior arises because of the modulation of the carrier densities in the fin and extension regions of the device with the applied biases. Underlapped FinFETs show the strongest atypical behavior; hence, the circuit design aspects need to be reevaluated [4, 6]. This chapter provides an in-depth review of FinFET technology, offering insights into key challenges and considerations involving device–circuit interactions in low-voltage digital circuits. A typical device structure for FinFETs to report the results in this review chapter is illustrated in Figure 13.1.

13.2 Advent of FinFETs

In 1965, Gordon Moore, Intel's co-founder, presented an influential theory [7] that predicted that the number of transistors in integrated circuits would double approximately every two years. This projection, now famously known as "Moore's Law," highlighted the trend of increasing transistor density over time, and subsequently became a fundamental guiding principle for the semiconductor industry. To sustain this growth in transistor numbers and maintain device performance, MOSFETs were aggressively scaled down. Engineers have implemented scaling techniques such as reducing the gate oxide thickness, minimizing the junction depth of the source and drain, and adjusting the substrate doping densities. However, these scaling strategies have introduced new technical challenges. For instance, while reducing the gate dielectric thickness enhanced gate control over the channel, it also resulted in increased leakage currents owing to tunneling through the thinner gate oxide layer. Additionally, increased body doping improved control but caused mobility degradation due to heightened scattering. Turning off scaled transistors completely in the OFF state also became difficult because of various short-channel effects (SCEs) such as higher subthreshold leakage currents, increased drain-induced barrier lowering (DIBL), gate-induced drain leakage (GIDL), hot carrier effects, and enhanced gate tunneling currents.

Numerous advancements in the fabrication techniques have been introduced to address these scaling challenges. Notably, in 2003, the 90-nm technology node marked the deployment of uniaxially strained silicon technology to boost device mobility [8]. This technique involves embedding a silicon–germanium (SiGe) layer in the PMOS source and drain regions, generating compressive stress in the channel, and thereby significantly enhancing hole mobility. For NMOS devices, tensile stress is applied to the channel through a silicon nitride (SiN) capping layer, which results in increased electron mobility. These strain-engineered technologies proved instrumental in enabling further scaling down to the 45 nm node. Additional enhancements include replacing traditional polysilicon gates with metal gates to reduce the effects of polysilicon depletion and introducing high-k dielectrics as gate insulators to reduce gate leakage. Despite these developments, moving beyond the 32-nm node requires a shift from planar to nonplanar architectures, as traditional scaling techniques have reached their limits.

In the ongoing pursuit of continued scaling in semiconductor technology, researchers have proposed a range of novel device architectures. As the dimensions of planar devices continue to shrink, maintaining an effective electrostatic control has become increasingly difficult. This has led to the development of new device architectures designed to enhance gate control, while simultaneously minimizing the interaction between the source and drain regions, which has proven to be crucial for maintaining device performance. Among these advancements, extremely thin body architectures (ETSOI), also referred to as ultrathin-body silicon-on-insulator (SOI) or fully depleted SOI (FDSOI) devices, have been introduced specifically to improve short-channel properties [9]. These innovative devices feature an ultrathin channel formed within a thin silicon film, which significantly enhances short-channel effects (SCEs) by effectively terminating the electric field lines in the underlying buried oxide (BOX) layer. Despite these advantages, the performance of ETSOI devices is still limited by challenges such as the increased parasitic source/drain junction capacitance, which poses a significant hurdle that must be addressed to fully realize their potential in advanced applications.

Multiple-gate architectures such as double-gate MOSFETs (DGMOSFETs) and vertical FinFETs provide better SCE control by maintaining electrostatic integrity. The double-gate structure, first proposed in 1984 [10], features a fully depleted SOI sandwiched between two gates, significantly reducing short-channel effects. Vertical DGMOSFETs with self-aligned gate structures have emerged as an excellent alternative. The first vertical double-gate MOSFET, known as DELTA [11], was fabricated in 1989. This concept evolved into FinFETs developed at the University of California, Berkeley [12]. With each generation, chip designers achieved a 0.7× reduction in transistor specifications, resulting in performance improvements, footprint reductions, power consumption improvements, and cost reductions. Leading foundries, such as Intel, TSMC, and Samsung, have successfully commercialized FinFETs at advanced nodes (22-/14-/10-nm, 16-/10-/7-nm, and 14-/10-/7-nm, respectively). However, beyond the 5 nm node, FinFETs are expected to face electrostatic integrity issues, necessitating alternative architectures. Vertically stacked nanosheet transistors (NSFETs), also known as multibridge channel MOSFETs (MBCFETs) by Samsung, are considered natural successors to FinFETs. NSFETs/MBCFETs feature multiple horizontally stacked channels surrounded by gates, offering excellent electrostatic control and optimized drive per footprint,

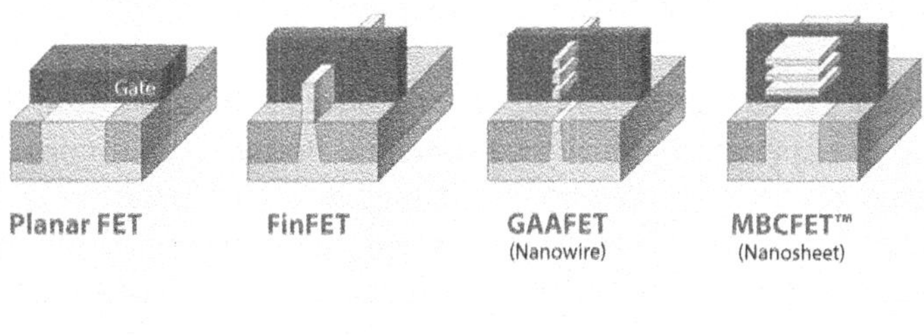

Figure 13.2 Trajectory of architecture in CMOS technologies [Courtesy: Samsung].

making them promising candidates for future technological nodes. Figure 13.2 illustrates the advent of multi-gate architectures from planar devices.

13.3 Critical Device-Circuit Co-Design Challenges in Low-Voltage Domain for FinFETs

In today's technological landscape, there is a pressing strategic need for computer systems that are not only cost-effective, but also exhibit significantly lower energy requirements. This need arises from the growing demand for devices that operate efficiently, while minimizing their impact on energy consumption and overall power usage. As power requirements continue to be downscaled in many applications, the importance of developing low-power and low-voltage solutions has become increasingly evident.

One promising approach to achieving these low-power objectives is the utilization of FinFET technology, which offers exceptional subthreshold performance. FinFETs are designed to improve the electrostatic control over the channel, allowing for enhanced performance at lower voltage levels. This unique characteristic enables the optimization of low-power circuits, making FinFET technology particularly well suited for applications where energy efficiency is a top priority. By leveraging the advantages of FinFETs, engineers and designers can create circuits that not only consume less power but also maintain high-performance levels, thereby addressing the growing demand for sustainable and energy-efficient computing solutions in various domains. As the industry continues to evolve, the integration of FinFET technology into the design of low-power circuits will play a crucial role in meeting the strategic need for energy-efficient computer systems in the future.

Commercial foundries more commonly use Si-bulk FinFETs owing to their better compatibility with existing process integration methods than SOI FinFETs. The critical parameters of a FinFET are the gate length (L_G), fin height (H_{fin}), fin width (W_{fin}), and fin thickness (T_{Si}). The extension region in a FinFET is a part of the silicon fin that extends beyond the gate; this region is unavoidable in manufacturing. Because the doping density cannot change abruptly from extremely high levels in the source and drain pads to nearly intrinsic levels in the fin region, extension regions are necessary. Maintaining a lightly doped body is crucial for controlling mobility degradation, corner effects, and random dopant fluctuations. However, lower fin doping may also result in a higher parasitic series resistance and capacitance for FinFETs. These factors significantly affect the circuit performance and must be addressed during the design stage.

Moreover, unlike planar transistors, the transistor width of FinFETs is discretized. For a double-gate FinFET with 'n' fins, the effective width is 2 * n * height of the fin. Thus, it is not possible to precisely adjust the total width as with planar transistors. Multiple fins are connected in parallel to approximate the required width; however, this approach significantly increases the fringing capacitance [13]. The parasitic capacitance in FinFETs is almost 40% higher than that observed in planar devices, owing to their structure. If the layout is not optimized, the increased capacitance can negate the performance improvements gained from scaling. Therefore, the FinFET-based circuit design requires a nuanced perspective to account for these effects.

A thorough understanding of the device-level phenomena in FinFETs is essential, particularly when the supply voltage is lowered from its nominal value. This understanding provides insights into how device-specific behaviors influence circuit performance.

Traditional approaches to designing digital circuits typically assume that the capacitance of the inverter input terminal (C_{in}) and its parasitic capacitance (C_p) remain consistent for an input voltage change (ΔV_{in}) [14]. These capacitances are not expected to be significantly affected by the load capacitance (C_L). However, various research [4, 15, 16] have shown that the capacitances of FinFET inverters are influenced by circuit parameters, unlike those of devices. The term extension transistor-induced capacitance shielding effect (ETICS) was coined to describe this phenomenon. The parasitic capacitances arising from the FinFET extension regions exhibit a shielding effect, as discussed in [15]. If the impact of capacitance shielding was not considered, the delay estimation of any circuit was reported to be erroneous. The authors in [5] described a novel phenomenon called 'drag,' where the input output transitions of a FinFET gate varied slowly for a duration dependent on the size and load of the buffer. This effect was attributed to the shielding of the gate-spacer-fin extension parasitic transistors. This introduces significant errors in the delay and power calculations of the FinFET-based multistage circuits. Because the duration of parasitic capacitance shielding depends on the circuit parameters, a strong relationship exists between the characteristics of the voltage transition at the input and the parasitic capacitance of the device. A low-power circuit design approach requires circuits to operate at a lower than the nominal supply voltage. Moreover, in scenarios where the primary focus is on minimizing the power consumption rather than maximizing the performance, designers often choose to operate at near-threshold voltage levels. This approach allows for significant energy savings, which are particularly advantageous for battery-operated and energy-efficient applications. In addition, designers may selectively apply near-threshold voltage supply voltages instead of nominal supply voltages to specific circuit blocks that are not critical for the overall functionality of the application. By doing so, they can effectively reduce the power usage in non-essential areas. Once the need for higher performance arises, these specific blocks can later be restored to their nominal supply voltage, thereby ensuring that the system operates optimally when required, while maintaining energy efficiency during less demanding tasks. As the applied supply voltage varies in a certain circuit, the behavior exhibited by the parasitic capacitances of the FinFET device also changes. When the supply voltages decrease by more than a certain value, the FinFET capacitances diverge from the ETICS behavior. The duration of this capacitance shielding depends on the circuit parameters, resulting in a strong interdependence between the voltage transition characteristics and the input and parasitic capacitances of the device. Hence, for low-voltage circuits, a careful understanding of device circuit codesign issues is required. This was extensively explored in [5]. Understanding and modeling the behavior of FinFET device capacitances across supply voltages is essential for optimizing cell design. The gate-to-drain extension parasitic capacitance of a FinFET significantly affects the circuit operation at various supply levels. Part of the gate-to-drain outer-fringing capacitance is modulated by bias conditions owing to carrier modulation in the low-doped region. This distributed capacitance is represented as C_{gd1}, C_{gd2}, ... C_{gdx}, as illustrated in Figure 13.3. C_{if} denotes the gate-to-drain inner-fringing capacitance and C_{gc} represents the total gate-to-channel capacitance. The components are detailed as follows.

In recent years, the industry has focused heavily on enhancing the energy efficiency of computing devices and extending battery life, particularly for IoT devices, wearable sensors, and biomedical applications. Reducing the

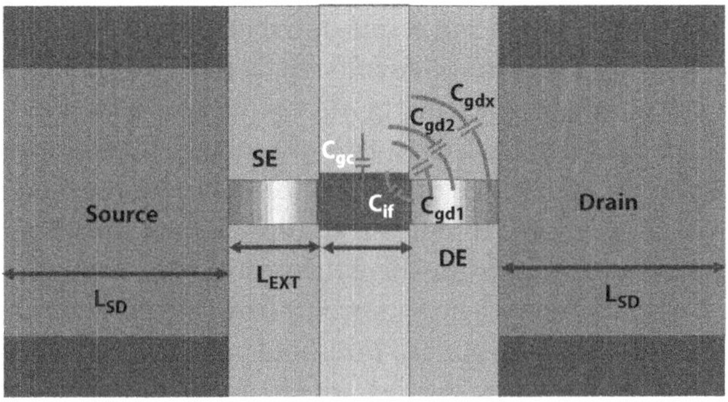

Figure 13.3 A 2D cut of a FinFET is shown highlighting the important capacitances impacting the delay calculation of a FinFET-based logic.

supply voltage has long been a popular method for optimizing the circuit power efficiency, a field widely explored under near-threshold computing for planar devices. FinFETs have demonstrated superior performance compared with planar transistors at scaled voltages, with research showing their near-threshold scalability. Research indicates that FinFETs exhibit 8.6 times greater energy efficiency [17] and a 4.8 times improvement in gain at near-threshold computing (NTC) compared with planar MOSFETs [18]. However, systematic design exploration of FinFETs at near-threshold voltages remains limited. A significant challenge in designing low-voltage FinFET circuits is the proper sizing of the pull-up and pull-down networks, which are critically constrained by the discretized effective width of the FinFETs themselves. In this context, the width of the transistor can only be increased by incorporating additional fins, which complicates the task of achieving a balanced rise and fall within a logic chain. The difficulty of this process is further compounded by the unique characteristics of the voltage transitions observed at both the input and output terminals of the FinFET inverter. This behavior leads to unusual capacitance patterns at the terminals, making it even more challenging to optimize circuit performance in low-voltage applications.

Traditional circuit design methods do not fully consider the impact of the supply voltage on the sizing methodology, making them valid only at nominal supply voltages. It is important to analyze how inverter capacitances vary with voltage for circuits intended to function at reduced supply voltages. While some researchers have extended the logical effort method to FinFETs, they often overlook the impact of circuit-level parameters on the effective capacitance. The unavoidable extension region in a FinFET affects the device's behavior depending on its operating region (subthreshold or super-threshold), influencing the logic gate capacitances. Design methodologies must account for the supply voltage and transition times to achieve efficient designs. This approach allows designers to select appropriate supply voltages, which is particularly crucial for low-voltage standard cells where not all cells operate at the same voltage.

The Logical Effort methodology of the circuit design estimates the minimum delay in a planar CMOS circuit with an optimal path length. This method allows designers to choose the best design among logically equivalent circuits, determine the size of transistors for each logic gate, and determine the optimal number of stages in a logic path. This methodology was applied to FinFETs operating across various supply voltages, including super-threshold, near-threshold, and subthreshold regions in [19]. However, the authors assumed the gate's net effective capacitance to be directly proportional to the number of parallel-connected fins, neglecting the significant impact of the input transition time and supply voltage. Consequently, delay estimation may not accurately reflect the actual delay.

In another study [20], the impact of the stage size, mode of operation (whether gates work independently or together), and fin width on circuit power consumption was analyzed, covering both active and standby power. A semi-analytical current-source modeling approach for FinFETs, which accounted for process variability, was introduced in [21]. This model uses curve fitting to correlate current and capacitance values with applied biases and process parameters, storing these points in look-up tables (LUTs). An enhanced semi-analytical approach called the multiple-input switching current source model was also discussed and further developed for other FinFET-based logic gates. In addition, the impact of stack sizing on FinFET logic gates and a framework for delay optimization were explored. A comprehensive investigation is necessary to establish the relationship between the currents and capacitances with device parameters such as the number of fins, extension length, and doping profile. The literature has not adequately addressed an appropriate sizing methodology that integrates these device-level effects, barring a few studies. The development of a precise sizing methodology tailored for circuit designers that utilizes readily accessible device-level parameters presents an intriguing challenge. This design can also pinpoint the optimal performance point of a circuit across the supply voltage spectrum.

Furthermore, several applications with ultralow power requirements prioritize energy efficiency over high performance. These scenarios benefit from operating at minimal supply voltage. Scaling down the supply voltage has emerged as a well-established strategy to effectively reduce energy consumption in digital circuits. According to Shannon's channel-capacity theorem [3], the fundamental threshold for the supply voltage is set to 36 mV for traditional planar CMOS circuits. However, FinFET devices, which possess nearly ideal subthreshold characteristics, demonstrate significant advantages over their planar counterparts in ultralow-voltage applications. Moreover, it is essential to determine the fundamental supply voltage limit specifically for FinFET-based logic circuits, as this limit is notably lower than that established for conventional planar CMOS technology. This discussion is particularly relevant in the context of low-power operation using FinFETs, where understanding these limits can lead to more efficient designs. In addition, the effects of temperature variability and various device design parameters on this fundamental voltage limit are critical considerations that cannot be overlooked. In the area of ultralow-voltage operation, a novel physics-based, semi-empirical current equation has been introduced for FinFETs, which is valid for supply voltages below 100 mV, as discussed in [2]. This current model is of considerable importance to circuit designers because it simplifies the process of performing quick calculations, making it a valuable tool for optimizing circuit designs in ultralow-voltage scenarios compared to traditional planar devices.

13.4 Inverter Capacitances in Low-Voltage Region of Operation

Quantitatively, the behavior of FinFETs at reduced supply voltages lies between that of traditional planar MOSFETs and FinFETs operating at standard voltage levels depending on the specific supply voltage applied [4]. Therefore, it is essential to understand the various behaviors that FinFET capacitances exhibit when small changes in the supply voltage occur. This understanding is particularly important when designing FinFET standard cells for near-threshold operations, as any miscalculation in estimating these parasitic behaviors can lead to inaccurate delay predictions. Consequently, the accurate modeling of the parasitic effects of FinFET devices is critical. Additionally, it is clear that changes in input capacitance (C_{in}) observed at low supply voltages cannot be explained solely by the

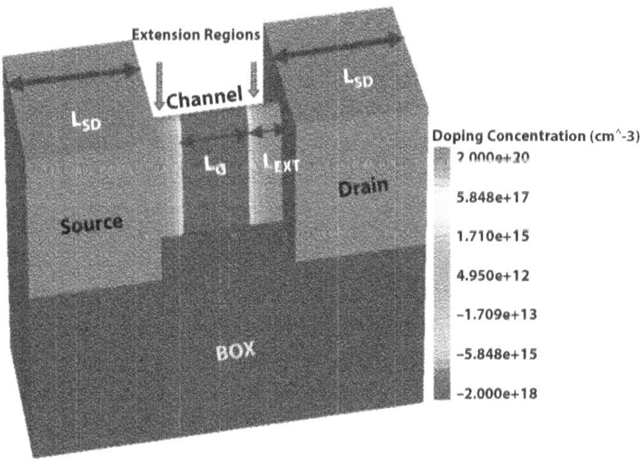

Figure 13.4 Showcases a 3D simulation of a 14-nm SOI FinFET conducted in Synopsys TCAD, illustrating the doping profile in the source-drain pads, source and drain extension regions, and the fin calibrated with the data in [22].

Table 13.1 FinFET design parameters.

Design parameter	Value (in nm)	Design parameter	Value
Length of gate	16	Width of Fin (H_{FIN})	25 nm
Height of fin	25	EOT	1.1 nm
Width of fin	8	Pad Doping	2×10^{20} cm^{-3}
Source drain pad length	40	Length of extension region	12–16 nm

well-known ETICS phenomenon [15]. In [4], researchers proposed a theory to explain the observed increase in input capacitance under low supply voltage conditions. Figure 13.4 illustrates the FinFET transistor used to gather these findings, with the key design parameters listed in Table 13.1. The simulations utilize proper physical models that are carefully chosen to align the simulation results with the experimental data [4].

A set of two cascaded FinFET inverters having one fin each, as shown in Figure 13.5, driven by a ramp input Vin with a transition time (t_{rin}) at the input node was examined in [4]. The first inverter acted as the Device Under Test (DUT). The input capacitance of the DUT is denoted as C_{in}. This investigation spanned over various supply voltages, ranging from the nominal 1 V down to 0.25 V, encompassing operations from near-threshold to super-threshold regions. To ascertain the effective input capacitance C_{in} of the inverter, the ratio of the charge Q_{in} to the change in input voltage Vin was calculated. To determine the charge Q_{in}, the input current of the inverter, I_{in}, during the 10%–90% or 20%–80% transition of the input

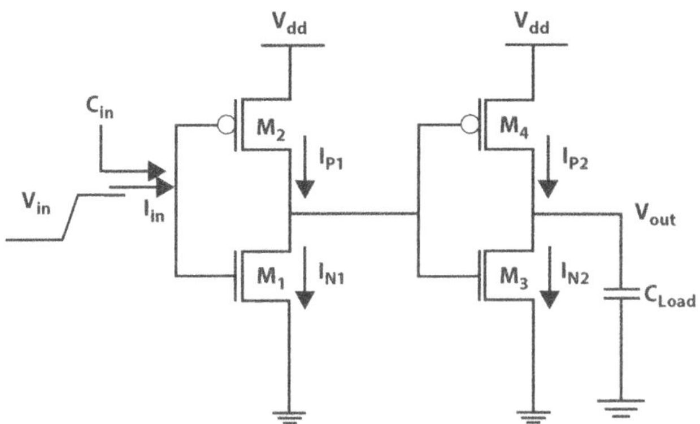

Figure 13.5 A pair of cascaded inverters is employed to analyze and report the dependence of FinFET parasitic effects.

voltage was integrated. This setup ensures a thorough exploration of the inverter behavior across different supply voltages, shedding light on its performance in diverse operating regimes. The methodological approach adopted in this experiment facilitates a comprehensive understanding of how the input capacitance varies with the supply voltage, which is essential for designing efficient and adaptable FinFET-based circuits.

Traditional circuit design techniques rely on the fundamental postulate of Logical Effort that the values of C_{in} and C_p for any logic gate, such as an inverter, remain constant for a given change in input voltage ΔV_{in} and output load. This consistency was observed in planar devices across repeated experiments. However, FinFETs exhibit a notable increase in the exhibited input capacitance when the transition time varies, while the output load remains unchanged. A 100 times variation in the input transition resulted in a remarkable increase of 72% in the input capacitance. Figures 13.6(a,b) show the variation in the input capacitance of the FinFET inverter and the parasitic capacitance of the inverter when the load being driven in the next stages is varied. These substantial variations can be attributed to the principles outlined in ETICS.

ETICS arises because the source and drain extension regions of a FinFET act as distributed parasitic transistors. This transistor action is different from the main transistor action and occurs between the gate, dielectric material of the spacers, and the fin extension region [14]. The low-doped part of the fin, where the density is significantly modulated by the applied drain bias, plays a significant role here. This parasitic transistor screens out a part of the fringing field capacitance of the gate and drain extension region when the bias between the gate and drain terminal V_{gd} is below a certain threshold during

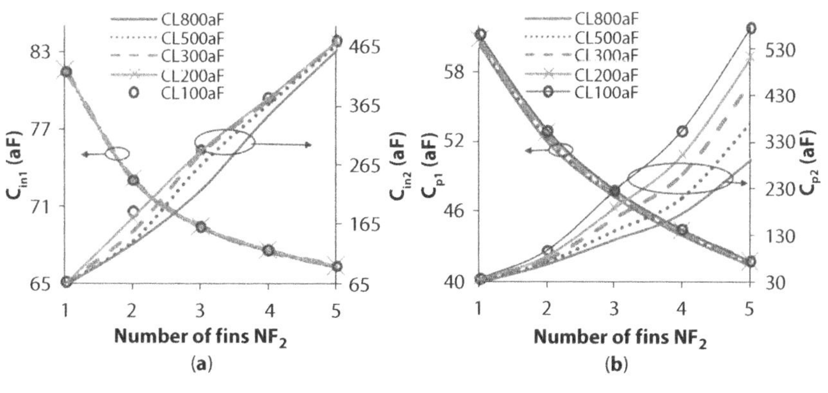

Figure 13.6 The variation seen in the (a) effective input capacitance C_{in1} and C_{in2} and (b) effective parasitic capacitance C_{p1} and C_{p2} with the fins NF_2 in the second stage of a two-stage inverter chain realized with FinFETs is driving a variable output load C_L [15].

the input and output node transitions. In other words, this parasitic transistor has its own threshold voltage, which differs from that of the transistor. This phenomenon originates from the modulation of the gate-to-drain capacitance (C_{gd}), which is a part of the outer fringing capacitance stemming from the gate fringing field lines terminating in the low-doped extension region. This modulation is influenced by the terminal voltages owing to changes in the carrier concentration in the source/drain extension region, which heavily depends on the applied voltage.

The Miller capacitance, which contributes to the total C_{gd}, varies based on the off/on state of the parasitic transistor, which is determined by the carrier density in the extension region during the transitions. When these parasitic transistors are in the off state, a portion of the Miller capacitance is shielded, and does not significantly contribute to the overall gate capacitance. This behavior effectively reduces the effect of the gate-extension fringing field capacitance on the net capacitance of the logic gate. Consequently, the presence of ETICS leads to variations in the gate capacitance based on the operating conditions of the FinFET, highlighting the importance of accounting for this effect in circuit design to ensure accurate modeling and performance prediction.

Owing to this phenomenon of ETICS, the input and parasitic capacitances of an inverter significantly increase when two specific conditions are simultaneously met. These conditions are as follows: (a) the Parasitic Drain Extension (DE) transistor is in the ON state and (b) there is a substantial "overlap" between the input and output transitions. This overlap refers to the period during which the changes V_{in} coincide with changes in the V_{out}. When the parasitic DE transistor is ON, it effectively contributes to the overall capacitance because the concentration of electrons in the DE region becomes significant, thus increasing the capacitance between the gate and drain. Simultaneously, when there is a large overlap in the input and output transitions, capacitive coupling between these nodes is maximized, leading to a further increase in the capacitance values.

In essence, the combined effect of the ON state of the parasitic DE transistor and the substantial overlap of the input and output transitions results in a marked increase in the effective input and parasitic capacitances. This dual condition significantly affects the performance of the inverter because increased capacitances can affect the switching speed and power consumption. Therefore, it is crucial to consider these factors when designing circuits that utilize the FinFET technology to ensure accurate performance predictions and efficient operation. The impact of ETICS on the capacitance values highlights the importance of advanced modeling techniques that can accurately capture these effects and guide the design process for high-performance low-power circuits.

As discussed in [14], initially, the number of fins in the second stage of the inverter is augmented from NF = 1 to NF = 5; traditionally, one would anticipate that the values of the 1st stage effective capacitances C_{p1} would remain relatively stable, while C_{in1} might experience a slight increase due to input 1 and input 2 capacitive coupling (C_{gd1}). However, the TCAD simulations in [14] revealed that both C_{in1} and C_{p1} decreased, contrary to expectations. This phenomenon can be attributed to the ETICS. The decrease in C_{in1} and C_{p1} with increasing NF2 is a consequence of the reduced Miller effect between the output and input nodes during the period when the parasitic transistor is ON. As NF2 increases, one typically expects C_{in2} and C_{p2} to increase proportionally. Nevertheless, a much larger increase in the values of C_{in2} and C_{p2} was observed, as depicted in Figure 13.6(b). This unconventional behavior can also be attributed to the ETICS. Additionally, it was observed that the decreases in C_{in1} and C_{p1} remained constant with increasing NF2 for different C_L values. However, owing to ETICS, the variation in the values of C_{in2} and C_{p2} (Figure 13.6(b)) with increasing NF2 differed for different C_L values. This discrepancy arises because C_{in2} and C_{p2} are heavily influenced by voltage transitions at the input–output nodes of Inv2, and consequently, C_L. Consequently, for a given NF2, the decrease in C_{p2} with increasing C_L stems from a reduction in the Miller coupling of output–input nodes during the period when the parasitic transistor is ON. As the load capacitance increased further, the impact of the ETICS diminished because the output voltage started to decline after the input voltage transition was nearly complete (for a rising input transition). Hence, the Miller coupling of output–input nodes remain relatively unchanged with further increases in the load capacitance.

In Figures 13.7(a–c), the electron density in the fin and extension regions of the transistor is shown. It is evident from the electron density profile that the threshold voltage of these extension transistors surpasses that of the main transistor formed between the gate, oxide, and the channel. Figure 13.7(a) illustrates the electron density along the fin when the gate voltage was varied in steps of 200 mV, ranging from 0 V to 1 V. This figure illustrates the changes in the electron density corresponding to the incremental increases in the gate voltage. The electron density along the fin when the gate voltage was varied from 200 mV to 600 mV, highlighting the behavior of the device during low-voltage operation, is shown in Figure 13.7(b). This range emphasizes how the electron density is affected under lower-voltage conditions. However, this does not follow the ETICS phenomenon. Figure 13.7(c) highlights the electron density along the fin when the gate voltage is varied from 500 mV to 1 V, showing the device's performance at higher supply voltages. This part emphasizes how the electron density shifts during higher-voltage operation. It was observed that for

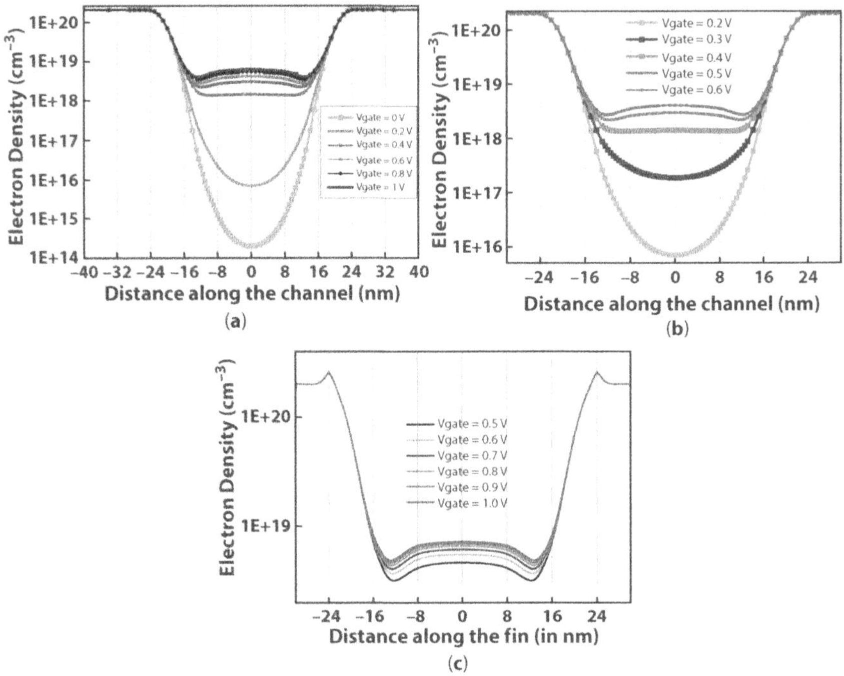

Figure 13.7 (a) The concentration of electrons in the fin and extension region is shown when V_G ranges from 0 V to 1 V. (b) The distribution density of electrons along the transistor's fin when the V_G is varied from 200 mV to 600 mV. (c) The distribution density of electrons along the transistor's fin when the gate voltage is varied from 500 mV to 1 V.

supply voltages above 500 mV, the threshold voltages of the extension transistors were higher. Conversely, for supply voltages below 500 mV, the electron density is greater in the extension transistor, indicating that current is predominantly used to charge the extension transistor. This highlights a distinct behavior in the electron density and current distribution, depending on whether the device operates at higher or lower supply voltages.

For supply voltages above 0.5 V (half of nominal V_{DD}), the concentration of electrons within the extension region is lower than that in the primary channel for a given gate input voltage. Consequently, the threshold voltage of the parasitic extension transistor is higher than that of the main transistor, which contributes to the ETICS. This led to an increase in the input capacitance as the input transition time increased. This corresponds to the case exhibited in Figure 13.7(c). However, when the supply voltage dropped further (V_{DD} <500 mV), the electron density became almost similar in both the main and extension regions. This occurs at a supply voltage of V_{DD} = 0.4 V. Reducing the supply voltage below this level results in a higher electron density in the extension region compared with the main channel transistor. Although this is an opposing physical effect, an increase in C_{in} is still evident in Figures 13.7(a,b), although the net percentage increase was reduced to 32% (for V_{DD} = 400 mV). This net rise continues to decline as the supply voltage approaches the FinFET threshold voltage, indicating that this change cannot be solely attributed to ETICS.

The observed phenomenon can be explained as follows: In the drain extension region, owing to the naturally low doping, the gate-to-drain capacitance C_{gd} increases with an increase in the electron density in the extension region, as illustrated in Figure 13.7(c). As the gate voltage increased gradually, the electron density in the main channel did not increase significantly. Under lower supply voltages, the current charges the extension transistor for a longer duration compared to the nominal voltage operation, where the charge predominantly supplies the main transistor. Consequently, this effect grows with an increase in the input transition time, as shown in Figures 13.8(a,b). In general, the input capacitance C_{in} shows an increasing trend with increasing input transition time. However, this increase was less pronounced when the supply voltage was reduced, resulting in a significantly smaller net percentage increase at lower supply voltages. This observation is crucial for understanding the behavior of FinFET devices under varying operating conditions, and for optimizing circuit designs that operate across a range of supply voltages. Therefore, the results in Figures 13.7, 13.8 therefore indicate the fundamental cause of the increase in input capacitance of a FinFET inverter with input transition time and its dependence on the supply voltage V_{DD}. Identifying this

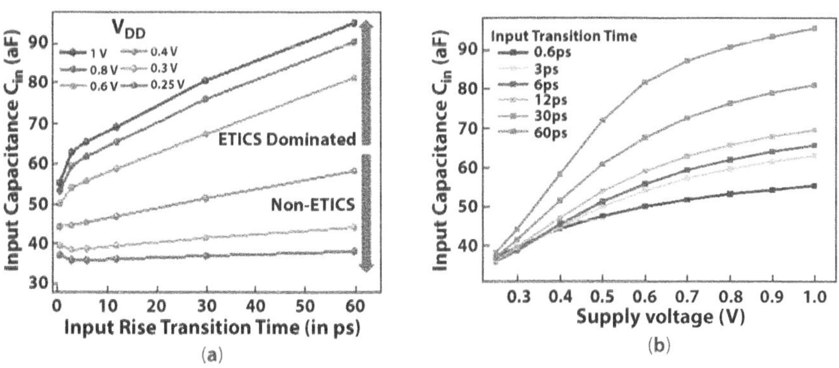

Figure 13.8 (a) Variation in the input capacitance of the device under test (DUT) as a function of input transition time for different supply voltage levels. (b) Variation in input capacitance relative to supply voltage when the input transition time is held constant.

transition point is essential for integration into circuit design to accurately predict delays in circuits that operate at multiple supply voltages.

Hence, instead of logical effort-based traditional circuit design methodologies that do not account for these extension-dependent effects leading to inaccurate delay estimations, a more comprehensive approach is required. These design methodologies were revised to incorporate the impact of significant carrier density modulation in extension regions to accurately model circuit delays. This is particularly critical for the design of standard cells operating at lower supply voltages. Because standard cells are typically designed to operate across a range of supply voltages rather than a fixed one, it is essential to address the varying changes in C_{in} of the inverter at different values of V_{DD}. The effective fan-out (FO) of the circuit varies depending on the V_{DD}, and proper modeling of this effect during the circuit design phase can mitigate the related concerns in delay estimation. Moreover, the standard cell design for low-supply voltage operation must also consider the varying input capacitance owing to different supply voltages. This variation can significantly affect the circuit performance, making it crucial to modify conventional design methodologies. The accurate modeling of this effect ensures reliable and efficient circuit operation across different supply voltage levels.

13.5 Minimum Supply Voltage for FinFET Logic Gates

Significant continuous advancements have been made in the design of deeply scaled low-voltage circuits. These developments are driven by applications that require extremely low power consumption without sacrificing functionality. Such applications include wearable devices, biomedical equipment, and IoT applications. Reducing the supply voltage has proven to be an effective method for reducing total energy usage. Lowering the supply voltage reduces dynamic power consumption in a quadratic manner, albeit at the cost of performance. This performance trade-off is often manageable in such applications as long as essential functionality is retained. Traditionally, high-performance FETs have achieved a balance between power efficiency and leakage current by adjusting both supply and threshold voltages. However, increasing leakage currents, such as sub-threshold and gate leakages, sets limits on the extent of this scaling. For planar CMOS circuits, the theoretical minimum energy per operation is set to 36 mV at 300 K. This limit was also confirmed by multiple studies that utilized standard subthreshold current models.

FinFETs providing superior gate control over the channel compared with planar MOSFETs are better suited for subthreshold operation. They deliver a higher current drive per unit area of silicon and exhibit lower drain-induced barrier lowering (DIBL), as well as reduced degradation in the sub-threshold slope. Consequently, establishing the minimum supply voltage limits for FinFET-based logic circuits is essential. Unlike planar MOSFETs, FinFETs demonstrate unique behavior at ultralow voltages owing to significant carrier concentration modulation in the drain extension (DE) region influenced by the terminal gate and drain bias voltages.

Various current models have been proposed for FinFETs operating at supply voltages below the transistor threshold voltage. However, these models are often computationally intensive because they rely on analytical solutions of Poisson's equation. Many of the current models published in the literature predominantly consider transport mechanisms governed by drift-diffusion. To accurately model the device's electrostatics, Poisson's equation is often solved. Taur *et al.* developed a methodology to describe the electrostatic potential in the channel for double-gate (DG) devices, as detailed in [23, 24]. A unified model based on the Pao–Sah integral for DG MOSFETs was also derived in [25]. Charge-based I–V models were developed in [26]. Such models avoid the use of transcendental equations, such as those used in surface-potential-based models. These models are particularly valid for describing the drain current in double-gate MOSFETs without the short-channel effects. In [27], an I–V model for a short-channel double-gate MOSFETS was proposed. This model was valid in the subthreshold region and incorporated short-channel effects (SCE) by adding corrections using fitting parameters. The literature in [28, 29] explored analytical solutions to 2-D Poisson's equation using conformal mapping. For a double-gate MOSFET with channel lengths less than 10 nm, the impact of quantum-mechanical confinement becomes significant. This leads to an increase in the threshold voltage of the transistor, the gate capacitance, and the carrier mobility in the channel. A compact model for drain current that includes these effects was developed in [30]. This model accounts for the difference in carrier injection from the source and drain, considering ballistic operation, as discussed in [31]. The work in [32] addressed DG MOSFETs with asymmetry owing to differences in the flat-band voltages of the two gates.

A semi-empirical formula for calculating the drain current and voltage relationship in similar asymmetric devices was discussed in [33]. A comprehensive model covering both the sub-V_{TH} and near-V_{TH} regions for FinFETs was introduced in [34]. The work in [35] presented a unified surface-potential-based model to estimate the drain current and terminal charges while accounting for ballistic transport effects. Over the years, several researchers have proposed various models specifically for subthreshold currents in FinFETs. In [36], a model based on an analytical solution of the three-dimensional Poisson equation was developed for subthreshold current estimation. Additional two-dimensional Poisson equation-based analytical models for the subthreshold current in underlap double-gate FinFETs were described in [37] and [38]. For circuit design purposes, the BSIM-CMG model implemented in Verilog-A remains a widely used option, as noted in [39].

Expanding on these foundational studies, the models not only improve our understanding of device behavior at the subthreshold level, but also contribute to the development of more accurate and practical circuit design tools. They highlighted the importance of considering various physical phenomena such as quantum mechanical effects, ballistic transport, and the influence of asymmetric gate structures, which become increasingly significant as the device dimensions continue to shrink.

Overall, the progression from basic electrostatic models to sophisticated charge and current models reflects the complexity of modern transistor behavior, particularly in nanoscale devices. The continuous refinement and validation of these models are crucial for advancing semiconductor technology and optimizing the performance of ultralow-power applications. For circuit designers, a current model that provides insights into the device's physics and is sufficiently simple for first-estimate calculations is preferable. The BSIM-CMG model is popular for circuit design. However, the notable influence of the intrinsic, bias-sensitive, low-doped areas within the drain extension on current conduction in the deep subthreshold region of FinFETs has received limited attention in all these discussions. To address this key gap, researchers in [2] developed a semi-empirical current model based on the fundamental physics of carriers crossing an energy barrier. This variability-aware I–V model is valid for very low operating voltages and intuitively accounts for the bias-dependent modulation of carriers in the extension's low doped regions.

Applications requiring deeply-scaled power consumption can achieve substantial power savings by lowering the supply voltage. Therefore, it is essential to examine the fundamental limit of V_{DD} scaling for FinFETs to design these circuits optimally. Utilizing the proposed deep subthreshold current model from [2] with a validated heuristic, the minimum supply voltage requirement for a FinFET CMOS inverter is shown to be well below the standard 36 mV limit for MOSFETs. Additionally, the impact of temperature variations and device design factors, such as extension length and channel doping, on this fundamental limit was explored to establish a practical range of scaled V_{DD} for various logic circuits. The authors in [2] proposed the following physics-based semi-empirical current model (1) for ultralow supply voltage operation.

$$I_{net} - I_o * \exp\left(\frac{q(V_{GS}+\beta V_{DS})}{\eta * KT}\right) * \left(1-\exp\left(-\frac{\gamma * q V_{DS}}{KT}\right)\right) \quad (13.1)$$

This model is valid for bias voltages less than 100 mV. This is based on the key postulates of the Virtual Source Model [40]. It develops on the probability of carriers crossing the electrostatic barrier between the source/drain and channel. Parameter η denotes the subthreshold slope of the device, and β accounts for the DIBL factor in Eq. (13.1), they are empirically adjusted according to the technology node. The term γ is an important parameter and is added to Eq. (13.1) to account for the faster current saturation of the drain current in FinFETs compared to planar devices. A comparison of the model developed in [2] with the conventional subthreshold IV model is shown in Figure 13.9. It is evident that the current saturation in the deep subthreshold region with the applied drain bias is not predicted by the conventional I–V model.

It is evident from Figure 13.9 that the drain current in FinFETs reaches saturation more quickly as the applied drain-to-source bias is increased across the FinFET. This behavior can be understood by analyzing how the low-doping region in the drain extension (DE) influences the current flow. By nature, a region with a low electron concentration forms at the boundary between the channel and drain extension region. This low doped region is

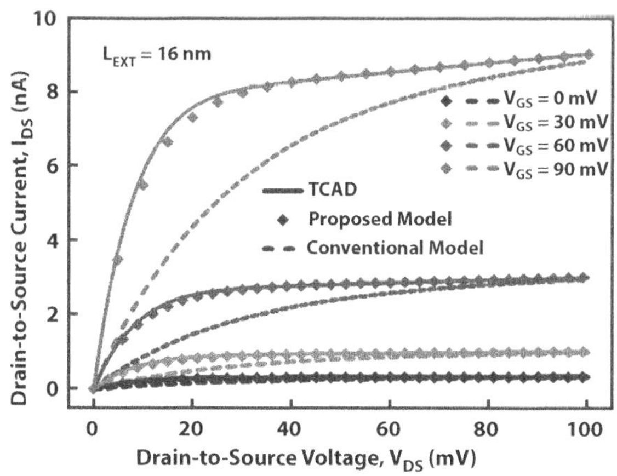

Figure 13.9 Comparison of the deep subthreshold model discussed in [2] with conventional planar current model.

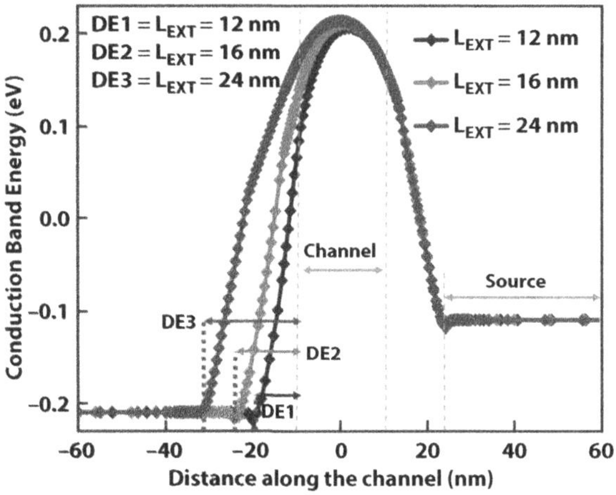

Figure 13.10 The energy level corresponding to the conduction band energy along the channel (fin and extension regions and S/D pads) varies with increasing DE. An increased slope indicating the electric field strength is observed for shorter extension lengths, indicating faster saturation.

bias-dependent. As the V_{DS} increased, this low-concentration area expanded. To sustain the current flow, the limited electrons within the channel due to the ultralow bias must move extremely rapidly toward the drain edge. This accelerated movement shifts the virtual source point [40] closer to the source extension region, as shown in Figure 13.10. Essentially, the result is a greater electric-field gradient within the channel, which moves the sparse charge carriers through the channel more quickly, leading to earlier current saturation at a lower V_{DS}. A factor γ is introduced to model the current saturation effect. Excluding this factor from the model would cause the predicted current to saturate at a higher V_{DS} than that observed. Furthermore, increasing the channel doping results in a higher doped channel compared to the low-doped DE region, which subsequently delays the onset of saturation of the drain current with the applied drain bias.

Using the proposed current model applicable for deeply scaled supply voltages, the authors investigated the theoretical limit of the V_{DD} required for FinFET-based digital gates. It was established that the theoretical limit for a FinFET inverter can be given by

$$V_{DD}(\min) = \frac{2V_T}{\gamma}\ln\left(1+\eta\gamma\right) \quad (13.2)$$

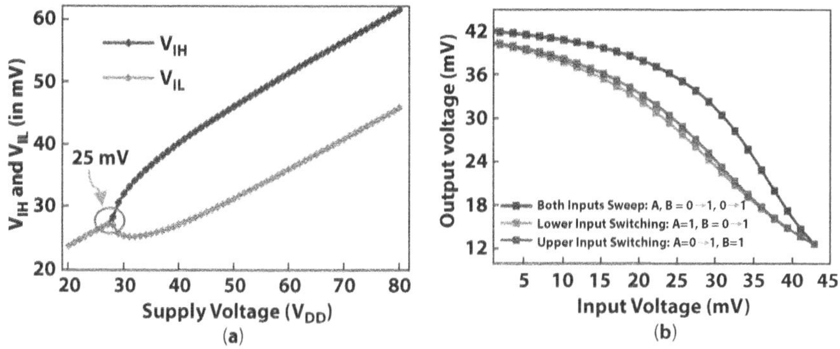

Figure 13.11 (a) Critical V_{IH} and low V_{IL} voltages of an inverter with respect to the V_{DD} show that the FinFET inverter's critical input voltages become indistinguishable at 25 mV. (b) NAND gate VTC at the derived lowest supply voltage.

When calculating this value for the inverter in question, a minimum supply voltage of 25 mV is required to drive the FinFET inverter at room temperature. Figure 13.11(a) indicates that at the minimum supply voltage, the critical input levels V_{IH} and V_{IL} of the inverter become indistinguishable. This limit was further investigated for the NAND gate. Considering the worst-case delay performance, it was observed that a supply voltage of 43 mV is required for the NAND operation realized using this FinFET technology. The voltage transfer characteristics at the lowest limit are shown in Figure 13.11(b). A variability study from the perspective of the device design and temperature was also extended in this study.

13.6 Conclusion

This chapter reviews recent developments in the area of low-voltage circuit design using FinFETs. It first delves into the need to switch to FinFETs and their origin. The critical challenges of FinFET-based digital circuits have been extensively presented. This chapter emphasizesthe peculiar-extension region-dependent carrier density modulation in the transistor to report several interesting device-level phenomena. As the industry switches towards low-voltage design, conventional design methodologies, without considering this critical device circuit issue fail to provide an optimum design. Hence, the device and circuit designs in the low-voltage domain coexist rogether and are interrelated. The origin of this novel phenomenon that explains the change in the FinFET-based logic gate's C_{in} with variations in the input transition time when operating at low supply voltages is tracked in the existing literature. It is also elucidated how this effect is fundamentally different from that observed when designing logic gates at nominal supply voltages. Given its critical influence on the fanout of a path, it is essential to model this effect and adapt the traditional circuit design methodologies for FinFETs accordingly. Consideration of this phenomenon is particularly important when designing standard cells for low-voltage operations. These standard cells operate at various V_{DD}s with varying optimizing power and performance. Hence, modeling V_{DD}-dependent effects ensures accurate performance predictions and efficient circuit functions across various supply voltages. As the voltage continues to scale down for the operation of circuits in the deep subthreshold region for ultra-low power circuits, another interesting phenomenon due to carrier modulation in the extension region appears. This phenomenon is explained by the fundamental physics of the chapter and is modeled into the device's current model. This results in the early onset of saturation of the current with the applied drain-to-source bias. Interestingly, this device-level phenomenon has interesting implications for the performance of logic gates. It is established in this chapter that FinFETs can support faithpul logic operation, albeit slow at even less scaled supply voltages than planar devices. Finally, the variability of this theoretical and fundamental limit for the inverter and NAND gate was discussed.

References

1. Crupi, F., Alioto, M., Franco, J., Magnone, P., Togo, M., Horiguchi, N., Groeseneken, G., Understanding the Basic Advantages of Bulk FinFETs for Sub- and Near-Threshold Logic Circuits From Device Measurements. *IEEE Trans. Circuits Syst. II Express Briefs*, 59, 7, 439–442, July 2012, doi: 10.1109/TCSII.2012.220017.
2. Yadav, S., Chauhan, N., Tyagi, S., Sharma, A., Banchhor, S., Joshi, R., Pratap, R., Bulusu, A., A Physical Insight into Variation Aware Minimum V_{DD} for Deep Subthreshold Operation of FinFET. *Semicond. Sci. Technol.*, 36, 12, 125002, 2021.
3. Meindl, J.D. and Davis, J.A., Fundamental limit on binary switching energy for terascale integration (TSI). *IEEE J. Solid-State Circuits*, 35, 10, 1515–1516, 2000, doi: 10.1109/4.871332.
4. Yadav, S., Chauhan, N., Pandey, A., Pratap, R., Bulusu, A., Behaviour of FinFET Inverter's Effective Capacitances in Low-Voltage Domain. *2021 25th International Symposium on VLSI Design and Test (VDAT)*, pp. 1–5, 2021.
5. Pandey, A., Kumar, H., Manhas, S.K., Dasgupta, S., Anand, B., Atypical Voltage Transitions in FinFET Multistage Circuits: Origin and Significance.

IEEE Trans. Electron Devices, 63, 3, 1392–1396, March 2016, doi: 10.1109/TED.2016.2520303.

6. Banchhor, S., Kumar, K.D., Dwivedi, A., Anand, B., A New Aspect of Saturation Phenomenon in FinFETs and Its Implication on Analog Circuits. *IEEE Trans. Electron Devices*, 66, 7, 2863–2868, July 2019, doi: 10.1109/TED.2019.2914867.
7. Moore, G.E. and Moore, G.E., Cramming More Components Onto Integrated Circuits. *Electron. Mag.*, 38, 8, 114–117, April 1965.
8. Ghani, T., *et al.*, A 90nm high volume manufacturing logic technology featuring novel 45nm gate length strained silicon CMOS transistors. *IEEE International Electron Devices Meeting 2003*, pp. 11.6.1–11.6.3, 2003, doi: 10.1109/IEDM.2003.1269442.
9. Xiong, R., Jang, P., Colinge, Self-aligned ground-plane FDSOI MOSFET. *2002 IEEE International SOI Conference*, pp. 23–24, 2002, doi: 10.1109/SOI.2002.1044401.
10. Sekigawa, T. and Hayashi, Y., Calculated threshold-voltage characteristics of an XMOS transistor having an additional bottom gate. *Solid State Electron.*, 27, 8–9, 827–828, 1984, ISSN 0038-1101, doi:10.1016/0038-1101(84)90036-4.
11. Hisamoto, D., Kaga, T., Kawamoto, Y., Takeda, E., A fully depleted lean-channel transistor (DELTA)-a novel vertical ultra thin SOI MOSFET. *International Technical Digest on Electron Devices Meeting*, pp. 833–836, 1989, doi: 10.1109/IEDM.1989.74182.
12. Hisamoto, D., Lee, W.-C., Kedzierski, J., Takeuchi, H., Asano, K., Kuo, C., Anderson, E., King, T.-J., Bokor, J., Hu, C., FinFET-a self-aligned double-gate MOSFET scalable to 20 nm. *IEEE Trans. Electron Devices*, 47, 12, 2320–2325, Dec. 2000, doi: 10.1109/16.887014.
13. Manoj, C.R., Sachid, A.B., Yuan, F., Chang, C., Rao, V.R., Impact of Fringe Capacitance on the Performance of Nanoscale FinFETs. *IEEE Electron Device Lett.*, 31, 1, 83 85, Jan. 2010, doi: 10.1109/LED.2009.2035931.
14. Sutherland, I., Sproull, B., Harris, D., *Logical Effort: Designing Fast CMOS Circuits*, Morgan Kaufmann, San Mateo, CA, USA, 1999.
15. Pandey, A., Raycha, S., Maheshwaram, S., Manhas, S.K., Dasgupta, S., Saxena, A.K., Anand, B., Effect of Load Capacitance and Input Transition Time on FinFET Inverter Capacitances. *IEEE Trans. Electron Devices*, 61, 1, 30–36, Jan. 2014, doi: 10.1109/TED.2013.2291013.
16. Pandey, A., Garg, P., Tyagi, S., Ranjan, R., Bulusu, A., A modified method of logical effort for FinFET circuits considering impact of fin-extension effects 2018. *19th International Symposium on Quality Electronic Design (ISQED)*, pp. 189–195, 2018, doi: 10.1109/ISQED.2018.8357286.
17. Pinckney, N., Shifren, L., Cline, B., Sinha, S., Jeloka, S., Dreslinski, R.G., Mudge, T., Sylvester, D., Blaauw, D., Near-threshold computing in FinFET technologies: Opportunities for improved voltage scalability. *2016 53nd ACM/EDAC/IEEE Design Automation Conference (DAC)*, pp. 1–6, 2016, doi: 10.1145/2897937.2898049.

18. Pinckney, N., Jeloka, S., Dreslinski, R., Mudge, T., Sylvester, D., Blaauw, D., Shifren, L., Cline, B., Sinha, S., Impact of FinFET on Near-Threshold Voltage Scalability. *IEEE Des. Test*, 34, 2, 31–38, April 2017, doi: 10.1109/MDAT.2016.2630303.
19. Lin, X., Wang, Y., Pedram, M., Joint sizing and adaptive independent gate control for FinFET circuits operating in multiple voltage regimes using the logical effort method. *2013 (ICCAD)*, pp. 444–449, 2013, doi: 10.1109/ICCAD.2013.6691155.
20. Cui, T., Wang, Y., Lin, X., Nazarian, S., Pedram, M., Semi-analytical current source modeling of FinFET devices operating in near/sub-threshold regime with independent gate control and considering process variation. *2014 19th Asia and South Pacific Design Automation Conference (ASP-DAC)*, pp. 167–172, 2014, doi: 10.1109/ASPDAC.2014.6742884.
21. Amelifard, B., Hatami, S., Fatemi, H., Pedram, M., A Current Source Model for CMOS Logic Cells Considering Multiple Input Switching and Stack Effect. *2008 Design, Automation and Test in Europe*, pp. 568–573, 2008, doi: 10.1109/DATE.2008.4484737.
22. Lin, C.-H., Greene, B., Narasimha, S., Cai, J., Bryant, A., Radens, C., Narayanan, V., Linder, B., Ho, H., Aiyar, A., Alptekin, E., An, J.-J., Aquilino, M., Bao, R., Basker, V., Breil, N., Brodsky, M., Chang, W., Clevenger, L., Chidambarrao, D., Christiansen, C., Conklin, D., DeWan, C., Dong, H., Economikos, L., Engel, B., Fang, S., Ferrer, D., Friedman, A., Gabor, A., Guarin, F., Guan, X., Hasanuzzaman, M., Hong, J., Hoyos, D., Jagannathan, B., Jain, S., Jeng, S.-J., Johnson, J., Kannan, B., Ke, Y., Khan, B., Kim, B., Koswatta, S., Kumar, A., Kwon, T., Kwon, U., Lanzerotti, L., Lee, H.-K., Lee, W.-H., Levesque, A., Li, W., Li, Z., Liu, W., Mahajan, S., McStay, K., Nayfeh, H., Nicoll, W., Northrop, G., Ogino, A., Pei, C., Polvino, S., Ramachandran, R., Ren, Z., Robison, R., Saraf, I., Sardesai, V., Saudari, S., Schepis, D., Sheraw, C., Siddiqui, S., Song, L., Stein, K., Tran, C., Utomo, H., Vega, R., Wang, G., Wang, H., Wang, W., Wang, X., Wehelle-Gamage, D., Woodard, E., Xu, Y., Yang, Y., Zhan, N., Zhao, K., Zhu, C., Boyd, K., Engbrecht, E., Henson, K., Kaste, E., Krishnan, S., Maciejewski, E., Shang, H., Zamdmer, N., Divakaruni, R., Rice, J., Stiffler, S., Agnello, P., High performance 14nm SOI FinFET CMOS technology with 0.0174μm2 embedded DRAM and 15 levels of Cu metallization. *2014 IEEE International Electron Devices Meeting*, pp. 3.8.1–3.8.3, 2014, doi: 10.1109/IEDM.2014.7046977.
23. Taur, Y., An analytical solution to a double-gate MOSFET with undoped body. *IEEE Electron Device Lett.*, 21, 5, 245–247, May 2000.
24. Taur, Y., Analytic solutions of charge and capacitance in symmetric and asymmetric double gate MOSFETs. *IEEE Trans. Electron Devices*, 48, 12, 2861–2869, Dec. 2001.
25. He, J., Xi, X., Lin, C.H., Chan, M., Niknejad, A., Hu, C., A non-charge-sheet analytic theory for undoped symmetric double-gate MOSFET from the exact

solution of Poisson's equation using SSP approach, in: *Proc. of the Workshop Compact Modeling, NSTI-Nanotech*, Boston, MA, pp. 124–127, May 2004.
26. Sallese, J.M., Krummenacher, F., Pregaldiny, F., Lallement, C., Roy, A., Enz, C., A design oriented charge-based current model for symmetric DG MOSFET and its correlation with the EKV formalism. *Solid State Electron.*, 49, 3, 485–489, Mar. 2005.
27. Liang, Y.P. and Taur, Y., A 2-D analytical solution for SCEs in DG MOSFETs. *IEEE Trans. Electron Devices*, 51, 9, 1385–1391, Sep. 2004.
28. Osthaug, J., Fjeldly, T.A., Iñíguez, B., Closed-form 2D modeling of sub-100 nm MOSFETs in the subthreshold regime. *J. Telecommun. Inf. Technol.*, 1, 70–79, 2004.
29. Kolberg, S. and Fjeldly, T.A., 2-D modeling of nanoscale DG SOI MOSFETs in the subthreshold regime. *J. Comput. Electron.*, 5, 217–222, 2006.
30. Baccarani, G. and Reggiani, S., A compact double-gate MOSFET model comprising quantum mechanical and nonstatic effects. *IEEE Trans. Electron Devices*, 46, 8, 1656–1666, Aug. 1999.
31. Jiménez, D., Sáenz, J.J., Iniguez, B., Suñé, J., Marsal, L.F., Pallarès, J., A unified compact model for the ballistic quantum wire and quantum well MOSFET. *J. Appl. Phys.*, 94, 2, 1061–1068, Jul. 15, 2003.
32. Lu, H. and Taur, Y., An analytic potential model for symmetric and asymmetric DMOSFETs. *IEEE Trans. Electron Devices*, 53, 5, 1161–1168, May 2006.
33. Roy, A.S., Sallese, J.M., Enz, C.C., A closed-form charge-based expression fordrain-current in symmetric and asymmetric double gate MOSFET. *Solid State Electron.*, 50, 4, 687–693, Apr. 2006.
34. Lin, X., Wang, Y., Pedram, M., Stack sizing analysis and optimization for FinFET logic cells and circuits operating in the sub/near-threshold regime. *Fifteenth International Symposium on Quality Electronic Design*, Santa Clara, CA, pp. 341–348, 2014.
35. Dasgupta, A., Agarwal, A., Khandelwal, S., Chauhan, Y.S., Compact Modeling of Surface Potential, Charge, and Current in Nanoscale Transistors Under Quasi-Ballistic Regime. *IEEE Trans. Electron Devices*, 63, 11, 4151–4159, Nov. 2016.
36. Havaldar, D.S., Katti, G., DasGupta, N., DasGupta, A., Subthreshold current model of FinFETs based on analytical solution of 3-D Poisson's equation. *IEEE Trans. Electron Devices*, 53, 4, 737–742, April 2006, doi: 10.1109/TED.2006.870874.
37. Chattopadhyay, A., Kundu, A., Sarkar, C.K., Two-dimensional modeling of the underlap graded-channel FinFET. *J. Comput. Electron.*, 19, 688–699, 2020. https://doi.org/10.1007/s10825-020-01458-w.
38. Narendar, V., Rai, S., Tiwari, S., A two-dimensional (2D) analytical surface potential and subthreshold current model for the underlap dual-material double-gate (DMDG) FinFET. *J. Comput. Electron.*, 15, 1316–1325, 2016. https://doi.org/10.1007/s10825-016-0899-x.

39. *BSIM-CMG 110.0.0 Technical Manual*, Department of Electrical Engineering and Computer Sciences, University of California, Berkeley, 2015. [Online] Available: http://bsim.berkeley.edu/models/bsimcmg.
40. Khakifirooz, A., Nayfeh, O.M., Antoniadis, D., A Simple Semiempirical Short-Channel MOSFET Current–Voltage Model Continuous Across All Regions of Operation and Employing Only Physical Parameters. *IEEE Trans. Electron Devices*, 56, 8, 1674–1680, Aug. 2009, doi: 10.1109/TED.2009.2024022.

14

A Novel Low-Power Approach of 8-Bit Vedic Multiplier Using Reversible Logic Gates

Aruru Sai Kumar[1]*, K. Sarangam[2†], P. Ravi Sankar[3], K. Nishanth Rao[4] and Yashika Gaidhani[5]

[1]School of Electronics Engineering, VIT-AP University, Amaravati, Andhra Pradesh, India
[2]Department of Electronics and Communication Engineering, National Institute of Technology, Warangal, India
[3]Department of CSE, SAHE University, Vijayawada, Andhra Pradesh, India
[4]Department of ECE, MLR Institute of Technology, Hyderabad, India
[5]Department of Electronics Engineering, Yeshwantrao Chavan College of Engineering, Wanadongri, Nagpur, India

Abstract

Reversible logic gates are crucial in quantum computing and low-power digital circuit designs because they uniquely conserve information. Conventional combinational circuits face issues such as information loss and energy inefficiency, which reversible computing addresses by ensuring that input bits are maintained in the output, thereby improving the overall efficiency. This study explores the construction and utilization of a full adder employing gates such as the Peres, Toffoli, and Feynman gates. The proposed solution is both energy-efficient and low-power, aiming to minimize ancilla inputs, quantum cost, and garbage outputs. Multipliers, critical for arithmetic operations, digital signal processing, and other computing tasks, require effective design and implementation to enhance circuit performance. Vedic multipliers, based on ancient Indian mathematical principles have been particularly noted for their efficiency and speed. This study examines and compares 8-bit Vedic multiplier architectures that use different adder structures, such as the Carry Lookahead Adder and the Ripple Carry Adder. The goal is

**Corresponding author*: asaikumar.nitw@gmail.com
†*Corresponding author*: sarangam_7@nitw.ac.in

Ekta Goel, Archana Pandey, Shiromani Balmukund Rahi and Arun Samuel (eds.) Handbook of Advanced Semiconductor Field Effect Transistors, (305–322) © 2026 Scrivener Publishing LLC

to evaluate the performance, delay, and logic levels of these multipliers relative to the existing designs. The proposed adder design demonstrates a 75% reduction in garbage count, while the multiplier shows a 57% improvement in delay compared with previous study. Validation was performed using the Vivado 2022.2.

Keywords: Vedic multiplier, reversible logic gate, garbage outputs, ripple carry adder, delay, carry look ahead adder

14.1 Introduction

Reversible logic gates play a crucial role in both quantum computing and energy-efficient computation. Reversible logic gates differ from traditional logic gates in that they do not lose information or dissipate energy as heat. Instead, the output states of the reversible logic gates uniquely identify the input states. This feature ensures that no information is lost during the computation, theoretically leading to zero-energy dissipation. This concept is based on the principle of reversibility in physics, which posits that physical processes can advance in both the forward and reverse orientations without a corresponding increase in entropy. It is consistent with the fundamental goal of developing more sustainable and efficient computational systems [1–5].

The primary motivation behind the development of reversible logic gates is their potential to minimize the power consumption. In conventional digital circuits, energy loss occurs primarily because of the erasure of bits of information. Landauer's principle quantifies this energy loss, stating that for every bit of information erased, a minimum amount of energy is dissipated as heat. Reversible logic gates circumvent this issue by ensuring that no information is erased, thereby significantly reducing or even eliminating energy loss. This property renders them particularly attractive for applications where energy consumption is a vital factor, such as large-scale data centers and portable electronic devices [6–8].

Quantum computers utilize the laws of quantum mechanics to perform tasks that are impossible for conventional computers. Reversible logic is intrinsically compatible with quantum computing because quantum processes are linear, implying an inherent reversibility. The development of quantum circuits requires the use of reversible gates, which include Toffoli and Fredkin gates, among other fundamental building blocks. In addition to enabling the preservation of quantum information, these gates also facilitate the implementation of intricate quantum algorithms that can address problems at an exponentially faster rate than conventional algorithms [9–11].

Reversible logic gates have significant implications for fault-tolerant computing. In classical computing, errors during the computation can lead to incorrect results and require additional resources for error detection and correction. If an error occurs, it can be traced back to its source by reversing the computational process. This property enhances the reliability and robustness of computational systems, making reversible logic an appealing choice for critical applications in which accuracy and fault tolerance are paramount [12].

The development and deployment of reversible logic gates pose unique challenges. One of the primary challenges is ensuring that the gates remain efficient and practical, while preserving reversibility. Traditional logic circuits often rely on several irreversible operations, and converting them into reversible operations can lead to increased circuit complexity and resource requirements. Researchers are actively exploring various approaches to optimizing reversible logic gate design, including the use of minimal ancilla bits (extra bits used to achieve reversibility) and the development of efficient synthesis algorithms to create reversible circuits with minimal overhead.

14.1.1 Fundamental Reversible Logic Gates

These basic reversible logic gates form the foundation of reversible computing, enabling the construction of complex logic circuits that preserve information and minimize energy dissipation. Each gate has unique properties and applications, making it a valuable tool for designing efficient and powerful computational systems.

Feynman Gate: The Feynman gate, illustrated in Figure 14.1, is a basic reversible logic gate. The system had two inputs and two outputs. The first input is directly replicated in the first output, but the second is the result of performing an exclusive OR operation for A and B.

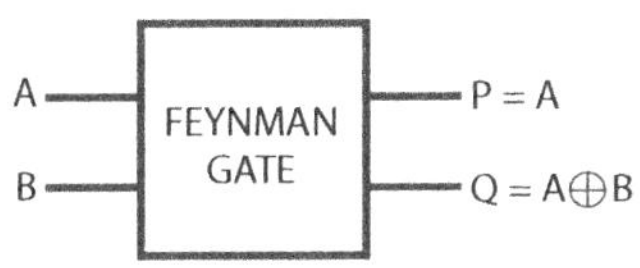

Figure 14.1 Feynman gate.

Toffoli Gate: It is a reversible logic gate that has the capability to do any computation in a reversible manner. The initial two inputs were immediately replicated in the initial two outputs. The third is the outcome of performing the XOR operation on C and the logical AND operation on A and B, as illustrated in Figure 14.2.

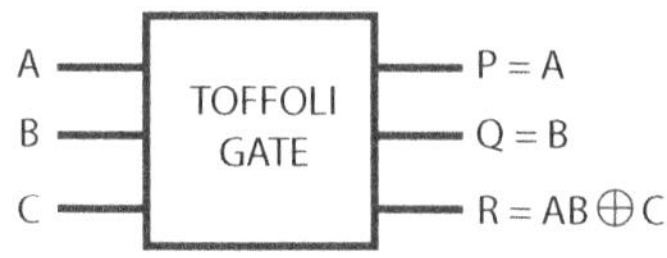

Figure 14.2 Toffoli gate.

Peres Gate: It is an additional significant reversible logic gate as shown in Figure 14.3. This can be interpreted as a hybrid of Toffoli and Feynman gates. The first output is identical to the first input, the second outcome is the exclusive OR of the initial two inputs, and the third is the XOR of the third input and the AND of A, B.

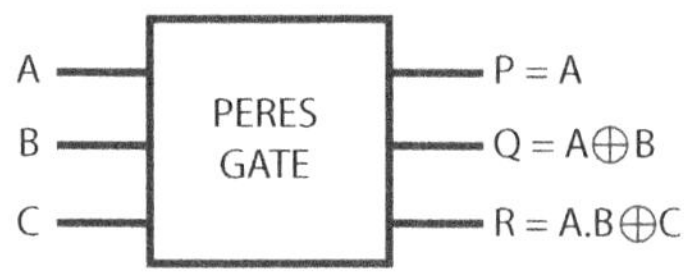

Figure 14.3 Peres gate.

Fredkin Gate: As shown in Figure 14.4, it operates on three bits, with the first acting as a control sign. The first bit acts as a control signal. If the control bit is 0, then the other two bits pass through unchanged. If the control bit is 1, the last two bits are interchanged. By preserving the information, the Fredkin gate offers a promising path towards reducing the energy consumption associated with traditional computing.

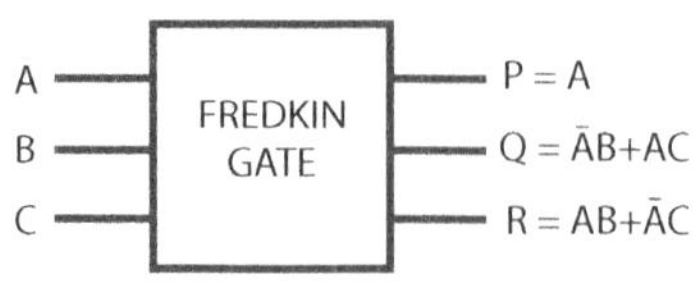

Figure 14.4 Fredkin gate.

TSG Gate: The TSG illustrated in Figure 14.5, is also known as the Toffoli-Swap-NOT gate, is a highly adaptable gate which combines the functionalities of the Swap gate, Toffoli and NOT gates enabling it to execute many logical operations simultaneously, hence enhancing the efficiency of reversible circuits. By combining multiple logical functions into a single reversible gate, the TSG gate becomes an asset in designing sophisticated computational systems.

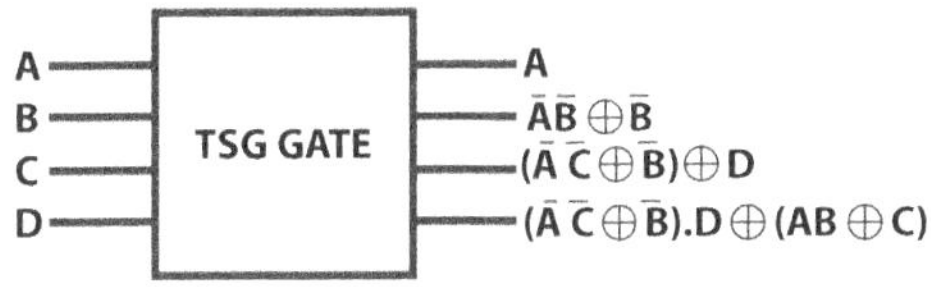

Figure 14.5 TSG gate.

The following is an expansion of the structure of this research paper: *Section 14.2* of this study outlines the comprehensive review of existing research, while *Section 14.3* introduces a novel technique for developing a reversible 8-bit Vedic multiplier. The performance of the proposed method is evaluated and analyzed in *Section 14.4*. Finally, *Section 14.5* presents the key findings and conclusions of the study.

14.2 Literature Survey

In the process of designing efficient adder units, a 4 × 4 reversible "TSG" gate was proposed by Thapliyal [13]. The demonstrated gate is utilized in the construction of three different types of carry adders, namely the RCA, the BCD, and the carry look-ahead algorithms. A key feature of the proposed circuit is its capacity to perform the functions of a reversible full adder, which is a crucial component. Beginning from this point onward, it is now feasible to create a reversible full adder using only a single gate. Adder designs that make use of the suggested gate have been shown to be significantly more efficient and optimized than literature-based competing techniques. This finding has been confirmed in several studies. An effective method to illustrate this is to create a link between the quantity of garbage and the variety of gates installed. This study introduces novel suggestions for a reversible two-bit CLA that preserves nanoscale parity.

Sarangam *et al.* [14] proposed the method of opts for low-power architectures, including quantum computing and the Internet of Things. Typical combinational logic circuits are inefficient and consume considerable energy. This is less of an issue with reversible computing, which preserves the input bits in the output. This study presents a low-power reversible full adder optimized for energy-efficient applications. A better comparison between energy-efficient full adders in reversible circuits and conventional CMOS circuits is recommended. The commonly used measurement for the loss of energy in CMOS devices is 0.5 CV_{dd}^{2}, which is similar to 0.5 CV_{Thr}^{2} in reversible logic. The amount of energy lost was $(V_{Thr}/V_{dd})^2$. In addition, the size of the ancillae and trash output was reduced by 16.67%. There were also 16.67% fewer transistors than previously.

Shukla *et al.*, [15] presented a unique method for implementing a reversible four-bit multiplier circuit. This approach has several superior characteristics. The fundamental principles of Vedic mathematics form the basis of Vedic multipliers. The multiplier is highly efficient because it only takes one step to compute the sum of all partial products. By employing a reversible methodology, the implementation of the mentioned multiplier will result in the creation of digital system multiplier circuits that are efficient, rapid, and devoid of any loss. Several criteria have been studied to determine the circuit's performance. These include quantum cost (QC), gate count (TC), CI, GO, and suggested multiplier design. Applications that require both high performance and low power consumption can take advantage of the benefits offered by the 4-bit reversible Vedic multiplier proposed by Saligram *et al.* [16]. One method for evaluating the aforementioned architecture is the total reversible logic implementation cost (TRLIC). Using this approach yields a TRLIC of 33% compared with the related literature.

14.3 Proposed Methodology

14.3.1 Reversible Full Adder

A full adder is a digital device specifically intended to execute the addition of three bits: two significant bits and one carry bit derived from prior addition. The process yields two results, total and carry-out, both of which are crucial for adding numerous binary values. The sum can be obtained by performing an exclusive OR operation on the three input bits, whereas the carry-out is determined by applying a combination of OR and AND operations to the input bits. Specifically, carry-out occurs when any two or all three input bits are high. The full adder's functionality is crucial in building arithmetic logic units and complex binary addition operations, as it allows for the cascading of multiple adders to handle larger binary numbers efficiently.

The equations for Sum and Carrry with A, B, and C_{in} as inputs are:

$$\text{Sum:A}\wedge\text{B}\wedge\text{Cin} \tag{14.1}$$

$$\text{Carry:(A}\wedge\text{B) Cin + A\&B} \tag{14.2}$$

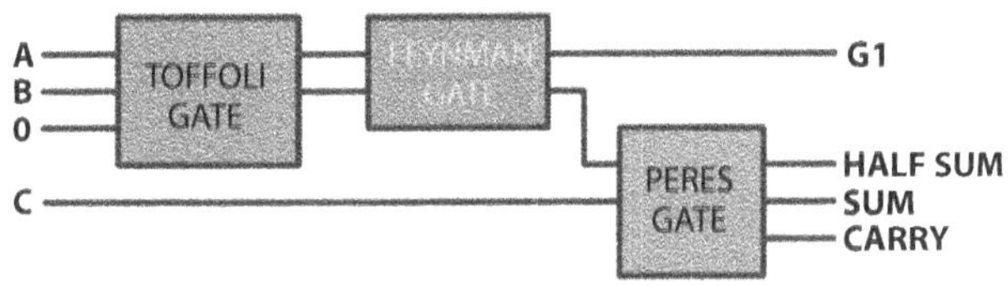

Figure 14.6 Full adder design using Toffoli, Feynman, and Peres gates.

Gates such as the Toffoli, Peres, and Feynman gates can be used to make a full adder. The Toffoli gate uses three inputs: A, B, and 0. The initial outcomes A and B are fed into a subsequent Feynman gate, whereas the third output is utilized as the third input for the Peres gate. The Feynman gate produces two outputs: A, which is considered a garbage output, and A⊕B. The Peres gate receives the XOR result of A and B from the Feynman gate as its first input and Cin as its second input. The Peres gate subsequently produces three outputs: the half sum (A⊕B), the sum result (A⊕B⊕C), and the carry as (A ⊕ B)Cin + A·B. This configuration enables the efficient and reversible computation of binary additions, as illustrated in Figure 14.6.

14.3.2 Ripple Carry Adder

RCA is a basic electronic circuit that is utilized to carry out binary addition by connecting many full adders in a chain, with each adder matching a certain bit location of the binary numbers that are being added. Each full adder in the binary system processes two binary digits and a carry-in from the previous bit position, generating a total and carry-out. Processing occurs from the LSB to MSB. As the "ripple" effect spreads across the chain, the carry-in from one full adder acts as the carry-out for the following. This straightforward design ensures that the addition operation is performed sequentially from the LSB to the MSB, culminating in the final sum and carry-out. Figure 14.7 depicts the configuration of the RCA with the claimed full-adder blocks.

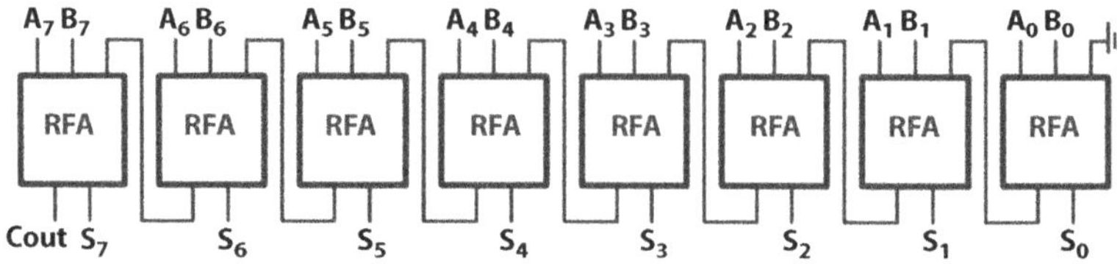

Figure 14.7 8-bit Ripple Carry adder using proposed full adder.

14.3.3 Carry Look Ahead Adder

The CLA shown in Figure 14.8 is a digital circuit that enhances the speed of binary addition by minimizing the time delay resulting from carry propagation in conventional adders such as the ripple carry adder. The CLA adder employs generate (G) and propagate (P) signals to forecast the carry outputs for every bit position concurrently rather than relying on the sequential ripple carry method. The generate signal (G) indicates that a bit pair will produce a carry regardless of the incoming carry, whereas the propagate signal (P) indicates that a bit pair will pass the incoming carry to the next higher bit. By calculating these signals for all bit positions in parallel, the CLA adder can determine the carry for each bit position without waiting for the previous positions, significantly accelerating the addition process. This parallel computation of carries allows the CLA to achieve significantly faster computation than RCA, especially for large bit widths, making it ideal for high-speed arithmetic operations in complex digital systems.

Initially, a carry generates and propagates a block, referred to as the PGA, as presented in Figure 14.9. The unit responsible for producing Si, Pi, and Gi is referred to as PGA. The PGA block was constructed using Feynman and Peres gates. The inputs to the Peres gate are A_i, B_i, and a constant input of 0 produces three outputs: the first output A_i is considered as the garbage output, the second output is fed into the Feynman gate, and the third output represents the carry generate (G_i). The Feynman gate receives the second output as an input, along with C_i, and produces carry propagate (P_i) and sum (S_i) as outputs. The generate function $G_i = A_i, B_i$ generates a carry if both the bits are equal to 1. The propogate function $P_i = A_i \oplus B_i$ is used to propagate a carry if either of the inputs is 1. The CLA design consists of four Feynman gates, six Peres gates, a TSG gate, and an R2 gate. In reversible circuit design, creating copies of certain bits for computation introduces fan-out, which is not permitted in reversible computing. To address this limitation, a special gate is required for the bit copying. The Feynman gate is the preferred solution for this task in a reversible circuit.

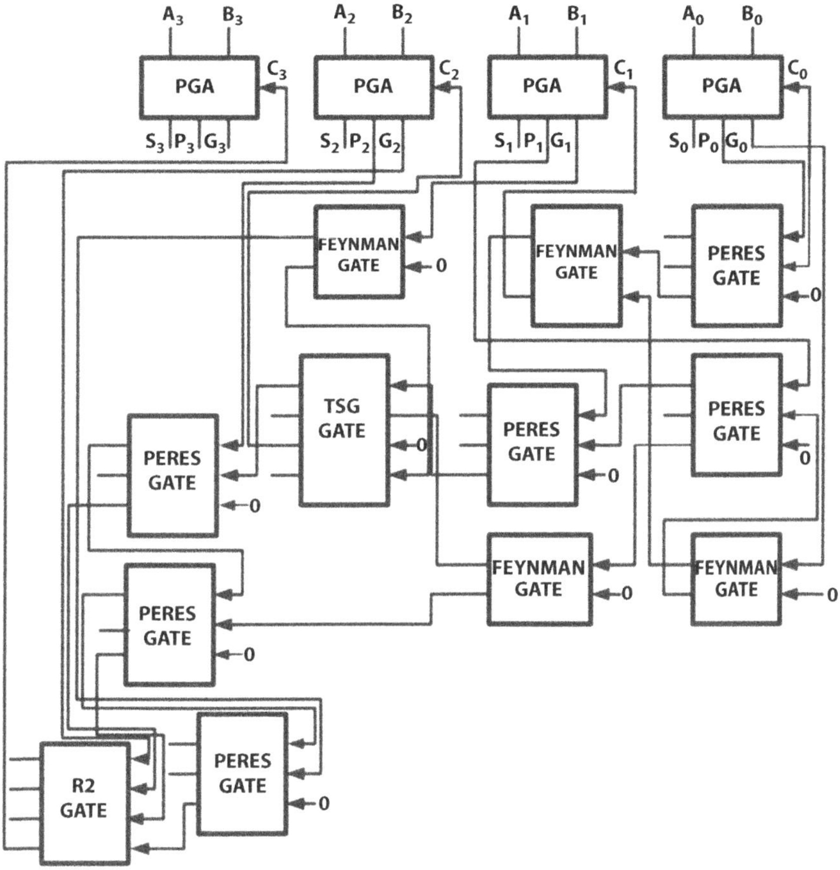

Figure 14.8 4-bit CLA design using reversible logic gates.

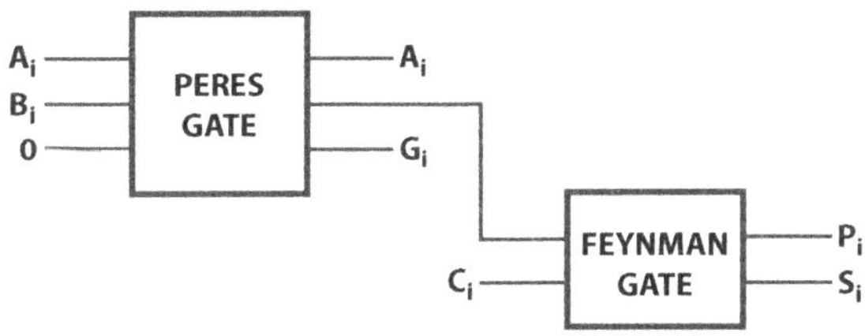

Figure 14.9 Carry generate and propagate block.

14.3.4 Proposed Reversible 8-Vedic Multiplier

The Vedic multiplier, which utilizes the Urdhva Triambakam technique, is a highly efficient algorithm derived from ancient Vedic mathematics, specifically designed for fast and parallel multiplication. This method, which translates to "Vertically and Crosswise," simplifies the multiplication process by breaking it down into smaller steps that can be executed simultaneously. In digital circuit design, this parallelism allows for the rapid generation of partial products, significantly reducing the computation time compared to traditional methods. The Urdhva Triambakam technique is particularly well-suited for binary multiplication, making it an ideal choice for modern digital systems where efficiency and speed are paramount.

The importance of using the Urdhva Triambakam approach in Vedic multipliers is its scalability, which allows it to handle numbers of any size easily. This renders the approach adaptable for a broad spectrum of purposes, ranging from basic arithmetic operations to intricate jobs in digital signal processing and cryptography. Additionally, the reduced number of computation steps and the ability to perform them in parallel results in lower power consumption and faster processing speeds, making the Vedic multiplier a valuable tool for the development of rapid processors as well as other digital systems. The Urdhva Triambakam technique showcases the enduring relevance of ancient mathematical principles in the advancement of modern technology.

14.3.4.1 2-Bit Vedic Multiplier

The Urdhva Triambakam (Vertically and Crosswise) technique is employed by a 2-bit Vedic multiplier, as illustrated in Figure 14.10, to effectively multiply two 2-bit binary numbers. Three critical stages comprise the procedure: Initially, the least significant bits (LSBs) of the two numbers multiply each other vertically to generate the LSB of the final product. Next, crosswise multiplication of the bits is performed, where the sum of these products forms the next bit in the final output. Finally, the most significant bits (MSBs) were multiplied vertically to yield the MSB of the product. Consider both two-bit numbers A[1:0] and B[1:0] for 2×2 multiplication. The process starts by multiplying LSBs (A0XB0), and the result is stored as s0 ($s0 = A0 \times B0$). Next, the crosswise products $A0 \times B1$ and $A1 \times B0$ were computed, and their sum was stored as c1s1 ($c1s1 = A1 \times B0 + A0 \times B1$). The carry bit(c1) from step is then added to the product of the most significant bits ($A1 \times B1$) to complete the multiplication.

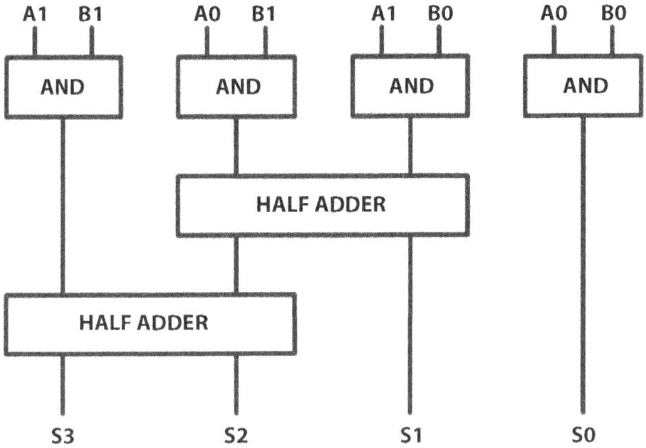

Figure 14.10 2-bit Vedic multiplier.

14.3.4.2 4-Bit Vedic Multiplier

A 4-bit Vedic multiplier, depicted in Figure 14.11, utilizing the Urdhva Triambakam (vertically and crosswise) technique, multiplies two 4-bit binary numbers by breaking down the operation into a series of parallel and straightforward steps. The process involves generating partial products through the vertical and crosswise multiplication of each bit, which are then combined to form the final product. Specifically, the LSBs are multiplied first, followed by the crosswise products, and finally, the MSBs. Each operation produces partial results that are summed to produce an 8 bit output. A 4-bit Vedic multiplier handles the multiplication of two 4-bit binary numbers, A[3:0] and B[3:0]. The process involves computing several partial products by multiplying bits vertically and crosswise; the initial product S0 is obtained by A0 × B0, followed by C1S1 from A1 × B0 + A0 × B1, etc. Subsequent carries C2, C3, and so on are added to the products of the remaining bit combinations.

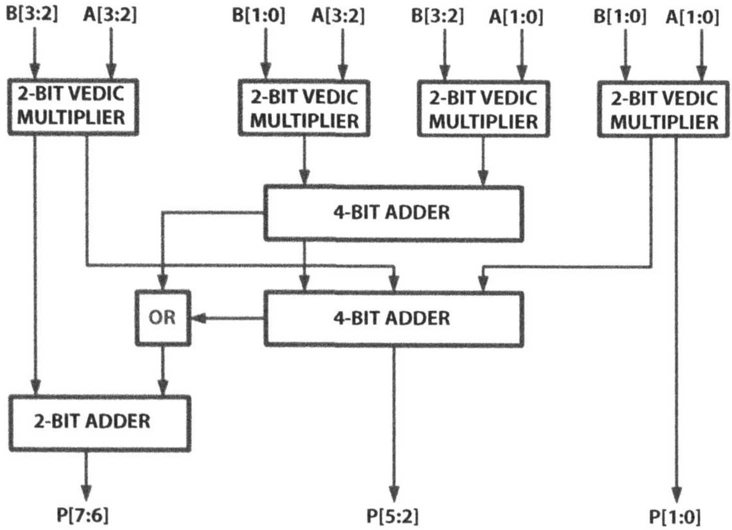

Figure 14.11 4-bit Vedic multiplier.

14.3.4.3 8-Bit Vedic Multiplier

An 8-bit Vedic multiplier shown in Figure 14.12, using the Urdhva Triambakam technique, multiplies two 8-bit numbers, where A = $A_7A_6A_5A_4A_3A_2A_1A_0$ and B = $B_7B_6B_5B_4B_3B_2B_1B_0$, by decomposing multiplication into a series of parallel, simpler operations. This process involves generating partial products for each bit pair through vertical and crosswise multiplication. The ultimate 16-bit outcome is obtained by aligning and summing these partial products. Specifically, the multiplication of LSBs generates initial partial products, crosswise multiplications handle intermediate results, and MSBs produce higher order products. For 8-bit multiplication, this approach involves dividing the numbers into 4-bit units. For example, the 8-bit number 10110011 can be separated into 1011 and 0011, and then a 4-bit Vedic multiplier is used to perform the vertical and crosswise multiplication of these units. This procedure simultaneously produces four partial products. An n-bit RCA was employed to combine partial products 2 and 3 at an identical stage. The two MSBs of partial product 1 are employed for the addition of partial product 3, which is subsequently processed using an n-bit RCA. The higher-order MSBs of this sum act as a carry for partial product 4, which is also added by using an n-bit ripple carry adder. This sequence results in an 8-bit final output, with each partial product calculated using n/2-bit multipliers.

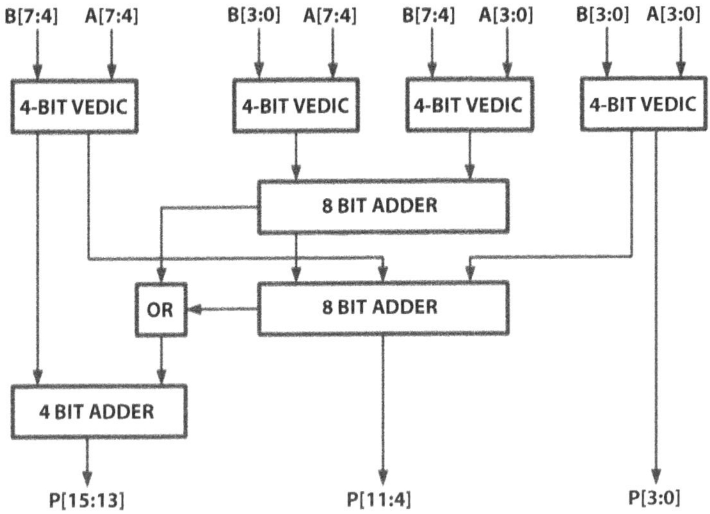

Figure 14.12 8-bit Vedic multiplier.

The multiplier employs RCA and CLA as adder units for the addition operations. A 4-bit and 8-bit RCA were constructed to facilitate the addition process within the multiplier circuit. The 8-bit CLA is implemented by combining two 4-bit CLA circuits.

14.4 Result Analysis

The proposed work was carried out using Vivado 2022.2, with the source code developed in Verilog HDL. To synthesize and simulate the Verilog code in Xilinx Vivado 2022.2, start by creating a new project, assigning it a suitable name, and selecting the RTL Project type. Upload Verilog source files and designate the top module of the design. Next, initiate the synthesis process by selecting "Run Synthesis" from the Flow Navigator. Once synthesis was complete, review the results. For simulation, create a test bench for your design, add it to the project, and set it as the top module. To begin the behavioral simulation, choose "Run Behavioral Simulation" from the Flow Navigator, and then analyze the results in the waveform viewer.

This study focused on employing reversible logic gates to design efficient adder and multiplier architectures. The simulation outputs of Full adder, 8-bit RCA, 4-bit CLA, 4-bit Vedic multiplier, 8-bit Vedic multiplier are illustrated in Figures 14.13, 14.14, 14.15, 14.16, and 14.17, respectively. The RTL Schematic for 8-bit Vedic multiplier is shown in Figure 14.18.

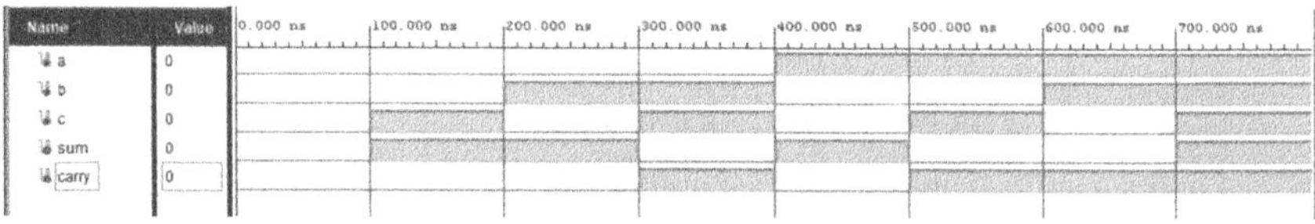

Figure 14.13 Simulation waveform of full adder.

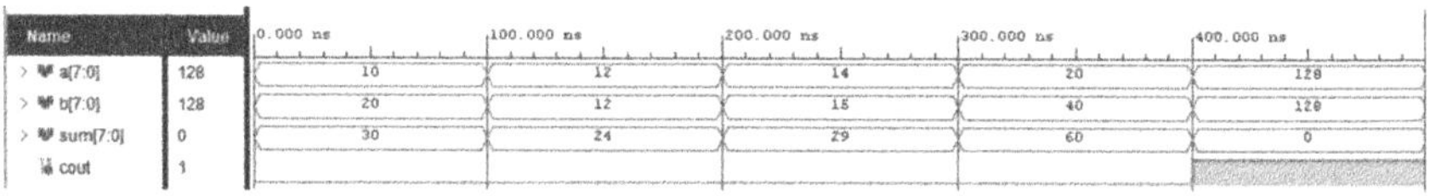

Figure 14.14 Simulation waveform of 8-bit Ripple Carry adder.

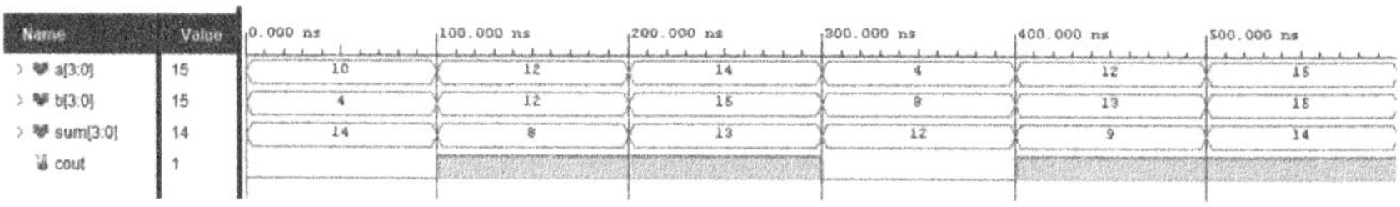

Figure 14.15 Simulation waveform of 4-bit Carry look ahead adder.

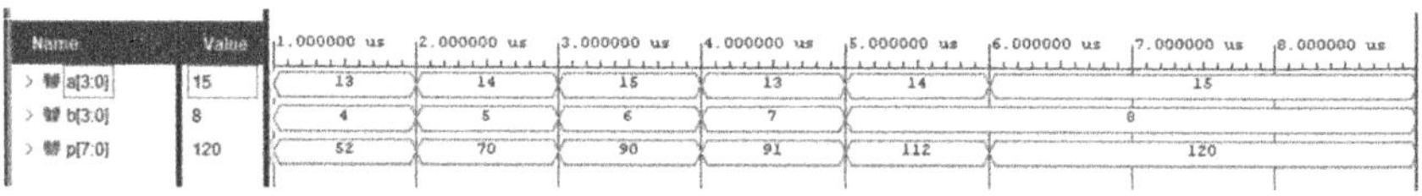

Figure 14.16 Simulation waveform of 4-bit Vedic multiplier.

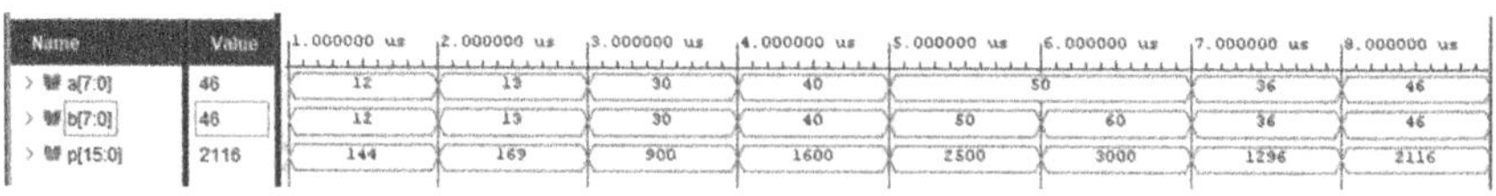

Figure 14.17 Simulation waveform of 8-bit Vedic multiplier.

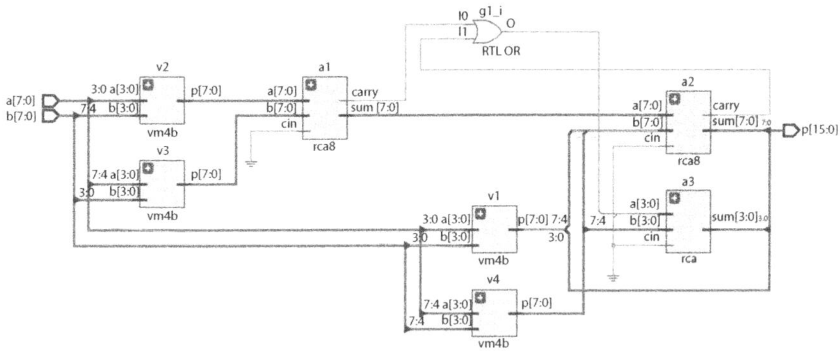

Figure 14.18 RTL schematic of 8-bit Vedic multiplier.

Table 14.1 Comparative analysis of proposed full adder with other designs.

Design	Garbage output	Ancilla inputs	Transistor count
Existing adder [17]	4	3	88
Existing adder [18]	2	1	–
Existing adder [19]	5	4	80
Existing adder [20]	5	4	64
Proposed adder	1	1	44

Table 14.2 Comparative analysis of 8-bit Vedic multiplier against existing architectures.

Multipliers	Logic levels	Delay (ns)
Array Multiplier [21]	19	12.934
Wallace Tree Multiplier [22]	12	7.815
Dadda Multiplier [23]	16	9.745
Existing Vedic Multiplier 1 [24]	15	0.130
Existing Vedic Multiplier 2 [25]	13	7.762
Proposed Vedic Multiplier using Reversible logic	10	5.527

Various input combinations are employed to rigorously test the designed circuits. Table 14.1 provides a comprehensive comparative analysis of the reversible full adder against existing adder designs. The proposed adder is efficient, utilizing only one ancilla input and producing only one garbage output while using 44 transistors. Table 14.2 provides an extensive analysis of the 8-bit Vedic multiplier against various other multiplier architectures. The proposed design required only 10 logic levels and achieved a delay of 5.527 ns.

14.5 Conclusion

This study implements an 8-bit Vedic multiplier using a full adder designed with reversible gates in Vivado 2022.2, with Verilog HDL as the source code. Utilizing Vedic mathematics, particularly the Urdhva-Tiryakbhyam sutra, the proposed design effectively managed multiplication operations with a reduced gate count and lower quantum cost. Comparative analysis with traditional multipliers reveals the Vedic Multiplier's advantages in terms of speed and resource efficiency. The proposed adder achieved a remarkable 75% reduction in garbage output and a 50% decrease in constant inputs. Additionally, the Vedic multiplier demonstrates 47% lower logic level utilization and 57% improvement in delay (ns) compared to the array multiplier. Therefore, this approach offers a promising path for the development of high-performance, energy-efficient digital systems.

References

1. Toffoli, T., Reversible computing, in: *Automata, Languages and Programming*, ser. Lecture Notes in Computer Science, vol. 85, J. de Bakker and J. van Leeuwen (Eds.), pp. 632–644, Springer Berlin Heidelberg, Berlin, 1980.
2. Pujar, J., Raveendran, S., Panigrahi, T., Vasantha, M.H., Kumar, Y.B.N., Design and Analysis of Energy Efficient Reversible Logic based Full Adder. *2019 IEEE 62nd International Midwest Symposium on Circuits and Systems (MWSCAS)*, USA, pp. 339–342, 2019.
3. Aditya, M., Kumar, Y.B.N., Vasantha, M.H., Reversible full/half adder with optimum power dissipation. *2016 10th International Conference on Intelligent Systems and Control (ISCO)*, Coimbatore, India, pp. 1–4, 2016.
4. Vasudevan, D.P., Lala, P.K., Di, J., Parkerson, J.P., Reversible-logic design with online testability. *IEEE Trans. Instrum. Meas.*, 55, 2, 406–414, 2006.
5. Ganesh, C., *et al.*, A Novel Design of Area Efficient Full Adder Architecture using Reversible Logic Gates. *7th International Conference on Devices, Circuits, and Systems (ICDCS)*, 2024.

6. Kumar, A.S. and Naresh Kumar Reddy, B., An Efficient Real-Time Embedded Application Mapping for NoC Based Multiprocessor System on Chip. *Wirel. Pers. Commun.*, 128, 2937–2952, 2023.
7. Sarangam, K., *et al.*, Design and Investigation of the 22 nm FinFET Based Dynamic Latched Comparator for Low Power Applications. *Trans. Electr. Electron. Mater.*, 25, 2, 1–14, 2024.
8. Kumar, A.S. and Hanumantha Rao, T.V.K., Scalable benchmark synthesis for performance evaluation of NoC core mapping. *Microprocess. Microsyst.*, 79, 103272, November 2020.
9. Kumar, A.S. and Hanumantha Rao, T.V.K., An Adaptive Core Mapping Algorithm on NoC for Future Heterogeneous System-on-Chip. *Comput. Electr. Eng.*, 95, 107441, October 2021.
10. Kumar, A.S., *et al.*, Nanosheet Field Effect Transistor Device and Circuit Aspects for Future Technology Nodes. *ECS J. Solid State Sci. Technol.*, 12, 8, 083009, 2023.
11. Kumar, A.S., *et al.*, Nanosheet FET for Future Technology Scaling, in: *Integrated Devices for Artificial Intelligence and VLSI*, pp. 25–47, 2024.
12. Kumar, A.S., *et al.*, Device Analysis of Vertically Stacked GAA Nanosheet FET at Advanced Technology Node. *2023 3rd International Conference on Advances in Computing, Communication, Embedded and Secure Systems (ACCESS)*, India, pp. 274–279, 2023.
13. Thapliyal, H., *et al.*, A novel reversible TSG gate and its application for designing reversible carry look-ahead and other adder architectures. *Proc. Asia-Pac. Conf. Adv. Comput. Syst. Architect.*, pp. 805–817, 2005.
14. Sarangam, K., *et al.*, Design and Energy Dissipation Analysis of Full-Adder for Low Power Applications Using Reversible Logic Gates. *3rd International Conference on Emerging Frontiers in Electrical and Electronic Technologies (ICEFEET- 2023)*, NIT Patna, 2023.
15. Shukla, V., Singh, O.P., Mishra, G.R., Tiwari, R.K., A novel approach for reversible realization of 4 × 4 bit vedic multiplier circuit, in: *Advances in VLSI, Communication, and Signal Processing: Select Proceedings of VCAS 2018*, Springer Singapore, pp. 733–745, 2020.
16. Saligram, R., *et al.*, Optimized reversible vedic multipliers for high speed low power operations. *2013 IEEE Conference on Information & Communication Technologies*, Thuckalay, India, pp. 809–814, 2013.
17. Parhami, B., Fault-tolerant reversible circuits, in: *IEEE Fortieth Asilomar Conference on Signals, Systems and Computers*, pp. 1726–1729, 2006.
18. Chiwande, S.S. and Dakhole, P.K., Design and Analysis of Low Power Full Adder using Reversible Logic. *2022 6th International Conference on Electronics, Communication and Aerospace Technology*, Coimbatore, India, pp. 146–150, 2022.
19. Bruce, J., *et al.*, Efficient adder circuits based on a conservative reversible logic gate, in: *Proceedings IEEE Computer Society Annual Symposium on VLSI*, pp. 74–79, 2002.

20. Kaur, P. and Dhaliwal, B.S., Design of fault tolearnt full adder/subtarctor using reversible gates, in: *International Conference on Computer Communication and Informatics (ICCCI)*, pp. 1–5, 2012.
21. Baugh, C.R. and Wooley, B.A., A Two's Complement Parallel Array Multiplication Algorithm. *IEEE Trans. Comput.*, 22, 12, 1045–1047, 1973.
22. Nagaraj, S., Thyagarajan, K., Srihari, D., Gopi, K., Design and Analysis of Wallace Tree Multiplier for CMOS and CPL Logic. *2018 International Conference on Computation of Power, Energy, Information and Communication (ICCPEIC)*, pp. 006–010, 2018.
23. Naresh, K., Sai, Y.P., Majumdar, S., Design of 8-bit Dadda Multiplier using Gate Level Approximate 4:2 Compressor. *2022 35th International Conference on VLSI Design and 2022 21st International Conference on Embedded Systems (VLSID)*, pp. 269–274, 2022, doi: 10.1109/VLSID2022.2022.00059.
24. Chandrashekara, M.N. and Rohith, S., Design of 8 Bit Vedic Multiplier Using Urdhva Tiryagbhyam Sutra With Modified Carry Save Adder. *2019 4th International Conference on Recent Trends on Electronics, Information, Communication Technology (RTEICT)*, pp. 116–120, 2019, doi: 10.1109/RTEICT46194.2019.9016965.
25. Kumar, A.S., Siddhesh, U., Sai kiran, N., Bhavitha, K., Design of High Speed 8-bit Vedic Multiplier using Brent Kung Adders. *2022 13th International Conference on Computing Communication and Networking Technologies (ICCCNT)*, Kharagpur, India, pp. 1–5, 2022.

15

64-Bit High Speed Parallel Prefix Adder Architectures

B. Harish[1]* and M.S.S. Rukmini[2]

[1]*Department of Electronics and Communication Engineering, Vignan's Lara Institute of Technology and Science, Guntur, India*
[2]*Department of Electronics and Communication Engineering, Vignan's Foundation for Science Technology and Research, Guntur, India*

Abstract
The increasing demand for real-time digital signal processing applications has called for the development of fast and efficient parallel prefix adder (PPA) designs. This paper examines various PPA architectures created for 64-bit applications, as well as their performance compared to traditional adders. All design entities were verified using Xilinx ISE, and the synthesis results were subsequently analyzed. It was observed that the 64-bit Pipelined Parallel prefix Brent-Kung adder (PPPBKA) exhibited the highest operating speed amongst the tested adders, including regular and parallel prefix adders. Furthermore, the power delay product (PDP) of the Pipelined Brent Kung adder was found to be lower than those of the other adders. The simulation and synthesis were conducted using Xilinx ISE.

Keywords: PPA, PPPBKA, PDP, FPGA, verilog

15.1 Introduction

15.1.1 Introduction to Parallel Prefix Adders

Addition is one of the most fundamental and widely employed arithmetic operations, playing a critical role in determining the speed and efficiency of digital devices. In modern microprocessors, digital signal processors, and application-specific integrated circuits (ASICs), the binary adder is an

**Corresponding author*: 17harish@gmail.com

Ekta Goel, Archana Pandey, Shiromani Balmukund Rahi and Arun Samuel (eds.) Handbook of Advanced Semiconductor Field Effect Transistors, (323–336) © 2026 Scrivener Publishing LLC

indispensable component, serving as the building block for all arithmetic operations. As such, the performance of adders greatly influences the overall speed and accuracy of digital systems. This has spurred significant research into optimizing adder architectures in very large-scale integration (VLSI) design [1–3].

Over the past decade, numerous adder architectures have been proposed and studied to improve binary addition performance. Particular focus has been given to parallel prefix adders (PPAs), which are well-suited for high-speed digital signal processing applications. Some notable examples of parallel prefix adders include the Kogge-stone adder, Ling adder, Brent-kung adder, Sparse Kogge-stone adder, Knowles adder, Sklansky adder, spanning tree adder, and Lynch adder. A brief overview of parallel prefix adders is provided in Section 15.1 [4–6].

In Section 15.2, the implementation of parallel prefix adders and other adder architectures in 64-bit is discussed. Section 15.3 presents the results and discussion, while Section 15.4 concludes with a summary of the findings and implications for future research. Throughout this report, attention is paid to the technical details and theoretical underpinnings of adder architectures, demonstrating a deep understanding of the subject matter. The objective is to contribute meaningfully to the ongoing efforts in optimizing adder efficiency in VLSI system design.

15.1.2 Parallel Prefix Adders

Prefix addition, as its name implies, is an iterative method of addition that begins at the least significant bit (LSB) and progressively moves up to the most significant bit (MSB). Parallel prefix addition (PPA), on the other hand,

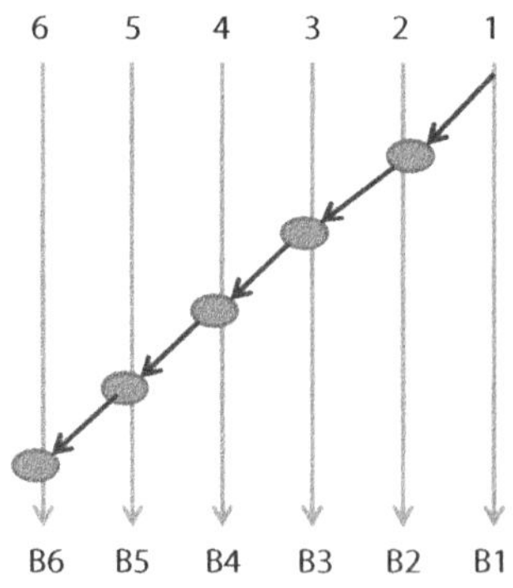

Figure 15.1 Serial addition [3].

refers to the simultaneous execution of addition operations. Figure 15.1 illustrates serial addition, where the result of each stage is dependent upon the previous stage, leading to an increase in the overall time required for the final sum. In contrast, Figure 15.2 shows parallel addition, where the result of each stage does not necessarily depend on the previous stage, resulting in a reduction in the time required to generate the final sum at a logarithmic rate. The parallel prefix addition process consists of three distinct stages, as depicted in Figure 15.3 [7–10].

In parallel prefix addition the propagate (P) and generate (G) bits are combined in a different way to generate a required carry input for each bit of addition. Figures 15.4 and 15.5 consist of black and gray cells and their logic circuit respectively [11–13].

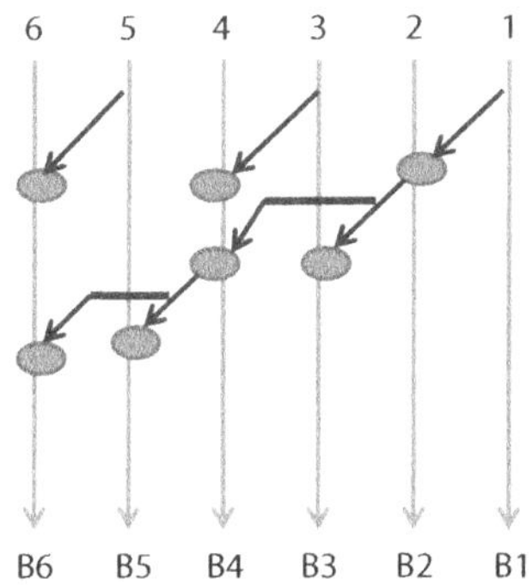

Figure 15.2 Parallel addition [3].

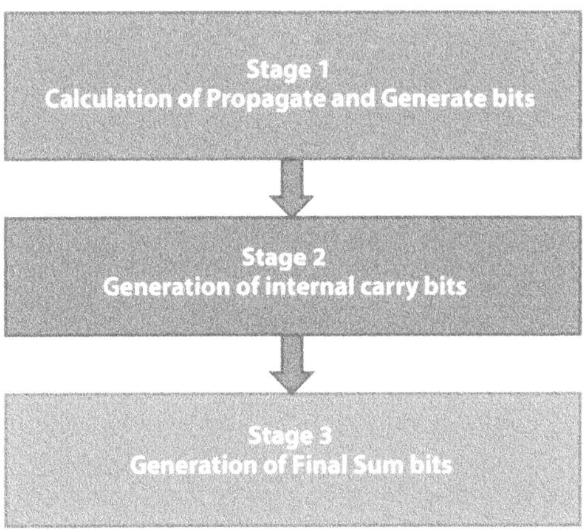

Figure 15.3 Steps in parallel prefix addition [3].

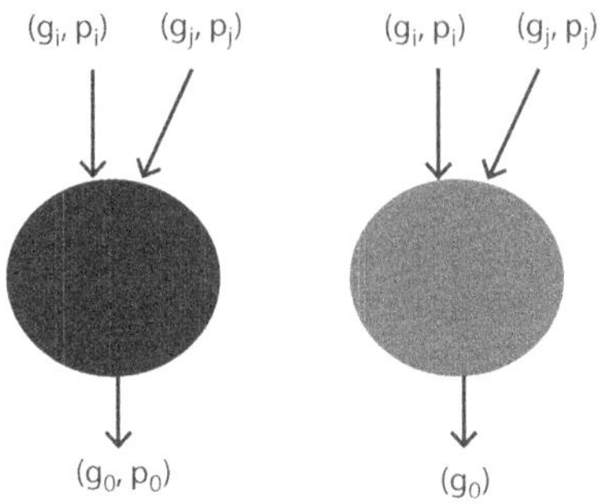

Figure 15.4 Black and gray cells [3].

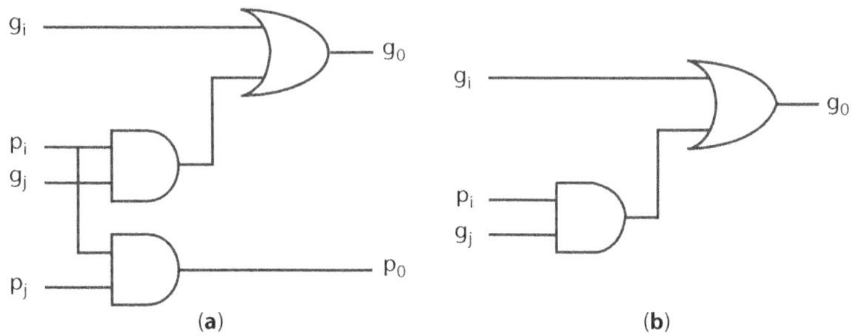

Figure 15.5 Logic circuits of (a) black and (b) gray cells [3].

The following are the equations produced by black and gray cells:
The black cell output logic equations:

$$G = Gi + Pi\ Gj \tag{15.1}$$

$$P = Pi\ Pj \tag{15.2}$$

The gray cell output logic equations:

$$G = Gi + Pi\ Gj \tag{15.3}$$

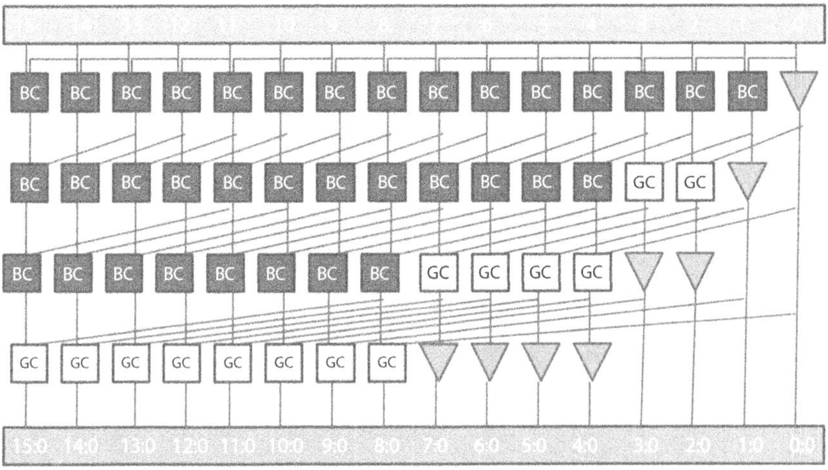

Figure 15.6 Kogge Stone adder architecture [10].

15.2 Implementation of PPA in 64-Bit

15.2.1 Kogge-Stone Adder

KSA is the first parallel prefix form of carry look-ahead adders. Kogge and Stone demonstrated that this adder has superior performance in terms of speed of operation in IC applications. Figure 15.6 depicts the KSA structure. The KSA is the quick adder with a wide area and low fan-out, in which each cell leads P and G bits that are determined using the equations (15.1), (15.2), and (15.3) [14].

15.2.2 Sparse Kogge Stone Adder

The Sparse Kogge-Stone (SKS) adder, as depicted in Figure 15.7, was first introduced by Sreenivasa Goud. This architecture consists of two distinct blocks, with the lower half dominated by numerous RCAs and the upper half characterized by gray and black cells. Compared to the traditional Kogge-Stone (KS) adder, the SKS adder exhibits a more substantial speed enhancement. However, this improvement comes at a cost in terms of increased power dissipation and area occupation. The SKS adder contains 11 black cells and 3 gray cells in total. The lower half of the architecture is comprised of 4 stages, each containing 4 RCAs. In this section, a total of 16 RCAs is present to accommodate the input of two 16-bit operands. The intermediate carriers generated in the first block are then passed on to the RCAs in the lower part of the architecture [9].

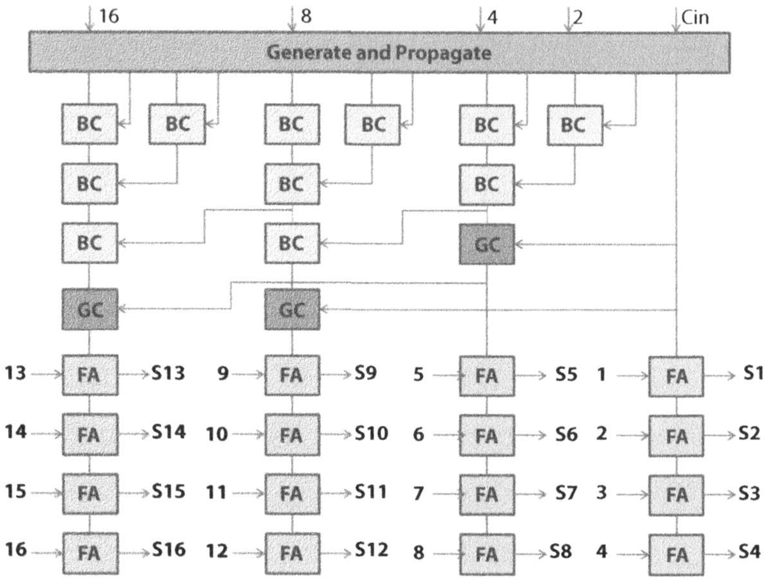

Figure 15.7 Sparse Kogge Stone adder architecture [18].

While the time required for intermediate carry generation in the first block is reduced due to the presence of RCAs in the lower part, the time taken for final carry generation is higher. The SKS adder thus strikes a balance between speed and complexity, offering a viable solution for addressing the ever-growing demands of high-performance digital systems.

15.2.3 Spanning Tree Adder

Lynch and Swartzlander's seminal work introduced the Spanning tree adder (STA), a hybrid of lookahead carry and prefix adder techniques. As depicted in Figure 15.8, this innovative adder exhibits superior performance compared to its constituent components. The STA employs a 4-input Manchester chain block tree to generate carry-in signals for selective addition at 8, 16, 24, 32, 40, 48, and 56-bit positions. This architecture consists of 22 black cells and 3 gray cells in its upper portion, supplemented by four stages of ripple carry adders (RCAs).

For the specified section of the carry, the STA utilizes two conventional 8-bit ripple carry adders to generate sums corresponding to Ci = 1 and Ci = 0. A Manchester carry chain then tests these signals and selectively activates the appropriate adder stage. In certain configurations, the Manchester carry chain may be ready before the total sum for the chosen carriage position.

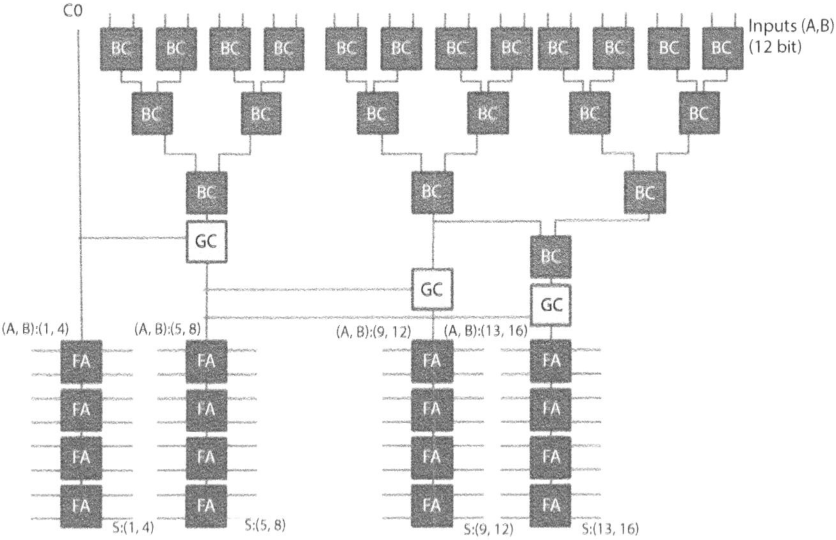

Figure 15.8 Spanning tree adder architecture [18].

However, all three adders - the STA, Recursive hybrid adder, and Spanning tree adder variable - employ three-level Manchester carry chains. The distinction lies in the intermediate outputs of the STA during its second and third stages, which enable the initiation of preemptive charging signals [15].

This professional presentation highlights the meticulous design of the Spanning Tree Adder, emphasizing its unique hybrid nature, superior performance, and adaptability across various carry positions. The use of precise terminology and clear explanations of the adder's architecture and operation demonstrate the authors' expertise in the field of digital circuit design.

15.2.4 Brent-Kung Adder

Richard Brent and Hsiang Kung introduced the Binary-Coded Key (BCK) adder in 1982, as depicted in Figure 15.9. Compared to the Karnaugh-Shoup (KS) adder, the BCK adder operates at a faster speed, exhibits less wiring complexity, and occupies less physical space. In the BCK adder, prefixes for the 2-bit groups are computed. These prefix trees are employed once more in the computation of the eight-bit group and so forth, to measure the other prefix for the four-bit groups. The architecture's stage count is given by 2^log2N - 1.

Despite using fifteen grey cells and thirty-six black cells, the sixteen-bit BCK adder utilizes only eleven grey cells and fourteen black cells. This leads to a

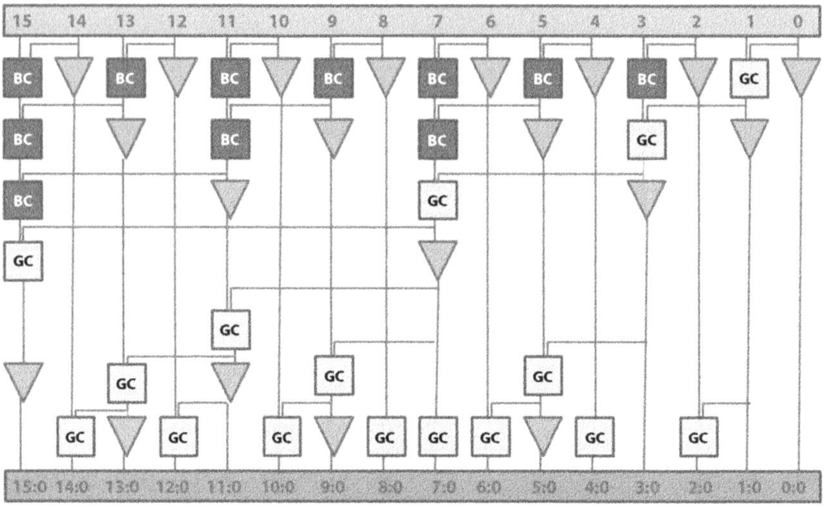

Figure 15.9 16-bit Brent-Kung adder architecture [9].

simpler construction and lower space requirements as compared to the KS adder. Furthermore, the BCK adder operates at a higher frequency than its predecessor. In summary, the BCK adder offers significant advantages in terms of speed, complexity, and space consumption over the KS adder [16, 17].

15.2.5 16-Bit Pipelined Brent-Kung Adder

Figure 15.10 presents the 16-bit Brent Kung adder (PBKA), which employs 16-bit registers. The design utilizes gray cells, black cells, and an additional four-stage pipeline register to enhance its functionality. Compared to the non-pipelined BK adder, the PBKA architecture is more complex and requires more chip area due to the pipelined registers. The high-speed addition capability of the pipelined BK adder is achieved through the implementation of pipelined stages. As shown in Figure 15.10, there are four pipelined stages, resulting in a clock speed that is four times higher than in an equivalent non-pipelined circuit. The lack of synchronization components in a non-pipelined circuit necessitates all computations to be generated internally.

The maximum pipeline clock rate for pipelined circuits is determined by the reciprocal of the maximum delay. Furthermore, an increase in the maximum frequency is possible by breaking down the overall complicated logic

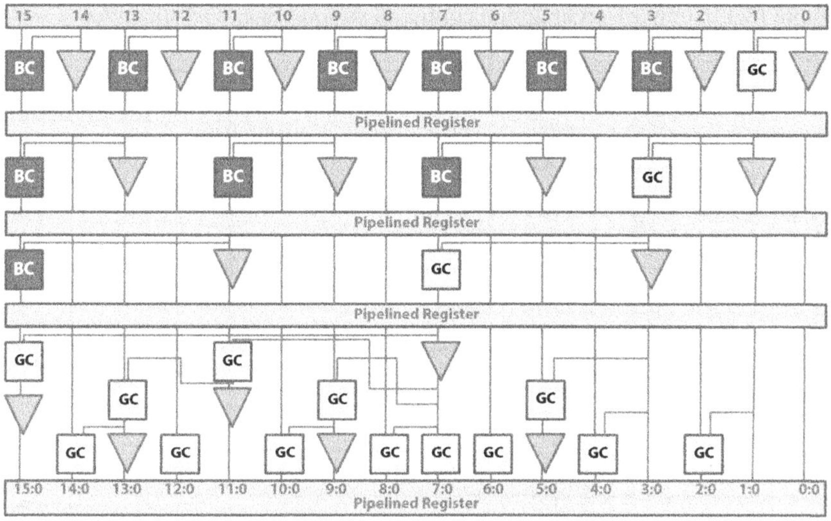

Figure 15.10 16-bit pipelined Brent-Kung adder architecture [1].

into smaller, simpler stages that still require synchronization components. Table 15.1 illustrates that the proposed model outperforms other PPAs and traditional adders in terms of operating speed.

Results and Discussion

Various parallel prefix adder topologies are meticulously examined. Particularly, the pipelined parallel prefix Brent Kung adder exhibits the fastest operation speed compared to all other adders discussed. Furthermore, the power delay product (PDP) of the pipelined Brent Kung adder is remarkably lower than those of its counterparts. A detailed comparison of design entities such as delay, area, and power dissipation among 16 and 64-bit adder topologies is presented in Table 15.1.

The simulation results for the pipelined Brent Kung adder (64-bit) and its corresponding Register Transfer Level (RTL) schematic are illustrated in Figures 15.11 and 15.12, respectively. Lastly, Figures 15.13 and 15.14 provide a comprehensive comparison of delay and power delay product (PDP) for each of the analyzed adders.

Table 15.1 Comparison of design entities of various 16 and 64-bit adder topologies [3].

S. no	Adder name	Delay(ns)		Device utilization (LUTs)		Power dissipation (mW)		PDP (nJ)	
		16 bit	64 bit	16 bit	64 bit	16 bit	64 bit	16 bit	64 bit
1	Ripple carry adder	3.64	95.7	24	96	266	1197	0.97	114.6
2	Carry look-ahead adder	4.30	37.12	24	96	253	1138	1.09	42.2
3	Carry skip adder	4.58	92.12	29	116	249	1109	1.14	102.2
4	Kogge stone adder	3.56	28.51	48	179	268	1323	0.95	37.7
5	Sparse kogge stone adder	3.39	43.22	44	159	322	1545	1.09	66.8
6	Spanning tree adder	3.75	57.8	35	126	310	1446	1.16	83.6
7	Brent kung adder	3.05	18.97	32	128	283	1413	0.86	26.8
8	Pipelined Brent Kung adder	0.73	3.21	48	189	428	3656	0.31	11.7

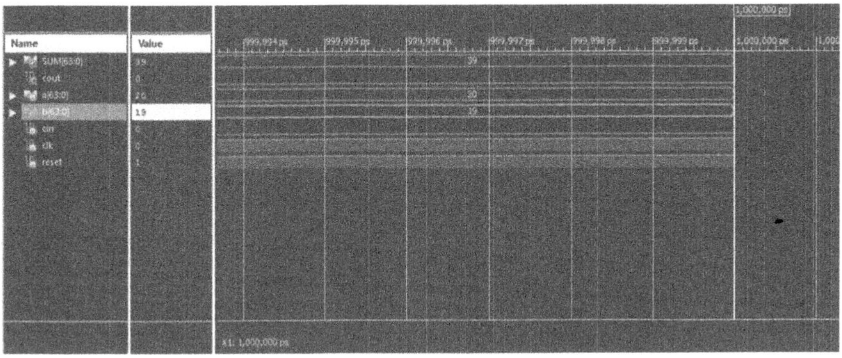

Figure 15.11 Simulation results of 64-bit pipelined Brent-Kung adder architecture [2].

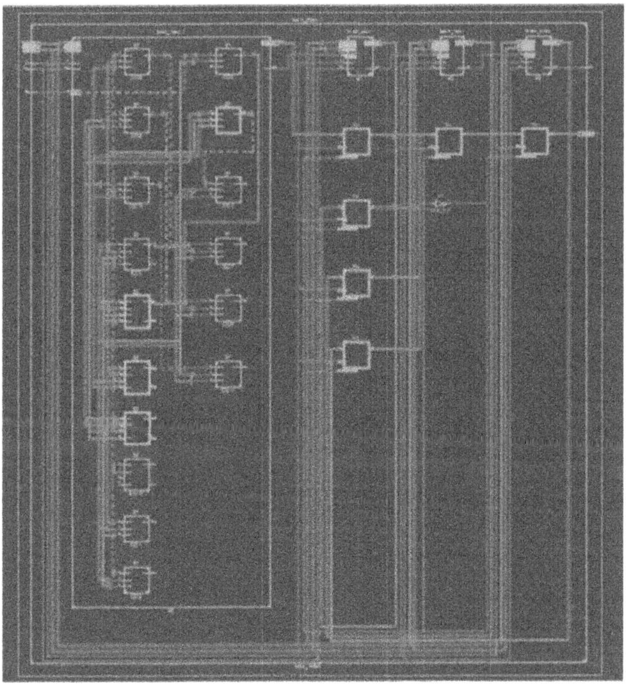

Figure 15.12 RTL schematic of 64-bit pipelined Brent-Kung adder architecture [5].

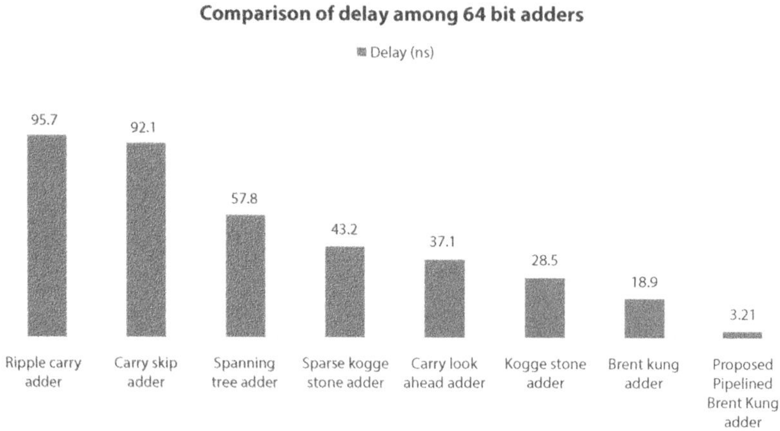

Figure 15.13 Delay observed for various adders (64-bit) [5].

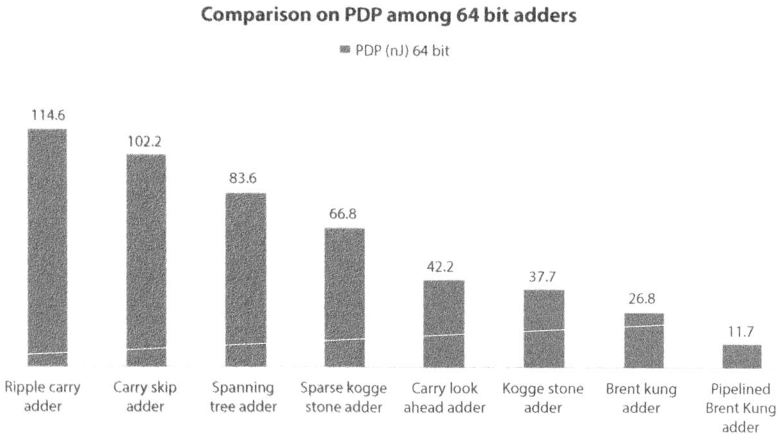

Figure 15.14 Comparison of PDP observed for various adders (64-bit) [5].

Conclusion

The speed of operation is a critical factor in the design of Very Large-Scale Integration (VLSI) systems, particularly in digital integrated circuit (IC) applications. In this study, it was found that the pipelined version of the Brent Kung (BK) adder exhibits significantly higher operating speeds when compared to other adders. A thorough analysis of delay, area, and power consumption revealed that the speed of operation of the Pipelined BK adder is improved by 73% when compared to the existing BK adder. Furthermore, the performance of the Pipelined BK adder (PBKA) was also found to be 56% faster than the conventional BK adder.

The pipelined nature of the PBKA allows for efficient implementation of functions such as the discrete Fourier transform (DFT) and fast Fourier transform (FFT). The adder can be easily scaled to accommodate 128-bit inputs, making it a versatile building block for high-performance digital systems. In conclusion, the Pipelined Brent Kung adder proves to be an invaluable addition to the toolbox of professionals designing VLSI digital ICs.

References

1. Harish, B., Rukmini, M.S.S., Sivani, K., Design of MAC unit for digital filters in signal processing and communication. *Int. J. Speech Technol.*, 11, 1–5, 2021.
2. Harish, B., Sivani, K., Rukmini, M.S.S., Performance comparison of various CMOS full adders, in: *2017 International Conference on Energy, Communication, Data Analytics and Soft Computing (ICECDS)*, 2017, August, IEEE, pp. 3789–3792.
3. Harish, B., Sivani, K., Rukmini, M.S.S., Design and Performance Comparison among Various Types of Adder Topologies, in: *2019 3rd International Conference on Computing Methodologies and Communication (ICCMC)*, 2019, March, IEEE, pp. 725–730.
4. Harish, B. and Rukmini, M.S.S., Ultra high-speed full adder for bio-medical applications. *Int. J. Reconfigurable Embed. Syst.*, *10*, 1, 25–31, 2021.
5. Harish, B., Sivani, K., Rukmini, M.S.S., Design of high-speed efficient parallel prefix Brent-Kung Adder architecture. *J. Green Eng.*, *10*, 7, 3508–3519, 2020.
6. Harish, B., Sivani, K., Rukmini, M.S.S., Implementation of frequency efficient multiplexer-based CORDIC on FPGA. *Int. J. Adv. Sci. Technol.*, *29*, 05, 753–762, 2020.
7. Barik, R.K., Pradhan, M., Panda, R., Time-efficient signed Vedic multiplier using redundant binary representation. *J. Eng.*, *2017*, 3, 60–68, 2017.

8. Beaumont-Smith, A. and Lim, C.C., Parallel prefix adder design, in: *Proceedings 15th IEEE Symposium on Computer Arithmetic. ARITH-15 2001*, 2001, June, IEEE, pp. 218–225.
9. Brent, R.P. and Kung, H.T., A regular layout for parallel adders. *IEEE Comput. Archit. Lett.*, *31*, 03, 260–264, 1982.
10. Kogge, P.M. and Stone, H.S., A parallel algorithm for the efficient solution of a general class of recurrence equations. *IEEE Trans. Comput.*, *100*, 8, 786–793, 1973.
11. Paliwal, P., Sharma, J.B., Nath, V., Comparative study of FFA architectures using different multiplier and adder topologies. *Microsyst. Technol.*, 7, 1–8, 2019.
12. Patali, P. and Kassim, S.T., High throughput FIR filter architectures using retiming and modified CSLA-based adders. *IET Circuits Devices Syst.*, *13*, 7, 1007–1017, 2019.
13. Pudi, V. and Sridharan, K., Low complexity design of ripple carry and Brent–Kung adders in QCA. *IEEE Trans. Nanotechnol.*, *11*, 1, 105–119, 2011.
14. Yizhen, L., Lin, L., Jun, W., The application of pipeline technology: An overview, in: *2011 6th International Conference on Computer Science & Education (ICCSE)*, 2011, August, IEEE, pp. 47–51.
15. Nagendra, C., Irwin, M.J., Owens, R.M., Area-time-power tradeoffs in parallel adders. *IEEE Trans. Circuits Syst. II: Analog Digit. Signal Process.*, *43*, 10, 689–702, 1996.
16. Carlson, D.A. and Sugla, B., Limited width parallel prefix circuits. *J. Supercomput.*, *4*, 2, 107–129, 1990.
17. Gizopoulos, D., Psarakis, M., Paschalis, A., Zorian, Y., Easily testable cellular carry lookahead adders. *J. Electron. Test.*, *19*, 3, 285–298, 2003.
18. Shashikala, B.N., P.R., Prahallad, P., R.S.K., Ramyashree, M.V., Design and implementation of Kogge-stone, Sparse Kogge-stone and Spanning tree adder. *2023 International Conference on Smart Systems for applications in Electrical Sciences (ICSSES)*, Tumakuru, India, pp. 1–6, 2023, doi: 10.1109/ICSSES58299.2023.10200644.

16

Design and Implementation of High-Performance Adaptive Baud Rate Generator for IoT Applications

B. Harish*, N. Jahnavi, M. Brammani, Md. Karishma and N. J. L. S. Manasa

Department of Electronics and Communication Engineering, Vignan's Lara Institute of Technology and Science, Guntur, Andhra Pradesh, India

Abstract

In the rapidly evolving landscape of Internet of Things (IoT), efficient communication protocols play a pivotal role. One critical aspect is the baud rate, which determines the rate at which data can be transmitted over a communication channel. Traditional UART systems require pre-agreed baud rates. Once set, changing the baud rate during communication is difficult. Because of this, irrespective of the state of the receiver, the amount of data being transmitted will be same. This may lead to either data loss or inefficient use of the receiver.

In this method of design, that is, when the baud rate is prefixed, the power consumption and delay are high. When compared among the different baud rates like 9600, 19200, 38400, the power consumption is 0.452W and the delay is 2.194ns when baud rate is prefixed to 9600 bps. The area consumption is up to 20 LUT's. The results are more efficient when baud rate is 9600 bps, when compared with others.

This project introduces an innovative solution to mitigate the aforementioned limitation. It involves the design of a high-performance adaptive baud rate generator, which dynamically adjusts the baud rate based on the state of the receiver. In the presented design, the power consumed and delay are 0.181W and 0.766ns respectively, which are reduced significantly when compared to the existing design. These achieved parameters establish the adaptive baud rate generator as well-suited for applications demanding low power consumption and fast communication.

**Corresponding author*: 17harish@gmail.com

Ekta Goel, Archana Pandey, Shiromani Balmukund Rahi and Arun Samuel (eds.) Handbook of Advanced Semiconductor Field Effect Transistors, (337–364) © 2026 Scrivener Publishing LLC

Keywords: UART, baud rate generator, IoT applications, low power consumption

16.1 Introduction

16.1.1 Overview of IoT Applications and Communication Requirements

The Internet of things, known as IoT in acronym form, is a system of linked physical objects containing sensors and other technological gadgets that allow them to collect and transmit data. Connecting to the web these devices are made up of wearables, industrial machines among others which enables them to interact with each other and with the world [1].

Applications for IoT:
IoT is applied in many areas some of which include:

i. **Smart cities:** Smart traffic systems, environmental monitoring; connected municipal utilities (water supply and electricity), intelligent waste management.
ii. **Smart homes:** Appliances controllable from afar, security systems; automatic lights plus temperature regulation.
iii. **Wearables and healthcare:** Wearable fitness technology, health monitors; medical apparels that are smart.
iv. **Industrial automation:** Remote-monitoring production lines; Optimizing production lines through remote maintenance of equipment.
v. **Retail and supply chain:** Enhancing logistics.

Communication Requirements for IoT:
When setting up an IoT project, you should think about how all the devices will talk to each other and share information. It's not as simple as just connecting them, there are a bunch of different factors to consider. Let's go over some of the key ones:

i. **Network technology:** There are a few options here, each with its own strengths and weaknesses: Cellular (like NB-IoT and LTE-M) is good for covering a wide area and using low power, which is perfect for remote sensors and wearables. Wi-Fi is great for high-bandwidth applications in local networks, like smart home gadgets. Bluetooth low energy (BLE) is perfect for short-range stuff like wearables and tracking assets. Low Power Wide Area Networks (LPWAN) is designed for long-range communication with low power usage, ideal for industrial and smart city projects.

ii. **Data transfer rate:** Think about how fast you need the data to move between devices. For simple sensor readings, you don't need a lot of speed, but for streaming video, you'll want something faster.

iii. **Security:** This is super important, because you should make sure your data is private and secure while it's being transmitted. Picking the right communication protocol depends on all these factors to make sure your IoT project can share information efficiently and reliably. So, when you're setting up your IoT project, just keep these things in mind and you'll be able to choose the right network tech and security measures for your needs [2–5].

16.1.2 Importance of Baud Rate Generation in IoT Devices

Baud rate generation is super important for making sure IoT devices can communicate with each other as efficiently as possible. Here's why:

i. **Data transmission speed:** The baud rate determines how fast data signals zip between devices. A higher baud rate means faster data exchange, which is great for apps like industrial monitoring or remote control that need real-time data.

ii. **Power consumption:** The baud rate is linked to how much power a device uses. Lower baud rates use less power, which is perfect for battery-powered IoT devices where conserving energy is key.

iii. **Network optimization:** By adjusting the baud rate, developers can manage network congestion. Picking the right baud rate for the amount of data being sent helps keep communication smooth and avoid overwhelming the network with too many signals.

iv. **Bandwidth requirements:** The baud rate helps decide how much bandwidth is needed for data transmission. Lower baud rates work better with narrower bandwidths, which can be useful for apps with limited bandwidth to work with.

Basically, baud rate generation is all about finding the right balance between data speed, power efficiency, and network optimization in IoT devices. By choosing the right baud rate, devices can make sure their devices can communicate well while also considering factors like battery life and network limitations [6–9].

16.2 Fundamentals of Baud Rate Generation

Baud rate is an important concept in data transmission, especially for serial communication in IoT devices. Here's a rundown of what it means and why it matters.

16.2.1 Definition of Baud Rate

Baud rate is like the speed limit of a road. It tells you how many things happen on a communication channel (like a wire or a radio signal) in one second. Those things could be voltage changes, pulses, or anything else that helps transmit data.

Baud Rate refers to the rate at which signal elements or changes occur per second when data passes through a transmission medium. Baud rate is crucial for determining how fast data is sent or received. It quantifies the number of signal elements (such as voltage shifts or pulses) per second [9–12].

16.2.2 Significance of Baud Rate

Baud rate is key because it affects a lot of stuff about how data travels from point A to point B:

i. **Data transfer rate:** Baud rate is related to bit rate (the number of bits that travel per second), but it's not the same. In simple systems, where one thing happening means one bit gets sent, baud rate and bit rate are the same. But with more complicated ways of encoding data (like turning one thing into zeroes and ones), a single thing can sometimes stand for more than one bit. So, baud rate is like the speed limit sign, but the actual speed you go might be higher or lower depending on how you're driving (aka encoding your data).
ii. **Power consumption:** Higher baud rates need more juice to send all those signal events through the air. For battery-powered

IoT devices, figuring out the best baud rate for your data needs, helps you save battery life.

iii. **Network efficiency:** If too many things are happening on the network at once, it can get all clogged up and stuff won't work right. Setting your baud rate right for the amount of data you're sending helps keep the network humming along smoothly.

iv. **Cable length:** In some cases, higher baud rates can be more sensitive to signal problems over long cables. This is important for wired connections in IoT deployments.

16.2.3 Basics of Serial Communication Protocols

Serial communication protocols are like languages that devices use to talk to each other one bit at a time. They tell devices how to format, send, and receive data over a serial communication channel. There are three common protocols used in IoT devices: UART, SPI, and I2C. Let's look at the basics of each one.

UART:

UART (Universal Asynchronous Receiver Transmitter) is an asynchronous protocol, meaning it doesn't need a separate clock signal to send data. It's often used for general-purpose serial communication between devices, like connecting microcontrollers to computers for programming and debugging.

Here's what you need to know about UART:

i. It's not too hard to understand and use.
ii. It uses start and stop bits to frame each data byte.
iii. You can set UART to work with different baud rates and data formats (number of bits per character).
iv. UART is often used for general-purpose communication.

SPI:

SPI (Serial Peripheral Interface) is a synchronous protocol, which means all the devices involved share a clock signal to send data. It's used for connecting microcontrollers to peripherals like displays, memory chips, and sensors.

Here's what you need to know about SPI:

i. SPI is faster than UART because it's synchronous.

ii. SPI uses separate lines for data input (MOSI), data output (MISO), clock (SCK), and slave select (SS) for device selection.
iii. SPI is well-suited for short-distance communication within a single device or system.

I2C:
I2C (Inter-Integrated Circuit) is another synchronous protocol, like SPI. It's used for communication between low-power integrated circuits on a single board.

Here's what you need to know about I2C:

i. I2C is great for low-power applications.
ii. I2C only needs two wires for serial communication (SCL for clock and SDA for data).
iii. You can connect multiple slave devices to a single master device on the same bus with I2C.

When deciding which protocol to use, you should consider things like:

i. How fast you need the communication to be (SPI is faster than UART).
ii. How many devices you need to connect (I2C is good for multiple slave devices).
iii. Power consumption (I2C is best for low-power situations).
iv. How complex the communication needs to be (UART is easier to use than SPI and I2C).

Understanding these basic principles will help you choose the right serial communication protocol for your IoT project.

16.2.4 Role of Baud Rate Generators in Communication Systems

Baud rate generators are super important in communication systems, especially in serial communication, which is used a lot in IoT devices. They're like the conductor of an orchestra, making sure everything's working together and data is transmitted smoothly. Here's how they work:

i. **Clock signal generation:** The baud rate generator makes a precise clock signal that keeps track of time. This clock decides when the "symbol changes" (like when the voltage shifts or a pulse happens) on the communication channel.

Detailed description of various protocal has shown in Table 16.1.

Table 16.1 Comparison among various protocols.

Protocol/ standard	Max. nodes	Max. bit rate [kbps]	Max. length [m]	Advantages	Disadvantages	Notes
UART	2			1. Full-duplex 2. Uses only two wires 3. Can provide both synchronous and asynchronous communication	1. Can connect only two devices 2. Controller should adjust settings to the controlled device 3. Usually slower compared to I2C and SPI	1. Configurable baud rate 2. Can use only one wire if one-sided communication is required 3. Uses internal IC clock system 4. Basic IC to IC or device to device communication
I2C	127 or 1023	5000	Undefined	1. Uses only two wires 2. Multi-master and multi slave 3. More slaves do not require more pins 4. More reliable than UART	1. Slower compared to SPI 2. More complex hardware than SPI 3. Slaves should have addresses defined	1. Requires pull-up resistors 2. Basic IC to IC communication
SPI	Depends on SS pins	Up to 10000	Undefined	1. Full-duplex 2. Very simple drivers 3. More reliable than UART 4. Very fast serial communication 5. Multi slave communication	1. Uses three or more wires (traces) 2. More slaves increase pin count 3. Only one master 4. Slaves cannot communicate freely with each other	1. Basic IC to IC communication 2. Low-resolution display/ image data
1-Wire	2^48	16.3	300	1. Requires only one wire 2. Parasitic power configuration 3. Very low power consumption 4. Very simple few pin devices	1. Only Maxim devices have this serial communication	1. Requires Pull-up

(Continued)

Table 16.1 Comparison among various protocols. (*Continued*)

Protocol/ standard	Max. nodes	Max. bit rate [kbps]	Max. length [m]	Advantages	Disadvantages	Notes
CAN	128	1000	500	1. Very robust 2. Multi slave 3. Error detection	1. Quite expensive	1. 120 Ω impedance transmission line 2. Uses differential pair 3. Used primarily on automotive electronics
LIN	16	19.2	40	1. Very cheap 2. Requires only one wire 3. Can have up to 15 slaves	1. Slow data rate	1. Used primarily on automotive electronics
RS-485	256	Up to 10000	1330	1. High achievable data rate 2. High achievable communication distance 3. High receiver sensitivity	1. Higher power consumption 2. Pretty complex hardware	1. Uses one or two differential pairs 2. Device to device communication
RS-232	2	128	15	1. Cheap	1. Slow data rate 2. Modern devices rarely have this connection or do not use this standard 3. Low receiver sensitivity	1. It can be seen often in many previous-generation devices 2. There are a lot of converters 3. Device to device communication

ii. **Data rate control:** By changing the frequency of the clock signal, the baud rate generator controls how fast data is sent, which is called the baud rate (symbol changes per second). A higher baud rate means faster data transmission, but it uses more power.

iii. **Synchronization:** In protocols like SPI and I2C, the baud rate generator makes sure data is transferred at the same time between devices. It ensures both the sender and receiver are on the same page about when data bits are sent and received.

iv. **Error prevention:** A stable and accurate baud rate helps prevent errors in data transmission. If the clock signal is inconsistent, data bits can get mixed up or corrupted.

Here's an analogy: Imagine a game of telephone where a group of people pass along a message. The baud rate generator is like the person in charge of keeping everyone in sync and passing the message along at a steady pace. This keeps the information clear and reduces misunderstandings.

16.3 Requirements and Challenges in IoT Baud Rate Generation

16.3.1 Specific Demands of IoT Applications on Baud Rate Generation

IoT applications are all about connectivity, and that means a lot of devices talking to each other at different speeds. The baud rate, or how many bits of data are transmitted per second, is a key factor in making sure everything works smoothly. But with so many devices and scenarios, there are some special challenges we need to consider when it comes to setting the baud rate.

One of the biggest challenges is power efficiency. Since many IoT devices run on batteries, we need to make sure the baud rate isn't using too much power. This means finding ways to adjust the baud rate based on the amount of data being sent, or using a lower baud rate when it's not needed as much.

Another important thing to think about is how fast the data needs to get from Point A to Point B. If you've got a real-time application, like a self-driving car, you'll want a higher baud rate to minimize lag time. But if you've got a sensor that just needs to send a little bit of data occasionally, like a temperature sensor in a greenhouse, you can get away with a lower baud rate to save power. When you've got a bunch of devices all trying to talk to each other at the same time, like in a wireless sensor network, things can get crowded pretty quick. That's why it's important to be able to manage the baud rate to avoid data collisions and keep everything running smoothly.

Different IoT devices have different capabilities when it comes to processing power and communication interfaces. The baud rate generator needs to be flexible enough to work with all of them. A low-power sensor with a basic interface might not be able to handle a super-high baud rate. Security is another thing to think about. In some applications, being able to quickly change the baud rate can help make it harder for unauthorized devices to eavesdrop on your communication [13–16].

So, to wrap it up, getting the baud rate just right in IoT can be a real balancing act. You've got to consider:

i. How fast the data needs to get from point A to point B.
ii. How much power the devices can handle.
iii. How well the network can handle all the traffic.
iv. And how well the devices can actually communicate with each other.

If you can find that sweet spot, you'll be well on your way to having a reliable, efficient IoT system that gets the job done without wasting energy or resources.

16.3.2 Challenges in Designing High-Performance Adaptive Baud Rate Generators

Designing high-performance adaptive baud rate generators for IoT apps can be tricky. Here's how:

i. Speed and power:
We need to balance fast switching and low power use. A wide range of baud rates means the generator needs to handle them all without using too much juice. That might call for some creative circuit design and power management strategies.

ii. Sync and accuracy:
We also need to make sure all the devices are on the same page, like literally. Rapidly changing baud rates can mess with sync, leading to errors in data transmission. To avoid this, we need to find a balance between fast switching and keeping the signal clean.

iii. Protocol flexibility:
Different serial comm protocols (like UART, SPI, and I2C) have different needs for baud rate generators. Our generator needs to be versatile and work with all of them.

iv. Real-world challenges:
Noisy channels from electromagnetic interference can make it hard for our generator to keep the baud rate steady. We need to make sure our generator is strong enough to filter out noise and keep things running smoothly.

v. Security and cost:
Cost can be a big issue for IoT devices with limited resources. Designing a high-performance generator that's also cheap can be tricky. We need to find a way to make it work without breaking the bank.

vi. It's complicated:
Making all these things work together can lead to a complex design for our generator. Finding ways to simplify things without sacrificing performance is a challenge we're still working on.

Overall, improving baud rate generators for IoT apps requires advances in integrated circuit design, low-power electronics, and communication protocols. This will help make our devices more efficient and reliable in an ever-changing world.

16.3.3 Factors Influencing Baud Rate Stability and Accuracy in IoT Environments

In IoT setups, there are various things that can affect baud rate stability, which is super important for making sure data transmission between devices is reliable. Here's a breakdown of some key influences:

i. Clock Source:
Crystal oscillators: These are like the most common and precise sources of clock signals for baud rate generation, but they can be affected by temperature changes and aging, causing the baud rate to be all over the place.

RC oscillators: These are cheaper and simpler, but they're way less accurate than crystal oscillators. Temperature changes can seriously mess with their frequency, making them not the best choice for high baud rate stability needs.

ii. Circuit Design and Implementation:
Component quality: The quality of components in the baud rate generator circuit can affect how stable it is. Using high-grade parts with tight tolerances helps keep clock jitter (like little fluctuations) in check and makes sure symbol changes happen at the right time.

Power supply noise: Fluctuations in the power supply can cause noise on the clock signal, which can make the baud rate all wonky. Using proper power management techniques and filtering circuits helps keep the power supply clean.

iii. Electromagnetic Interference (EMI):
Noise from other devices: In crowded IoT setups where lots of devices are close together, EMI can mess with the clock signal. Using shielding techniques and making sure all the circuits are properly grounded can help reduce EMI and keep the baud rate steady.

Environmental factors: External things like lightning strikes or strong radio signals can also cause EMI and make the baud rate unstable. In those situations, extra shielding or using error correction techniques might be needed.

iv. Software Implementation:
Baud rate configuration: The software that controls the baud rate generator needs to be accurate and reliable to make sure it sets the baud rate right. Using calibration routines and error checking can help make sure there aren't any config mistakes.

Synchronization algorithms: The algorithms used for syncing up devices during baud rate changes can impact stability. Using robust sync techniques helps make sure all the devices are in sync at the new baud rate, with minimal overhead.

v. Aging and Component Degradation:
Over time, electronic components can wear out, which can cause slight changes in their characteristics. This can make the clock frequency drift, messing with the baud rate stability. Using components with high stability ratings or recalibrating regularly can help minimize this issue.

By understanding these factors and using appropriate design techniques, developers can make baud rate generators in IoT devices way more stable and accurate. This helps make sure data transmission is reliable, which means a more efficient and reliable IoT ecosystem overall [17].

16.4 State-of-the-Art Techniques in Baud Rate Generation

16.4.1 Review of Existing Baud Rate Generation Architecture

In the context of data communication, baud rate generation refers to the process of generating the timing signals necessary for transmitting and receiving data at a specific rate. The architecture for baud rate generation can vary depending on the specific hardware or software implementation.

During the review process, it is important to consider factors such as accuracy, stability, power consumption, and compatibility with different communication protocols. Additionally, the architecture should be able to handle various baud rates efficiently and support error detection and correction mechanisms. To provide more specific information, one important aspect to consider is the trade-off between hardware and software implementations. In hardware-based architectures, dedicated circuits or microcontrollers are often used to generate the baud rate. This approach can provide precise timing and high-speed communication, but it may require additional hardware components and can be less flexible for different baud rate requirements.

On the other hand, software-based architectures utilize software algorithms to generate the baud rate. This approach offers more flexibility and can be easily adjusted to different baud rates without requiring hardware changes. However, it may introduce some timing inaccuracies or overhead due. One important aspect to consider is the trade-off between hardware and software implementations.

In hardware based architectures, dedicated circuits or microcontrollers are often used to generate the baud rate. This approach can provide precise timing and high-speed communication, but it may require additional hardware components and can be less flexible for different baud rate requirements.

When reviewing an existing baud rate generation architecture, it's important to evaluate factors such as performance, resource utilization, and compatibility with the target system. Additionally, considering the specific requirements of your project, such as the desired baud rate range and any constraints on power consumption or timing accuracy, will help in assessing the suitability of the existing architecture.

16.4.2 Adaptive Techniques for Baud Rate Adjustment

Adaptive techniques refer to methods that dynamically adjust the baud rate based on certain conditions or requirements. These techniques are commonly used in communication systems to optimize data transmission and reception.

i. One popular adaptive technique is called Automatic Baud Rate Detection. This technique allows the receiver to automatically detect the baud rate of the incoming data stream. It works by analyzing the timing and transitions of the received signals to determine the baud rate. This is particularly useful when dealing with data streams from different sources with varying baud rates.
ii. Another adaptive technique is called Adaptive Baud Rate Control. This technique adjusts the baud rate during the data transmission process to optimize the communication performance. It takes into account factors such as noise levels, signal quality, and channel conditions to dynamically adapt the baud rate.
iii. This helps to maintain reliable and efficient data transfer, especially in situations where the communication environment is prone to changes or disturbances.
iv. Adaptive techniques for baud rate adjustment are crucial in ensuring reliable and efficient data communication. By adapting the baud rate based on real-time conditions, these techniques help to optimize data transmission, improve error detection and correction, and enhance overall system performance.

16.4.3 Techniques for Power Optimization in Baud Rate Generators

Power optimization techniques aim to reduce the power consumption of the baud rate generator while maintaining its functionality and performance.

One common technique is Clock Gating, which involves selectively disabling the clock signal to unused or idle parts of the baud rate generator. By gating the clock, unnecessary power consumption can be minimized, especially in situations where the baud rate generator is not actively generating or receiving data.

Another technique is Voltage Scaling, which involves adjusting the supply voltage of the baud rate generator based on the required performance level.

By reducing the supply voltage during periods of lower activity or lower baud rates, power consumption can be significantly reduced without sacrificing performance.

Additionally, using Low-Power Design methodologies such as optimizing circuit designs, reducing leakage currents, and employing power-saving circuit techniques like sleep modes or power gating can further contribute to power optimization in baud rate generators.

These techniques, along with proper system-level power management strategies, can help reduce power consumption in baud rate generators without compromising their functionality and performance. Important points about power optimization in baud rate generators:

i. **Clock Gating:** This technique involves selectively disabling the clock signal to unused or idle parts of the baud rate generator to minimize unnecessary power consumption.
ii. **Voltage Scaling:** By adjusting the supply voltage of the baud rate generator based on the required performance level, power consumption can be reduced during periods of lower activity or lower baud rates.
iii. **Low-Power Design:** Optimizing circuit designs, reducing leakage currents, and implementing power-saving circuit techniques like sleep modes or power gating can contribute to power optimization.
iv. **System-Level Power Management:** Considering overall system-level power management strategies can help in effectively reducing power.

16.5 High-Performance Adaptive Baud Rate Generators

16.5.1 Introduction to Adaptive Baud Rate Generator (Proposed System)

An adaptive baud rate generator is an important part of serial communication systems, especially when the conditions of transmission are all over the place. Its main job is to change the baud rate, which is the speed at which data moves through a communication channel. Unlike fixed baud rate generators, adaptive ones can adjust the speed based on the parameters like signal strength, interference, or how far apart the devices are that are talking to each other. This flexibility is important for making sure data can get from point A to point B smoothly and efficiently.

Especially in wireless comm systems or places where the signal conditions can change a lot, adaptive baud rate generators are like superheroes. They can adjust to whatever is going on and keep the whole system running smoothly. This helps boost performance, make things more reliable, and just makes serial communication in general better in all kinds of areas like telecommunication, industrial automation, consumer electronics, and beyond.

16.5.2 Design Considerations and Constraints

A baud rate generator is a crucial component in communication systems that determines the rate at which data is transmitted. Here are some important considerations and constraints to keep in mind.

i. **Accuracy:** The baud rate generator should provide accurate timing to ensure reliable data transmission. It needs to generate the correct number of pulses per second to match the desired baud rate.
ii. **Stability:** The generator should maintain a stable and consistent output frequency over time. Any fluctuations or variations in the generated baud rate can lead to communication errors.
iii. **Flexibility:** It's important to consider the flexibility of the baud rate generator. Different communication protocols may require different baud rates, so the generator should be able to support a wide range of baud rate options.
iv. **Power Consumption:** The design of the baud rate generator should take into account power consumption. It should be efficient and minimize power usage to optimize overall system performance.
v. **Noise Immunity:** The generator should be designed to minimize the impact of noise and interference on the generated baud rate. This can be achieved through proper shielding and filtering techniques.
vi. **Compatibility:** Consider the compatibility of the baud rate generator with other components and systems in the communication setup. It should be able to interface seamlessly with other devices and ensure smooth data transmission.

16.5.3 Block Diagram

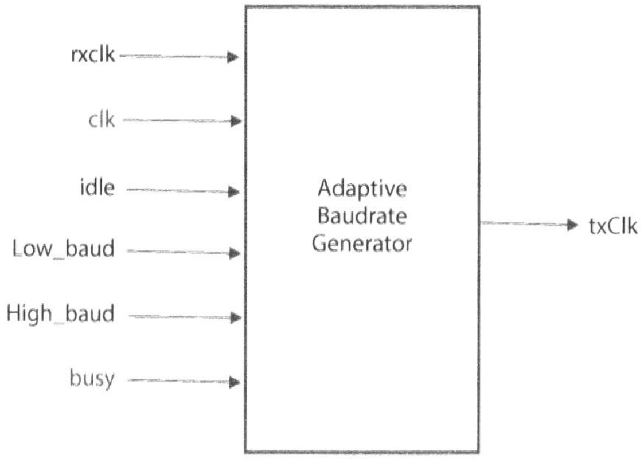

Figure 16.1 Block diagram of adaptive baudrate generator.

The block diagram of adaptive baud rate generator is depicted above in Figure 16.1. It consists of the following signals which are:

i. Rxclk (Receiver clock)
ii. Clk (Clock)
iii. Idle
iv. Low_baud
v. High_baud
vi. Busy
vii. TxClk (Transmitter clock)

Only the txClk signal is output signal. Remaining all are the input signals.

16.5.3.1 Working Principle of Adaptive Baud Rate Generator

In the proposed adaptive baud rate generator, we consider the different states of the receiver, and change the baud rate according to the condition of the receiver. In the existing baud rate generator irrespective of the state of the receiver, the data will be sent with respect to the baud rate fixed originally. If that is the case, then that leads to either inefficient use of the receiver or loss of data and noise occurrences.

So, to avoid those circumstances, we consider the state of the receiver to adjust the baud rate. We consider four states of the receiver which are:

i. Default state
ii. Free state
iii. Partially busy state
iv. Busy state

When the receiver is in default state, the baud rate assigned will be 9600 bps. But when the receiver is in free state and is capable of receiving a greater number of bits, then the baud rate will be adjusted to 19200 bps. In another case, if the receiver is partially busy and can only accept a few bits of data, then the baud rate will be adjusted to 4800 bps. And finally, if the receiver is busy and is incapable of receiving any more data, then the baud rate is adjusted to 0 bps and in this case, there is no communication taking place as there is no data being transmitted.

This adaptation allows for improved reliability and flexibility, especially in scenarios where the communication channel's characteristics may vary or change over time.

16.6 Results and Discussion

16.6.1 Result for Baud Rate Generator (Existing System)

i. The below Figure 16.2 shows the baud rate value is 9600. For 9600 baud rate the static power is 0.092W, dynamic power is 0.0360W and total power consumed is 0.128W.

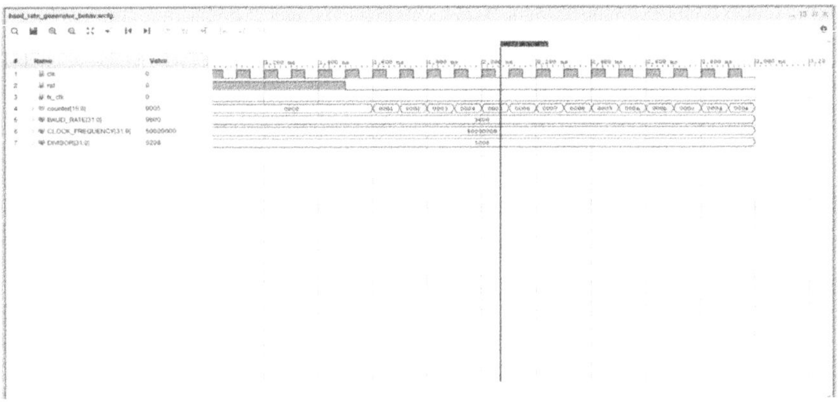

Figure 16.2 Baud rate = 9600.

The Area Consumed: 20 (LUT's).

Delay: 2.106 ns

ii. The below Figure 16.4 shows the baud rate value is 12600. For 19200 baud rate the static power is 0.092W and dynamic power is 0.361 and total power consumed is 0.453W.

The Area Consumed: 20 (LUT'S).

Delay: 2.270 ns

Schematic diagram for boud rate for case 19200 is shown in Figures 16.3, 16.5, 16.7 to 16.9.

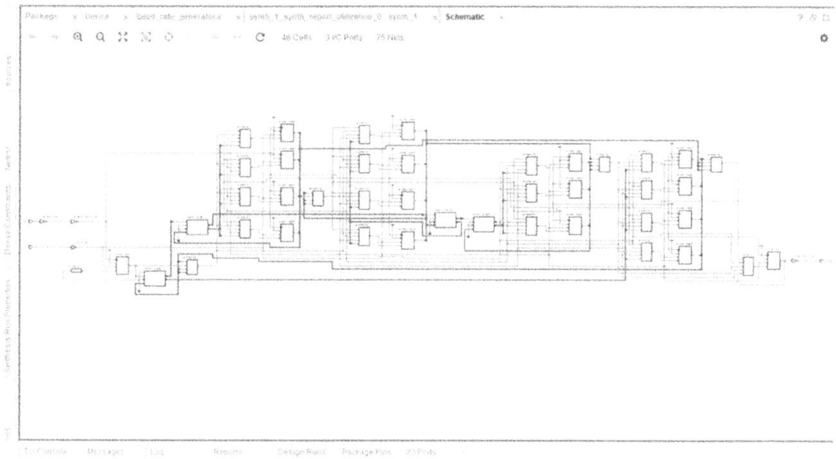

Figure 16.3 Schematic diagram when baudrate = 9600.

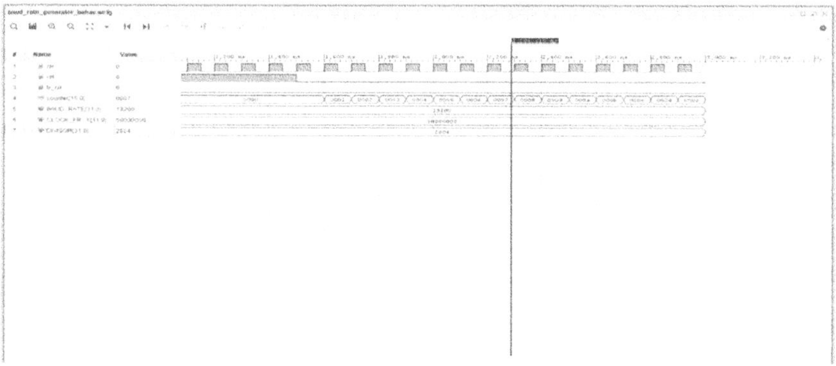

Figure 16.4 Baud rate = 19200.

iii. The below Figure 16.6 shows the baud rate value is 38400. For 38400 baud rate the static power is 0.092W and dynamic power is 0.348W and total power consumed is 0.44W.

The Area Consumed: 20 (LUT's).

Delay: 3.306 ns

From the above comparison, we can say that the baud rate generator is more efficient when baud rate is fixed to 9600 bps.

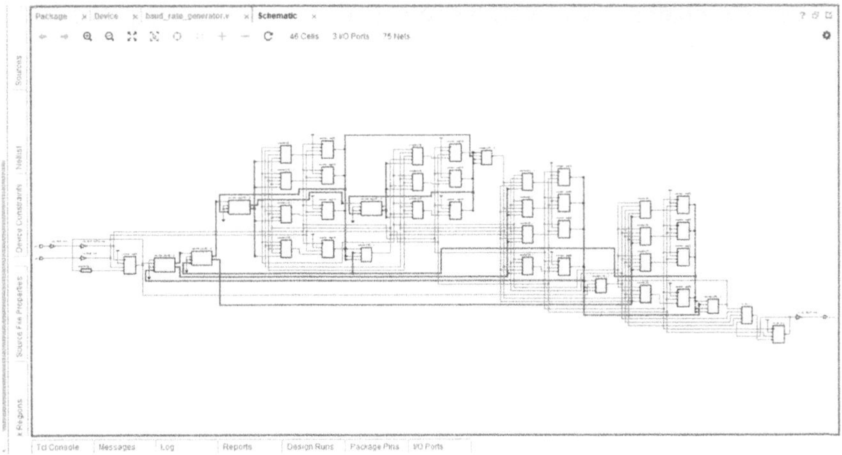

Figure 16.5 Schematic diagram when baudrate = 19200.

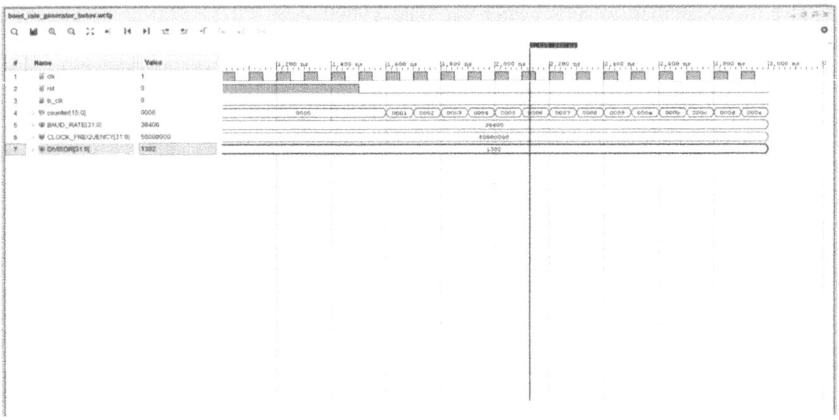

Figure 16.6 Baud rate = 38400.

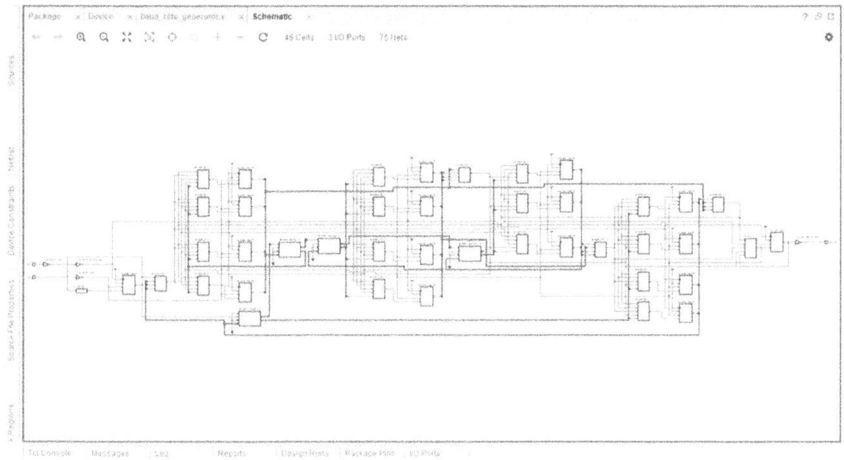

Figure 16.7 Schematic diagram when baud rate = 38400.

16.6.2 Result for Adaptive Baud Rate Generator (Proposed System)

In this proposed system, we took a case where we consider 4 states of the receiver which are default state, free state, partially busy state and fully busy state. We represent them as idle, High_baud, Low_baud, busy parameters.

i. When the idle state is active, it means that the receiver is in default state, and the baud rate is set to 9600 bps.
ii. When the High_baud is active, it means that the receiver is in free state, and the baud rate is set to 19200 bps.
iii. When the Low_baud is active, it means that the receiver is in partially busy state, and the baud rate is set to 4800 bps.
iv. When the busy is active, it means that the receiver is in busy state, and the baud rate is set to 0 bps.

For the adaptive baud rate generator, the delay is 0.766 ns and the power consumed is 0.181W. When compared between both the existing system and the proposed system, we can say that the performance of the adaptive baud rate generator is better.

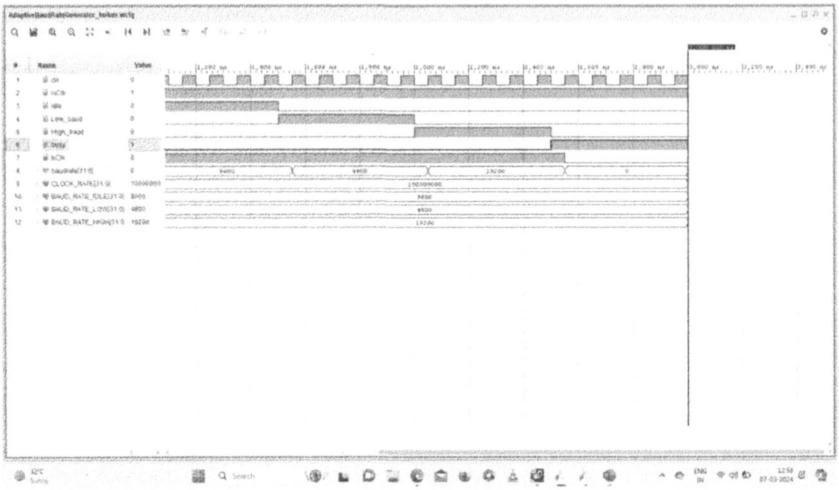

Figure 16.8 Adaptive baud rate generator.

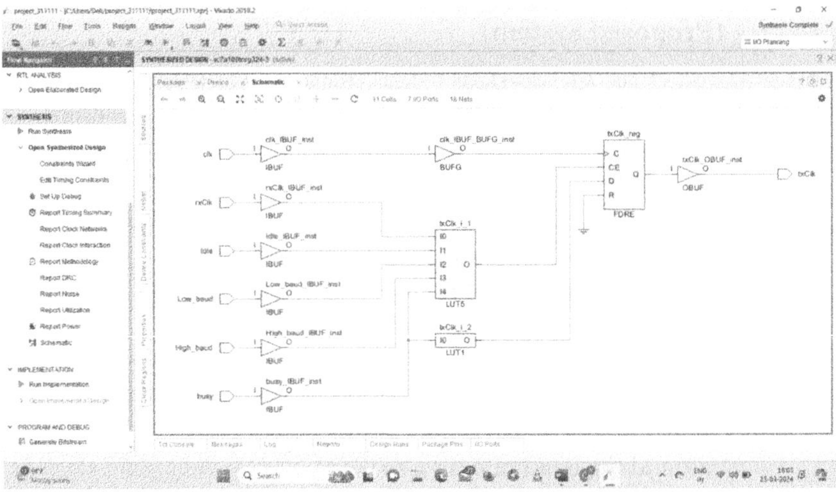

Figure 16.9 Schematic diagram of adaptive baud rate generator.

16.6.3 Comparison Table

Table 16.2 Comparison of area, delay and power dissipation of baud rate generator and adaptive baud rate generator.

Architecture	Baud rate generator (existing)	Adaptive baud rate generator (proposed)
Area	30 (LUT's)	5 (LUT's)
Propagation delay	2.194 (ns)	0.766 (ns)
Power Dissipation	4.52 x 10^5 (μW)	1.81 x 10^5 (μW)

As we can see in Table 16.2, all the parameters are having better values in the proposed adaptive baud rate generator when compared to the existing baud rate generator. We can say that the proposed adaptive baud rate generator is highly efficient because of its low power dissipation, low area occupation and less propagation delay when compared to the existing generator.

16.7 Future Directions and Challenges

Emerging trends in IoT communication are totally shaping the scene of interconnected devices and systems. Like, one major trend is the integration of 5G technology, which brings super-fast connectivity and low latency, enabling advanced IoT applications in all sorts of industries. Another trend is edge computing, which allows data processing to happen closer to where it's collected, reducing latency and improving system efficiency. Blockchain technology is also getting really popular for enhancing security and trust in IoT ecosystems by using decentralized and tamper-resistant data storage. And let's not forget about AI and machine learning, which enable IoT devices to analyze data on the spot and make decisions without having to rely on humans all the time.

Besides those, there are also other trends like LPWAN technologies, like Lora WAN and NB-IoT, which let devices communicate over long distances with super-low power consumption. Standardization efforts are also important for making sure that different IoT devices and platforms can work together seamlessly, fostering innovation and collaboration.

Edge AI is a thing now too, where AI algorithms are run directly on IoT edge devices to enable real-time data processing and decision-making. And of course, cybersecurity solutions are super important for protecting IoT networks and data from all the hackers and bad guys out there.

Digital twins, which are like virtual representations of physical objects or systems, are also becoming a big deal for monitoring and optimizing things in real-time. And environmental monitoring IoT solutions are super important for tracking and managing environmental conditions, which is super important for sustainability efforts.

These trends, like said, are all part of the ongoing evolution and maturation of IoT communication, driving efficiency, innovation, and value across various industries and applications. It's pretty cool to see how it's all coming together.

16.8 Conclusion

Real-world scenarios for IoT baud rate era are all over the put, like, in so numerous distinctive places and circumstances. It's not fair one thing, it's everything. Like, in industry, they utilize it for their machines and stuff, to make beyond any doubt they can conversation to each other and not have any issues. And in ranches, they utilize it to keep track of their crops and make beyond any doubt they're watered right. And indeed, in our homes, like with our keen apparatuses and stuff.

But the cool thing around baud rate era in IoT is that it can adjust to diverse circumstances; like, in a manufacturing plant, it can alter the speed of the information transmission based on how active the arrange is or if there's obstructions or something. That way, the machines can still have conversation to each other and get the work done. And in horticulture, it can alter based on the separate between sensors and the central center or cloud server, so the information gets there quick and doesn't get lost.

Here are a few cases of how baud rate era is utilized in the genuine world:

1. In agribusiness, sensors on the cultivate screen soil dampness, temperature, and mugginess, and they send that information back to a central center or cloud server. The baud rate era makes beyond any doubt that the information gets there quick and effective, so ranchers can make savvy choices approximately watering their crops and stuff.
2. In industry, sensors and actuators keep an eye on machines and generation forms. They send information back to control

frameworks, and the baud rate era makes beyond any doubt that data gets there in real-time, so the individuals in charge can make changes and optimize things. This makes a difference as it decreases downtime and makes the entire operation more efficient.

3. IoT devices such as smart interior controllers, security cameras, and lighting at home communicate with one another, a central hub, or even our phones.
4. In healthcare, IoT gadgets like wellness trackers and farther wellbeing screens collect information from patients and send it to healthcare suppliers or cloud-based stages. The baud rate era makes beyond any doubt that information gets there quick, so specialists can keep an eye on patients and give superior care.
5. And indeed, in cities, IoT sensors and gadgets are utilized to track activity, discuss quality, squander administration, and open security. The baud rate era changes how these devices communicate with one another and with central control centers, allowing cities to make far better decisions and operate more effectively.

Designing a high-performance adaptive baud rate generator (ABRG) for the Internet of Things (IoT) is important for smooth communication between devices. It lets the ABRG adjust its baud rate based on changing conditions, ensuring optimal data transmission and reception. This innovation is going to make IoT devices way more efficient and connected. The ABRG's adaptability is key here. By automatically adjusting the baud rate based on the environment, it can handle varying signal strengths and interference without any issues. That's pretty cool. Plus, it's also energy-efficient, which means IoT devices can run longer on their batteries. Another awesome thing about the ABRG is that it makes IoT applications more reliable. It can adapt to different network conditions, so data transfer stays consistent and there's less risk of losing important information. Overall, the ABRG is a game-changer for IoT, making communication faster, more efficient, and way more reliable. In conclusion, the high-performance adaptive baud rate generator is a total must-have for IoT applications. It's going to make our connected world even better by improving device efficiency, boosting reliability, and enabling seamless communication. The future looks bright with this technology.

In summary, baud rate era is a crucial component of IoT since it influences how all these unique devices and systems communicate with one another and cooperate in any situation.

References

1. Zhang, Y., Yu, Z., Yang, X., Design and Implementation of Configurable UART with Self-Test Function, in: *Presented at the 5th International Conference on Electronic Engineering and Informatics (EEI)*, 2023.
2. Li, J., Yang, L., Feng, X., Xing, Y., Chen, Z., Wan, P., UART Controller with FIFO Buffer Function Based on APB Bus, in: *IEEE 16th International Conference on Anti-counterfeiting, Security, and Identification (ASID)*, 2022.
3. Gupta, A.K., Raman, A., Kumar, N., Ranjan, R., Design and Implementation of High-Speed Universal Asynchronous Receiver and Transmitter (UART), in: *7th International Conference on Signal Processing and Integrated Networks (SPIN)*, 2020.
4. Harutyunyan, S., Kaplanyan, T., Kirakosyan, A., Momjya, A., Design And Verification of Auto configurable UART Controller, in: *IEEE 40th International Conference on Electronics and Nanotechnology (ELNANO)*, 2020.
5. Sonwane, R., Deshmukh, A., Choudhary, S., UART (Universal Asynchronous Receiver Transmitter) for Serial Data Communication: Design and Implementation on FPGA Platform, in: *2023 2nd International Conference on Futuristic Technologies (INCOFT)*.
6. Wang, W., Optimization of UART Communication Protocol Based on Frequency Multiplier Sampling Technology and Asynchronous FIFO, in: *2023 IEEE 2nd International Conference on Electrical Engineering, Big Data and Algorithms (EEBDA)*, Changchun.
7. Mishra, S., Ray, A., Singh, M., Venkatesan, S., Anand, A.S., Automated Hardware Auditing Testbed for UART and SPI based IoT Devices, in: *2023 10th International Conference on Internet of Things: Systems, Management and Security (IOTSMS)*.
8. Sivaranjani, P., Sasikala, S., Lavanya, A., Keerthana, M., Design and Analysis of UART Protocol with Sec-Ded and Implementation on FPGA, in: *2023 14th International Conference on Computing Communication and Networking Technologies (ICCCNT)*.
9. Kamath, A., Mendez, T., Ramya, S., Nayak, S.G., Design and Implementation of Power-Efficient FSM based UART, in: *1st International Conference on Artificial Intelligence, Computational Electronics and Communication System (AICECS 2021)*, Manipal, India, 28–30 October 2021.
10. Kumar, K., Kaur, A., Ramkumar, K.R., Effective data transmission with UART on Kintex-7 FPGA, in: *2020 12th International Conference on Computational*

Intelligence and Communication Networks (CICN), Bhimtal, India, pp. 492–497, 2020.
11. Sowmya, K.B., Gomes, S., Tadiparthi, V.S.R., Design of UART Module using ASMD Technique, in: *2020 5th International Conference on Communication and Electronics Systems (ICCES)*.
12. Priyanka, B., Gokul, M., Nigitha, A., Poomica, J.T., Design of UART Using Verilog and Verifying Using UVM, in: *2021 7th International Conference on Advanced Computing and Communication Systems (ICACCS)*.
13. Govil, A., Karnwal, A., Sindhu, G., Singh, A., Shukla, S., Design and Implementation of UART Using FPGA Board, in: *2022 International Journal for Research in Applied Science & Engineering Technology (IJRASET)*.
14. Pham-Thai, N., Ho-Ngoc, B., Do-Duy, T., Truong, P.Q., Phan, V.-C., A novel multichannel UART design with FPGA-based implementation. *Int. J. Comput. Appl. Technol.*, 67, 4, 358–369, 2021.
15. Harutyunyan, S., Kaplanyan, T., Kirakosyan, A., Khachatryan, H., Configurable Verification IP for UART, in: *2020 IEEE 40th International Conference on Electronics and Nanotechnology (ELNANO)*, pp. 234–237, 2020 Apr 22.
16. Kashyap, B. and Ravi, V., Universal Verification Methodology Based Verification of UART Protocol. *J. Phys. Conf. Ser.*, 1716, 1, 012040, 2020 Dec 1.
17. Yang, L., Linan, L., Liangguang, Z., Xiaowei, H., Design and Realization of Universal UART Module Based on Verilog HDL [J]. *Electron. Des. Eng.*, 28, 08, 174–179, 2020.

17

Biomedical Applications in VLSI Field

Jyoti Kandpal[1]*, Divya Sharma[2] and Ekta Goel[3]

[1]*Department of Electronics and Communication Engineerining, Graphic Era Hill University, Dehradun, India*
[2]*Department of Electronics and Communication Engineering, Birla Institute of Technology Mesra, Ranchi JH, India*
[3]*National Institute of Technology, Warangal, India*

Abstract
In today's world, healthcare costs are rising due to an aging population, while governments struggle to allocate sufficient funds. As a result, patients face significant challenges in accessing medical services, leading to distress and hardship. This chapter explores key advancements in medical and biomedical sensing using VLSI technology. With continuous progress in VLSI, its integration into biomedical applications is set to further transform healthcare delivery and enhance patient outcomes.

Keywords: VLSI, bio sensors, CMOS, nano materials

17.1 Introduction

The health care industry is in a "state of emergency" as a result of recent tragic events that have disrupted the ongoing delivery of service. The growing lack of confidence between patients and health bionetworks may impede the desired advancement of the Indian healthcare industry and the country's overall health. The ageing population and growing costs have made life miserable for the general populace. The government must take the lead in reviewing the health care system's essential components, including assurance handling, adherence to treatment and care norms and recommendations, cost management, reimbursement models, integrity, and the use of new technologies. In this case, Wi-Fi transmission may be used in conjunction

**Corresponding author*: jayakandpal27@gmail.com

Ekta Goel, Archana Pandey, Shiromani Balmukund Rahi and Arun Samuel (eds.) Handbook of Advanced Semiconductor Field Effect Transistors, (365–382) © 2026 Scrivener Publishing LLC

with innovative methods of providing healthcare that reduce costs, alleviate patient suffering, and need fewer personnel [1, 2].

Very large-scale integration (VLSI) technology may be used to build this wireless structure, making it suitable for biogenic applications. In the field of biomedicine, VLSI design results in smaller chips with improved speed and range. The CMOS fabrication technique, the architecture of the medical implant communication system (MICS) receiver for critical medical operations, the field programmable gate array (FPGA) execution of semantic networks, the neuro-fuzzy system, the neuromorphic computing approach, and the neural net performance in analogue hardware and digital network are all examined in this analysis for the VLSI employment of neural networks. Additionally, the benefits and drawbacks of various categories and techniques are discussed here.

With ageing populations and growing prices, health care is getting more and more expensive in the modern period, and the government is unable to pay for it. Patients are experiencing trauma as a result of this issue, and they are finding it difficult to access medical care. Here, wireless communication lessens medical expenses and suffering for the patient. Very large-scale integration, or VLSI, is a concept that may be applied to the implementation of this wireless system in biomedical applications. In neurology, the use of VLSI techniques aids in circuit reduction, area, and speed enhancement.

We are living in a moment of progress, but we are also going through growing pains right now. Furthermore, the health care industry is limited by three elements: awareness, utilization, and allocation. There is currently a large disparity between these variables, which results in a broken system with insufficient access to medical treatment. Even though the medical industry faces challenges with accessibility, privacy, and the ageing trend, technological advancements to address health issues have advanced quickly thus far. This is where wireless technology enters the picture, lowering costs, manpower, and pain points for patients.

Additionally, modern wireless gadgets that are implanted in patients' bodies use precise technology to monitor patients and their health; data access is simple, costs are reduced, and bed space is preserved. The goal of a multi-objective optimization issue is to find the optimal solution while simultaneously optimizing for each of the objective functions.

17.2 Role of VLSI in Biomedical Application

Very Large-Scale Integration (VLSI) plays a crucial role in the advancement of biomedical applications by enabling the development of highly sophisticated and efficient medical devices and systems. The implementation of new technology ends this strain. In medical services, the use of wireless technology helps save costs, minimize patient stress, and ease patients' issues during doctor visits. To minimize power consumption in the medical industry, wireless technology has mostly expanded with new applications, such as the use of RF transceivers, IF mixers, and other circuitry parts, to lessen patient suffering throughout daily medical services [3, 4].

VLSI design in the biomedical field has several advantages. VLSI design can produce a reduction in the size of the chips, enhance the range, and improve the speed of biomedical devices. Reduced circuit size, improved area, and faster devices are all benefits of using VLSI techniques in neurology. Because integrated neural networks operate on low-cost, low-power, and low-area VLSI designs, this is extremely important for biomedical systems. VLSI systems applied to medical imaging systems can provide significant benefits, such as improved performance, reduced size, and increased efficiency. The VLSI implementation in biomedical applications can lead to advancements in various areas, including neural network-based systems, medical imaging, and healthcare monitoring devices. A typical CMOS biological system's sensor, actuators, recognition components, analog/digital interface circuits, microfluidic structures, and biological samples are shown in Figure 17.1 [5].

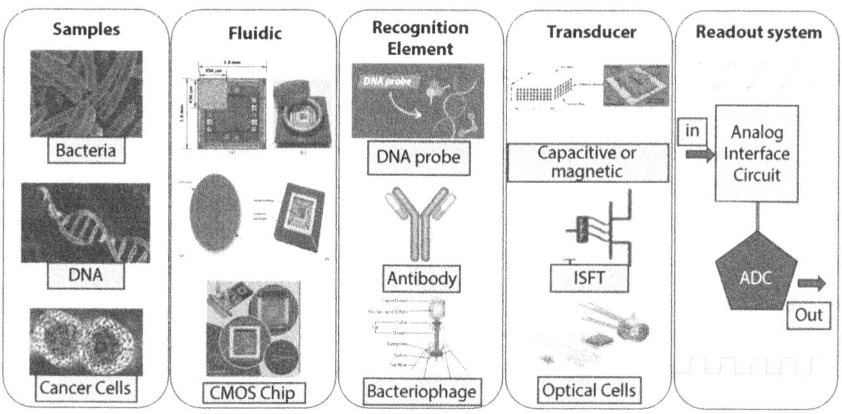

Figure 17.1 CMOS biosensor illustration [5].

Here are several key contributions of VLSI technology in the biomedical field [6–13]:

1. Medical imaging:
   - CT scanners and MRI machines: VLSI chips are used in the signal processing units of CT scanners and MRI machines, enhancing image quality and processing speed.
   - Ultrasound devices: VLSI technology helps improve the resolution and accuracy of ultrasound images through advanced signal processing techniques.
2. Wearable health monitors:
   - Fitness trackers and smartwatches: VLSI circuits are integral to wearable health monitoring devices, allowing for the very accurate assessment of vital indicators such blood oxygen levels, heart rate, and activity level.
   - Continuous glucose monitors: These devices benefit from VLSI technology for real-time monitoring of blood glucose levels in diabetic patients.
3. Implantable medical devices:
   - Pacemakers and defibrillators: VLSI chips in these devices ensure reliable heart rhythm monitoring and intervention.
   - Cochlear implants: Advanced VLSI circuits enable better sound processing and clearer hearing for users.
4. Telemedicine:
   - Remote monitoring systems: VLSI technology supports the development of compact and energy-efficient sensors for remote patient monitoring, transmitting health data to medical professionals in real time.
   - Portable diagnostic devices: Handheld diagnostic tools for point-of-care testing leverage VLSI circuits for quick and accurate results.
5. Neural interfaces:
   - Brain-machine interfaces (BMIs): VLSI is crucial in developing interfaces that can decode neural signals and translate them into commands for prosthetic devices, aiding individuals with disabilities.
   - Neuroprosthetics: These devices use VLSI circuits to provide sensory feedback and improve motor control in prosthetic limbs.

6. Drug delivery systems:
   - Microelectromechanical systems (MEMS): VLSI technology enables the creation of MEMS for controlled drug delivery, allowing precise dosages to be administered over time.
7. Lab-on-a-chip:
   - Microfluidic devices: VLSI technology is key in the implementation of lab-on-a-chip systems, which miniaturize laboratory processes such as DNA analysis, blood testing, and cell sorting onto a single chip.
8. Personalized medicine:
   - Genomic sequencing: VLSI circuits facilitate faster and more cost-effective genomic sequencing, aiding in the development of personalized treatment plans based on individual genetic profiles.

The integration of VLSI in biomedical applications enhances the functionality, reliability, and accessibility of medical technologies, improving the results for patient and the advancement of healthcare.

17.3 Application

This research explores the diverse applications of VLSI in biomedical fields, highlighting its impact on medical imaging, wearable health monitors, implantable devices, telemedicine, neural interfaces, drug delivery systems, lab-on-a-chip technologies, and personalized medicine.

a) Analog front-end (AFE) circuits

The development of electronic systems has long depended on analogue front-end (AFE) circuits, which serve as a bridge between analogue signals and digital signal processing operations. These circuits have a wide range of uses, including biological signal collection systems, radio receivers, and sensors.

By serving as an interface between several generic analogue and digital system chambers, the system servers provide hardware modularity *via* shared AFE TOBs. AFEs for health monitoring are available from a number of well-known semiconductor providers, such as Texas Instruments (ADS1298), AFE4400, and AFE4490. Additionally, analogue devices sell a CN0209 for use in test and measurement applications, while Atmel provides analogue front-ends for smart meters [14].

b) Circuits for bioimpedance test

According to Ohm's law, the measurement of an impedance usually uses voltammetry, which can be implemented using a bioimpedance measurement integrated circuit (IC). These VLSI-based bioimpedance measurement ICs provide the necessary analog front-end circuitry, such as current sources, voltage measurement units, and signal conditioning, to accurately capture the bioimpedance signals [15].

VLSI-based bioimpedance measurement techniques have found applications in various healthcare domains, including:

- Body composition analysis: VLSI ICs enable the development of compact, wearable devices for segmental body composition determination, which is particularly beneficial for disabled individuals.
- Tissue health monitoring: Bioimpedance measurement can be used to assess tissue health, such as detecting changes in hydration or detecting the onset of pressure ulcers.
- Impedance spectroscopy and tomography: VLSI circuits can be utilized in electrical impedance spectroscopy (EIS) and electrical impedance tomography (EIT) systems, which provide valuable information about the physiological state of tissues.
- Cardiovascular monitoring: Impedance-based techniques, such as impedance cardiography (ICG), can be used for non-invasive monitoring of cardiovascular parameters, like stroke volume and cardiac output.

c) Capacitive based circuits

Capacitive coupling is a prevalent phenomenon in VLSI circuits, where adjacent nets can induce electrical signals in each other due to the capacitive coupling between them. This capacitive coupling can lead to logic failures and degradation of timing in VLSI circuits, a phenomenon known as crosstalk [16].

One application of capacitive-based circuits in VLSI is the switched-capacitor (SC) circuit technique. The basic SC circuit operation is based on two-phase non-overlap clocks, which allow for the transfer of charge between capacitors. These SC circuits have various applications in VLSI, such as the implementation of charge pump DC/DC converters, where the model and measurement results are consistent.

Another application of capacitive-based circuits in VLSI is the implementation of synaptic dynamics in neuromorphic ICs. The VLSI implementation of a conductance-based multi-synapse using switched-capacitor circuits can effectively capture the synaptic dynamics.

While the most accurate approach to analyze capacitive coupling is to use a transistor-level circuit simulator, this method is computationally inefficient and not suitable for large-scale VLSI circuits. Therefore, alternative techniques, such as the use of capacitive-based circuits, are often employed to address the challenges posed by capacitive coupling in VLSI design.

d) Circuits for new sensing devices and microelectrodes

VLSI technology has been widely applied in the development of new sensing devices and microelectrodes for various biomedical applications. This technology enables the integration of thousands or even millions of electronic components onto a single chip, leading to significant advancements in the field of biosensing and neural interfacing [17].

VLSI-based sensing devices offer several advantages, including miniaturization, improved performance, and increased power efficiency. These compact and integrated circuits are essential for the realization of wearable, portable, and implantable sensing devices that can continuously monitor various biomolecules, neural signals, and physiological parameters.

One key opportunity lies in the development of low-power VLSI designs for IoT (Internet of Things) devices. Power efficiency is crucial for these battery-powered devices, as they need to operate for extended periods without recharging. VLSI engineers can design energy-efficient circuits to extend the battery life of IoT-based sensing devices, making them more practical and user friendly.

Furthermore, VLSI technology enables the integration of various sensors onto a single chip, reducing the size and complexity of the overall system. This integration of sensor functionalities on a single VLSI chip opens up possibilities for new applications and use cases in the field of biomedical monitoring and neural interfacing.

One such application is the development of implantable neural interfacing microsystems, which consist of a microelectrode array (MEA), an analog front-end for recording and stimulation, and a wireless communication module. These sophisticated VLSI-based implants can provide high-density and high-resolution neural recordings, enabling neuroscientists to study brain functions and potentially treat neural disorders.

Additionally, VLSI-based biosensing interfaces and microelectrodes have been utilized for various applications, such as amperometric and voltametric sensing, neural interfacing, and molecular biology analysis. The integration of these VLSI circuits with advanced materials and micro/nano-structures has led to the development of highly sensitive, selective, and accurate sensing platforms for a wide range of biomedical applications.

e) Circuits for implantable and wearable devices

Implantable medical devices rely on VLSI circuits to perform a variety of critical functions, such as:

- Sensing physiological parameters (e.g., neural activity, glucose levels)
- Electrical stimulation for therapy delivery (e.g., deep brain stimulation, cardiac pacing)
- Data processing and storage
- Wireless communication for remote monitoring and control
- Power management and efficient energy harvesting

The miniaturization and low-power capabilities of VLSI have been crucial in developing implantable devices that can be safely and unobtrusively integrated within the human body. These devices often require custom-designed application-specific integrated circuits (ASICs) to meet the stringent size, power, and performance requirements [18].

Wearable medical devices have also benefited significantly from VLSI technology. These devices can provide continuous health monitoring, early disease detection, and real-time feedback to patients and healthcare providers. Key VLSI-enabled features in wearable devices include:

- Compact and flexible circuit design for comfortable and unobtrusive wearability
- Low-power sensor interfaces for extended battery life
- Wireless communication for data transmission and remote control
- Integrated data processing and analytics for intelligent monitoring

The integration of VLSI circuits in wearable devices has enabled the development of a wide range of medical applications, such as remote patient monitoring, early disease detection, and personalized healthcare.

f) Low-power and Low-voltage VLSI circuits

The proliferation of IoT devices and the need for energy-efficient and environmentally responsible operations have led to a growing emphasis on the design and optimization of low-power VLSI circuits for biomedical applications. The low-power, low-voltage operation of electronic devices is crucial for biomedical applications such as hearing aids, pacemakers, and EEG monitors. These devices often have stringent power and resource constraints, necessitating the development of area and power-efficient VLSI architectures that can perform analysis and processing in a power-constrained environment. VLSI design techniques play a crucial role in enhancing the capabilities of each transistor, ensuring excellent performance, low power consumption, and improved operations in integrated circuits (ICs) for biomedical applications. Voltage and frequency scaling are two examples of architectural options that can significantly cut power consumption without sacrificing performance. Optimizing transistors, particularly through voltage scaling, can result in substantial energy savings and increased device lifespan, which is crucial for implantable biomedical devices. Additionally, the unique requirements of biomedical applications, such as maintaining lower power consumption and addressing heat generation, necessitate the implementation of advanced cooling systems and the integration of techniques like gating and voltage scaling to achieve energy efficiency [19].

The design of low-power voltage-controlled oscillators (VCOs) is of particular importance for biomedical applications, as they are used in Phase Locked Loop circuits, clock circuits, and frequency generation circuits. For biomedical applications like pacemakers, the frequency should be low, corresponding to the heartbeat. By employing techniques like current-starved VCO design, the normal frequency of oscillations can be reduced from gigahertz to hertz range, enabling the use of VCOs in low-power, low-frequency biomedical devices.

Overall, the application of VLSI design techniques in the development of low-power, low-voltage circuits is crucial for enabling energy-efficient and sustainable biomedical devices, addressing the growing demand for compact, portable, and implantable medical technologies.

g) Micro-energy harvesting

Energy harvesting is the process of extracting energy from the ambient environment to power stand-alone systems. Micro-energy harvesting refers to the extraction of small amounts of energy, typically in the micro-watt range, from sources such as vibration, solar, thermal, and electromagnetic fields.

VLSI circuits play a crucial role in the design and implementation of micro-energy harvesting systems.VLSI techniques allow for the integration of various energy harvesting, power management, and energy storage components onto a single chip, enabling the creation of compact and efficient micro-energy harvesting systems. One application of VLSI in micro-energy harvesting is the design of inductor-less on-chip micropower management systems for light energy harvesting applications. These systems utilize VLSI techniques to integrate power conversion, regulation, and storage components on a single chip, enabling efficient energy harvesting and power management for a wide variety of applications. Another application of VLSI in micro-energy harvesting is the design of low-voltage energy harvesting SoC (System-on-Chip) systems for ultra-low-power biomedical applications. VLSI techniques are employed to integrate energy harvesting, power management, and sensor interface circuitry on a single chip, enabling the development of self-powered and miniaturized biomedical devices. VLSI-based micro-energy harvesting systems offer several advantages, such as reduced size, improved power efficiency, and enhanced integration capabilities, making them well-suited for a variety of applications, including wearable electronics, wireless sensor networks, and IoT (Internet of Things) devices [20].

h) Circuits and systems in clinical applications

VLSI technology has had a significant impact on the development of advanced medical devices and healthcare technologies. The integration of VLSI circuits has enabled the creation of compact, sophisticated, and cost-effective medical devices that can enhance patient care and improve treatment outcomes. One key application of VLSI in the medical field is the development of wireless biosensing and body implant systems. VLSI-based circuits can be used to design compact, energy-efficient, and reliable wireless sensors for *in-vivo* monitoring of various physiological parameters, such as neural activity, glucose levels, and cardiovascular data. These integrated circuits enable wireless data transmission and can be seamlessly integrated into implantable medical devices, providing continuous and real-time monitoring of the patient's health [21].

VLSI technology has also been instrumental in the advancement of neural interface circuits, which are critical for various clinical applications, including neural recording, neurochemical sensing, and neural stimulation. These integrated circuits can be used to develop high-density, low-power neural interfaces that can effectively record and stimulate neural activity, enabling the development of neuroprosthetic devices and brain-computer interfaces. Furthermore, VLSI-based circuits have found applications in the field of biosensing interfaces, where they are used to develop advanced sensor interfaces for amperometric, potentiometric, and impedimetric biosensors. These integrated circuits can enhance the performance of these biosensors, improving their sensitivity, selectivity, and signal-to-noise ratio, which is crucial for accurate and reliable biomedical measurements. It has also been utilized in the development of circuits for molecular biology applications, such as label-free medical diagnostics, nucleic acid analysis, and digital microfluidic biochips. These integrated circuits enable the miniaturization and integration of various biological and chemical analysis functionalities, leading to the creation of portable, cost-effective, and high-throughput diagnostic devices.

In addition to these specific applications, VLSI circuits have also been employed in various medical imaging modalities, such as X-ray, CT, MRI, and ultrasound, to enhance the performance, speed, and cost-effectiveness of these imaging systems. The integration of VLSI circuits in medical imaging devices can lead to improved patient care by expediting diagnosis and treatment. Overall, the integration of VLSI technology in the development of circuits and systems for clinical applications has been transformative, enabling the creation of advanced, compact, and cost-effective medical devices that can significantly improve patient outcomes and the overall quality of healthcare.

i) Circuits for cells, DNA, bacteria, viruses

- ***VLSI circuits for cellular and DNA analysis***
  VLSI technologies, such as CMOS (Complementary Metal-Oxide-Semiconductor) processes, have been utilized to create massively parallel integrated circuits with high sensitivity for cellular and DNA analysis. These circuits can be used for applications like DNA sequencing, gene expression profiling, and high-throughput screening of biological samples.
- ***VLSI circuits for bacterial and viral detection***
  VLSI-based integrated circuits have also been employed for the detection and analysis of bacteria and viruses. These circuits can integrate various sensing elements, such as electrochemical or optical sensors, to identify the presence and characteristics of specific bacterial or viral agents. This enables rapid and accurate diagnosis, which is crucial in various healthcare and disease control scenarios.

- ***VLSI-based biomedical diagnostics and monitoring***
  VLSI technology has also been instrumental in the development of advanced biomedical diagnostic and monitoring devices. These devices can integrate various sensing modalities, signal conditioning circuits, and data processing capabilities on a single chip, enabling on-site or point-of-care analysis of biological samples. This includes applications such as biosensors, wearable health monitoring systems, and implantable devices for continuous monitoring of physiological parameters.

j) Brain interfaces

Very-large-scale integration (VLSI) has emerged as a crucial technology in the development of brain-computer interfaces (BCIs). BCIs aim to establish a direct communication channel between the human brain and external devices, enabling individuals with disabilities to regain functionalities or control electronic devices using their neural signals [22]. VLSI techniques have facilitated the design of complex integrated circuits for BCIs, including specialized hardware and neuromorphic architectures that mimic the brain's neural networks. These VLSI-based systems can efficiently process and transmit neural data, overcoming challenges such as power efficiency and implant size that are critical for reliable and effective wireless BCI implants. VLSI-based BCI systems often incorporate advanced spike sorting techniques to classify neural spikes with high accuracy while maintaining the power and size constraints of the implant. Spike sorting is a crucial step in extracting meaningful information from the recorded neural signals [22]. Furthermore, VLSI has enabled the development of modular and customizable neural prostheses, where specialized circuits can be integrated to address specific needs or applications. This modularity allows for greater flexibility and personalization in BCI design and implementation. VLSI-based neural signal processors have also been designed to efficiently process and analyze neural data from implantable microelectrode arrays (MEAs), which provide high-resolution recording of neural activities. These VLSI-based processors can be integrated into wireless BCI systems to enable reliable data transmission and processing. In addition to neural signal processing, VLSI has been utilized in the development of advanced prosthetic limbs and rehabilitation devices. The integration of VLSI chips with sensors, stimulators, and control modules has enabled the creation of more sophisticated and user-friendly assistive technologies.

k) Internet of Things for remote healthcare

VLSI technology has played a vital role in the advancement of remote healthcare systems powered by the Internet of Things (IoT). These VLSI-enabled solutions have revolutionized patient care by enabling real-time monitoring of health parameters through various sensors and microcontrollers. VLSI-powered diagnostic devices can enable rapid analysis of biological samples, such as blood or saliva, providing healthcare professionals with immediate insights into a patient's condition. This capability is especially valuable in emergency settings, ambulances, and remote healthcare environments where timely diagnosis is critical. Portable imaging devices leveraging miniaturized VLSI technology can bring advanced diagnostics to remote areas, significantly improving healthcare access and revolutionizing the delivery of medical services. VLSI-based remote patient monitoring systems use IoT technology to enable continuous health monitoring, allowing healthcare providers to track patient data and receive alerts in real-time. This architecture facilitates remote patient care, especially for individuals with chronic conditions or those in need of long-term monitoring. The wireless structure of these remote monitoring systems can be constructed using VLSI concepts, making them suitable for biomedical applications. This wireless design enables the transmission of patient data to remote locations for further analysis and diagnosis. VLSI-enabled sensors have revolutionized healthcare by enabling continuous patient monitoring through wearable devices. These smart devices integrate multiple sensors, data processing capabilities, and wireless communication, allowing for comprehensive health tracking and early intervention. In the field of electrocardiogram (ECG) monitoring, VLSI-based architectures have been developed to enhance power management and optimize performance, enabling long-term and battery-operated cardiac monitoring. These solutions leverage techniques such as dynamic transmission mode adjustment, slope-based feature extraction, and lossless data encoding to maximize battery life and data reliability. VLSI technology has also enabled the development of energy-efficient and resource-optimal microcontroller units (MCUs) for wireless body sensor networks (WBSNs), facilitating the integration of multiple sensors and advanced signal processing capabilities while maintaining low power consumption [23, 24]. Table 17.1 and Table 17.2 summarize the impact of nanomaterial in biomedical application.

Table 17.1 Nanomaterials for biomedical applications [24].

	Nanomaterials	**Applications**
Biosensor	The carboxy lated SWCNTs	Electrochemical SARS-CoV-2 biosensor
	The graphene sponge (GS)	Electrochemical glucose biosensor
	The graphene oxide (GO)	Fluorescent RNAi biosensor
	The carbon dot (CD) and GO	Fluorescent organophosphate pesticide biosensor
Drug delivery	Chitosan/agarose/graphitic carbon nitride	MCF7 cancer cell treatment
	Gelatin/carbon dot	Antibacterial drug release
	Silica nanoparticle/CNT-DNA	HeLa cancer cell treatment
	Mesoporous silica-modified CNT	D2A1 cancer cell treatment
Tissue engineering	VA-CNT	Neural stem cell differentiation
	rGO/GelMA	Peripheral nerve regeneration
	GT-PDA-PCNTs	Skeletal muscle regeneration
	HPU/CNT fiber	Artificial muscle actuation

Table 17.2 Different carbon-based nanomaterials for various cancer therapy [25].

Carbon-based nanomaterials	***In vitro***	**Therapy**
NY-ESO-1, CpG-ODNs with MWCNT	Dendritic cells	Immuno response
Magnetic ferrite nanoparticles filled	CNTSKOV3 cells	Imaging and therapy
PEG functionalized MWCNTS	U87, U373MG, NHA	Brain tumor therapy
Carbon Nanotubes (CNT)	-	Microbeam radiation therapy

(Continued)

Table 17.2 Different carbon-based nanomaterials for various cancer therapy [25]. (*Continued*)

Carbon-based nanomaterials	***In vitro***	**Therapy**
CNT	-	Microbeam radiation therapy
Multi Walled Carbon Nanotube (MWCNT)	PANC-1	Pancreatic cancer
MWCNT	HeLa	Photothermal therapy
Single Walled Carbon Nanotube (SWCNT)	4T1	Chemo-photothermal therapy
SWCNT	4T1	Photothermal
Porphyrin immobilized NanoGO	U87MG, HBMEC, ACBRI376	Photothermal
Nano graphene sheet		Photothermal
Double network structured GO	HCT116	Chemo-photothermal
(PEG-g-PDMA-HA)@rGO	MDAMB-231, A549	Photothermal
GO decorated Ru(II)-PEG complex	A549	Photodynamic-photothermal
Iron oxide-GO	HeLa	Chemo-photothermal
GO (188Re)-modified Fe3O4/silica	-	Chemo-photothermal
(HA)-modified Q-Graphene	A549, MRC-5	Chemotherapy
CuS-GO	HeLa	Chemo-photothermal
RGO-PEG	U87	Chemo-photothermal photodynamic
GQD-Ce6-HA	A549	Photodynamic
UCNP-GQD	4T1	Photodynamic
7Gd-encapsulated Graphene Carbon	SSC-7	Photodynamic
GQDs	BT-474, MCF-7	-
Graphene Quantum Dots (GQDs)	PANC-1, A-549, HepG2	
GQDs	RG2	Chemotherapy
GQDs	La29, HaCaT, Mia-Pa-Ca-2	Photothermal photodynamic
GQDs	SW620, HCT116	Radiotherapy

17.4 Conclusion

VLSI technology has become indispensable in the biomedical field, driving innovation and significantly enhancing the effectiveness of medical devices and systems. Its wide-ranging applications, including medical imaging, wearable health monitors, implantable devices, telemedicine, neural interfaces, drug delivery systems, lab-on-a-chip technologies, and personalized medicine, underscore its broad impact. As VLSI technology continues to advance, its integration into biomedical applications is poised to further revolutionize healthcare delivery and improve patient outcomes. Several factors, such as increasing lifespans, rising obesity rates, and an aging population, are dramatically escalating healthcare costs. Developments in semiconductor technology and IC design methodologies have resulted in the creation of microsystems that include sensing and processing capabilities. These systems have the potential to complement, augment, or perhaps replace conventional diagnostic and treatment processes. At previously unheard-of low prices, integrated biomedical systems based on VLSI technology provide surprisingly effective means of prompt diagnosis, therapy, and illness management. VLSI technologies and integrated microsystems also facilitate a number of revolutionary biomedical applications that save lives, save quality of life (QoL) for an aging global population using minimally-invasive surgery with less morbidity, offer low-cost preventative healthcare monitoring/preventative solutions for everyday use, provide effective chronic disease management tour de force minimizes the need for hospitalizations/hospital visits belies application sophistication point-of-care diagnosis enables early detection/treatment which may be too late later on high-throughput bio sequencing/bio screening unveils new discoveries digital brain-machine interface where synergy born from deep understanding of human intelligence is explored. VLSI technology plays a key role in biomedical microsystems to realize the functions such as interfacing to sensors and actuators, operating effectively with energy sources (mainly renewable), miniaturization-based high level of integration, intelligence embedding, and connectivity.

References

1. Shin, M., Lim, J., Park, Y., Lee, J.Y., Yoon, J., Choi, J.W., Carbon-based nanocomposites for biomedical applications. *RSC Adv.*, *14*, 10, 7142–7156, 2024.
2. Maiti, D., Tong, X., Mou, X., Yang, K., Carbon-based nanomaterials for biomedical applications: a recent study. *Front. Pharmacol.*, *9*, 1401, 2019.

3. K., *VLSI circuits for biomedical applications*, Artech House, USA, 2008.
4. Sowmya, N. and Rout, S.S., A Review on VLSI Implementation in Biomedical Application, in: *Innovations in Bio-Inspired Computing and Applications.* IBICA 2019. Advances in Intelligent Systems and Computing, vol. 1180, A. Abraham, M. Panda, S. Pradhan, L. Garcia-Hernandez, K. Ma (Eds.), Springer, Cham, 2021, https://doi.org/10.1007/978-3-030-49339-4_14.
5. Forouhi, S., Dehghani, R., Ghafar-Zadeh, E., CMOS based capacitive sensors for life science applications: A review. *Sens. Actuators A: Phys.*, *297*, 111531, 2019, https://doi.org/10.1016/j.sna.2019.111531.
6. Barragán-Montero, A., Javaid, U., Valdés, G., Nguyen, D., Desbordes, P., Macq, B., Lee, J.A., Artificial intelligence and machine learning for medical imaging: A technology review. *Physica Med.*, *83*, 242–256, 2021, https://doi.org/10.1016/j.ejmp.2021.04.016.
7. Auepanwiriyakul, C., Waibel, S., Songa, J., Bentley, P., Faisal, A.A., Accuracy and acceptability of wearable motion tracking for inpatient monitoring using smartwatches. *Sensors*, *20*, 24, 7313, 2020, https://doi.org/10.3390/s20247313.
8. Seidman, S.J., Guag, J., Beard, B., Arp, Z., Static magnetic field measurements of smart phones and watches and applicability to triggering magnet modes in implantable pacemakers and implantable cardioverter-defibrillators. *Heart Rhythm*, *18*, 10, 1741–1744, 2021, https://doi.org/10.1016/j.hrthm.2021.06.1203.
9. Abdellatif, M.M. and Mohamed, W., Telemedicine: An IoT Based Remote Healthcare System. *Int. J. Online Biomed. Eng.*, *16*, 6, 72–81, 2020, https://doi.org/10.3991/ijoe.v16i06.13651.
10. Salahuddin, U. and Gao, P.X., Signal generation, acquisition, and processing in brain machine interfaces: a unified review. *Front. Neurosci.*, *15*, 728178, 2021, doi: 10.3389/fnins.2021.728178.
11. Chircov, C. and Grumezescu, A.M., Microelectromechanical systems (MEMS) for biomedical applications. *Micromachines*, *13*, 2, 164, 2022, https://doi.org/10.3390/mi13020164.
12. Venkatesalu, S., Dilliyappan, S., Kumar, A.S., Palaniyandi, T., Baskar, G., Ravi, M., Sivaji, A., Prospectives and retrospectives of microfluidics devices and lab-on-A-chip emphasis on cancer. *Clin. Chim. Acta*, 552, 117646, 2024.
13. Elemento, O., The future of precision medicine: towards a more predictive personalized medicine. *Emerg. Top. Life Sci.*, *4*, 2, 175–177, 2020, https://doi.org/10.1042/ETLS20190197.
14. Wang, Z., Li, W., Chen, C., Sun, C., Chen, W., A multichannel reconfigurable EEG acquisition system design with felt-based soft material electrodes, in: *2018 IEEE International Instrumentation and Measurement Technology Conference (I2MTC)*, 2018, May, IEEE, pp. 1–6, https://doi.org/10.1109/I2MTC.2018.8409883.
15. Zarifi, M.H., Frounchi, J., Asgarifar, S., Baradaran Nia, M., FPGA implementation of a fully digital demodulation technique for biomedical application. *2008*

Canadian Conference on Electrical and Computer Engineering, Niagara Falls, ON, Canada, pp. 001265–001268, 2008, doi: 10.1109/CCECE.2008.4564742.
16. Maheshwari, S., Serb, A., Papavassiliou, C., Prodromakis, T., An adiabatic capacitive artificial neuron with RRAM-based threshold detection for energy-efficient neuromorphic computing. *IEEE Trans. Circuits Syst. I Regul. Pap.*, *69*, 9, 3512–3525, 2022, https://doi.org/10.1109/TCSI.2022.3182577.
17. Mariani, F., Gualandi, I., Schuhmann, W., Scavetta, E., Micro-and nano-devices for electrochemical sensing. *Microchim. Acta*, *189*, 12, 459, 2022, https://doi.org/10.1007/s00604-022-05548-3.
18. Garrido-Regife, L., Rivero-Antúnez, P., Morales-Flórez, V., The dispersion of carbon nanotubes in composite materials studied by computer simulation of Small Angle Scattering. *Phys. B Condens. Matter*, *649*, 414450, 2023, https://doi.org/10.1016/j.physb.2022.414450.
19. Sajotra, D., Dhariwal, S., Mishra, R.S., Gupta, R., Lamba, V.K., Design and Simulation of Low Power Voltage Controlled Oscillator for Biomedical Applications. *2018 International Conference on Intelligent Circuits and Systems (ICICS)*, Phagwara, India, pp. 138–142, 2018, doi: 10.1109/ICICS.2018.00038.
20. Chalasani, S. and Conrad, J.M., A survey of energy harvesting sources for embedded systems, in: *IEEE SoutheastCon 2008*, 2008, April, IEEE, pp. 442–447, https://doi.org/10.1109/SECON.2008.4494336.
21. Huang, Z. and Inoue, Y., A sub-100nA power management system for wireless structure health monitoring applications, in: *2011 IEEE International Symposium of Circuits and Systems (ISCAS)*, 2011, May, IEEE, pp. 2897–2900.
22. Yúfera, A., Huertas, G., Calvo, B., Design and Application of Biomedical Circuits and Systems. *Electronics*, *9*, 11, 1920, 2020.
23. Pulimamidi, R., Leveraging IoT Devices for Improved Healthcare Accessibility in Remote Areas: An Exploration of Emerging Trends. *Internet Things Edge Comput. J.*, *2*, 1, 20–30, 2022, https://thesciencebrigade.com/iotecj/article/view/158.
24. Kandpal, J., Baluti, G., Ranjan, P., Integration of Artificial Intelligence with Opto-VLSI for the Biomedical and Healthcare Industries, in: *Opto-VLSI Devices and Circuits for Biomedical and Healthcare Applications*, pp. 161–175, CRC Press, Boca Raton, Florida, USA, 2023.
25. McNamara, K. and Tofail, S.A., Nanoparticles in biomedical applications. *Adv. Phys.: X*, *2*, 1, 54–88, 2017.

18

Self-Powered Biosensor Field-Effect Transistors

Archana Pandey* **and Shradha Saxena**

Dept. of ECE, Jaypee Institute of Information Technology, Noida, India

Abstract
Biosensors have revolutionized various fields by providing rapid, sensitive, and portable means of detecting biological analytes. They find applications in healthcare, environmental monitoring, food safety, and biodefense, among others.

As traditional biosensors often face challenges related to power consumption and limited operational lifetimes, the incorporation of self-powering mechanisms becomes a transformative solution. In recent years, a significant focus has been directed towards the development of self-powered biosensor FETs, which incorporates energy harvesting mechanisms to enable autonomous operation without the need for external power sources. This chapter provides a comprehensive overview of the need of FET biosensors, working principle, need of energy harvesting, latest advancements in self-powered biosensor FET technology, fabrication techniques, and applications.

Keywords: Biosensor, energy harvesting, FET, self-power

18.1 Introduction to Biosensors

18.1.1 Definition and Importance

Biosensors are analytical instruments that integrate a biological element with a physicochemical detector to detect and measure specific biological molecules. They are essential in diverse areas, including medical

**Corresponding author*: archana.pandey@jiit.ac.in

Ekta Goel, Archana Pandey, Shiromani Balmukund Rahi and Arun Samuel (eds.) Handbook of Advanced Semiconductor Field Effect Transistors, (383–400) © 2026 Scrivener Publishing LLC

diagnostics, environmental monitoring, food safety, and bioprocess control [1, 2]. The biological component, often referred to as the bioreceptor, can be enzymes, antibodies, nucleic acids, or cells that interact specifically with the target analyte. This interaction produces a signal that is converted into a measurable response by the physicochemical detector, providing valuable data for further analysis.

18.1.2 Overview of Traditional Biosensors

Traditional biosensors can be broadly categorized into electrochemical, optical, thermal, and piezoelectric biosensors. Each type relies on a different transduction principle to convert the biological interaction into a measurable signal. Figure 18.1 demonstrates various types of biosensors. Figure 18.2 demonstrates electrochemical biosensing.

Optical Biosensors- These biosensors measure changes in light properties, such as absorbance, fluorescence, or surface plasmon resonance, induced by the analyte-bioreceptor interaction. Optical biosensors are widely used in laboratories for DNA sequencing and protein analysis. Figure 18.3 demonstrates the working principle of optical biosensing [4–7].

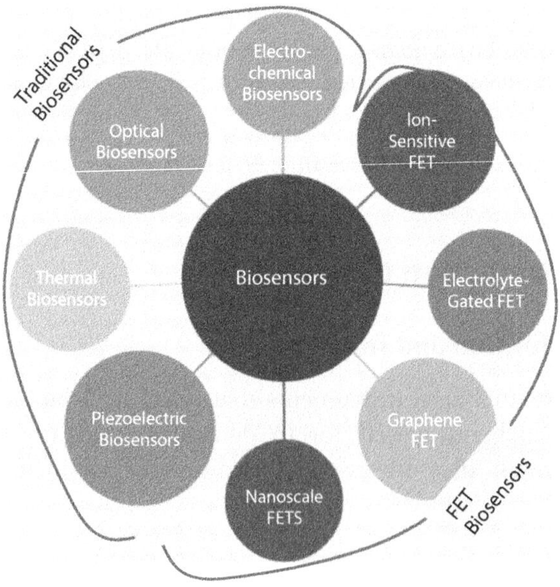

Figure 18.1 Various types of biosensors.

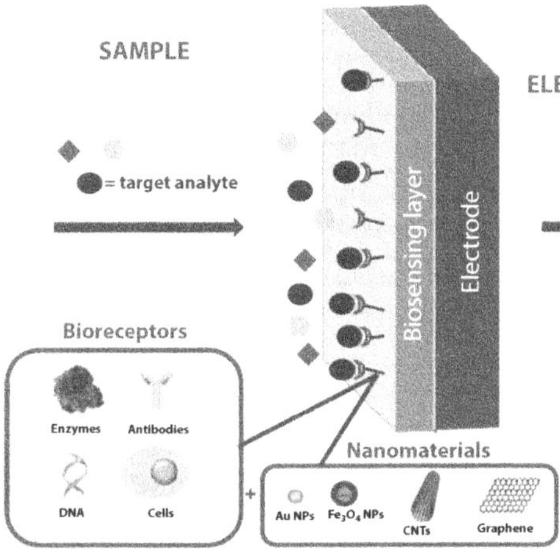

Figure 18.2 Schematic diagram of electrochemical biosensing [3].

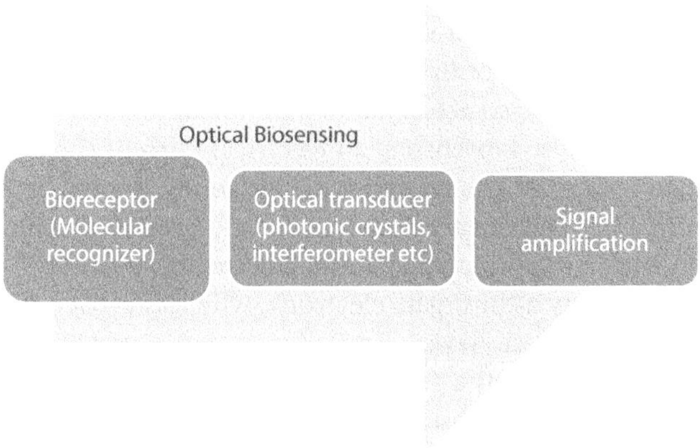

Figure 18.3 A flow chart of the process of optical biosensing.

Thermal Biosensors-
These detect changes in temperature resulting from the biochemical reaction between the analyte and the bioreceptor. Although less common, thermal biosensors are employed in specific applications where heat generation is significant [8, 9].

Piezoelectric Biosensors-
These sensors measure changes in mass or mechanical properties through variations in the oscillation frequency of a piezoelectric crystal. They

are particularly useful in detecting small molecules and pathogens. Piezoelectricity is a mechanism which works on the principle that the material produces voltage when is in mechanical strain. This is valid for the reverse case as well, that when material is put in mechanical strain, it produces voltage. Aluminum phosphate (berlinite), aluminum nitride, zinc oxide, crystalized topaz (Al2SiO4(F, OH)2), barium and lead titanate, gallium orthophosphate, quartz (SiO2), tartrate tetrahydrate (Rochelle salt), polyvinylidene fluoride, polylactic acids can be mentioned as typical piezoelectric materials [10, 11].

18.2 Field-Effect Transistor (FET)-Based Biosensors

18.2.1 Working Principle

FETs were developed in 1970s. Since then, FET devices have been used as a biosensor for various biosensing applications due to their ability to detect various analytes. Field-effect transistor (FET)-based biosensors are a class of biosensors that utilize the field-effect transistor as the transduction element. The basic structure of an FET includes a source, drain, and gate terminal. In these biosensors, an electrical current travels through a semiconductor channel connecting two electrodes, known as the Source and Drain (S-D). When charged molecules are present on the surface of FET biosensors, the conductance of the channel between the Source and Drain is regulated by a third gate electrode, which is capacitively coupled through a thin dielectric layer. Typically, an oxide material like silicon dioxide (SiO2) is used as the gate dielectric for insulation. The device also features a p-type silicon substrate (bulk), where applying a negative gate potential leads to an accumulation of holes (the primary charge carriers), increasing the channel's conductance. In contrast, a positive gate potential depletes holes, decreasing conductance. The adsorption of molecules on the surface of the semiconductor channel either modifies its local surface potential or directly dopes the channel, leading to changes in conductance. In biosensing applications, the gate is typically functionalized with a bioreceptor, which interacts with the target analyte. This interaction alters the charge distribution at the gate, modulating the current flowing between the source and drain [12].

This makes FETs highly promising as sensing devices, offering customizable configurations, enhanced sensitivity, and real-time detection capabilities. Moreover, advanced chemical techniques are employed to address the issue of nonspecific absorption, preventing surface fouling when exposed to complex media like human serum or blood [13]. In solution-gated FET biosensors, analytes are detected within an aqueous environment. The semiconducting nanowire channels are immersed in a flow or sensing chamber that confines the solution. To prevent current leakage caused by ionic conduction, the source and drain electrodes are insulated using materials such as poly (methyl methacrylate), poly dimethylsiloxane) (silicone rubber), SiO2 thin films, or SU8 passivation layers [14–17].

The working principle of FET biosensors revolves around the modulation of the channel conductance by the electric field generated by the bioreceptor-analyte interaction. When the target analyte binds to the bioreceptor on the gate, it induces a change in the surface potential. This change influences the electric field across the FET channel, altering its conductivity. The resulting change in current or voltage is measured and correlated to the concentration of the analyte, providing a quantitative analysis. Figure 18.4 shows a flow chart of the biosensing process in FET based biosensors.

Bioreceptor interacts with the target analyte and makes changes in the charge distribution of the gate. This may impact the effective work function of the gate. For example, let us take an example of a platinum gate of the FET, which comes in contact with the target analyte. Let us assume that on the introduction of target analyte work function of the platinum material

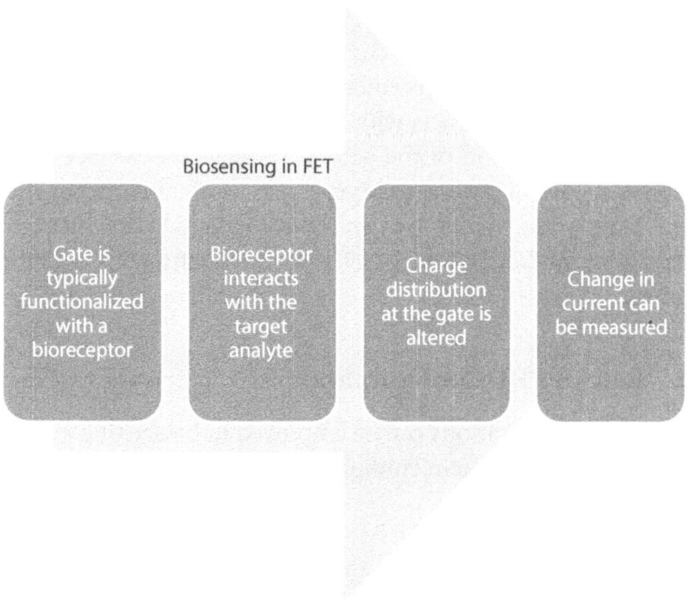

Figure 18.4 A flow chart of the biosensing process in FET based biosensors.

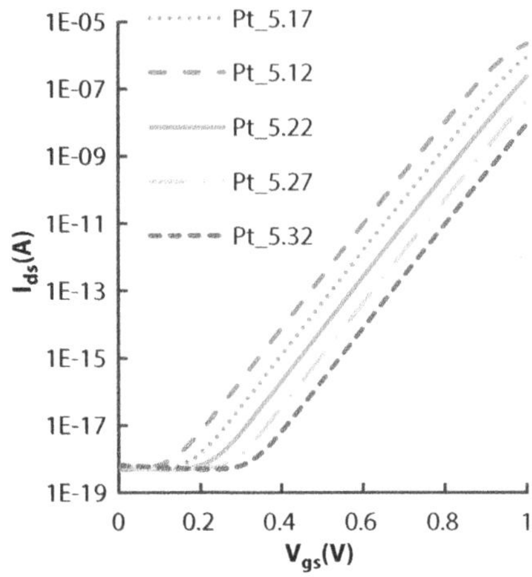

Figure 18.5 Current-voltage characteristics of FET as obtained from the TCAD Sentaurus simulation; work function of the gate material is varied from 5.17 to 5.32.

changes from 5.17 to 5.32. We have analyzed the IV characteristics of the FET device using TCAD simulations with the help of calibrated simulation set up of Synopsis TCAD senaturus tool [18]. In Figure 18.5, as the work function of the gate varies from 5.17 to 5.32, we can see that there is a significant change in the IV characteristics of the FET. Y axis shows the drain current and X axis shows gate to source voltage. Figure 18.5 represents that with the help of the measured change in IV characteristics, the type or amount of the analyte can be predicted.

As the bioreceptor interacts with the target analyte and makes changes in the charge distribution of the gate, it changes the surface potential of the channel in FET as shown in Figure 18.6 as obtained from the TCAD sentaurus simulation results. We can see in the Figure 18.6 that the surface potential curve shifts downwards as the work-function of the gate material is increased.

18.2.2 Role of FET Configurations

Different FET configurations can be employed in biosensing, each offering unique advantages. The main configurations include:

Metal-Oxide-Semiconductor FET (MOSFET)-
MOSFETs are widely used due to their high sensitivity and compatibility with silicon-based technology [12]. The gate of the MOSFET can be functionalized with various bioreceptors to detect specific analytes.

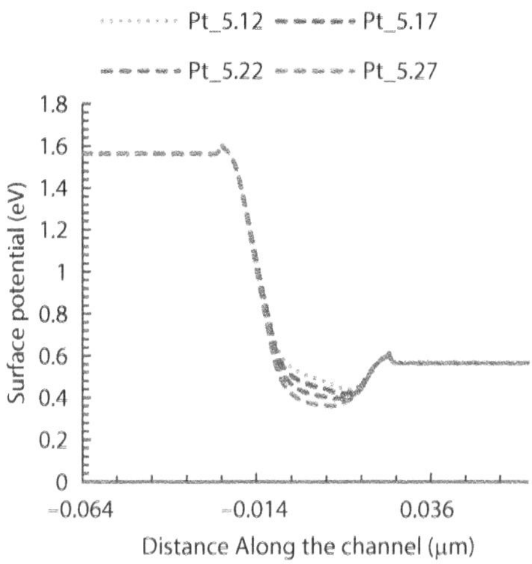

Figure 18.6 Surface potential plotted with respect to the distance along the channel for various values of the gate- work-function.

Figure 18.7 (a) shows the two-dimensional cross-section of a SOI-MOSFET, Figure 18.7 (b) shows three dimensional MOSFET structure and Figure 18.7 (c) shows three dimensional FinFET structure obtained from TCAD sentaurus tool. Novel FETs such as FinFET, Nanowire FET and Nanosheet FET have large surface to volume ration [19–23]. Therefore, sensitivity of these devices is better than the traditional FETs [24–29].

Ion-sensitive FET (ISFET)
ISFETs are designed to detect ions in solution, making them suitable for pH sensing and detecting ionic species in biological samples. They are commonly used in medical diagnostics and environmental monitoring .

Electrolyte-gated FET (EGFET)-
EGFETs use an electrolyte as the gate medium, which allows for direct interaction with the analyte solution. This configuration is beneficial for biosensing applications in aqueous environments [31, 32].

Graphene FET (GFET)-
Graphene-based FETs utilize graphene as the channel material. Graphene's high surface area and excellent electronic properties make GFETs highly sensitive and capable of detecting low concentrations of analytes [33–35].

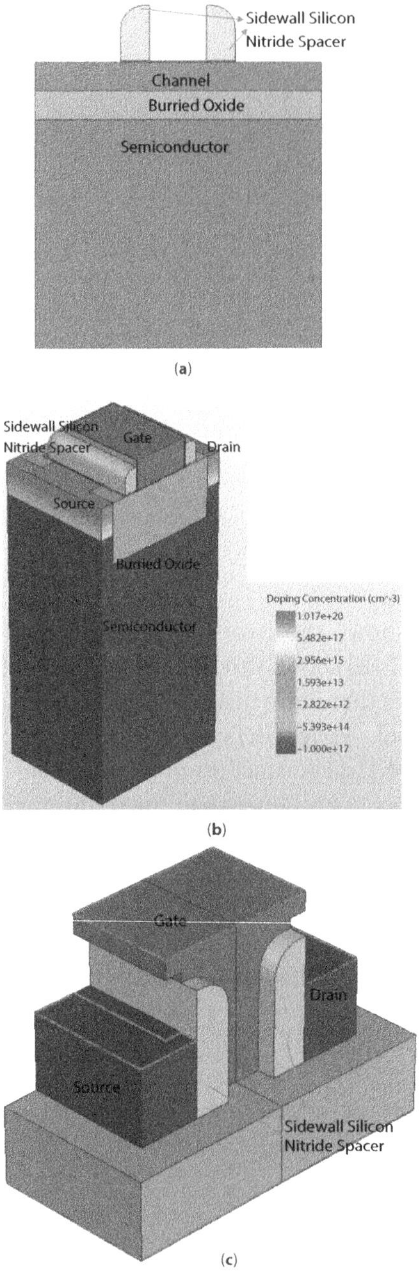

Figure 18.7 (a) Two-dimensional cross-section of a SOI-MOSFET taken from TCAD sentaurus tool. (b) A 3D MOSFET taken from TCAD sentaurus tool. (c) Three-dimensional cross-section of a FinFET device taken from TCAD sentaurus tool.

Figure 18.8 shows some of the most frequently used FET device configurations in biosensor applications. Here the substrate is silicon (Si), and the dielectric material used is silicon dioxide (SiO2). Figure 18.8 (a) shows a bottom gate FET, here the substrate is silicon (Si), and the dielectric material used is silicon dioxide (SiO2). Figure 18.8 (b) Electrolyte-Gated Field-Effect Transistor (EG-FET): This device has a planar configuration where the gate is positioned in the same plane as the source and drain. The gate is insulated by an electrolyte solution. Figure 18.8 (c) Electrochemical Transistor (ECT): In this setup, the gate is an external electrode immersed in the electrolyte solution. Figure 18.8 (d) Ion-Sensitive Field-Effect Transistor (ISFET): Here, the gate electrode is replaced by a reference electrode, commonly made of silver/silver chloride (Ag/AgCl). Figure 18.8 (e) Chemically Sensitive Field-Effect Transistor (ChemFET): In this type of FET, the gate electrode is separated from the source and drained by an electrolytic solution, with a semi-permeable membrane positioned at the gate interface.

Nanowires-based FETs-

A wide variety of nanomaterials has been used as channel materials for constructing metal-oxide semiconductor FETs, ion-selective FETs, and nanowire (NW)-based FET biosensors. These channels are in direct contact with the environment, providing better control over surface charge. Significant efforts have been made to enhance FET designs by incorporating various metals like Ag, Pt, Au, and Cu, as well as semiconductors such as ZnO, SnO2, Si, GaN, TiO2, In2O3, InP, and polymer NWs to investigate

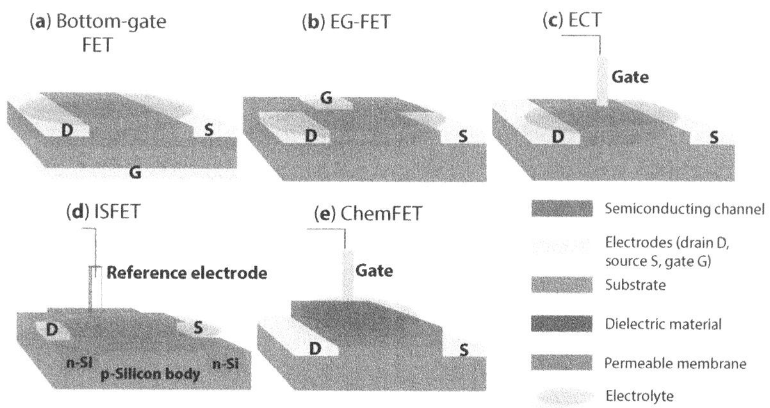

Figure 18.8 Some of the most common structures of FET devices that are used as biosensors, (a) Bottom-gate FET (b) EG-FET (c) ECT (d) ISFET (e) ChemFET [36].

biomolecular interactions and overcome the physical limitations of FET technology. NW fabrication methods are generally categorized into bottom-up and top-down approaches. The top-down method is essential for its ability to precisely align and control the orientation of NWs, though it cannot achieve the same level of miniaturization as the bottom-up method. Notably, NW-based FET biosensors effectively address several limitations, making them highly suitable for FET-based biosensor fabrication [23–30].

18.2.3 Sensitivity and Selectivity

The sensitivity of FET-based biosensors is primarily attributed to the high surface-to-volume ratio of the transducer and the efficient modulation of the channel conductance by the electric field. The nanoscale dimensions of the FET channel enable the detection of minute changes in charge distribution, making these sensors highly sensitive to low concentrations of analytes.

Selectivity in FET biosensors is achieved through the functionalization of the gate with specific bioreceptors that interact selectively with the target analyte. The choice of bioreceptor—such as antibodies, enzymes, or nucleic acids—determines the specificity of the sensor. Additionally, advanced surface modification techniques can enhance selectivity by reducing non-specific binding and improving the affinity of the bioreceptor for the analyte [37, 38].

18.3 Label-Free Detection with FET Biosensors

18.3.1 Comparison with Traditional Methods

Traditional biosensing methods often require labeling the analytes with fluorescent, radioactive, or enzymatic tags to facilitate detection. While these methods can be highly sensitive, they involve additional steps for labeling, which can be time-consuming, costly, and may alter the properties of the analyte [14].

Label-free detection with FET biosensors eliminates the need for such labels, simplifying the assay process and preserving the native state of the analyte. This approach leverages the inherent electrical properties of the analyte-bioreceptor interaction to produce a measurable signal. As a result, label-free FET biosensors offer a more direct, rapid, and cost-effective means of detection [24–27].

18.3.2 Rapid Response Times

One of the significant advantages of FET biosensors is their rapid response time. The direct interaction between the analyte and the bioreceptor at the gate surface produces an immediate change in the electric field, which is promptly reflected in the FET's current or voltage output. This real-time response capability is particularly valuable in applications requiring quick decision-making, such as point-of-care diagnostics and environmental monitoring [30, 31].

18.3.3 Advantages in Biosensing

FET biosensors offer several advantages over traditional biosensing methods:

High Sensitivity- The ability to detect low concentrations of analytes due to the high surface-to-volume ratio and efficient electric field modulation [37].
Label-Free Detection- Eliminates the need for additional labeling steps, reducing complexity, cost, and potential alterations to the analyte [32, 33].
Real-Time Monitoring- Provides immediate feedback, enabling real-time analysis and rapid response to changes in analyte concentration.
Miniaturization and Integration- The small size of FET sensors allows for the development of compact, portable devices and the integration into lab-on-a-chip systems for multiplexed analysis.

18.4 Need for Self-Powering Mechanisms in Biosensors

18.4.1 Challenges in Power Consumption

As biosensors become more advanced and integrated into various applications, power consumption emerges as a critical challenge. Many biosensors, particularly those with continuous monitoring capabilities, require a constant power supply to function effectively. This dependence on external power sources can limit their portability and practicality, especially in remote or resource-limited settings [39].

18.4.2 Limited Operational Lifetimes

Biosensors powered by conventional batteries or external power sources face limitations in operational lifetimes. Batteries have a finite energy capacity and require periodic replacement or recharging, which can be inconvenient and impractical for long-term or continuous monitoring applications. The need for frequent maintenance can also increase the overall cost and complexity of the biosensing system [39].

18.4.3 Importance of Autonomous Operation

To overcome these challenges, there is a growing need for biosensors with self-powering mechanisms that enable autonomous operation. Self-powered biosensors can harness energy from the environment or the biological system itself, eliminating the dependence on external power sources. This capability enhances the portability, convenience, and sustainability of biosensors, making them more suitable for a wide range of applications, including wearable devices, implantable sensors, and remote environmental monitoring systems [39–43].

18.5 Energy Harvesting in Biosensor FET Technology

18.5.1 Piezoelectric Mechanisms

In addition to piezoresistive and capacitive FET pressure sensors, piezoelectric materials are commonly used in pressure sensor fabrication due to the charge they generate under mechanical stress. When stress is applied to certain materials, electrical dipole moments are created within them. The neutral state of the material is largely due to parasitic effects that cause recombination of the piezoelectric charges. Piezoelectric sensors are highly sensitive and have a fast response time, making them ideal for detecting dynamic pressures, such as vibrations from sound waves. Moreover, the self-generated electrical dipoles make piezoelectric materials promising candidates for low-power or self-powered sensing devices. However, these sensors are not effective at detecting static pressure or forces with minimal variation. To overcome this limitation, piezoelectric sensors are often paired with other pressure sensors or electronic devices, typically incorporating amplifiers like transistors. Combining FETs and piezoelectric sensors has become a standard method in fabricating flexible pressure sensors, improving their sensitivity.

Piezoelectric energy harvesting is particularly useful in applications where mechanical movements, such as bodily motions or environmental vibrations, are readily available [10]. Similarly, environmental sensors can harness energy from vibrations in industrial settings or natural phenomena, enabling long-term, self-sustained operation in remote or hazardous locations [11].

Triboelectric techniques [39] biofuels and solar cell integration are other energy harvesting mechanisms which are useful for environmental monitoring applications, where sensors can be deployed in the field and powered by ambient sunlight [44–46].

Integration of energy harvesting in FET biosensor platforms, their design and implementation and performance evaluation is a real challenge [39, 40].

18.6 Applications of Self-Powered Biosensor FETs

Point-of-Care diagnostics [47], Environmental monitoring [36], Food safety analysis [1–3], Wearable applications [2] are the trending applications of the self-powered biosensor FETs. They play a crucial role in various fields such as medical diagnostics, environmental monitoring, food safety, and bioprocess control.

18.7 Conclusion

The integration of energy harvesting mechanisms with FET biosensors represents a significant advancement in biosensor technology. Self-powered FET biosensors offer numerous advantages, including high sensitivity, label-free detection, real-time monitoring, and autonomous operation. These sensors have the potential to revolutionize point-of-care diagnostics, environmental monitoring, food safety analysis, and wearable applications.

References

1. Bunney, J., Williamson, S., Atkin, D., Jeanneret, M., Cozzolino, D., Chapman, J., Power, A., Chandra, S., The Use of Electrochemical Biosensors in Food Analysis. *Curr. Res. Nutr. Food. Sci.*, 5, 83–95, 2017.

2. Pantelopoulos, A. and Bourbakis, N.G.A., Survey on Wearable Sensor-Based Systems for Health Monitoring and Prognosis. *IEEE Trans. Syst. Man Cybern. Syst.*, 40, 1–12, 2010.
3. Dridi, F., *et al.*, Nanomaterial-based electrochemical biosensors for food safety and quality assessment, in: *Nanobiosensors*, pp. 167–204, Academic Press, Amsterdam, Netherlands, 2017.
4. Zhao, Y., Tong, R.-j., Xia, F., Peng, Y., Current status of optical fiber biosensor based on surface plasmon resonance. *Biosens. Bioelectron.*, 142, 111505, 2019.
5. Wang, X.-d. and Wolfbeis, O.S., Fiber-optic chemical sensors and biosensors (2015–2019). *Anal. Chem.*, 92, 1, 397–430, 2019.
6. Monfared, Y.E., Overview of recent advances in the design of plasmonic fiber-optic biosensors. *Biosensors.*, 10, 7, 77, 2020.
7. Singh, A.K., Mittal, S., Das, M., Saharia, A., Tiwari, M., Optical biosensors: A decade in review. *Alexandria Eng. J.*, *67*, 673–691, 2023.
8. Vasuki, S., Varsha, V., Mithra, R., Dharshni, R.A., Abinaya, S., Dharshini, R.D., Sivarajasekar, N., Thermal biosensors and their applications. *Am. Int. J. Res. Sci. Technol. Eng. Math*, *1*, 262–264, 2019.
9. Yao, N., Wang, J., Zhou, Y., Rapid determination of the Chemical Oxygen Demand of water using a thermal biosensor. *Sensors*, 14, 6, 9949–9960, 2014.
10. Pohanka, M., Overview of Piezoelectric Biosensors, Immunosensors and DNA Sensors and Their Applications. *Materials*, 11, 3, 448, 2018. https://doi.org/10.3390/ma11030448.
11. Lian, Y., He, F., Wang, H., Tong, F., A new aptamer/graphene interdigitated gold electrode piezoelectric sensor for rapid and specific detection of staphylococcus aureus. *Biosens. Bioelectron.*, 65, 314–319, 2015.
12. Sadighbayan, D., Hasanzadeh, M., Ghafar-Zadeh, E., Biosensing based on field-effect transistors (FET): Recent progress and challenges. *TrAC, Trends Anal. Chem.*, 133, 116067, 2020.
13. Hideshima, S., Wustoni, S., Kobayashi, M., Hayashi, H., Kuroiwa, S., Nakanishi, T., Osaka, T., Effect of Human Serum on the Electrical Detection of Amyloid-β Fibrils in Biological Environments Using Azo-Dye Immobilized Field Effect Transistor (FET) Biosensor. *Sens. Bio-Sens. Res.*, 17, 25–29, 2018.
14. Gutierrez-Sanz, O., Andoy, N.M., Filipiak, M.S., Haustein, N., Tarasov, A., Direct, Label-Free, and Rapid Transistor-Based Immunodetection in Whole Serum. *ACS Sens.*, 2, 1278–1286, 2017.
15. Chu, C.-H., Sarangadharan, I., Regmi, A., Chen, Y.-W., Hsu, C.-P., Chang, W.-H., Lee, G.-Y., Chyi, J.-I., Chen, C.-C., Shiesh, S.-C., *et al.*, Beyond the Debye Length in High Ionic Strength Solution: Direct Protein Detection with Field-Effect Transistors (FETs) in Human Serum. *Sci. Rep.*, 7, 5256, 2017.
16. Chen, P.-C., Chen, Y.-W., Sarangadharan, I., Hsu, C.-P., Chen, C.-C., Shiesh, S.-C., Lee, G.-B., Wang, Y.-L., Editors' Choice—Field-Effect Transistor-Based Biosensors and a Portable Device for Personal Healthcare. *ECS J. Solid State Sci. Technol.*, 6, Q71–Q76, 2017.

17. Kao, W.-C., Chen, Y.-W., Chu, C.-H., Chang, W.-H., Shiesh, S.-C., Wang, Y.-L., Lee, G.-B., Detection of C-reactive Protein on an Integrated Microfluidic System by Utilizing Field-Effect Transistors and Aptamers. *Biomicrofluidics*, 11, 044105, 2017.
18. Synopsys, *Sentaurus Device User Guide*, K-2015.06, Synopsys, Santa Clara, CA, 2015.
19. Gao, X.P.A., Zheng, G., Lieber, C.M., Subthreshold Regime Has the Optimal Sensitivity for Nanowire FET Biosensors. *Nano Lett.*, 10, 547–552, 2010.
20. Pandey, A., Recent Trends in Novel Semiconductor Devices. *Silicon*, 14, 9211–9222, 2022. https://doi.org/10.1007/s12633-022-01694-8.
21. Archana Pandey, S.R., Maheshwaram, S., Manhas, S.K., Dasgupta, S., Saxena, A.K., Anand, B., Effect of load capacitance and input transition time on FinFET inverter capacitances. *IEEE Trans. Electron Devices*, 61, 1, 30–36, 2014.
22. Narula, M.S. and Pandey, A., Performance Evaluation of Stacked Gate Oxide/High K Spacers based Gate All Around Device Architectures at 10nm Technology Node. *Silicon J.*, 14, 5, 2397–2407, April 2022. doi.org/10.1007/s12633-022-01685-9.
23. Narula, M.S. and Pandey, A., A Comprehensive Review on FinFET, Gate All Around, Tunnel FET: Concept, Performance and Challenges, in: *8th International Conference on Signal Processing and Communication (ICSC)*, Noida, India, pp. 554–559, 2022, doi: 10.1109/ICSC56524.2022.10009504.
24. Singh, A. and Pandey, A., Gate-All-Around Nanosheet FET Device Simulation Methodology Using a Sentaurus TCAD, in: *Advanced Field-Effect Transistors*, pp. 232–259, CRC Press, Boca Raton, 2023, ISBN- 9781003393542.
25. Patolsky, F., Zheng, G., Lieber, C.M., Fabrication of Silicon Nanowire Devices for Ultrasensitive, Label-Free, Real-Time Detection of Biological and Chemical Species. *Nat. Protoc.*, 1, 1711–1724, 2006.
26. Stern, E., Klemic, J.F., Routenberg, D.A., Wyrembak, P.N., Turner-Evans, D.B., Hamilton, A.D., LaVan, D.A., Fahmy, T.M., Reed, M.A., Label-Free Immunodetection with CMOS-Compatible Semiconducting Nanowires. *Nature*, 445, 519–522, 2007.
27. Huang, Y.-W., Wu, C.-S., Chuang, C.-K., Pang, S.-T., Pan, T.-M., Yang, Y.-S., Ko, F.-H., Real-Time and Label-Free Detection of the Prostate-Specific Antigen in Human Serum by a Polycrystalline Silicon Nanowire Field-Effect Transistor Biosensor. *Anal. Chem.*, 85, 7912–7918, 2013.
28. Lin, C.-H., Hung, C.-H., Hsiao, C.-Y., Lin, H.-C., Ko, F.-H., Yang, Y.-S., Poly-Silicon Nanowire Field-Effect Transistor for Ultrasensitive and Label-Free Detection of Pathogenic Avian Influenza DNA. *Biosens. Bioelectron.*, 24, 3019–3024, 2009.
29. Zhang, G.-J., Zhang, L., Huang, M.J., Luo, Z.H.H., Tay, G.K.I., Lim, E.-J.A., Kang, T.G., Chen, Y., Silicon Nanowire Biosensor for Highly Sensitive and Rapid Detection of Dengue Virus. *Sens. Actuators B Chem.*, 146, 138–144, 2010.

30. Lin, C.-H., Hsiao, C.-Y., Hung, C.-H., Lo, Y.-R., Lee, C.-C., Su, C.-J., Lin, H.-C., Ko, F.-H., Huang, T.-Y., Yang, Y.-S., Ultrasensitive Detection of Dopamine Using a Polysilicon Nanowire Field-Effect Transistor. *Chem. Commun.*, 44, 5749–5751, 2008.
31. Wu, C.-C., Ko, F.-H., Yang, Y.-S., Hsia, D.-L., Lee, B.-S., Su, T.-S., Label-Free Biosensing of a Gene Mutation Using a Silicon Nanowire Field-Effect Transistor. *Biosens. Bioelectron.*, 25, 820–825, 2009.
32. Ma, S., Li, X., Lee, Y.-K., Zhang, A., Direct Label-Free Protein Detection in High Ionic Strength Solution and Human Plasma Using Dual-Gate Nanoribbon-Based Ion-Sensitive Field-Effect Transistor Biosensor. *Biosens. Bioelectron.*, 117, 276–282, 2018.
33. Ma, S., Lee, Y.-K., Zhang, A., Li, X., Label-Free Detection of Cordyceps Sinensis Using Dual-Gate Nanoribbon-Based Ion-Sensitive Field-Effect Transistor Biosensor. *Sens. Actuators B Chem*, 264, 344–352, 2018.
34. Xu, S., Zhang, C., Jiang, S., Hu, G., Li, X., Zou, Y., Liu, H., Li, J., Li, Z., Wang, X., *et al.*, Graphene Foam Field-Effect Transistor for Ultra-Sensitive Label-Free Detection of ATP. *Sens. Actuators B Chem.*, 284, 125–133, 2019.
35. Thakur, B., Zhou, G., Chang, J., Pu, H., Jin, B., Sui, X., Yuan, X., Yang, C.-H., Magruder, M., Chen, J., Rapid Detection of Single E. coli Bacteria Using a Graphene-Based Field-Effect Transistor Device. Biosens. *Bioelectron*, 110, 16–22, 2018.
36. Lei, Y.-M., Xiao, M.-M., Li, Y.-T., Xu, L., Zhang, H., Zhang, Z.-Y., Zhang, G.-J., Detection of Heart Failure-Related Biomarker in Whole Blood with Graphene Field Effect Transistor Biosensor. *Biosens. Bioelectron.*, 91, 1–7, 2017.
37. Elli, G., Hamed, S., Petrelli, M., Ibba, P., Ciocca, M., Lugli, P., Petti, L., Field-Effect Transistor-Based Biosensors for Environmental and Agricultural Monitoring. *Sensors*, 22, 11, 4178, 2022. https://doi.org/10.3390/s22114178.
38. Lee, M.S., Choi, J.H., Kim, D.H., Removal of Non-Specifically Bound Proteins Using Rayleigh Waves for Enhanced Biosensor Performance. *Sensors*, 22, 11, 4096, May 2022.
39. Stock, D., Muntze, G.M., Figge, S., Eickhoff, M., Ion Sensitive AlGaN/GaN Field-Effect Transistors with Monolithically Integrated Wheatstone Bridge for Temperature- and Drift Compensation in Enzymatic Biosensors. *Sens. Actuators B Chem*, 263, 20–26, 2018.
40. Grattieri, M. and Minteer, S.D., Self-powered biosensors. *ACS Sens.*, *3*, 1, 44–53, 2018.
41. Baingane, A. and Slaughter, G., Self-Powered Electrochemical Lactate Biosensing. *Energies*, 10, 1582, 2017.
42. Chen, C.-H., Lee, P.-W., Tsao, Y.-H., Lin, Z.-H., Utilization of self-powered electrochemical systems: metallic nanoparticle synthesisand lactate detection. *Nano Energy*, 42, 241–248, 2017.

43. Moreira, F.T.C., Sale, M.G.F., Di Lorenzo, M., Towardstimely Alzheimer diagnosis: A self-powered amperometric biosensor for the neurotransmitter acetylcholine. *Biosens. Bioelectron.*, 87, 607–614, 2017.
44. Ruff, A., Pinyou, P., Nolten, M., Conzuelo, F., Schuhmann, W.A., Self-Powered Ethanol Biosensor. *ChemElectroChem*, 4, 890–897, 2017.
45. Jiang, Y., Liang, P., Liu, P., Yan, X., Bian, Y., Huang, X., Acathode-shared microbial fuel cell sensor array for water alert system. *Int. J. Hydrogen Energy*, 42, 4342–4348, 2017.
46. Jiang, Y., Liang, P., Liu, P., Miao, B., Bian, Y., Zhang, H., Huang, X., Enhancement of the sensitivity of a microbial fuel cellsensor by transient-state operation. *Environ. Sci. Water Res. Technol.*, 3, 472–479, 2017.
47. Schievano, A., Colombo, A., Cossettini, A., Goglio, A., D'Ardes, V., Trasatti, S., Cristiani, P., Single-chamber microbial fuel cells as on-line shock-sensors for volatile fatty acids in anaerobic digesters. *Waste Manage*, 71, 785–791, 2018.

19

Vertical Tunneling FETs (V-TFETs): A Novel Approach in Biosensing Technology

Sourav Das[1], Ekta Goel[2] and Kunal Singh[1]*

[1]Solid State Electronics and VLSI Lab, Department of ECE, NIT Jamshedpur, Jharkhand, India
[2]National Institute of Technology, Warangal, India

Abstract
This chapter introduces and explores a number of aspects pertaining to tunnel field-effect transistors (TFETs) for dielectric-modulated (DM) label-free biosensors. An overview of DM TFET biosensors has been given in brief. Principles and design perceives are highlighted in the discussion of the dielectric modulation idea in TFETs. The incorporation of implanted nanogaps into TFET geometries using computer-aided design (TCAD) tool and suitable physics-based simulation models will be explored. Through TCAD simulations, the sensitivity of gate-engineered TFET is investigated under several conditions. Lastly, the state of the proposed V-TFET is pointed out, and an overview map showing the sensitivities of the major efforts in DM label-free biosensors is presented.

Keywords: TFET, V-TFET, biosensor, biomolecules, dielectric modulation, sensitivity

19.1 Introduction

Because the semiconductor industry's capital are based on Moore's Law, metal oxide semiconductor field-effect transistors (MOSFETs) were mandatory to reduce the size of their devices to accommodate a lot of transistors on a given chip. Short-channel effects (SCEs) were shown to be a sign of MOSFET performance degradation as the device dimensions were reduced to the nanoscale dimension. Threshold voltage roll-off,

**Corresponding author*: kunalsingh.ece@nitjsr.ac.in

Ekta Goel, Archana Pandey, Shiromani Balmukund Rahi and Arun Samuel (eds.) Handbook of Advanced Semiconductor Field Effect Transistors, (401–418)

drain-induced barrier lowering, and mobility deterioration were caused by the inversion charge sharing between the source and drain regions in short-channel MOSFETs [1–3]. Because low-power purposes required lower supply voltages and low-off current requirements, these issues restricted MOSFET progress.

Thus, the semiconductor industry has been searching for new devices that can provide performance that is better than MOSFETs and successfully resolve scaling challenges. In recent decades, various devices have been proposed by researchers and industry players as potential substitutes for MOSFETs in low-power uses. A large number of devices operate on different principles from MOSFETs. In its 2015 publication "Beyond CMOS," the International Technology Roadmap for Semiconductors detailed the developing devices according to charge/non-charge entities, material composition, and design [4]. This includes different types of devices such as spin field-effect transistor (FET) [5, 6], graphene FET [7, 8], tunnel field-effect transistor (TFET) [9, 10], carbon nanotube FET [11, 12], nanowire FET [13, 14], and negative gate capacitance FET [15, 16]. Because their basic fabrication processes are comparable to those of MOSFETs, as well as their facility to attain sub-60-mV/dec subthreshold swing (SS) and lower I_{OFF} than MOSFETs, TFETs have attracted a specific interest in low-power applications among these devices. TFETs operate through an interband tunneling mechanism as opposed to MOSFETs, which use thermionic emission, leading to low SSs and off currents. Numerous TFET structures have been put forward, so far, in an effort to boost on-currents and enhance performance. Heterojunction TFET [17, 18], III-V TFET [19], triple material gate TFET [20, 21], and cylindrical TFET [22] are some of the broadly used structures. There are several low-power applications for TFETs, including memory and digital circuitry [23, 24]. Over the lateral TFET, the vertical TFET (V-TFET) offers several advantages, one of which being a steeper subthreshold slope as tunneling is directly regulated by the gate voltage.

The steep subthreshold slope makes it possible to operate at low power and low OFF current. The tunneling arises vertical to the Si-Oxide interface when the gate voltage strikes a particular threshold, enabling the gate voltage to directly regulate the tunneling. As a result, this device may have a smaller SS and a greater ON current than the lateral gated p-i-n construction [25, 26]. When biomolecules are present, the biosensor's sensitivity is determined in relation to a reference value. As dielectric-modulated DM biosensors, a variety of TFET shapes have been reported, and evaluations of their sensitivities depending on device parameters are available [27–29].

19.2 Types of Biosensors

Biosensors are basically categorized on the basis of their transduction methodology and detecting method. The biological substances used in the detection process include proteins, drugs, enzymes, biological tissues, antibodies, and bacteria [30]. The targeted biomolecules are examined by superimposing them on the detecting element, which causes some physiochemical processes to start. These reactions produce byproducts, which are then used as inputs by the transducer elements.

Assuming the transduction process, the various physiochemical responses induced by the sensing devices are divided into four [31] most important types and a few subclasses [31], as shown in Figure 19.1. The biosensors listed in Figure 19.1 facilitate the detection of a broad range of biomolecules. When target biomolecules react, the electrical characteristics of the sensor change, which is essentially an electrochemical biosensor works. The sensor uses the observed change as the corresponding measurement parameter, and it is further categorized into three types: conductometric, potentiometric, and amperometric. Several types of biomolecules found in the human body, such as DNA, biotin, uricase, protein, glucose, and hemoglobin, are detected by the electrochemical biosensor.

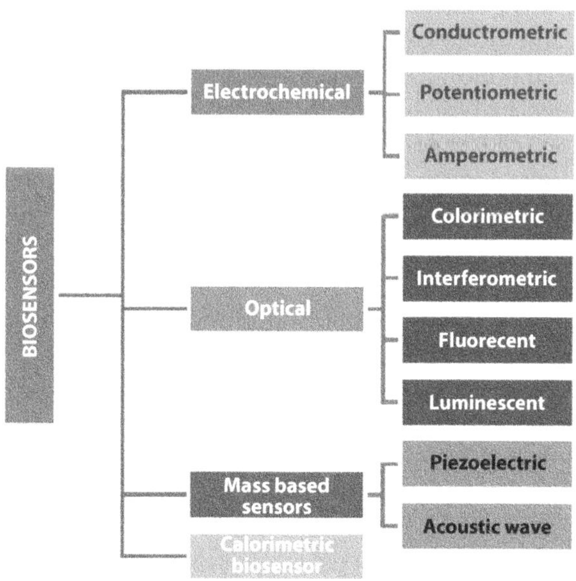

Figure 19.1 Categorization of biosensors [31].

Because optical biosensors require less sample preparation in order to detect target biomolecules, they are an extremely effective substitute for conventional analytical type biosensors. An optical biosensor senses biomolecules such as toxins, cancer cells, and tumor biomarkers by utilizing the interaction between the analyte and optical fields. The fundamental idea of a reaction to a change in mass is used by mass-based biosensors. These sensors are widely used in MEMS systems; in particular, the piezoelectric base sensor has attracted significant attention. The acoustic wave sensor and the piezoelectric belong to this group and have excellent applications in the identification of DNA, glucose, and living things. Basically, three key factors—such as accuracy, specificity, and ease of fabrication—should be taken into account while creating any reliable and precise biosensor. Mass-based and calorimetric biosensors have low response times and very complicated designs.

19.3 Comparison of FET- and TFET-Based Biosensors

Researchers from all over the world have recently become highly fascinated in FET-based biosensors because of their excellent characteristics, which include label-free detection, compact size, fast response, and reliability [28, 32, 33]. The oxide layer of the FET is used in conjunction with the bio-receptors and bio-recognition element to recognize certain biomolecules. Although FET-based biosensors offer several advantages, they also face significant challenges. While FET-based biosensors offer numerous benefits, they also face significant challenges such as I. the miniaturization process's SECs and scaling issues and II. the theoretical limit on the lowest probable SS (SS > 60 mv/dec) [33, 34]. All of these problems cause the device's sensitivity and performance to decline, and the thermionic electron emission in FETs causes a large power dissipation. Because of band-to-band tunneling of the carrier and sharp SS, researchers are turning their attention to the emerging technology of FET-based biosensors or TFET-based biosensors, which have least power and excellent features [32, 33, 35, 36]. Another key feature of biosensors is response time; a fast response requires a SS that is least desirable. Because TFETs may fulfil SS (SS < 60 mV/dec) lower than CFETs, there has been a lot of study recently focused on producing biosensors based on TFET technology.

19.4 Dielectric Modulation in TFETs: Principle and Design

19.4.1 Working Principle

The basic configuration of TFET contains three regions the source, drain, and channel. The main difference between TFETs and conventional field-effect transistors (CFETs) is the doping of the source and drain. In a CFET, the source and drain are doped with similar doping elements (P-type or N-type), whereas, in a TFET, the source and drain are doped differentially. In TFETs, the channel area is typically intrinsic or mildly doped. The arrangement has a gated p-i-n diode-like appearance. The TFET barrier width is thin to facilitate charge carrier tunneling (high doping of the source and drain allows for potential tunneling at barrier junctions; most charge carrier tunneling happens at the source-channel junction due to the source's higher doping than the drain). The operation of a transistor under different biasing situations is described by its current characteristic, and with three electrodes—the gate, drain, and source—the TFET-based biosensor is equipped with a bio-recognition element in the channel, which is the area between the source and drain. The targeted biomolecules are interacted with by this bio-recognition element, which also senses their existence and monitors electrical impact. Next, the biosensor immediately converts the biological data into a signal that can be monitored. The three main steps in the functioning of a TFET-based biosensor are as follows: 1) alteration of the charge concentration at the channel's surface; 2) the change of charge results in an alteration of the effective gate voltage; and 3) an improvement in the drain current due to the gating effect's reduction of the effective tunneling length.

Figure 19.2 illustrates the bending of the energy bands that allows to tunneling, despite considering that the energy levels of the source/channel are not coincided prior to biomolecule trapping. The triangular tunnel barrier at the source-channel junction is modeled, and the tunneling probability of it is computed using the wentzel–kramers–brillouin (WKB) approximation as [38]:

$$T_E = \exp\left[-\frac{4\lambda\sqrt{2m^*\left(E_g^*\right)^{\frac{3}{2}}}}{3q\hbar\left(E_g^* + \Delta\varphi\right)}\right] \tag{19.1}$$

where λ is the screening tunneling length, E_g is the source-channel tunnel junction energy band gap, m^* is the effective mass, q is the electronic charge, $\Delta\varphi$ is the energy overlap of the bands at the tunnel junction, and $\hbar$ is the Planck's constant.

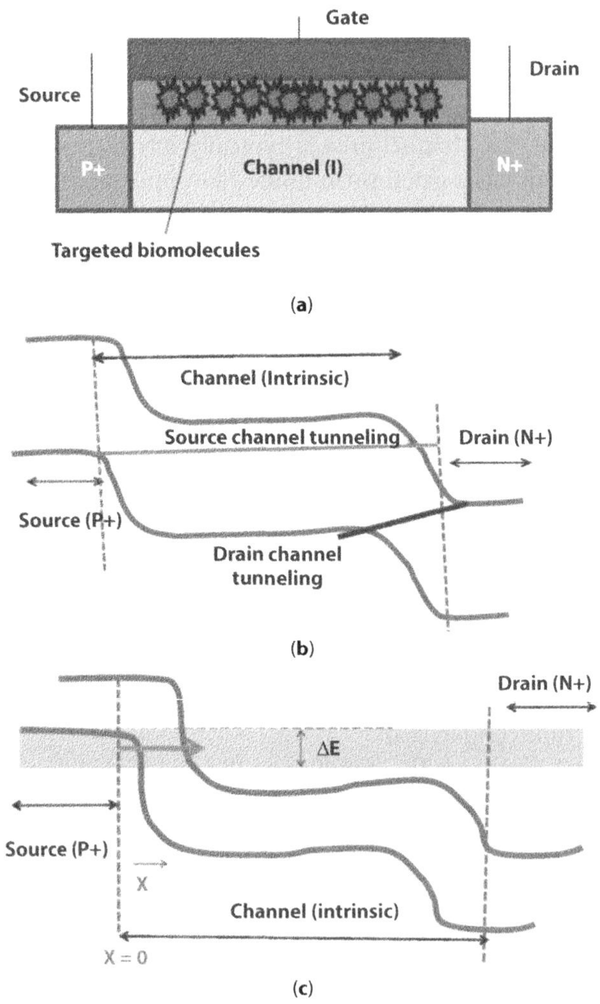

Figure 19.2 (a) 2D configuration of TFET (b) off state: energy band (c) on state: energy band distribution [37].

The definition of the screening length, λ, is

$$\lambda = \sqrt{\frac{\varepsilon_{si}}{\varepsilon_{ox}} t_{ox} t_{si}}$$

where ε_{ox} is the gate oxide's dielectric constant, ε_{si} is the semiconductor's dielectric constant, t_{ox} is the dielectric's thickness, and t_{si} is the device thickness. When there are lower dielectric constants at the tunnel junction than if there are greater dielectric constants at the tunnel junction, the tunnel width is more even at higher gate voltages due to the decreased gate-to-channel coupling [38, 39]. The energy bands at the channel-drain junction of a standard n-TFET are induced to form a tunneling barrier when the drain voltage is positive and the gate voltage is negative. This triggers the current to tunnel at this junction.

19.4.2 Design

As Section 19.4.1 clearly specified, in order to achieve the objectives of using dielectric modulation for biosensing, altered geometry must meet the following primary requirements.

- In the gate dielectric region, the biomolecules have to be kept immobile. The gate dielectric must include a nanogap in order to do this. With one gate dielectric having a fixed dielectric constant and the other carrying the dielectric constant of the biomolecules, the gate dielectric is thus formed of two areas; it appears to function similarly to a dual gate dielectric device.
- Biomolecules must be able to enter the cavity through the height of the implanted nanogap. For simulation investigations in nanoscale TFETs, a lowest possible height of 10–11 nm is typically taken into consideration.
- In the addition, a double-gate structure with mention design could be used for capturing additional biomolecules.
- The drain doping concentration is especially significant for TFETs that make use of the ambipolar behavior of the devices.

19.5 Literature Review

Because of the proposal of ion-sensitive FETs in the 1970s [40], Im *et al.* suggested a DM FET–based biosensor [41]. A nanowire TFET illustrating the gating effect for positively charged biomolecules was introduced by Sarkar and Banerjee [42]. However, Kim *et al.* confirmed partially filled nanogaps in real-world scenarios because of steric hindrance, and they suggested a metric called fill factor, which is the proportion of the nanogap that is occupied by biomolecules [43]. Similar simulation investigations for partially filled nanogaps in DM FET and PNPN TFET were presented by Narang *et al.* [7]. To further account for the effect of dielectric modulation in TFETs, Narang *et al.* provided an analytical model based on the Poisson equation [28]. Due to the effective dielectric constant's variation with position within the nanogap, partially filled nanogaps reduce the biosensor's response. As a biosensor, Ahangari reported on the use of a dual material gate nanowire junctionless TFET [44]. The sensitivity significantly improved as the nanogap length increased. However, K. Vanlalawmpuia [45] confirmed that, owing to the structural variations between the biosensors, the hetero-stacked source TFET biosensor has the maximum sensitivity when the step profile decreases, whereas the heterostructure V-TFET biosensor has the highest sensitivity when the step profile increases. The model by Devi *et al.* [46] takes into account the impacts of charge and dielectric constant of V-TFET, providing a generalized solution that works for both charged and neutral biological molecules.

19.6 Simulation Methodology for a DM TFET as a Label-Free Biosensor

By using physics-based models useful to the design, biosensor performance analysis is most conveniently performed on a computing platform. A broad range of industrial simulators exist, and the majority of them are fitted with features that allow users to define a geometry and feed it through successive iterations of specific models chosen from their libraries. In order to design the nanogap in a DM TFET, that area of the gate dielectric is replaced with an oxide dielectric material, whose dielectric constant can be changed to fit the requirements.

The oxide-semiconductor interface is where charges must be taken into consideration. biomolecules that carry charges. It is possible to accurately simulate the immobilization of biomolecules by adjusting the charge and dielectric constant. It is important to select the TFET models carefully. While considerable band bending at equilibrium is typically achieved by TFETs with high-source doping concentration, Fermi-Dirac statistics must be employed [47]. When dealing with highly doped geometries, bandgap narrowing is essential [47]. In order to develop an appropriate simulation environment for a device that works on the idea of interband tunneling, non-local band-to-band tunneling models are necessary [47]. Mobility models that depend on the field may be used [47].

19.7 Sensitivity Parameters

Any MOS-based device, including TFETs, can be evaluated by its electrical characteristics. The drain current and threshold voltage are the most commonly used constraints to determine sensitivity. Although the measurement of SS depends on the orders of the logarithmic scale along the drain current, SS may also be a parameter to figure out a sensitivity value. This value may not be constant because the drain current fluctuates in response to changes in the dielectric constant of the immobilized biomolecules. When the nanogap in a DM biosensor is empty of biomolecules, it is assumed to be filled with air and have a dielectric constant of 1, meaning there are no charges at the oxide-semiconductor interface. The mathematical expression for the drain current-based sensitivity is [43]

$$Sensitivity,\ S_1 = \frac{I_{D,K}}{I_{D,K=1}} \mid V_{GS} \tag{19.2}$$

The drain current rates while the nanogap is filled with biomolecules and when it is empty are denoted by $I_{D,K}$ and $I_{D,K=1}$. To obtain an accurate sensitivity value, the values required to be precise at the identical gate voltage. The threshold voltage is determined by the charge at the semiconductor-oxide interface and the gate dielectric constant. As biomolecules that have different dielectric constants become immobile, the shift in threshold voltage may be used as a sensitivity parameter. It can be stated numerically as

$$Sensitivity,\ S_T = V_{T,K=1} - V_{T,K} \tag{19.3}$$

where the threshold voltages $V_{T,K}$ and $V_{T,K=1}$ correspond to the situations in which the nanogap is filled with biomolecules and fully empty, correspondingly. This extraction principle states that the threshold voltage is the intercept on the gate voltage axis formed by the tangent to the drain current curve that corresponds to the highest value of $g_m = \frac{dI_D}{dV_{GS}}$ We have taken a specified range of gate voltage into consideration in all of the comparative scenarios, even though this extraction method may cause a change in threshold voltage with changes in gate voltage variation.

19.8 Non-Idealities in Dielectric-Modulated Biosensors

When assessing a TFET's biosensing capacity through simulation using a TCAD tool, it is fitting to imagine that the nanocavity is fully filled with biomolecules. In actuality, though, the embedded nanogap cannot be fully filled due to steric hindrance and probe placement concerns [29]. Therefore, partially filled nanogaps form.

19.8.1 Steric Hindrance

The biomolecules that become immobilized first in a steric hindrance situation prevent other biomolecules from entering. In certainty, there is a barrier that prevents the biomolecules from immobilizing in the nanogaps, which would cause partial hybridization. Various patterns of immobilization within the nanogap are considered, such as increasing, decreasing, convex, and concave placement profiles, in order to adjust for simulation [29].

19.8.2 Probe Placement

In the nanogap, probes or receptors are utilized to immobilize the biomolecules. The frequent insertion of probes into the nanogap for biomolecule immobilization could lead to partially filled nanogaps [29].

19.8.3 Fabrication Issues

A native oxide is first created in a DM biosensor, and it is then etched off to create the nanogap [28]. However, this approach is challenging, and there is the possibility that this process could lead to damage. In a TFET, the alliance of the gate with the source-channel tunnel junction and the tunneling phenomenon are extremely contingent on each other.

It is possible for the gate edge to come out of the junction during production. Depending on whether the gate moves in the direction of the source or the channel, there could be an overlap or an underlap as a result. When there is a gate-source overlap, the TFET's properties usually get better because the shifted gate has more control over the energy bands in the source-channel junction. The tunnel junction is least impacted by an underlap, though, as the gate edge shifts away from the junction in that scenario.

19.9 Impact of Charged Biomolecules on Sensitivity

First, we consider that biomolecules occupy the full embedded nanogap of the TFET and analyze the impact of various parameters on its sensitivity. The ratio of the area enclosed by immobilized biomolecules to the entire region of the nanogap is referred to as the fill factor of a nanogap, and it is typically stated as a percentage. The fill factor is 100% for a nanogap that is completely filled. For a metal-oxide-semiconductor configuration, the voltage balance equation is as follows:

$$V_G = \Psi_s + \phi_{MS} - \frac{QN_{bio}}{C'_{OX}} \tag{19.4}$$

where Q represents the electronic charge value, N_{bio} is the number of charges per unit area, C'_{OX} is the consequent capacitance per unit area, Ψ_s is the surface electrostatic potential, φ_{MS} is the contact potential, and V_G is the gate voltage.

The consequent capacitance derived from

$$C'_{OX} = \frac{K}{t_{OX}} \tag{19.5}$$

where t_{OX} denotes the dielectric thickness and K is the dielectric constant. With a fixed gate voltage V_G in Equation 19.4, an increase in the magnitude of a negatively charged biomolecule implies a decrease in Ψ_s. Therefore, as the amount of negative charges increases, the drain current drops. When V_G is fixed and N_{bio}'s value increases, Ψ_s must also be raised, which improves drain current and, thus, sensitivity [48].

19.10 Device Architecture and Simulation of Model

The 2-D sectional view of the suggested single-metal double-gate V-TFET (SMDG–V-TFET) biosensor based on Si is displayed in Figure 19.3.

The Silvaco TCAD technology is utilized to verify the authenticity of the proposed structure. Table 19.1 shows the relevant design parameters chosen for the SMDG–V-TFET simulation. Single metal with suitable work function of 4.3 is positioned along the two sides of the channel. To evaluate

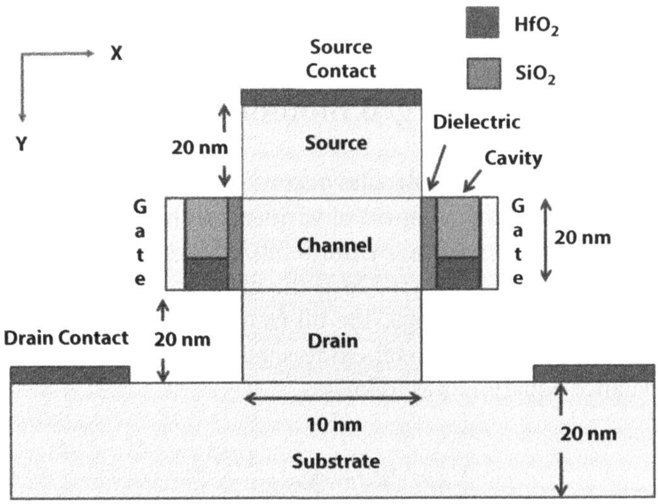

Figure 19.3 Schematics representation of SMDG–V-TFET biosensor.

Table 19.1 Design constraints of SMDG-V-TFET biosensor.

Parameters	Values
SiO_2 thickness (t_{SiO2})	0.62 nm
HfO_2 thickness (t_{HfO2})	3.38 nm
Channel doping (N_{CH})	5×10^{16} cm^{-3}
Source doping (p-type) (N_S)	5×10^{19} cm^{-3}
Drain doping (n-type) (N_D)	1×10^{18} cm^{-3}
HfO_2 dielectric constant (K_{Hfo2})	25
SiO_2 dielectric constant (K_{SiO2})	3.7

the performance of the proposed biosensor, different biomolecule dielectric constants (K = 1, 8, 12) [25] have been considered and examined with respect to drain characteristics, electric field and energy band diagram. The proposed biosensor's transfer curve moves to the left as K values increase. Similarly, the biosensor's threshold voltage drops and its on-state current increases. When K values increase, SS decreases proportionately. It is also observed that when K values change, OFF current nearly stays constant while ON current varies marginally displayed in Figure 19.4(a). Improving the doping profile will be accomplished by increasing the K immobilized biomolecule level. As a result, the steep doping profile at the tunneling junction has improved the electric field. Figure 19.4(b) displays the plot of various electric field variations with various biomolecules. The band on the channel side began to bend downward as the gate voltage increased, and this caused electrons to begin tunneling. Thus, electrons are able to tunnel from source to drain. In Figure 19.4(c), the energy band distribution is displayed.

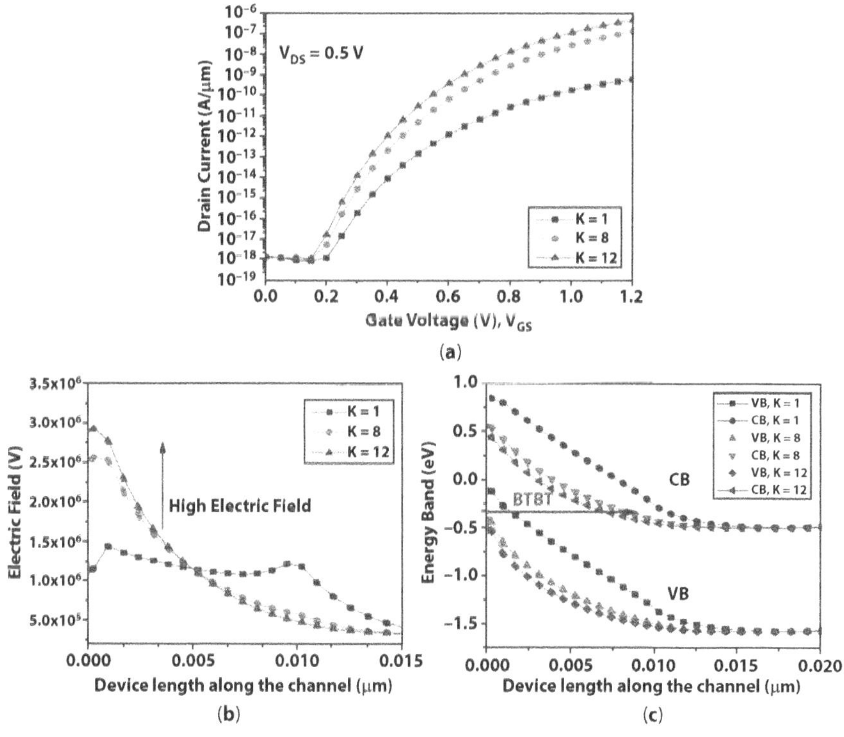

Figure 19.4 Comparison of (a) I_d-V_{gs} characteristics. (b) Electric Field distribution. (c) Energy band diagram for various values of the dielectric constant (K) of SMDG-V-TFET biosensor at V_{GS} = 1.2V and V_{DS}= 0.5V.

19.11 Conclusion

An overview of DM biosensors based on TFETs has been offered in this chapter. In addition to their scalability, transistors have emerged as one of the most important components for low-power appliances. This chapter covers the development and design of DM V-TFETs as biosensors, with a focus on simulation analysis. To design and construct a TFET-based DM biosensor, a number of challenges, including fabrication problems, probe positioning uncertainty, and steric hindrance, have been highlighted. To authenticate the results using the TCAD tool, appropriate physical-based models are required.

References

1. Sarangi, S., Bhushan, S., Santra, A., Dubey, S., Jit, S., Tiwari, P.K., A rigorous simulation based study of gate misalignment effects in gate engineered double-gate (DG) MOSFETs. *Superlattices Microstruct.*, 60, 263–279, 2013. https://doi.org/10.1016/J.SPMI.2013.05.009.
2. Choi, W.Y., Park, B.G., Lee, J.D., Liu, T.J.K., Tunneling field-effect transistors (TFETs) with subthreshold swing (SS) less than 60 mV/dec. *IEEE Electron Device Lett.*, 28, 743–745, 2007. https://doi.org/10.1109/LED.2007.901273.
3. Le Royer, C. and Mayer, F., Exhaustive experimental study of tunnel field effect transistors (TFETs): From materials to architecture, in: *Proc. 10th Int. Conf. Ultim. Integr. Silicon ULIS 2009*, pp. 53–56, 2009, https://doi.org/10.1109/ULIS.2009.4897537.
4. Gargini, P., The international technology roadmap for semiconductors (itrs): past, present and future. GaAs IC Symposium. *IEEE Gallium Arsenide Integrated Circuits Symposium.* 22nd Annual Technical Digest 2000. (Cat. No. 00CH37084). IEEE, 2000.
5. Shen, M., Saikin, S., Cheng, M.C., Spin injection in spin FETs using a step-doping profile. *IEEE Trans. Nanotechnol.*, 4, 40–44, 2005. https://doi.org/10.1109/TNANO.2004.840150.
6. Datta, S. and Das, B., Electronic analog of the electro-optic modulator. *Appl. Phys. Lett.*, 56, 665–667, 1990. https://doi.org/10.1063/1.102730.
7. Fiori, G. and Iannaccone, G., Simulation of graphene nanoribbon field-effect transistors. *IEEE Electron Device Lett.*, 28, 760–762, 2007. https://doi.org/10.1109/LED.2007.901680.
8. Echtermeyer, T.J., Lemme, M.C., Baus, M., Szafranek, B.N., Geim, A.K., Kurz, H., Nonvolatile switching in graphene field-effect devices. *IEEE Electron Device Lett.*, 29, 952–954, 2008. https://doi.org/10.1109/LED.2008.2001179.

9. Nirschl, T., Wang, P.F., Hansch, W., Schmitt-Landsiedel, D., The tunneling field effect transistor (TFET): The temperature dependence, the simulation model, and its application, in: *2004 IEEE International Symposium on Circuits and Systems (ISCAS)*, 3, III–713, IEEE, 2004, May.
10. Jiang, C., Liang, R., Xu, J., Investigation of Negative Capacitance Gate-all-Around Tunnel FETs Combining Numerical Simulation and Analytical Modeling. *IEEE Trans. Nanotechnol.*, 16, 58–67, 2017. https://doi.org/10.1109/TNANO.2016.2627808.
11. Wong, H.S.P., Beyond the conventional transistor. *IBM J. Res. Dev.*, 46, 133–168, 2002. https://doi.org/10.1147/RD.462.0133.
12. Guo, J., Datta, S., Lundstrom, M., A numerical study of scaling issues for Schottky-barrier carbon nanotube transistors. *IEEE Trans. Electron Devices*, 51, 172–177, 2004. https://doi.org/10.1109/TED.2003.821883.
13. Yang, F.L., Lee, D.H., Chen, H.Y., Chang, C.Y., Liu, S.D., Huang, C.C., Hu, C., 5nm-gate nanowire FinFET, in: *Digest of Technical Papers. 2004 Symposium on VLSI Technology, 2004*, pp. 196–197, IEEE, 2004, June.
14. Yoon, J.S., Kim, K., Rim, T., Baek, C.K., Performance and variations induced by single interface trap of nanowire FETs at 7-nm node. *IEEE Trans. Electron Devices*, 64, 339–345, 2017. https://doi.org/10.1109/TED.2016.2633970.
15. Salahuddin, S., Review of negative capacitance transistors, in: *2016 Int. Symp. VLSI Technol. Syst. Appl. VLSI-TSA 2016*, 2016, https://doi.org/10.1109/VLSI-TSA.2016.7480491.
16. Jo, J. and Shin, C., Impact of temperature on negative capacitance field-effect transistor. *Electron. Lett.*, 51, 106–108, 2015. https://doi.org/10.1049/EL.2014.3515.
17. Toh, E.H., Wang, G.H., Chan, L., Sylvester, D., Heng, C.H., Samudra, G.S., Yeo, Y.C., Device design and scalability of a double-gate tunneling field-effect transistor with silicon - germanium source. *Jpn. J. Appl. Phys.*, 47, 2593–2597, 2008. https://doi.org/10.1143/JJAP.47.2593/XML.
18. Ahish, S., Sharma, D., Kumar, Y.B.N., Vasantha, M.H., Performance enhancement of Novel InAs/Si hetero double-gate tunnel fet using Gaussian doping. *IEEE Trans. Electron Devices*, 63, 288–295, 2016. https://doi.org/10.1109/TED.2015.2503141.
19. Dewey, G., *et al.*, Fabrication, characterization, and physics of III–V heterojunction tunneling field effect transistors (H-TFET) for steep sub-threshold swing, in: *2011 International Electron Devices Meeting*, pp. 33–6, IEEE, 2011, December.
20. Bagga, N. and Dasgupta, S., Surface Potential and Drain Current Analytical Model of Gate All Around Triple Metal TFET. *IEEE Trans. Electron Devices*, 64, 606–613, 2017. https://doi.org/10.1109/TED.2016.2642165.
21. Bagga, N. and Sarkar, S.K., An Analytical Model for Tunnel Barrier Modulation in Triple Metal Double Gate TFET. *IEEE Trans. Electron Devices*, 62, 2136–2142, 2015. https://doi.org/10.1109/TED.2015.2434276.

22. Dash, S. and Mishra, G.P., A new analytical threshold voltage model of cylindrical gate tunnel FET (CG-TFET). *Superlattices Microstruct.*, 86, 211–220, 2015. https://doi.org/10.1016/J.SPMI.2015.07.049.
23. Luong, G.V., Strangio, S., Tiedemannn, A., Lenk, S., Trellenkamp, S., Bourdelle, K.K., Zhao, Q.T., Mantl, S., Experimental demonstration of strained Si nanowire GAA n-TFETs and inverter operation with complementary TFET logic at low supply voltages. *Solid State Electron.*, 115, 152–159, 2016. https://doi.org/10.1016/J.SSE.2015.08.020.
24. Mookerjea, S., *et al.*, Experimental demonstration of 100nm channel length In 0.53 Ga 0.47 As-based vertical inter-band tunnel field effect transistors (TFETs) for ultra low-power logic and SRAM applications, in: *2009 IEEE International Electron Devices Meeting (IEDM)*, pp. 1–3, IEEE, 2009, December.
25. Das, S., Kumar, B.B., Bundela, P., Singh, K., Performance assessment of Si based dual metal double gate vertical TFET biosensor. *Micro Nanostruct.*, 191, 207864, 2024. https://doi.org/10.1016/J.MICRNA.2024.207864.
26. Han, Z.F., Ru, G.P., Ruan, G., A simulation study of vertical tunnel field effect transistors, in: *Proc. Int. Conf. ASIC*, pp. 665–668, 2011, https://doi.org/10.1109/ASICON.2011.6157293.
27. Narang, R., Saxena, M., Gupta, R.S., Gupta, M., Dielectric modulated tunnel field-effect transistor-a biomolecule sensor. *IEEE Electron Device Lett.*, 33, 266–268, 2012. https://doi.org/10.1109/LED.2011.2174024.
28. Narang, R., Reddy, K.V.S., Saxena, M., Gupta, R.S., Gupta, M., A dielectric-modulated tunnel-FET-based biosensor for label-free detection: Analytical modeling study and sensitivity analysis. *IEEE Trans. Electron Devices*, 59, 2809–2817, 2012. https://doi.org/10.1109/TED.2012.2208115.
29. Narang, R., Saxena, M., Gupta, M., Comparative Analysis of Dielectric-Modulated FET and TFET-Based Biosensor. *IEEE Trans. Nanotechnol.*, 14, 427–435, 2015. https://doi.org/10.1109/TNANO.2015.2396899.
30. Mohanty, S.P. and Koucianos, E., Biosensors: A tutorial review. *IEEE Potentials*, 25, 35–40, 2006. https://doi.org/10.1109/MP.2006.1649009.
31. Reddy, N.N. and Panda, D.K., A Comprehensive Review on Tunnel Field-Effect Transistor (TFET) Based Biosensors: Recent Advances and Future Prospects on Device Structure and Sensitivity. *Silicon*, 13, 3085–3100, 2021. https://doi.org/10.1007/S12633-020-00657-1/METRICS.
32. Sarkar, D. and Banerjee, K., Fundamental limitations of conventional-FET biosensors: Quantum-mechanical-tunneling to the rescue, in: *Device Res. Conf. - Conf. Dig. DRC*, pp. 83–84, 2012, https://doi.org/10.1109/DRC.2012.6256950.
33. Kanungo, S., Chattopadhyay, S., Gupta, P.S., Sinha, K., Rahaman, H., Study and Analysis of the Effects of SiGe Source and Pocket-Doped Channel on Sensing Performance of Dielectrically Modulated Tunnel FET-Based Biosensors. *IEEE Trans. Electron Devices*, 63, 2589–2596, 2016. https://doi.org/10.1109/TED.2016.2556081.

34. Nair, P.R. and Alam, M.A., Screening-limited response of NanoBiosensors. *Nano Lett.*, 8, 1281–1285, 2008. https://doi.org/10.1021/NL072593I/ASSET/IMAGES/MEDIUM/NL-2007-02593I_0003.GIF.
35. Cao, W., Sarkar, D., Khatami, Y., Kang, J., Banerjee, K., Subthreshold-swing physics of tunnel field-effect transistors. *AIP Adv.*, 4, 67141, 2014. https://doi.org/10.1063/1.4881979/20818.
36. Khatami, Y. and Banerjee, K., Steep subthreshold slope n- and p-type Tunnel-FET devices for low-power and energy-efficient digital circuits. *IEEE Trans. Electron Devices*, 56, 2752–2761, 2009. https://doi.org/10.1109/TED.2009.2030831.
37. Kumar, M.J., Vishnoi, R., Pandey, P., Tunnel field-effect transistors (TFET): modelling and simulation, John Wiley & Sons, 2016.
38. Knoch, J. and Appenzeller, J., A novel concept for field-effect transistors - The tunneling carbon nanotube FET. *Device Res. Conf. - Conf. Dig. DRC*, 2005, 153–156, 2005. https://doi.org/10.1109/DRC.2005.1553099.
39. Knoch, J., Mantl, S., Appenzeller, J., Impact of the dimensionality on the performance of tunneling FETs: Bulk versus one-dimensional devices. *Solid State Electron.*, 51, 572–578, 2007. https://doi.org/10.1016/J.SSE.2007.02.001.
40. Bergveld, P., Thirty years of ISFETOLOGY: What happened in the past 30 years and what may happen in the next 30 years. *Sens. Actuators B Chem.*, 88, 1–20, 2003. https://doi.org/10.1016/S0925-4005(02)00301-5.
41. Im, H., Huang, X.J., Gu, B., Choi, Y.K., A dielectric-modulated field-effect transistor for biosensing. *Nat. Nanotechnol.*, 27, 2, 430–434, 2007. https://doi.org/10.1038/nnano.2007.180.
42. Sarkar, D. and Banerjee, K., Proposal for tunnel-field-effect-transistor as ultra-sensitive and label-free biosensors. *Appl. Phys. Lett.*, 100, 14, 2012.
43. Kim, C.H., Ahn, J.H., Lee, K.B., Jung, C., Park, H.G., Choi, Y.K., A new sensing metric to reduce data fluctuations in a nanogap-embedded field-effect transistor biosensor. *IEEE Trans. Electron Devices*, 59, 2825–2831, 2012. https://doi.org/10.1109/TED.2012.2209650.
44. Ahangari, Z., Performance assessment of dual material gate dielectric modulated nanowire junctionless MOSFET for ultrasensitive detection of biomolecules. *RSC Adv.*, 6, 89185–89191, 2016. https://doi.org/10.1039/C6RA17361F.
45. Vanlalawmpuia, K. and Bhowmick, B., Analysis of Hetero-Stacked Source TFET and Heterostructure Vertical TFET as Dielectrically Modulated Label-Free Biosensors. *IEEE Sens. J.*, 22, 939–947, 2022. https://doi.org/10.1109/JSEN.2021.3128473.
46. Wangkheirakpam, V.D., Bhowmick, B., Pukhrambam, P.D., N+ Pocket Doped Vertical TFET Based Dielectric-Modulated Biosensor Considering Non-Ideal Hybridization Issue: A Simulation Study. *IEEE Trans. Nanotechnol.*, 19, 156–162, 2020. https://doi.org/10.1109/TNANO.2020.2969206.

47. *Atlas User's Manual Silvaco Int. Softw.*, Silvaco Inc., Santa Clara, CA, USA, 2016.
48. Devi, W.V., Bhowmick, B., Pukhrambam, P.D., N+ pocket-doped vertical TFET for enhanced sensitivity in biosensing applications: Modeling and simulation. *IEEE Trans. Electron Devices*, 67, 2133–2139, 2020. https://doi.org/10.1109/TED.2020.2981303.

20

Micro-Electromechanical System (MEMS) and Field-Effect Transistor (FET)–Coupled Sensors

Shradha Saxena* **and Archana Pandey**

Electronics & Communication Engineering, Jaypee Institute of Information Technology, Noida, India

Abstract

Over the past decades, advancements in sensing and detection technologies have significantly impacted various sectors, including the food industry, medical field, pharmaceuticals, defense, and environmental monitoring. Recent progress in micro-electromechanical systems (MEMS) has enabled the miniaturization of sensor technology, allowing the integration of transducing devices and signal processing systems onto a single platform.

MEMS-based sensors are celebrated for their small size, biocompatibility, electrical nature, ease of fabrication, low cost, low power consumption, ultra-high sensitivity, fast detection speed, high accuracy, and real-time analysis. These advantages make MEMS sensors suitable for applications such as bioimplants, including retinal and neural implants. Common transducing techniques used in MEMS sensors include piezoelectric, piezoresistive, electrostatic, electromagnetic, electrothermal, and optical methods. However, these techniques often suffer from low output responses. To address this issue, the integration of MEMS sensors with field-effect transistors (FETs) has been proposed, enhancing the performance of MEMS-FET sensors. This integration allows changes in the output parameters of the FET (such as mobility, threshold voltage, on/off ratio, and source-drain currents) in response to MEMS inputs, enabling precise detection of various substances.

This chapter provides a detailed review of MEMS-FET sensors, starting with an in-depth discussion of the advantages of MEMS sensors over traditional sensors. It then explores the operating principles, advantages, disadvantages, and limitations

**Corresponding author*: shradha.saxena@jiit.ac.in

Ekta Goel, Archana Pandey, Shiromani Balmukund Rahi and Arun Samuel (eds.) Handbook of Advanced Semiconductor Field Effect Transistors, (419–438) © 2026 Scrivener Publishing LLC

of various transduction technologies. The chapter continues with a description of different configurations, classifications, and applications of MEMS sensors.

The focus will be on the working principle of MEMS-FET sensors, detailing the methodology of coupling FET with MEMS sensors. Advances in FETs and MEMS structures and their compatibility will be discussed. Special attention will be given to various types of MEMS-FET sensors, including force, accelerometer, pressure, and gas sensors, and their applications in fields, such as aerospace, navigation, unmanned aerial vehicles, explosive impact detection, and biomedical applications.

Finally, the chapter will highlight future perspectives in the field of MEMS-FET sensors, providing a comprehensive overview from the fundamentals of sensor technology to advanced MEMS-FET sensor development.

Keywords: MEMS, FET, biosensor

20.1 Introduction to Micro-Electromechanical System (MEMS)–Based Sensor

Over the past few decades, micro-electromechanical system (MEMS)–based sensors have gained popularity compared to existing sensors because of their feasibility for micro-scale miniaturization, batch fabrication, low power consumption, low cost, and high sensitivity. The use of MEMS-based sensors in various applications, from biomedical implants to the food industry, defense, structural health monitoring, environmental monitoring, and other sensing and detection sectors, has been well-documented in the literature. Furthermore, advancements aimed at improving the output response of MEMS-based sensors have led to their integration with well-established field-effect transistor (FET) technology. This chapter primarily focuses on recent developments in FET-based MEMS sensor technology.

20.1.1 Definition and Importance of MEMS

MEMS is a technology that integrates mechanical, electrical, and semiconductor disciplines on a single platform. In MEMS, micro-scaled mechanical systems, such as sensors, actuators, and other electronic components, are fabricated on a silicon wafer. The development of highly sensitive, small-scale sensors and actuators has become possible due to advancements in MEMS fabrication technology and the superior electrical and mechanical properties of silicon. MEMS sensors can sense and monitor various physical properties (such as pressure, acceleration, and displacement) and environmental properties (such as temperature, humidity, and gases).

They are primarily composed of three components: a sensing element to detect physical or environmental quantities, a transducing device to convert the detected parameter into its equivalent electrical signal, and an interpreting device, such as a computer, to analyze the electrical signal.

20.1.2 Types of MEMS Sensors

Depending on the sensing element and application, the broad range of MEMS has been divided into four categories: bio-MEMS, fluidic MEMS, radio-frequency (RF) MEMS, and optical MEMS [1]. In bio-MEMS, biosensors interact with biological proteins, cells, and other medical reagents, finding applications in drug delivery and medical diagnosis. In fluidic MEMS, sensors measure flow, pressure, acceleration, etc., and are mainly used in biological, chemical, and pharmaceutical applications. In RF MEMS, sensors are involved in detecting or transmitting radio frequencies and are utilized in tunable capacitors, antennas, and other high-frequency applications. Additionally, sensors involved in optimizing the functioning of optical switches and reflectors fall under the category of optical MEMS.

20.1.3 Sensing Mechanisms of MEMS Sensors

The output of MEMS sensors is primarily detected using piezoresistive, capacitive, and optical mechanisms [1]. Each technique has its own advantages and disadvantages.

20.1.3.1 Piezoresistive Sensing Mechanism

In this mechanism, the resistance of the piezoresistive material changes in response to variations in surface stress. While this technique is capable of detecting small changes in stress, the resistivity of the piezoresistive material is highly sensitive to elevated temperatures, which limits its applications. The piezoresistive sensing technique has been primarily used in biomarker detection, chemical detection, humidity sensing, and pressure sensing applications, among others.

20.1.3.2 Capacitive Sensing Mechanism

In this mechanism, a change in capacitance occurs because of a variation in the distance between one fixed and one movable plate, resulting in a corresponding change in output voltage. This mechanism offers advantages such as low temperature dependency, high sensitivity, and minimal power consumption. Due to these benefits, the capacitive sensing mechanism has been widely explored in various applications, including biomedical sensing, navigation, and RF MEMS.

20.1.3.3 Optical Sensing Mechanism

This technique operates by detecting differences between transmitted and received light. It is commonly used in communication systems and air pollutant detection. Compared to the other two mechanisms, optical sensing is relatively expensive. In addition to the aforementioned techniques, piezoelectric and thermal sensing methods have also been employed in various sensing applications.

20.1.3.4 Piezoelectric and Thermal Sensing Mechanisms

In piezoelectric sensing, an electrical charge is generated in response to stress or strain applied to a piezoelectric material. When pressure or force is applied, the mechanical stress or strain on the material produces an electrical output. In the thermal sensing technique, temperature changes are detected on the basis of variations in the thermal properties of the materials.

20.2 Introduction to Field-Effect Transistor (FET)–Based Sensors

In semiconductor technology, the FET is one of the fundamental building blocks of modern electronic devices. FET devices offer advantages such as gate-controlled channel current, low power consumption, low cost, and easy fabrication methods. In recent years, FET-based sensors have garnered significant attention from researchers because of their unique features, including label-free detection, high sensitivity, simple design, and rapid response, in addition to the inherent benefits of FET technology. Compared to existing sensing technologies, FET-based sensors excel at detecting low-concentration targets. They detect target adsorbates by monitoring changes in the conductance of the channel material.

20.2.1 Different Configurations and Working Principle of FET Sensors

The primary components of FET sensors include the source, drain, gate electrodes, gate oxide material, and the channel formed within the semiconductor

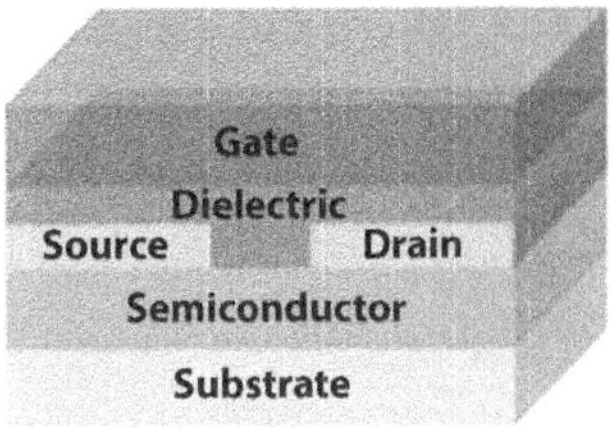

Figure 20.1 Basic structure of FET device [2] (Source: https://doi.org/10.3390/bios13040426).

layer between the source and drain as shown in Figure 20.1. According to the working principle of FETs, the device activates once the gate potential reaches the threshold voltage, which forces charge carriers to move, causing a drain current to flow along the channel between the source and drain electrodes. To harness the advantages of FET devices for sensing applications, researchers have developed electrolyte-gated (EG)–FET devices, where the gate oxide–insulating material is replaced by various electrolytes, such as polymers and ionic liquids. Similar to traditional FETs, the drain current in EG-FETs is controlled by the gate electrode via the electrolyte solution. The key feature of EG-FETs is their low operating voltage, typically less than 1V, and their high gate capacitance. In literature, various configurations of EG-FETs have been explored for sensing and detecting applications. To grasp the operation of EG-FETs sensors, it is essential to understand the fundamentals, advantages, and limitations of each FET configuration. By analyzing the requirements of specific sensing applications, the most appropriate EG-FET configuration can be selected. This section offers a concise overview of different EG-FETs sensor configurations. Depending on the relative positioning of the gate electrode and the semiconductor layer, these configurations are discussed in the subsequent subsections.

20.2.1.1 Top Gate Configuration

In this configuration, the gate electrode is positioned on the top side of the semiconductor layer and functions as the sensing component, instead of the channel. This configuration is characterized by low contact resistance and high mobility. The probe-based top gate configuration of EG-FET is illustrated in Figure 20.2(a), where the gate electrode probe is manually connected over the channel. This configuration is primarily used for sensing biological molecules [2].

20.2.1.2 Side Gate Configuration

In this configuration, the gate electrode is placed in the same plane as the channel, as shown in Figure 20.2(b). Because the gate is located on one side of the channel, this setup allows for simultaneous biasing of multiple semiconductor channels, making it suitable for multifunctional bio-sensing applications. The side gate configuration offers advantages such as precise control of the gate electrode and ease of fabrication [3].

20.2.1.3 Extended Gate Configuration

In this configuration, the gate electrode is extended off-chip to separate the sensing area from the transistor, as shown in Figure 20.2(c). The extended gate acts as a sensing layer and is connected to the transistor gate via a

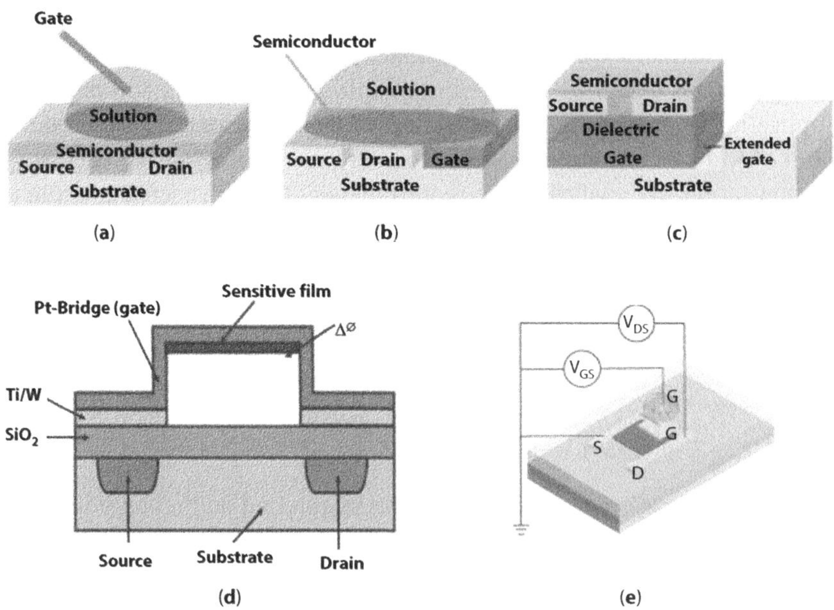

Figure 20.2 Schematics of different gate configurations of FET-based sensors [2, 3]: (a) top-gate, (b) side gate, (c) extended gate, (d) suspended gate, and (e) floating gate (Source [2]: https://doi.org/10.3390/bios13040426), Source [3]: https://doi.org/10.1186/s12951-023-02153-1).

metal line. This configuration is highly immune to environmental interference and offers advantages such as a simple design and easy fabrication [3].

20.2.1.4 Suspended Gate (SG) Configuration

This configuration is created by introducing an air gap between the gate oxide layer and the semiconductor channel layer [3]. The schematic of the suspended gate–FET (SG-FET) is shown in Figure 20.2(d). In this setup, the SG itself functions as the sensing layer. This configuration is used in various sensing applications, such as gas sensing and force sensing.

20.2.1.5 Floating Gate (FG) Configuration

In this configuration, an additional floating metal gate is placed in the solution between the main top controlling gate and the channel, as shown in Figure 20.2(e). The FG is electrically insulated from both the channel and the top gate [3]. In this setup, charge carriers are trapped around the FG for an extended period, allowing the channel conductivity to be controlled even in the absence of a power supply.

20.2.2 Sensing Mechanism in EG-FET Sensors

In FET-based sensors, three types of interactions between target adsorbates and channel material have been illustrated in the literature. One type is physical adsorption, which primarily occurs because of van der Waals forces between the target and the channel material.

Next is chemical adsorption, where direct bonding occurs between the target adsorbates and the channel material. Another possible interaction mechanism is a chemical reaction between the adsorbate and the channel. In sensing and detection applications, FET-based sensors demonstrate a wide range of capabilities, from gas sensing to the detection of heavy metal ions, proteins, and bacteria [4].

20.2.3 Limitation of FET-Based Sensors

FET-based sensors offer advantages such as high sensitivity, label-free detection, low power consumption, and compatibility with complementary metal-oxide-semiconductor (CMOS)-based fabrication processes. However, they also have limitations, including high packaging costs and a large chip area. Therefore, there is a need for technology that minimizes the chip area of FET sensors without compromising their overall cost and sensitivity.

20.3 Introduction of MEMS-FET Sensor

When comparing MEMS- and FET-based sensor technologies, it is evident that both have their own advantages and have contributed to revolutionary developments in sensing and detection, from the micro- to the nanoscale. Recently, MEMS-FET–based sensors have emerged, integrating MEMS devices with FET devices on a single platform. These sensors can operate in two ways: either MEMS acts as the sensing device and FET as the measuring device, or FET serves as the sensing device and MEMS as the supporting device. However, the first configuration, where MEMS acts as the sensing device and FET as the measuring device, has been found to be more efficient compared to the alternative.

20.3.1 Advantages of MEMS (Sensing)–FET (Measuring) Sensors

Studies have shown that conventional MEMS sensors, which use passive transduction methods (such as piezoresistive, capacitive, and piezoelectric), often suffer from a low signal-to-noise ratio (SNR). This issue can be addressed by incorporating a FET-based active transduction mechanism with MEMS sensors. The amplification properties of FET devices enhance the output SNR and facilitate the integration of MEMS sensors with CMOS-based readout circuits. Additionally, MEMS-FET sensors offer advantages such as low noise, reduced chip area, and low power consumption.

20.3.2 Importance of Suspended Gate (SG) Architecture of FET in MEMS-FET Sensors

Research indicates that the SG-FET configuration is particularly well suited for integration with MEMS sensor technology, resulting in highly sensitive MEMS-FET sensor devices. Literature reports various MEMS–SG-FET–based applications, including accelerometers, biosensors, and pressure sensors [4–7]. In these devices, mechanical inputs cause displacement of the SG, leading to a change in gate capacitance and, consequently, the drain current of the SG-FET. The measured drain current is directly proportional to the magnitude of the mechanical input.

20.4 Applications of MEMS-FET Sensors

In recent years, MEMS-FET–based sensors have emerged as a transformative technology, driving significant advancements in various sensing and detection applications. In the biomedical industry, it is now possible to measure biomarker parameters directly, eliminating the cumbersome and time-consuming process of traditional blood analysis. This chapter summarizes recent developments in diverse sensing applications utilizing MEMS-FET sensor technology.

20.4.1 MEMS-FET–Based Gas Sensors

Gas sensors are used across various industries, including chemical, biomedical, and aerospace, to detect gases and biomaterials. These industries typically require gas sensors that are compact, cost-effective, highly selective, sensitive, and responsive. While existing MEMS gas sensors meet many of these requirements, they still fall short in providing a high SNR.

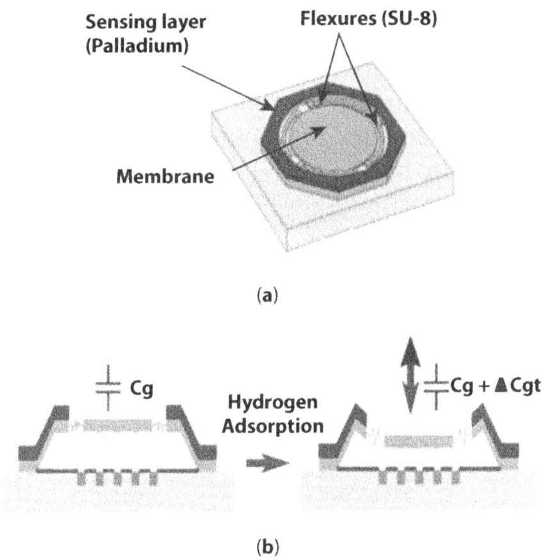

Figure 20.3 (a) Schematic and (b) working principle of ring-flexure-membrane (RFM) suspended gate–field-effect transistor (SG-FET)–based MEMS gas sensor [5] (https://doi.org/10.3390/mi14050944).

Consequently, research is ongoing into MEMS-FET–based gas sensors to address this limitation. In this context, J. Zacharias *et al.* [5] reported a ring-flexure-membrane (RFM) SG-FET–based MEMS gas sensor designed primarily for detecting hydrogen gas. The sensor features a suspended octagon-shaped palladium sensing material as the outer ring, which is connected to an SU-8–based inner circular membrane via four flexures, as shown in Figure 20.3(a). The RFM structure of the sensor supports the SG in a fingered FET configuration. Upon gas adsorption, a compressive force is applied to the outer ring, causing a uniform downward displacement of the inner ring and, consequently, the gate electrode, as illustrated in Figure 20.3(b). This downward displacement increases the gate capacitance, which, in turn, enhances the drain current response. The authors conducted a detailed modelling and simulation study of the RFM–SG-FET–based MEMS gas sensor and demonstrated that the proposed design offers an efficient transduction mechanism, resulting in a high SNR in the output response.

Similarly, B. Sharma *et al.* [6] fabricated a multilayer hydrogen gas sensor chip using MEMS technology, which incorporates dual FETs and an embedded micro heater on a single platform, as shown in Figure 20.4. One of the FETs serves as the sensing element, with its gate made of graphene-decorated palladium-silver (Pd-Ag) alloy (GPA) nanoparticles. The GPA-FET exhibits a high adsorption rate for hydrogen molecules (H_2). The other FET, which features a platinum (Pt) gate, serves as a reference and remains unaffected by changes related to H_2 particles. The embedded micro heater maintains a uniformly high working temperature around the

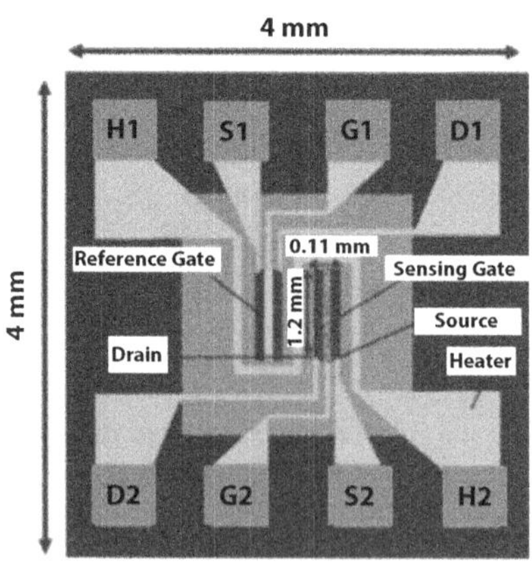

Figure 20.4 Top view of fabricated multilayer hydrogen gas sensor chip [4] (Source: https://doi.org/10.1038/s41598-018-24324-z).

sensing area while keeping the outer substrate temperature as low as possible. The authors reported achieving a minimum temperature of 102.4°C and a maximum temperature of 192.6°C with the embedded micro heater. This approach results in low power consumption, low thermal mass, and high temperature stability. The operating principle of the H_2 sensor involves a change in the work function of the gate material of the GPA-FET in the presence of H_2, whereas the work function of the reference FET remains unchanged. This difference in work function causes a corresponding shift in the gate threshold voltages, which, in turn, alters the drain current. The authors claimed that the GPA-FET achieves a high response to H_2 gas concentrations above 1 ppm within a temperature range of 25°C–254.5°C.

20.4.2 MEMS-FET–Based Accelerometer Sensors

MEMS-based accelerometer sensors are well-established technologies used to measure vibrations and are widely applied in defense, navigation, aerospace, industrial, and commercial sectors. Recent research aims to enhance the efficiency of MEMS-based accelerometers by leveraging the advantages of SG-FETs. In this context, P. Martha *et al.* [7] proposed a stepped architecture for the SG and reported a novel design for a z-axis accelerometer based on a MEMS and stepped SG (SSG)–FET array. The n-stepped segregated structure of the SG is shown in Figure 20.5(a). Figure 20.5(b) illustrates the four–flexible guided beam–based mechanical structure to which the stepped suspended (SS) gate FET is attached.

When input vibrations are applied, the SG displaces in the out-of-plane direction, resulting in a corresponding change in channel capacitance and, consequently, the drain current. The authors claim that their design achieves an enhanced stable driving range from 33.3% to 50% of the initial

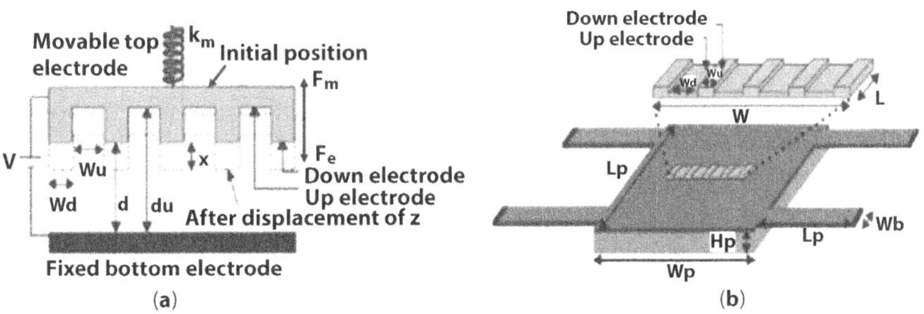

Figure 20.5 (a) Schematic of proposed stepped suspended gate and (b) stepped suspended gate FET–based MEMS accelerometer sensor [7] (Source: 10.1109/TED.2021.3101997).

air gap, with approximately a 90% increase in pull-in voltage. Additionally, the SSG-FET–based MEMS accelerometer exhibits a sensitivity of 38 mV/g at an input voltage of 3.3 V.

To further miniaturize MEMS–SG-FET to the micron level, P. Martha *et al.* [8] addressed the issues arising from short-channel effects (SCEs) in SG-FETs and introduced an efficient U-channel SG (USG) silicon-on-insulator (SOI) FET-based MEMS accelerometer. Literature indicates that reducing the channel length of SG-FETs below 10 μm diminishes the gate electrode's impact on the drain current, thereby affecting the performance of SG-FETs [9, 10]. Various techniques, such as U-channel FET [11], halo implantation [12], and lightly doped drain [13], have been suggested to mitigate SCE in sub-micron MOSFET technology. Among these, the U-channel FET architecture is noted for its efficiency, CMOS compatibility, and immunity to SCE effects.

In the proposed USG-SOIFET–based MEMS accelerometer, a U-shaped channel is formed on an SOI substrate, with the gate of the USG-SOIFET attached to the MEMS structure, as shown in Figure 20.6. The U-channel in the USG-SOIFET creates an air gap between the source and drain, reducing the impact of the drain electric field on the source voltage. It has been reported that the USG-SOIFET MEMS accelerometer achieves a channel length four times shorter than that of the SG-FET MEMS accelerometer for the same application. The sensitivity and bandwidth of the USG-SOIFET MEMS accelerometer are observed to be 4.185 μA/g and 100 Hz, respectively.

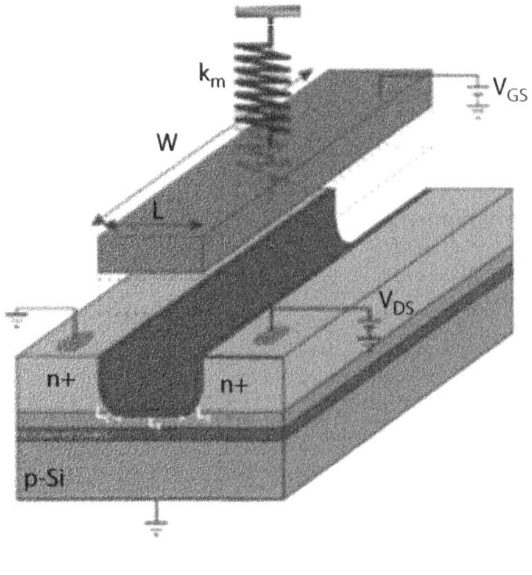

Figure 20.6 Schematic of U-channel suspended gate (USG) silicon-on-insulator (SOI) FET–based MEMS accelerometer [8] (Source: 10.1109/JSEN.2021.3060186).

20.4.3 MEMS-FET Force Sensor

MEMS force sensors are crucial in the biomedical industry, where they are used to study various biological processes and reactions. Micro force sensors are instrumental in measuring properties such as Young's modulus, locomotion traction, viscosity, and the survival rate of living cells. Recent research aims to analyze biomedical processes at the molecular level, highlighting the need for ultra-low MEMS force sensors capable of detecting forces below the nano-newton range. As a result, many researchers are focusing on movable gate (MG)–FET–based transduction methodologies rather than passive transduction mechanisms. The schematic of an MG-FET is shown in Figure 20.7. In an MG-FET, the MG can float either in the vertical (VMG-FET) or lateral (LMG-FET) direction over the channel area. In MEMS-MG-FET force sensors, the mechanical structure is attached to the MG of the FET. Movement of the gate electrode due to the sensed force changes the channel capacitance, and the resulting drain current reflects the sensed force. MEMS–LMG-FET–based force sensors offer high sensitivity to nano-newton input forces, a high SNR, and good compatibility with chip-level integration.

To further enhance the accuracy and capability of LMG-FETs, W. Gao *et al.* [14] proposed a novel decoupled sandwich structure for LMG-FETs and fabricated a MEMS–LMG-FET–based force sensor, as shown in Figure 20.8(a). The broad view of the sandwich structure is depicted in Figure 20.8(b), which includes a cross-axis decoupling gate array layer made of

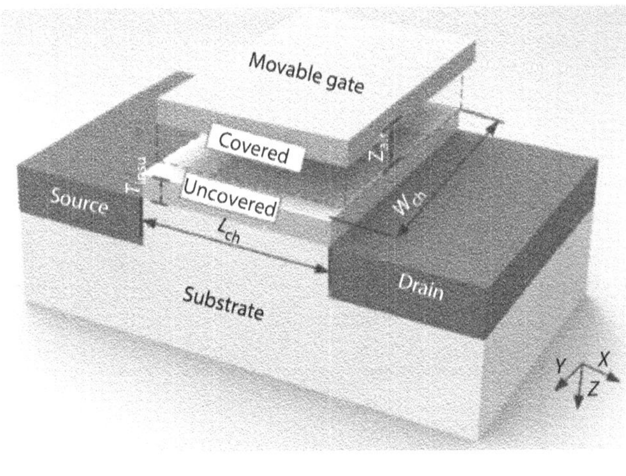

Figure 20.7 Architecture of movable gate (MG)–FET device [14] (Source: 10.1016/j.eng.2022.06.018).

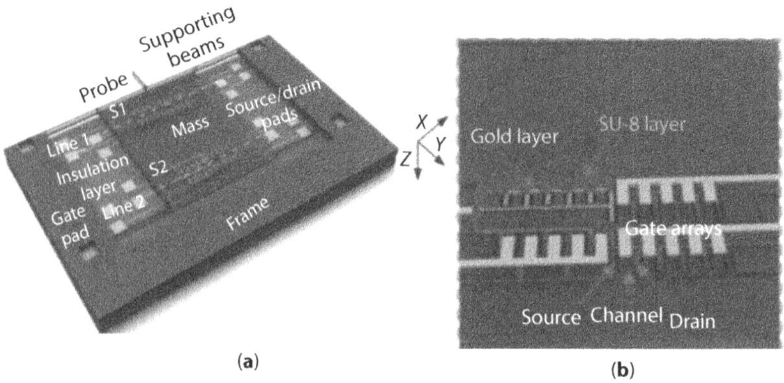

Figure 20.8 Proposed laterally movable gate (MG)–FET–based MEMS force sensor [14]: (a) chip level schematic and (b) broad view of decoupled sandwich structure of MG-FET (Source: 10.1016/j.eng.2022.06.018).

gold and two covered flexible layers of SU-8 photoresist material. The proposed sensor features a dual-differential decoupling sensing configuration, resulting in high sensitivity and improved anti-interference capability.

20.4.4 MEMS-FET–Based Microphone

In this technology, a flexible mechanical diaphragm structure serves as the moving gate of the SG-FET device. When deflected by input sound pressure, the diaphragm causes a change in the drain current. The overall

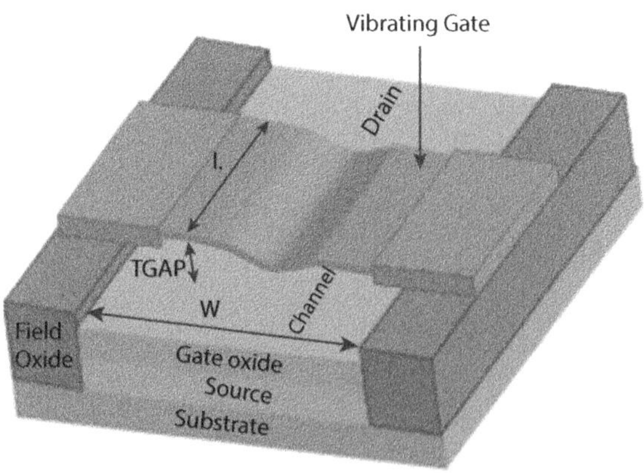

Figure 20.9 Three-dimensional schematic of MEMS-suspended gate FET–based microphone [15] (Source: 10.1016/j.proche.2009.07.218).

sensitivity of a MEMS–SG-FET–based microphone sensor can be increased by enhancing either the mechanical or electrical sensing components. Increasing mechanical sensitivity often introduces fabrication complexities. Therefore, improving sensitivity through electrical means is a simpler approach. In this context, M. C. Kaura *et al.* [15] utilized an SOI substrate-based SG-FET operating in the subthreshold region, where the drain current is exponentially proportional to the difference between the gate voltage and the threshold voltage. The structure of the proposed MEMS–SG-FET sensor is shown in Figure 20.9.

This design results in a high level of sensitivity, which can be further adjusted by tuning various device factors such as air gap, doping concentration, and silicon thickness.

20.5 Self-Powered MEMS-FET Sensors

In the current scenario, advancements in MEMS-FET sensor technology have led to significant progress across various industrial sectors. Sensors are now used in applications ranging from household settings to biomedical, defense, automotive, aerospace, and wearable technologies. In applications such as structural and health monitoring, continuous sensing is essential. However, modern sensor technology, which relies on batteries, has a limited lifespan. Therefore, there is a growing need for self-powered sensors that can recharge using ambient energy.

Recent research is focused on developing advanced self-powered MEMS-FET sensors for continuous monitoring applications. Ambient energy sources such as solar, thermal, wind, and RF can be used to harvest energy on a macro scale. At the micro level, electromagnetic, electrostatic, and piezoelectric transduction mechanisms are employed to harvest electrical energy from environmental vibrations. Recently, for biomedical applications, the triboelectric nanogenerator (TENG) transduction mechanism has gained attention [16]. TENG operates on the principle of triboelectric and electrostatic effects. When two different triboelectric materials come into contact and then separate or slide against each other, electrons transfer between the materials, creating electrostatic charges of opposite polarities on their surfaces. In the quest to make MEMS-FET sensor technology self-powered, TENG mechanisms show promise for integration with FET devices. TENG-based MEMS-FET sensors offer a platform for the continuous monitoring of temperature, humidity, and various physiological and physical parameters. Furthermore, to enhance the energy scavenging capability of these sensors, researchers are exploring hybrid energy harvesting techniques, which combine the advantages of multiple transduction mechanisms on a single platform.

20.5.1 Self-Powered MEMS-FET–Based Biosensor

R. Omar *et al.* [16] developed a self-powered, freestanding multifunctional microneedle (MMN) extended gate (EG) MEMS-FET sensor for wearable online health monitoring, using a hybrid energy harvesting technology combining TENG and solar cells. Microneedle (MN)–based MEMS sensors are effective in wearable biomedical devices for monitoring various biomarkers, such as potassium (K+), calcium (Ca2+), and sodium (Na+), in interstitial fluid [16, 17]. The MN is a tube-like structure created by inserting a poly-dimethylsiloxane (PDMS) mold into a polystyrene (PS) solution of dried dimethylformamide. The MN is then spray-coated with silver nanowires. Each MN in an array is designed with a specific sensing layer tailored to detect selective electrolytes like Na+, K+, Ca2+, and pH. The schematic of the MMN with different biomarkers and its scanning electron microscope (SEM) image are shown in Figures 20.10(a) and (b). In the MMN-EGFET biosensor, an array of MMNs is used as an extended gate for the FET device. The sensor's response is measured as a change in the FET's drain current. The self-powered MMN-EGFET wearable sensor includes components such as the MMN-EGFET biosensor, a hybrid energy harvesting module (combining TENG and solar cells), a rechargeable battery, and an Arduino board, all operating in a closed-loop system. The wearable MMN-EGFET biosensor is placed on the wrist of the wearer, with TENG (one triboelectric material) and solar cells attached to the inner arm, and the other triboelectric material positioned at the waist, as shown

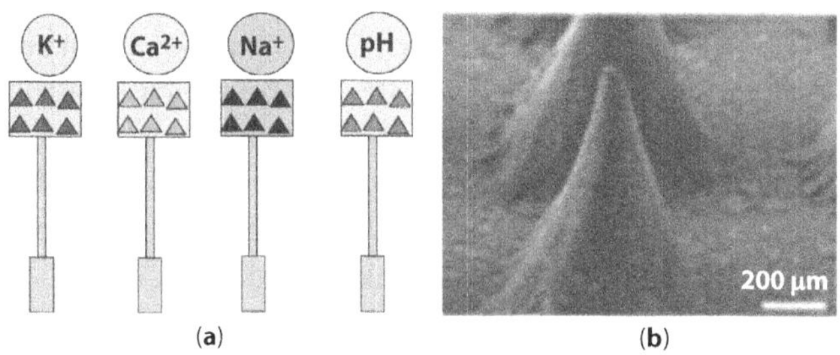

Figure 20.10 (a) Schematic of multifunctional microneedles (MMNs) with different biomarkers [16] and (b) scanning electron microscope (SEM) image of MNs (Source: 10.1016/j.snb.2023.134788).

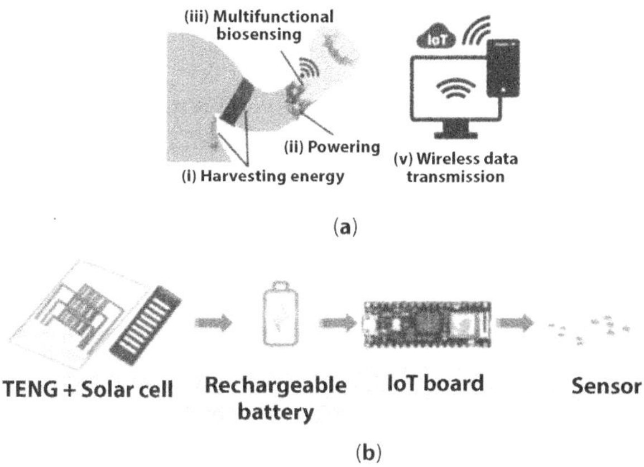

Figure 20.11 (a) Overview of online sensing and measuring analysis of self-powered multifunctional microneedles (MMN) and extended gate–FET–based wearable biosensor [16] (Source: 10.1016/j.snb.2023.134788).

in Figure 20.11(a). The combined body movement (TENG) and solar energy (solar cells) generate electricity sufficient to charge the rechargeable battery of the Arduino microcontroller, which is coupled with the biosensor and energy harvesting module. The microcontroller, programmed to measure biomarker data from the biosensor, communicates the data to phones or computers via Internet of Things technology. The closed-loop sensing and powering methodology is illustrated in Figure 20.11(b). The authors reported achieving average output values of 40 V and 4 μA from the TENG mechanism and about 4 V and 330 μA from the solar cells.

20.6 Future Direction and Challenges

In the literature, there is limited research on self-powered MEMS-FET–based sensors. While the development of self-powered electronic devices is a well-explored field, there remains significant scope for further investigation in self-powered MEMS-FET sensors. To date, only the TENG transduction mechanism has been extensively explored in conjunction with MEMS-FET sensors. A comprehensive analysis of MEMS-FET sensors using other energy harvesting mechanisms, such as biofuels and piezoelectric systems, is still needed. Additionally, to fully exploit the benefits of hybrid energy harvesting technologies, further analysis is required on coupling passive energy harvesting mechanisms with macro-scale energy transduction sources.

20.7 Conclusion

This chapter provides a comprehensive overview of MEMS-FET–based sensors. It covers fundamental concepts, advantages, and limitations of MEMS-FET technology and includes recent research on its applications in various sensing and detecting fields. The chapter aims to present the latest advancements in MEMS-FET sensor technology in a clear and engaging manner. It begins with an introduction to MEMS and FET technologies and then explores the application of MEMS-FET sensors in detecting gases, acceleration, force, sound, and biological parameters. Additionally, it addresses the development of self-powered MEMS-FET sensors and their potential impact on the biomedical industry.

References

1. Algamili, A.S., Khir, M.H.M., Dennis, J.O., *et al.*, A Review of Actuation and Sensing Mechanisms in MEMS-Based Sensor Devices. *Nanoscale Res. Lett.*, 16, 16, 1–21, 2021. https://doi.org/10.1186/s11671-021-03481-7.
2. Hao, R., Liu, L., Yuan, J., Wu, L., Lei, S., Recent Advances in Field Effect Transistor Biosensors:Designing Strategies and Applications for Sensitive Assay. *Biosensors*, 13, 4, 426, 2023. https://doi.org/10.3390/bios13040426.
3. Nguyen, T.T.H., Nguyen, C.M., Huynh, M.A., *et al.*, Field Effect Transistor based Wearable Biosensors for Healthcare Monitoring. *J. Nanobiotechnol.*, 21, Article 411, 1–29, 2023. https://doi.org/10.1186/s12951-023-02153-1.
4. Mao, S., Chang, J., Pu, H., Lu, G., He, Q., Zhang, H., Chen J., J., Two-Dimensional Nanomaterial-Based Field-Effect Transistors for Chemical and Biological Sensing. *Chem. Soc. Rev.*, 46, 22, 6872–6904, 2017. https://doi.org/10.1039/c6cs00827e.
5. Zacharias, J., Martha, P., Seena, V., Polymer Ring–Flexure–Membrane Suspended Gate FET Gas Sensor: Design, Modelling and Simulation. *Micromachines*, 14, 5, 944, 2023. https://doi.org/10.3390/mi14050944.
6. Sharma, B. and Kim, J.S., MEMS Based Highly Sensitive Dual FET Gas Sensor Using Graphene Decorated Pd-Ag Alloy Nanoparticles for H_2 Detection. *Sci. Rep.*, 8, 5902, 2018. https://doi.org/10.1038/s41598-018-24324-z.
7. Martha, P., Kadayinti, N., Seena, V., CMOS-MEMS Accelerometer With Stepped Suspended Gate FET Array: Design & Analysis. *IEEE Trans. Electron Devices*, 68, 10, 5133–5141, 2021. https://doi.org/10.1109/TED.2021.3101997.
8. Martha, P., Kadayinti, N., Seena, V., A CMOS-MEMS Accelerometer With U-Channel Suspended Gate SOI FET. *IEEE Sens. J.*, 21, 9, 10465–10472, 2021. https://doi.org/10.1109/JSEN.2021.3060186.

9. Kang, H.S., Lee, K.H., Yang, D.Y., You, B.H., Song, I.H., Micro-accelerometer Based On Vertically Movable Gate Field Effect Transistor. *Nano Micro Lett.*, 7, 3, 282–290, 2015. https://doi.org/10.1007/s40820-015-0041-9.
10. Gao, W., Jia, C., Jiang, Z., Zhou, X., Zhao, L., Sun, D., The Design And Analysis of a Novel Micro Force Sensor Based on Depletion Type Movable Gate Field Effect Transistor. *J. Microelectromech. Syst.*, 28, 2, 298–310, 2019. https://doi.org/10.1109/JMEMS.2019.2899621.
11. Muralidhar, R., Dennard, R.H., Ando, T., Lauer, I., Hook, T., Advanced FDSOI Device Design: The U-Channel Device For 7 nm Node And Beyond. *IEEE J. Electron Devices Soc.*, 6, 551–556, 2018. https://doi.org/10.1109/JEDS.2018.2809587.
12. Yeh, W.K. and Chou, J.W., Optimum Halo Structure for Sub-0.1 μm CMOSFETs. *IEEE Trans. Electron Devices*, 48, 10, 2357–2362, 2001. https://doi.org/10.1109/16.954477.
13. Ang, D.S. and Ling, C.H., A Unified Model for the Self-Limiting Hot-Carrier Degradation in LDD N-MOSFETS. *IEEE Trans. Electron Devices*, 45, 1, 149–159, 1998. https://doi.org/10.1109/16.658824.
14. Gao, W., Qiao, Z., Han, X., Wang, X., Shakoor, A., Liu, C., Lu, D., *et al.*, A MEMS Sensors Based on A Laterally Movable Gate Field-Effect Transistor (LMGFET) with A Novel Decoupling Sandwich Structure. *Engineering*, 21, 61–74, 2023. https://doi.org/10.1016/j.eng.2022.06.018.
15. Kaura, M.C., Pratapa, R., Bhat, N., Design of a high sensitivity FET integrated MEMS microphone. *Procedia Chem.*, 1, 875–878, 2009. https://doi.org/10.1016/j.proche.2009.07.218.
16. Omar, R., Yuan, M., Wang, J., *et al.*, Self-Powered Freestanding Multifunctional Microneedle-Based Extended Gate Device for Personalised Health Monitoring. *Sens. Actuators B Chem*, *398*, 134788, 2024. https://doi.org/10.1016/j.snb.2023.134788.
17. Tehrani, F., Teymourian, H., Wuerstle, B., Kavner, J., Patel, R., Furmidge, A., *et al.*, An Integrated Wearable Microneedle Array for the Continuous Monitoring of Multiple Biomarkers in Interstitial Fluid. *Nat. Biomed En.*, 6, 1214–1224, 2022. https:// doi.org/10.1038/s41551-022-00887-1.

21

Memory Design Using Conventional DRAM Unit Cell

Husien Salama[1*], Zina Guesmi[2], Faouzi Nasri[2,3], Billel Smaani[4], Khalifa Ahmed Salama[5] and Ahmed Gawa[6]

[1]Computer System Institute, Boston, USA
[2]Research Laboratory for Advanced Materials and Nanotechnology, ISSAT Kasserine, Kasserine, Tunisia
[3]Center for Research in Microelectronics & Nanotechnology, Sousse, Tunisia
[4]Abdelhafid Boussouf University Centre of Mila, Mila, Algeria
[5]Libyan Center for Engineering Research and Information Technology, Bani Walid, Libya
[6]University of Bridgeport, Bridgeport, USA

Abstract

This work examines key factors in memory design, such as the operating system, application size, and operation diversity, which influence memory usage and its type (virtual or real). Traditional dynamic random access memory (DRAM), using one MOSFET and one capacitor, can lead to space inefficiencies for multi-bit storage. To counter this, row and column logic are placed outside the memory array. Activating a row line selects the entire row, requiring a buffer for longer lines. On the other hand, control logic latches address signals, with the column address determining which bits to access once a row is selected. This chapter outlines the circuits needed for RAM implementation. A microarchitectural framework for protocol controller design is presented, breaking down memory functions into fundamental activities for effective protocol execution. Moreover, we introduced two memory designs: reduced-area memory (area: 1/8 and 1/4) and high-density memory (density: 4/16 bits/cell) using conventional DRAM unit cells based on the soft-error principles. We presented the design objectives and concept of the new memory designs, showing that the new memory designs are cost-effective without bringing in large-area overhead for error correction code (ECC) or error localization. We have found that, without the use of any proprietary tools, a useful ECC

**Corresponding author*: husien.salama@gmail.com

Ekta Goel, Archana Pandey, Shiromani Balmukund Rahi and Arun Samuel (eds.) Handbook of Advanced Semiconductor Field Effect Transistors, (439–464) © 2026 Scrivener Publishing LLC

current waveform function library has been developed to be the building blocks of exploratory computer-aided design (CAD) for DRAM data sensing.

***Keywords*:** Memory addressing, multi-core processor addressing, DRAM memory, nanobiotechnology, SRAM-based in-memory computing

21.1 Introduction

Dynamic random access memory (DRAM) conventional DRAM is a tried-and-true memory that many companies have manufactured over the past 27 years. The one transistor, one capacitor (1T- 1C) memory cell has gradually evolved to be the hub of what a conventional DRAM is known for. Like the other semiconductor components, the capacitor has also evolved in design and fabrication. Another strength of the DRAM memory is its ability to scale to smaller structures at a low cost. The DRAM has always been cheap, and the price per bit has always been less than other types of memory. There are some drawbacks like the requirement for refresh and read-before-write protocols. However, memory designers have managed to take advantage of developing high-speed technology when possible [1–3]. The once simple 1T-1C memory cell is now a multi-shaped, multi-layered, capacitor cell used in association with an array of circuits that are required to access the data on each cell. A revisionist approach seeks to move away from the generic spice and the IBM SPICE model used in modern design and fabrication and to implement a DRAM memory that can be optimized for timing and reduces power consumption while simultaneously providing a low cost of manufacturing. This memory has been named the simple DRAM and is under development at the Electrical and Computer Engineering Department of Concordia University.

Over the past 5 years, the use of submicron technologies at both the wafer and cell levels has allowed the doubling of the capacity of DRAM memory components every 3 years while, at the same time, maintaining a decrease in the cost of manufacturing. Consumer and workstation computing applications are now able to take advantage of the storage capacity that can only be supported by semiconductor technologies. For example, a 5-year-old 16-Mb DRAM component costs about 8–9 dollars to manufacture, whereas, today, a 64-Mb DRAM costs about that much. This paper will begin by introducing some basic information about the mechanism of a DRAM component. Information will be presented about the structure of the strobe, the sense amplifier, and the steering circuits. The presentation will move on to investigate the design of a high-speed memory by using state-of-the-art DRAM memory [4].

21.1.1 Definition and Basic Principles of DRAM

In the design of dynamic memory cells, the amount of charge required to place a specified charge on the storage node, usually, a poly gate, can be obtained by discharging the interconnected devices into a specified level. We measure the device drive, leakage, and delay through specific charge levels to calculate the transistor sizes. We first choose a predetermined amount of charge by dividing the desired probability of the state into deciles or another convenient number. We then design the transistor sizes based on the amount of charge that needs to be driven or discharged during the access operation. The storage node charge design is then modified to ensure that the transistor can drive to the appropriate levels. For a functional DRAM, we are forced to cell height and width, which dictates the minimal selector designs. The state probability and transistor sizing will affect the amount of sense amplifier current that must be dissipated within the read-write time and, thereby, the time constant, i.e., positioning of the capacitance on the data sheet. For large implementations, consideration has to be taken into account for tolerances in the transistor drive to ensure proper reading capabilities for the memory state [5–8].

We first describe the definition and basic principles of the DRAM. The DRAM consists of memory cells, a sense amplifier, a column decoder, a row decoder, and positive-MOS (PMOS) and negative-MOS (NMOS) transistors for the control and operation of the memory cells and the external interface. Each memory cell consists of a capacitor and a pass transistor for cell access. The basic operation of the memory consists of three different signal levels: recharge, write, and read. In the recharge mode, the bit lines of the sense amplifier are precharged to a high voltage. There is no read or write operation in the array during the precharge mode. In the write mode, when the word line is opened, the unselected memory cells are pushed to the high voltage on the bit line. Then, the low voltage on the bit line drives the bit line either to the low voltage, in case of writing 0, or high voltage, in case of writing 1, through the drive capability of the memory cell.

21.1.2 Historical Background and Development of DRAM Technology

It is worth noting that the proposed structures naturally utilize the main direction of the resulting reduced charge storage of the storage node. This charge does not need to drive the access transistor switch into the stored voltage and does receive little energy from the output amplifier. Encouraging devices become turned off more during storage operations.

Overall, it is vital to enable longer storage times [9]. The primary disadvantage of these oriented design elements must remain: The cell will have a lower capacitance if the storage node is capacitive. The following text details one possible direction for memory development based on the proposed user transplantation. We have not considered, for example, redistribution of the low capacitance until a designer proposes and explores a useful implementation. In this respect, we continue to have an open mind and believe that the proposed structure has great potential for achievement. For access dynamic range cell cells, the amount of current that an access transistor can carry is a critical parameter. When the access transistor is on, it must be able to drive the charge-restoring power switch to the storage node and provide enough energy to raise its value.

A DRAM cell consists of a single transistor and a storage node consisting of a metal-insulator-metal structure. When a DRAM cell is not being accessed, its value is read and used for a significant amount of time. During this time, the storage node discharges through the transistor, and, ultimately, it must be recharged by refreshing the cell. By reducing the charge storage node by a large factor, memory cells with significantly longer retention times can be designed. By increasing the cell capacitance by a large factor, memory chips with dense cell layouts can be designed. One way to accomplish this better than conventional memory cells begins by adding, in its simplest form, a gate across the access transistor so that it can heighten the discharge efficiency of the storage node. This first proposed memory design is a logical extension of a conventional memory cell. The gate is, in fact, the word line significantly of the memory cell, and activating this word line will drive the storage node's access transistor into a positive resistance region and drain the storage node's potential to the low supply voltage [10–14]. When the word line is deactivated and the storage node's access transistor is now at cutoff, the storage node will try to recover its value by being stealthily charged by the neighboring access transistor. Again, the circuit that connects the storage node to the access transistor should have high energy. The underlying storage node in conventional memory cells usually has one of the smallest transistors crispest, and, thus, it will contribute a great deal to the signal driving the access transistor. Very few organizations have proposed some form of positive feedback to increase the efficiency of the storage node.

21.1.3 Importance of DRAM in Modern Computing Systems

DRAM is an attractive memory because it provides the greatest density and lowest cost per bit of any existing memory technology. As described in the introduction, only DRAM and "natively logical" memories that borrow a portion of the DRAM concepts can achieve density-dominated access times [15, 16]. These "natively logical" memories (i.e., graphics frame buffers, register files, and microprocessor caches) achieve their low access time by sacrificing bit cell (i.e., bit storage) density. The DRAM organizational approach is the most area-efficient way to provide dense user-addressable memory. No structures exist that can satisfy latency-dependent memory access requirements in dense memories. The word lines necessarily provide wiring-dependent current paths in a dense memory to allow reasonable address decode energy per bit. Among the available "classical capacitor," DRAM organizations are one-transistor (n-MOS or pMOS) and three-transistor memory cells. Several capacitor-less scaling schemes have also been proposed for conventional bit-cell DRAM arrays. High-density, low-cost memory storage places high demands on the classical n-MOS DRAM bit cell because of its inherently small feature size. This chapter discusses bit-cell challenges and design techniques for the n-MOS memory cells of the commercial NDRO, NLT, and RAM1T asynchronous write, long refresh cycle (hours), low cost, wide 397-Mbit DDR DRAM, and the sub-2-GHz DDR-1T DRAM designs.

Modern computers utilize main memory storage that is in a separate silicon device chip (DRAM). The data is moved between the processor and main memory storage using logic that forms the memory system. However, several issues exist with this architecture. For example, data movement between the DRAM chip and processor requires a large amount of power and is the primary determiner of overall system price and performance [17–20]. Power consumption is a severe problem for portable systems because of the inherently large transfer signal power. Modern DDR designs specify memory arrays that consume tens of milliwatts. These powers result from signal swing, loading, and word-line driving. DRAM carries a significant fraction of portable and desktop systems. Any advances in DRAM power or performance will contribute directly to systems employing DRAM-based memories.

21.2 Conventional DRAM Unit Cell Structure

In a DRAM, whenever logic is written onto the data node, the amount of charge stored in the junction capacitance of the storage node is decreased. Therefore, the node voltage drops to 0 volts, a voltage lower than logic 0 [19]. This node must be restored to a high-voltage level, which means a "1" voltage level. Because FETs are power-saving switches, whenever the

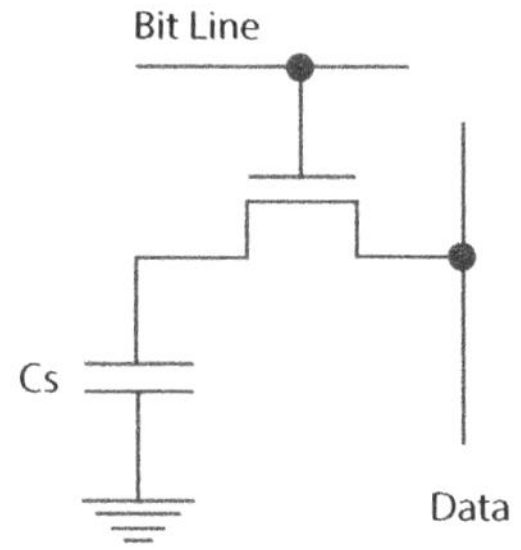

Figure 21.1 Conventional DRAM unit cell.

core power is suddenly turned off, LPC is also connected to the core power source, as shown in Figure 21.1.

The thick-film 1T-2C is essential to absorb the write margin reduction by the bit-line coupling. It requires a large die area but is suitable for high-speed write operation of advanced DRAMs. In this paper, the difference between the cells is revealed in terms of the capacitor limitation, charge loss write margin, and transistor scalability. Results prove that the charge loss write margin is not due to the thin-film effect but due to the onion layer error rate. It is easy to reduce the bit-line coupling-based write margin reduction at the turbo timing by the thick-film 1T-2C. The structure protects the transistor from the short-channel effect. The write margin at the turbo timing is easily swallowed by the thick-film 1T-2C, even when the write operation speed is very high [9].

Information in this section is about the cell architecture, thin capacitor structure, charge loss of the write operation using the thick capacitor structure, and read performance issues regarding transistor scalability affected by a thin capacitor. It assumes very high speed (GHz) write operation and the advanced soft-error rate is not the major concern. The difference between a 1T/1C cell and a 1T/2C cell DRAM is a capacitor structure in a unit cell [21].

21.2.1 DRAM Unit Cell Operation

This section explains the data reading and writing principles of the conventional DRAM unit cell used in existing memory. The output buffer operates as an inverter, and the sense line reads the voltage level [22].

DRAM is a basic memory type constituting computers and the fundamental storage unit of information stored in computers. The expansion of DRAM to mass storage created a big data boom and, consequently, led to an enormous amount of new research in DRAM technology. Because DRAMs are the principal storage element in computers, the memory design has a significant effect on the performance of computers. In addition, as research related to DRAM has entered the nano-scale era of DRAM technology, decisions regarding DRAM design are becoming more and more important. This chapter presents a memory design that combines the conventional DRAM unit cell currently used with SRAM characteristics and the operation of the conventional DRAM, data reading, and writing. Figure 21.2 shows the operation of DRAM [23].

DRAM read operation on a conventional cell. A conventional DRAM cell has enough capacitance to discharge fast. But, by using clever circuit techniques, the sense amplifier feedback has been restructured as a differential sensing scheme, resulting in amplification of the small signals induced across the bit lines due to the true and complement copies of the field-effect transistor (FET) turning ON and small currents flowing through the capacitance. Each true FET sits connected to a bit line, and its complement sits connected to the inverse bit line. When a cell multiplexer activates (also guaranteeing that no other cell in the same row multiplexes at the same time and allowing its gates to activate), the contents of that cell are sensed. The data is written back into that row at the end of sensing to

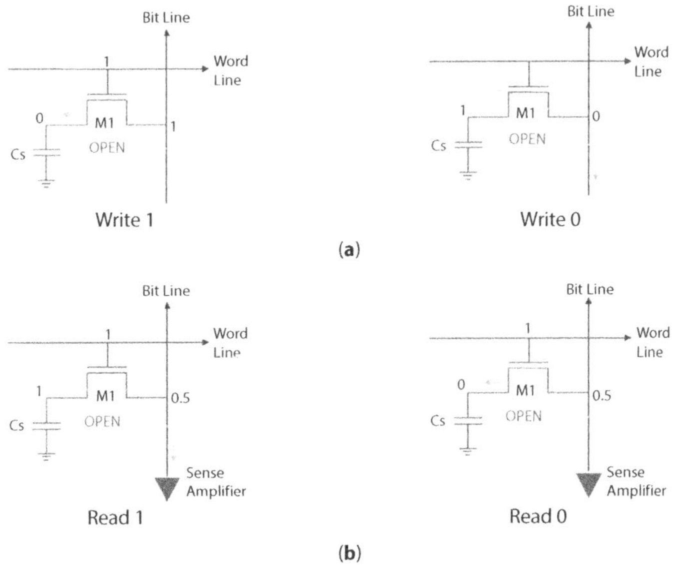

Figure 21.2 Operational steps of the DRAM memory cell.

return to its state before the sensing operation. A control circuit governs the read-and-write operations to obtain this functionality. Typically, it consists of a pre-decoder responsible for activating a small number of the total passed word lines and the associated other control signals, depending on the requested row address. This drives the word lines active, at very low resistance values because of the cell refresh scheme [24–26].

A conventional cell stores information using a single transistor and one capacitor. The transistor at the other end is connected to a sense amplifier that strengthens the signal before writing it to (or reading from) the bit line. The information in a certain cell is read by activating the corresponding word line, thus activating the transistors, and connecting the capacitor to the bit line to charge it. The other cells in its row also activate the corresponding word lines, but, because of a physical phenomenon between bit lines, only the cell of interest offers enough transition to activate the sense amplifier. A particular property of a capacitor is the discharging process after the first reading operation. This is one of the reasons why the information read from a cell in a certain row must be rewritten there after the read operation. This complicates the operation of this memory [27].

21.2.2 Overview of DRAM Unit Cell Components

DRAMs have storage elements protected by thin gates from possibly interfering with the charge that is strongly held. The transistor turn-off condition should never apply to these gates, due to how the memory cell operates. Therefore, care should be taken in the design of the generated external address signals such that address contention or "aliasing" is avoided. This means that the decoded word-line input paths must be the only externally addressable selected paths to memory cells. There may be more than one root to secondary addresses, but only one version can reach the respective memory. Differentiations must occur early in these address build paths (address matrix row and column) [28].

The memory unit design illustrated with conventional DRAM is considered a two-transistor and one-capacitor configuration. Each of the access transistors is connected to a row line and a column line which is orthogonal to each other in the word line direction and the bit line direction, respectively. The transistors of the same row and column are connected to a corresponding diffusion area, i.e., drain, source, and the memory cell. The address (column and row) decode is used to select the appropriate word and bit line. The column decode controls the compiled data and guides it to the actual bit line [29].

21.2.3 Capacitor and Transistor in a DRAM Unit Cell

The disadvantage is the memory access time, which is usually quite long since the data must be read from an entire row or column of the memory cells. There is a need for at least one transistor (select functionality) for each capacitor to connect the memory to a supply voltage and a data bus connected to the outside world. So, the area of the transistor should not be underestimated. There are thus many conflicting demands that limit accuracy in modeling. Moreover, modeling a memory under different traffic profiles (streaming accesses, random accesses, and repeated writes) is almost impossible, and shifting some regions of the memory operating to a different mode of working is very frequent. As shown in Figure 21.3, a single bit of data is stored as the presence or absence of an electrical charge on a tiny capacitor within a memory cell [30].

Inflation of the transistor size reduces the effective capacitance. Inflation has the effect of increasing the switching time and hence the memory access time, as well as the power consumption. The decrease in the effective capacitance may also lead to soft errors. It is the small size process of the capacitor that causes the most yield problems. The most common structure for a DRAM unit cell is a capacitor in series with a FET. The capacitor stores charge when its ferroelectric material is polarized, whereas the FET enables read-and-write operations (select functionality). The value of the capacitor is coupled to a single control line, which is both good and bad. The high density and the great number of memory cells allow for higher memory capacity at a modest cost. Furthermore, writing the memory is simple: just set an address on the address lines, specify the value on the control line, and, finally, activate the control line [31, 32].

In a DRAM, one bit of memory data is stored in a single DRAM unit cell. Each DRAM unit cell consists of a capacitor and a transistor.

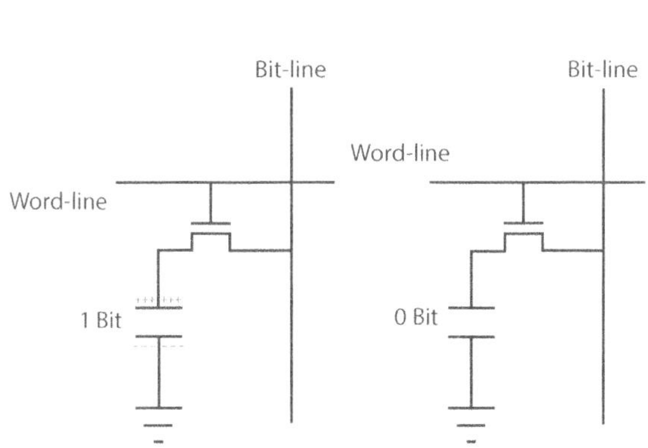

Figure 21.3 DRAM: One bit is stored as the presence or absence of charge on a capacitor.

The capacitance is the basic storage element in a DRAM and is represented by two equal and opposite charges in the capacitor plates. The charge then represents the two DRAM states, and the highest voltage indicates a logic level 1. The smallest physical size is always preferred for the capacitors as this will result in the smallest die and, consequently, in the higher amount of resulting DRAM chips. The size of the capacitors is limited by the stacking process, the tight layer-to-layer registration, the trench etches processes, and the need for a high aspect ratio to maximize the capacitance. There are tight ranges and strict electrical requirements attached to the value of the capacitor that can be easily met by varying the dimensions of the unit cell, capacitor, and transistor together [5].

21.2.4 Functionality and Operation of a DRAM Unit Cell

As a rule of design, a DRAM uses rechargeable storage capacitors that are charged to an essential voltage level to verify the logic state. Furthermore, as a rule of power design, the storage capacitor is reduced to a sub-fF level. Notably, the storage capacitor is detached from the bit line during a read operation. This task of reading is performed by an access transistor. The task of an access transistor during a read operation is twofold. First, it offers a way out of voltage across its drain and source terminals to the bit line and internal node to specify a logic value [33]. Second, it closes the way out of the charge stored in the storage capacitor to the internal node. Finally, the third functional block in a unit cell is a word line. This is used to control the opening and closing of an access transistor. Accountably, a unit cell is activated by turning on the appropriate word line. In general, the word lines are activated sequentially during a read operation in a DRAM.

When it comes to memory applications, DRAM is the most eminent candidate. That is because of its inherent disadvantages such as bulk storage, bit line limitations for large-sized RAM designs, favor with associating multiple bits to different addresses, and the facile nature of a unit cell. In this section, a brief introduction to the DRAM unit cell is given. A DRAM unit cell is broadly divided into three functional blocks: a storage capacitor, an access transistor, and a word line. Furthermore, this subsection also unveils the stylized operation of the unit cell [34].

21.2.5 Read-and-Write Operations

The power is dissipated by the read circuits and the write circuits with latches, initiated by the respective voltage levels of the write data select word. In the write operation, normalized, and provided on the data input lines. Signals are pre-charged to predetermine voltage levels. While all the latch and write row word lines are at a low-voltage level, the write drive data switches are turned on to apply the pre-charged voltages on the dedicated word lines and latches. A dummy read followed by the dummy write operation is required to complete the multilevel write or multilevel erase operation and refresh the conventional DRAM arrays. Only a small difference is sufficient for the word to be enabled for far programming voltage specified process. The different operation is caused by the different intermediate data voltage and signal levels provided by the reconfigurable classic [35, 36]. Figure 21.4 shows that 1-bit data storage is achieved using one transistor and one capacitor.

Cs<<CBL.

$$\Delta V = V_{BL} - V_{PRE} = V_{BIT} - V_{PRE} \frac{C_S}{C_S + C_{BL}} \tag{21.1}$$

To write a digital bit, the capacitor is charged by applying a pulse in WL and VDD in BL to erase the cell, and the capacitor is discharged VDD by applying 0 in BL. To sense (read) the device, ½ VDD is applied in the bit line, whereas short read pulse is applied in the WL.

The bit line (BL) voltage is sensed as a digital signal to detect the data during the read operation. The voltage differential between the trigger level voltage (VT) and the bit line voltage level determines the read margin. The sense amplifier must amplify and convert the differential voltage

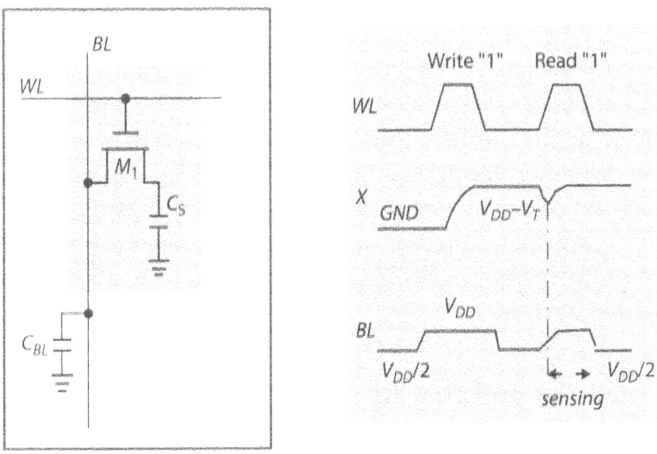

Figure 21.4 One-bit data storage is achieved using one transistor and one capacitor.

into the respective replacing voltage level compatible with the multilevel cell. The circuits for sensing, amplifying, and latching are available in the read circuit that completes the read operation. The word line voltage level of the encountered read word decides if the word is sensed, the bit is activated (active low), the read operation is enabled, and the multilevel data is latched into. Alternatively, the whole write unit is powered on in the state [37].

21.3 Design Considerations for Conventional DRAM

The inherent low static power offered by a capacitor-based memory cell encouraged researchers to develop design tools that would assist the design of these cells. The capacitor's low intrinsic leakage characteristics and the CMOS implementation used to access the contents also offered low standby power for the design. This was of prime concern in designing phosphor screens used as a display device in personal computers where images would remain on the screen for long periods without the need for memory cell refresh-in activity. Devices that depend on the capacitance effect, such as the EEPROM, Flash EEPROM, and the ROM, also offer lower power consumption and have been studied in detail [38–40]. The DRAM cell, on the other hand, offers the most compact structure and the presence of row decode is minimal. A key challenge, however, is the presence of complex drive circuitry and multiple probing points, giving rise to a higher bit-line capacitance.

To achieve larger amounts of on-chip memory and smaller memory access times in future microprocessor designs, the traditional SRAM-based memory cell technology is expected to be replaced by a DRAM-based memory design. A reduction in the process feature size allowed to production of denser and faster memories so far. Innovations in the memory design structure, however, were limited to those that would allow the fabrication of higher densities of the cell [41]. The primary developers of DRAM devised cell structures that were smaller and would require short access times, but they function on completely different principles when compared to the memory cells based on the capacitive effect and resistor-based latching devices.

21.3.1 Cell Size and Density in DRAM Design

The memory cell design depends on the technology node and tuning requirements such as the desired cell size, current leakage limits, retention time, refresh rate, and the desired operating frequency (read-and-write rate). Incorrect mode provides a technique for reading two adjacent cells to discharge or charge the columns and provides a way to allow low-voltage precoding. Statistical memory simulations provide a way to estimate memory size requirements based on types of applications. The areas consumed by dynamic random access memory (DRAM) settings, input/output (I/O), ECC, spacing, and timing require a careful study to achieve the lowest level of cost and the highest level of quality of service. The key goal of any memory block design is to maximize the memory capacity while minimizing all associated chip areas including the size of DRAM settings. Hence, finding the best point for a given technology node is the selection of the best bit-cell type.

Memory designs for an intellectual property block are realized as an array of capacitors and cell-select switches (transistors) [42]. Word line drivers and sense amplifiers complete the read-and-write circuits. Conventional DRAM designs are based on 1T1C unit cells in which a single transistor is used in combination with a single capacitor. DRAM memories, like static memory found in microprocessors and ASICs, are realized as an array of unit cells. 4Kx1 bit, 8Kx1 bit, and 16Kx1 bit are typical implementations [43].

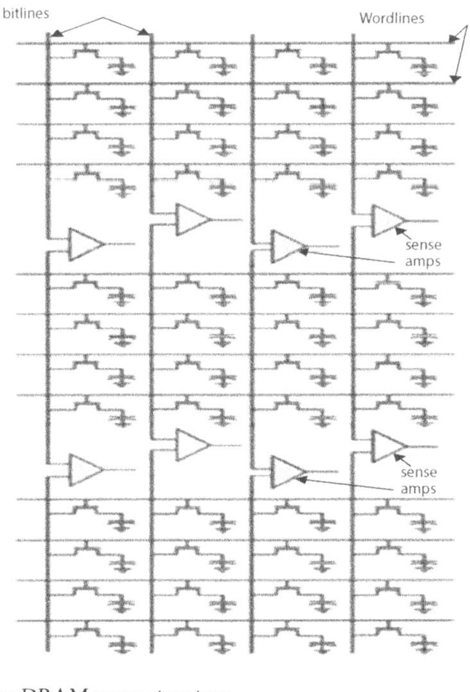

Figure 21.5 Bit-lines DRAM array structure.

Other types of non-volatile memories, (PROM, EPROM, EEPROM, and Flash) also adopt a similar design approach. A conventional 1T1C design is shown in the figure. The cell layer is used to connect nodes with different functions, separating those that are accessed by a set of different word lines. Data buses in the read-and-write paths are multiplexed in 16- or 32-bit-wide groups. Figure 21.5 shows the layout of an open-bit line of DRAM array structure.

21.3.2 Read-and-Write Operations in DRAM

The value of data in a DRAM unit cell can be read in two main steps: activation and column access. First, to read the value from the DRAM unit cell, the target row address is sent to the row decoder, and the row line is strobed high. The other row lines are not selected during the read operation. This step is called the activation process [44, 45]. The data is transferred to the standalone sense amplifier/row line. Second, the data is sent to the input/output pads to be read by the circuit. The sense amplifier isolation and data transfer time for the activation row line is the column access time. Data is read back from the DRAM unit cell to the sense amplifier for a refresh cycle if it is in one of the selected banks. Data is written back to the DRAM cell if it has undergone data corruption issues. A read operation can have a read disturbance. In a read disturbance, a value may be erroneously written back to the DRAM cell. The read operation is usually performed using a single read command. The latency between the successive read commands is the read cycle time [46].

With the trend in high-density memory designs, conventional DRAM memory is slower and more powerful. Future memory designs should focus on low read-and-write latencies to be useful for information processing. In this section, we discuss various conventional DRAM read-and-write operations.

21.3.3 Reading Data from a DRAM Cell

$$Vsh(k) = (K - 1)/2 ** (k - 1)\ V = (V - 1)/2 ** (k - 1)\ V\ (k > 2) \quad (21.2)$$

$$Vsl(k) = K/2 ** (k - 1)\ V = V/2 ** (k - 1)\ (k > 2) \quad (21.3)$$

In general, monitoring the capacitor charging progress requires generating an n-step staircase waveform, where n is the resolution of the structure detecting the state of the capacitor. This approach is also called an n-step

ADC (analog to digital converter). The detectors with a desired resolution can be constructed iteratively using the two-detector device, as shown in Figure 21.3, where the damp pass transistor and the sense capacitor are both identical structures [47]. The detectors have the same resistances and parasitic capacitances. Consider the circuit working in the floating-point representation of the signal. First, the source and the channel of the damp pass transistor, as well as the sensor capacitor, are discharged. Second, a high-level logic signal n (k), which is equal to ground, is changed to the high-level logic signal. After that, the device receives the following arguments in descending order: the source and the channel of the damp pass transistor are set to a high-level logic. As a result, a current flow from the bit line to the capacitor [48]. At the same time, the voltage on the sense capacitor increases until the comparator detects the change [48]. The procedure is not in the voltage level corresponding to the numerical value of the charge on the capacitor. Consequently, the adjacent level is determined. In the detector, two values of voltage upper limit, Vsl and Vsh, exist, which are determined by Equations (21.2) and (21.3).

In this section, we will describe the reading process in a DRAM system. Consider the capacitor and access power supply from a conventional DRAM unit cell. An AC voltage is applied to the bit line to copy the state of the capacitor. The signal charges the bit line by an amount of $\Delta V \approx Vdd/2 - Vt$, where $Vt > 0$ is the threshold voltage of a non-conducting pass transistor. There are two possible ways to read the integer value from the DRAM cell capacitor [49–51]. The first method is based on a binary search, and the amplifiers can determine the state of the capacitor in this way. The DRAM cell storing a state of the background is read, and the amplifiers map it into the binary value corresponding to the state. The probability that the reading result is incorrect is much less than 0.1. If the amplifiers make an erroneous decision, then the reading is repeated. However, applying a binary search means using many comparators because they need to detect a lot of discrete voltage levels.

21.3.4 Writing Data to a DRAM Cell

Moreover, it is observed that the bit line voltage is continuously decreasing during the write operation with an increase in the bit line current in the read case. This suggests that a leakage current flows through the device in the off-state, determined by the on-state and VDD. However, they are only a few turns, and the access transistor will turn off once the time derivative of the current is expressed as the capacitance times the sum of the bit line voltage and off-state rate. Only a small amount of charge is needed to switch the potential of the capacitive node to the bit line voltage satisfying the read condition for output, but there is a finite time constant for charge recovery [52].

The process of writing data into a DRAM cell involves a single access transistor with the characteristic curve of writing a voltage-dependent current into a capacitor. With the access transistor in the off state, at a low-voltage level on the bit line, an on-current flows into the capacitive node, which stores the charge. At a high-voltage level on the bit line, the off-current amounts to writing a threshold voltage in the opposite direction. The threshold in the write scenario is preset to a high voltage so that all cells on the bus will be written to the higher voltage [53].

In the subthreshold regime of operation, the operation starts from the turning of the word line (WL), and the voltage of the bit line (BL) is switched from 0V to the supply voltage (VSS). Subsequently, the bit line transfer gate (TG) is turned on to enable the flow of the drain current. If the bit line is recharged to the supply and the memory cell in its initial state has its intrinsic drain current enabled (both the pass transistor and the pull-down path are enabled), then a 1-kb data can be written into any unselected array cell, which provided two additional work lines.

21.3.5 Refresh Mechanisms in DRAM

The stored charge leaks out with time, and after a time interval called the retention time, the DRAM cell enters an illegal state. The POSTING bit in DDR4 and DDR5 variants of JEDEC specification is removed, where Write and Recharge can be done simultaneously causing bit flips. This row hammer vulnerable data is a fault in the data that is exposed due to adjacent fast data accesses, which in turn cause bit flips in the weak adjacent cell [54]. A known refresh mechanism is to periodically read the contents of each cell row and re-write it with the same data to store the charge and maintain the regular operation of the memory. The memory periodically activates this DRAM refresh, which wastes a significant amount of energy and degrades the normal memory performance. The concern is that normally refreshing less often than the retention interval causes a retention fault, in which a refresh failure results in data loss. Another rare event that could occur is read-disturb, a read-induced capture that erases valid data from the single cell itself. The probability of read-induced capture increases as the read when the data in the cells are at low-voltage states, where the probability of the capture is higher [55].

21.4 Challenges and Limitations of Conventional DRAM

The large memory capacity and high-speed processor increase the long refresh memory capacity based on the DRAM. The long refresh memory device is needed to maintain good data integrity. The abnormal data at high frequency causes the system to malfunction. A high refresh current leads to a low DRAM memory access time. When the refresh current is very large, the IC package overheats because of a high current's degradation. Moreover, a high current induces some noise and failure in the DRAM memory architecture. The poor power supply induces system data errors, and, then, error control code encoding, and correction are operated. The poor power supply affects the entire chip, and it induces long data retention from noise in the read data. An ECC error detection code, encoding, and error correcting latency are increased because of the lower data signal-to-noise ratio [56].

A large single-unit capacitor increases the area per bit because it has a limitation of the memory fabrication process. The capacitor size, especially the capacitor unit length, should be equivalent to one memory cell because of the selected transistor size or refresh time. A very small unit capacitor length causes a very high capacitor leakage current, and, then, refresh time increases to obtain a large read margin. The refresh time is limited by 64 ms in JEDEC standard. The refresh time of a trench cell capacitor exceeds 500 ms shortly based on the capacitor F-N tunneling leakage current. Therefore, the refresh time and large memory chip's area and static current are increased by the growth of trench cell capacitors. There are several retention enhancement techniques, such as DRAM device architecture, advanced technology, and ECC logic, which could suppress the retention time increment by the capacitor leakage current buildup [57, 58].

21.4.1 Issues with Scaling Down DRAM Technology

Regarding storage node capacitance, the original purpose of SKT was line-to-line capacitance reduction. The storage nodes in the high-temperature region include oxygen and nitrogen. However, the conventional SKT does not reduce the line-to-line capacitance [59]. Therefore, it was necessary to assign the assignment in consideration of the OEM requirement and to make the storage node independently. Afterward, the length of the storage node increased when a lower voltage operation was required. However, increasing the length of the storage node causes the degradation of the memory capacity and requires a main cell increase, which increases the unit cell area.

During the last several decades, the DRAM has been dynamically scaled in terms of active and passive dimensions. In recent years, in scaled-down DRAM technology, the length of the word line, bit-line, and storage node has been more aggressive, but certain issues have been found too. These issues include the capacitance value of a bit-line, the decrease of access time as the data sequence grows, and the power dissipation. When the channel length becomes a smaller size than the minimum feature size, the corner effect occurs, which increases drastically the threshold voltage. Therefore, it is required that the transfer length solves this problem in the same manner. However, the cross-point length of a bit-line is increased, which causes a decrease in the capacitance. The next technology should resolve this issue [60–63].

21.4.2 Power Consumption and Heat Dissipation Challenges

In conventional memory devices, the primary source of power consumption and heat generation comes from memory access: activate, recharge, and data bus charging. It is easy to understand why data bus charging and recharge allow the flow of data onto the memory's data output lines, as the dynamic power needs to be dissipated for every memory access (read or write). However, the activation needs to be activated to isolate the memory bank from other memory banks, whereas the signals are potent and charge a specific cell. In other words, it charges the access transistor that allows the connection between the memory cell and the word line control plurality. The access transistors in most memory designs are significantly larger than conventional CMOS logic gates [64].

A characteristic that is inherent to memory stacks is related to its power consumption and heat dissipation. While a 3D stacked design is beneficial in terms of performance and power consumption, it is also true that thermal management becomes more critical. Certainly, there are some known methodologies for dealing with heat arising from power density challenges that are also employed in 3D stacking, such as through-silicon vias and micro-channel cooling [65, 66].

21.4.3 Prospects and Potential Solutions

In the DDRAM, voltage swing is an important factor that influences DDRAM's access time. A smaller swing leads to a shorter time. Thus, voltage swings on bit lines shall be kept as small as possible.

Because the source follower load of a sensing amplifier can be regarded as a currency of the output data of the DDRAM array, it is indispensable to lessen it. This is one of the crucial challenges to be overcome. The current consumption of one-bit line is decreased in proportion to the number of divisions. Hence, if one of the divided bit line pairs is selected and transferred to the encoder, then current consumption can be excessively decreased. However, in this case, the area of the bit line unit becomes large, and the total length of bit lines also increases unrealistically. Alternatively, the use of the dummy bit line is a well-known technique to lessen the current consumption of DDRAM. Because the area overhead and the routing complexity of the dummy bit line depend heavily on its modality, it is important to devise the memory architecture with a less complex dummy bit line.

21.5 Conclusion and Future Directions

Compared to the basic fresh SRAM design (substitution and read/write-restore), our infrastructure additions increased the timing, energy, and area of the memory array by 52.3%, 44.22%, and 13.4%, respectively. However, we showed that, by employing heterogeneous slicing to exploit the best characteristic of different memory bits, we achieved 34.6% area, 34.8% delay, and 41.8% energy proximity scaling benefits on average, with very little degradation in error/slowdown, at the balance of "1.23RO+0.20RESET" on average on the maximum-throughput and corner-mode targets.

To validate our proposed memory designs, we designed a larger and more detailed functional DRAM memory using a functional soft error rate tool. Inserting read-and-write disturbances at all possible places on the bit-cell and reading soft-error occurrences using a pattern-aware fault-simulator, we showed that both the proposed designs are soft-error immune.

In this chapter, we introduced two memory designs: reduced-area memory (area: 1/8 and 1/4) and high-density memory (density: 4/16 bits/cell) using conventional DRAM unit cells based on the soft-error principles. We presented the design objectives and concept of the new memory designs, showing that the new memory designs are cost-effective without bringing in large-area overhead for ECC or error localization.

21.5.1 Summary of Key Points in DRAM Design

In this chapter, the ECAD framework for FPGA-based memory design is presented. Without the use of any proprietary tools, a useful EC current waveform function library is developed for being the building blocks of exploratory CAD for DRAM data sensing.

Designing DRAM is an interdisciplinary subject, involving process technology, CAD algorithms, reliability, testing, and FPGA/Xilinx tools from electrical engineering as well as system design and computer architecture. Various groups with different backgrounds have been working with memory design. Currently, the voltage scaling is one of the various design parameters. However, the conventional process technology and circuit design lead to the limitation of scaling Vdd directly. The DRAM design now wants to use various process technologies to have lower supply voltage as well. As a newer model for process technology adoption, the DRAM design is enhanced with newer knowledge from electrical engineering and other fields.

21.5.2 Emerging Trends in DRAM Technology Research

There has been a report that conventional transistor one capacitor cell DRAM may no longer be an option beyond the 60-nm generation. It has been predicted that, by the 45-nm generation, the length of the cell capacitor will be less than 90 nm. Conventional scaling will not be possible if it is designed with minimum geometry because the capacitor area is inversely proportional to the voltage noise and proportional to the threshold voltage. The reduction in the length will be inversely proportional to the scale factor because the capacitance C is proportional to k * A/d. This has made DRAM technologists search for alternatives. In the past, various attempts at transistor DRAM and three-dimensional capacitors have been made. Many of them have resulted in feasibility results, but compatibility with the cost structures of a DRAM is low. During the research for alternatives, it was realized that all the designs were skimping the problem by introducing a major process or design parameter. The parameter would typically not allow the design to scale for more than a generation. As the parameters reached a limit, it would replace the technology in existence. The authors surmise that any new DRAM design that emerges in the future will not increase the number of manufacturing steps or costs. In addition, it will have to scale and last beyond 7 years or beyond two DRAM generations. Any new design that is independent of this requirement will result in evolutionary death. Because neither nanotubes nor capacitors have demonstrated their independence and maturity currently, the authors foresee a long life for the current DRAM design.

21.5.3 Potential Advancements and Innovations in DRAM Design

One advantage of unit capacitor designs in these cases is that many advanced materials reported to have applications in DRAMs are used in the semiconductor fabrication process. As of October 2004, authors reported a new dielectric material, Gd2O3, that has a dielectric constant greater than 20 and etch selectivity comparable to that of SiO_2. Other activities include the use of carbon nanoclusters (5–7 A in size) as a floating gate in metal-insulator-metal capacitors (MIM) capacitors to allow an ultrahigh-density charge storage scheme. The use of Cu electrodes in MIM capacitors to suppress the resistance of common Al electrodes is also intended to allow advanced materials to be applied to DRAMs.

The previous sections presented issues and concerns in a DRAM array by assuming the performance and characteristics of a unit capacitor. Because many issues regarding the specific characteristics of conventional capacitors have been reported in previous studies, synchronization of their performance with that of the rest of the memory would be considered a challenge in a conventional DRAM. However, advancement in capacitor design has been ongoing. For example, authors have presented a three-dimensional stacked DRAM to suppress crosstalk between adjacent cells. They also discussed strategies for designing advanced capacitor units for DRAMs and new applications. Other activities have included increased dielectric materials, such as focusing on the dielectric constant, adhesivity, and thermal and mechanical properties of MIM capacitor units.

References

1. Khalil, A.S. and Collins, J.J., Synthetic biology: applications come of age. *Nature*, 11, 5, 367–379, 2010.
2. Bhati, I., *et al.*, Flexible Auto-Refresh: Enabling Scalable and Energy-Efficient DRAM Refresh Reductions, in: *ISCA*, 2015.
3. Choi, J., *et al.*, Multiple Clone Row DRAM: A Low Latency and Area Optimized DRAM, in: *ISCA*, 2015.
4. Ghose, S., *et al.*, Demystifying Complex Workload-DRAM Interactions: An Experimental Study, in: *SIGMETRICS*, 2019.
5. Pessl, P., *et al.*, DRAMA: Exploiting DRAM Addressing for Cross-CPU Attacks, in: *USENIX Security*, 2016.
6. Lee, D., *Reducing DRAM latency at low cost by exploiting heterogeneity*, arXiv preprint arXiv:1604.08041, 2016.

7. Zhang, X., *et al.*, Restore Truncation for Performance Improvement in Future DRAMSystems, in: *HPCA*, 2016.
8. Zhao, D., Xia, Z., Yang, T., Yang, Y., Zhou, W., Huo, Z., A Novel Capacitorless 1T DRAM with Embedded Oxide Layer. *Micromachines*, 13, 10, 1772, 2022.
9. Luo, H., Shahroodi, T., Hassan, H., Patel, M., Giray Yaglikci, A., Orosa, L., Park, J., Mutlu, O., CLR-DRAM: A Low-Cost DRAM Architecture Enabling Dynamic Capacity-Latency Trade-Off, in: *2020 ACM/IEEE 47th Annual International Symposium on Computer Architecture (ISCA)*, pp. 666–679. IEEE, 2020.
10. Husien, S., Saman, B., Heller, E., Gudlavalleti, R.H., Mays, R., Jain, F., Twin Drain Quantum Well/Quantum Dot Channel Spatial Wave-Function Switched (SWS) FETs for Multi-Valued Logic and Compact DRAMs. *Int. J. High Speed Electron. Syst.*, 27, 03n04, 141, 2018. 1840024.
11. Tayal, S., Smaani, B., *et al.*, Incorporating Bottom-Up Approach Into Device/Circuit Co-Design for SRAM-Based Cache Memory Applications. *IEEE Trans. Electron Devices*, 69, 11, 6127–6132, 2022. doi: 10.1109/TED.2022.3210070.
12. Vidya Wicaksana Putra, R., Abdullah Hanif, M., Shafique, M., EnforceSNN: Enabling resilient and energy-efficient spiking neural network inference considering approximate DRAMs for embedded systems. *Front. Neurosci.*, 16, 937782, 2022.
13. Ausavarungnirun, R., *Techniques for Shared Resource Management in Systems with Throughput Processors*, PhD diss., Carnegie Mellon University, 2017.
14. Smaani, B., Rahi, S.B., Labiod, S., Analytical Compact Model of Nanowire Junctionless Gate-All-Around MOSFET Implemented in Verilog-A for Circuit Simulation. *Silicon*, 14, 10967–10976, 2022. https://doi.org/10.1007/s12633-022-01847-9.
15. Patel, M., *Enabling Effective Error Mitigation in Memory Chips That Use On-Die Error-Correcting Codes*, arXiv preprint arXiv:2204.10387, 2022.
16. Seshadri, V. and Mutlu, O., *In-DRAM Bulk Bitwise Execution Engine*, arXiv preprint arXiv:1905.09822, 2019.
17. Singh Meena, J., Min Sze, S., Chand, U., Tseng, T.Y., Overview of emerging nonvolatile memory technologies. *Nanoscale Res. Lett.*, 9, 1–33, 2014.
18. Hassan, H., *Improving DRAM Performance, Reliability, and Security by Rigorously Understanding Intrinsic DRAM Operation*, arXiv preprint arXiv:2303.07445, 2023.
19. Salama, H., Saman, B., Gudlavalleti, R., Mays, R., Heller, E., Chandy, J., Jain, F., Compact 1-Bit Full Adder and 2-Bit SRAMs Using n-SWS-FETs. *Int. J. High Speed Electron. Syst.*, 29, 01n04, 2040013, 2020.
20. Nasri, F., Mzoughi, H., Salama, H., Salama, K.A., Numerical Analysis of Self-Heating Effect in HEMT Transistors, in: *2023 International Conference on Electrical, Communication and Computer Engineering (ICECCE)*, Dubai, United Arab Emirates, pp. 1–5, 2023, doi: 10.1109/ICECCE61019.2023.10442598.

21. Gudlavalleti, R.H., Saman, B., Mays, R., Salama, H., Heller, E., Chandy, J., Jain, F., A novel addressing circuit for SWS-FET based multi-valued dynamic random access memory array. *Int. J. High Speed Electron. Syst.*, 29, 2040009, 2020.
22. Nguyen, D.T., Ho, N.M., Wong, W.F., Chang, I.J., *OBET:* On-the-Fly Byte-Level Error Tracking for Correcting and Detecting Faults in Unreliable DRAM Systems. *Sensors*, 21, 24, 8271, 2021.
23. Nam, H., Baek, S., Wi, M., Jaemin Kim, M., Park, J., Song, C., Sung Kim, N., Ho Ahn, J., DRAMScope: Uncovering DRAM Microarchitecture and Characteristics by Issuing Memory Commands, in: *2024 ACM/IEEE 51st Annual International Symposium on Computer Architecture (ISCA)*, pp. 1097–1111, IEEE, 2024.
24. Abbas, H., Li, J., Shenp Ang, D., Conductive Bridge Random Access Memory (CBRAM): Challenges and Opportunities for Memory and Neuromorphic Computing Applications. *Micromachines*, 13, 5, 725, 2022.
25. Nasri, F. and Salama, H., Numerical Investigation of the Electrothermal Properties of SOI FinFET Transistor. *Int. J. High Speed Electron. Syst.*, 32, 02n04, 2350020, 2023.
26. Husien, S,., Smaani, B., Nasri, F., Tshipamba, A., Nanotechnology and quantum dot lasers. *J. Comput. Sci. Technol. Stud.*, 5, 1, 45–51, 2023.
27. Hyun Shin, H. and Chung, E.Y., In-DRAM Cache Management for Low Latency and Low Power 3D-Stacked DRAMs. *Micromachines*, 10, 2, 124, 2019.
28. Hong, S., Kim, D., Lee, J., Oh, R., Yoo, C., Hwang, S., Lee, J., *DSAC: Low-Cost Rowhammer Mitigation Using In-DRAM Stochastic and Approximate Counting Algorithm*, arXiv preprint arXiv:2302.03591, 2023.
29. Kim, S., Senol, D., Xin, H., Lee, D., Ghose, S., Alser, M., Hassan, H., Ergin, O., Alkan, C., Mutlu, O., *GRIM-filter: fast seed filtering in read mapping using emerging memory technologies*, 2017.
30. Kim, Y., Daly, R., Kim, J., Fallin, C., Lee, J.H., Lee, D., Wilkerson, C., Lai, K., Mutlu, O., Flipping Bits in Memory Without Accessing Them: An Experimental Study of DRAM Disturbance Errors, ACM SIGARCH Computer Architecture News, 42, 3, 361–372, 2014.
31. Gautam, S., Manhas, S., Kumar, A., Pakala, M., Yieh, E., Row Hammering Mitigation Using Metal Nanowire in Saddle Fin DRAM, in: *IEEE TED*, 2019.
32. Luo, H., Olgun, A., Yağlıkçı, A.G., Tuğrul, Y.C., Rhyner, S., Cavlak, M.B., Lindegger, J., Sadrosadati, M., Mutlu, O., RowPress: Amplifying Read Disturbance in Modern DRAM Chips, in: *ISCA*, 2023.
33. Cohen, Y., Tharayil, K.S., Haenel, A., Genkin, D., Keromytis, A.D., Oren, Y., Yarom, Y., HammerScope: Observing DRAM Power Consumption Using Rowhammer, in: *CCS*, 2022.
34. Beigi, M.V., Cao, Y., Gurumurthi, S., Recchia, C., Walton, A., Sridharan, V., A Systematic Study of DDR4 DRAM Faults in the Field, in: *IEEE International Symposium on High-Performance Computer Architecture (HPCA)*, 2023.

35. Hassan, H., Patel, M., Kim, J.S., Yağlıkçı, A.G., Vijaykumar, N., Ghiasi, N.M., Ghose, S., Mutlu, O., Crow: A Low-Cost Substrate for Improving Dram Performance, Energy Efficiency, and Reliability, in: *ACM/IEEE International Symposium on Computer Architecture (ISCA)*, 2019.
36. Hou, C.-S., Chen, Y.-X., Li, J.-F., Lo, C.-Y., Kwai, D.-M., Chou, Y.-F., A Built-In Self-Repair Scheme for DRAMs with Spare Rows, Columns, and Bits, in: *2016 IEEE International Test Conference (ITC)*, 2016.
37. Wang, Y., Ferraiuolo, A., Edward Suh, G., Timing Channel Protection for a Shared Memory Controller, in: *IEEE International Symposium on High-Performance Computer Architecture (HPCA)*, IEEE, 2014.
38. Anderson, D., Bevan, P., Lang, K., Liberty, E., Rhodes, L., Thaler, J., A High-Performance Algorithm for Identifying Frequent Items in Data Streams, in: *Proceedings of the 2017 Internet Measurement Conference*, pp. 268–282, 2017.
39. Kurmus, A., Ioannou, N., Neugschwandtner, M., Papandreou, N., Parnell, T., From Random Block Corruption to Privilege Escalation: A Filesystem Attack Vector for Rowhammer-like Attacks, in: *11th USENIX Workshop on Offensive Technologies (WOOT 17)*, 2017.
40. Oh, T.Y., Hoeju, C., Park,J-Y., Lee, K-W., Oh, S., Doo, S-Y., Kim, H-Kw., *et al.*, A 3.2 Gbps/pin 8 Gbit 1.0 V LPDDR4 SDRAM with integrated ECC engine for sub-1 V DRAM core operation. *IEEE J. Solid-State Circuits*, 50, 1, 178–190, 2014.
41. Wang, Y., *et al.*, Re-architecting the On-chip memory Sub-system of Ma-chine-Learning Accelerator for Embedded Devices, in: *ICCAD*, 2016.
42. Kotra, J., *et al.*, Hardware-Software Co-design to Mitigate DRAM Refresh Overheads: A Case for Refresh-Aware Process Scheduling, in: *ASPLOS*, 2017.
43. Rosenfled, P., *et al.*, DRAMSim2: A Cycle Accurate Memory System Simulator, in: *IEEE Computer Architecture Letters*, 2011.
44. Isen, C., *et al.*, ESKIMO-Energy Savings using Semantic Knowledge of Inconsequential Memory Occupancy for DRAM subsystem, in: *MICRO*, 2010.
45. Liu, S., *et al.*, Flikker: Saving DRAM Refresh-power through Critical Data Partitioning, in: *ASPLOS*, 2011.
46. Cui, Z., *et al.*, DTail: A Flexible Approach to DRAM Refresh Management, in: *ICS*, 2014.
47. Bhati, I., *et al.*, Flexible Auto-Refresh: Enabling Scalable and Energy-Efficient DRAM Refresh Reductions, in: *ISCA*, 2015.
48. Seshadri, V., *et al.*, Ambit: In-Memory Accelerator for Bulk Bitwise Operations Using Commodity DRAM Technology, in: *MICRO*, 2017.
49. Li, S., Tian, M., Gu, C., Wang, R., Wang, M., Xiong, X., Li, X., Huang, R., Wu, Y., BEOL compatible 15-nm channel length ultrathin indium-tin-oxide transistors with I on= 970 μA/μm and on/off ratio near 10 11 at V ds= 0.5 V, in: *2019 IEEE International Electron Devices Meeting (IEDM)*, pp. 3–5, IEEE, 2019.

50. Oota, M., Ando, Y., Tsuda, K., Koshida, T., Oshita, S., Suzuki, A., Fukushima, K., *et al.*, 3D-stacked CAAC-In-Ga-Zn oxide FETs with gate length of 72nm. in: *2019 IEEE International Electron Devices Meeting (IEDM)*, pp. 3–2, IEEE, 2019.
51. Bera, R., *et al.*, Hermes: Accelerating Long-Latency Load Requests *via* Perceptron-Based Off-Chip Load Prediction, in: *MICRO*, 2022.
52. Hamerly, G., Perelman, E., Lau, J., Calder, B., Simpoint 3.0: Faster and more flexible program phase analysis. *J. Instr. Level Parallelism*, 7, 4, 1–28, 2005.
53. Jacob, B., Ng, S.W., Wang, D.T., *Memory Systems: Cache DRAM Disk*, Morgan Kaufmann, San Mateo, CA, USA, 2007.
54. Nasri, F. and Salama, H., Numerical Elucidation of the Electrothermal Characteristics of Gate-All-Around FET Devices, in: *2023 IEEE 11th International Conference on Systems and Control (ICSC)*, Sousse, Tunisia, pp. 01–05, 2023, doi: 10.1109/ICSC58660.2023.10449698.
55. Rosenfeld, P., Cooper-Balis, E., Jacob, B., DRAMSim2: A cycle accurate memory system simulator. *IEEE Comput. Archit. Lett.*, 10, 1, 16–19, Jan. 2011.
56. Yourst, M.T., PTLsim: A cycle accurate full system x86-64 microarchitectural simulator, in: *Proc. ISPASS*, pp. 23–34, 2007.
57. Nair, P., Chou, C.C., Qureshi, M.K., A case for refresh pausing in DRAM memory systems, in: *Proc. HPCA*, 2013.
58. Ghosh, M. and Lee, H.-H., Smart refresh: An enhanced memory controller design for reducing energy in conventional and 3d die-stacked DRAMs, in: *Proc. MICRO*, 2007.
59. Isen, C. and John, L.K., ESKIMO–energy savings using semantic knowledge of inconsequential memory occupancy for DRAM subsystem, in: *Proc. MICRO*, 2009.
60. Kornaros, G., Papaefstathiou, I., Nikologiannis, A., Zervos, N., A fully programmable memory management system optimizing queue handling at multi gigabit rates, in: *Proc. Design Automation Conference*, 2003.
61. Reinhardt, S.K., Larus, J.R., Wood, D.A., Tempest and typhoon: user-level shared memory, in: *Proc. International Symposium on Computer Architecture*, 1994.
62. Carter, J., Hsieh, W., Stoller, L., Swanson, M., Zhang, L., Brunvand, E., Davis, A., Kuo, C.-C., Kuramkote, R., Parker, M., Schaelicke, L., Tateyama, T., Impulse: Building a smarter memory controller, in: *Proc. International Symposium on High-Performance Computer Architecture*, 1999.
63. Narayanan, R., *et al.*, Minebench: A benchmark suite for data mining workloads, in: *International Symposium on Workload Characterization*, 2006.
64. Stuecheli, J., Kaseridis, D., Daly, D., Hunter, H.C., John, L.K., The virtual write queue: coordinating DRAM and last-level cache policies, in: *Proc. Interntional Symposium on Computer Architecture*, 2010.

65. Tayal, S., Smaani, B., Rahi, S.B., Labiod, S., Ramezani, Z., eds. *Device Circuit Co-design Issues in FETs*, CRC Press, 2023.
66. Salama, H., Tshipamba, A., Ahmed, K., Modeling for CMOS circuit design, in: *Device Circuit Co-Design Issues in FETs*, pp. 1–20, CRC Press, 2023.

22

Ensuring Robustness: Reliability Analysis of 4H-SiC Trench MOSFETs in High-Performance Analog Applications

Ajay Kumar[1]*, Mandeep Singh Narula[1], Neha Gupta[2], Aditya Jain[3], Kaushal Kumar[4] and Amit Kumar Goyal[5]

[1]*Department of Electronics and Communication Engineering, Jaypee Institute of Information Technology, Noida, India*
[2]*Department of Applied Science, GNIOT Group of Institutions, Greater Noida – Uttar Pradesh, India*
[3]*Department of Electronics and Communication, Symbiosis Institute of Technology, Symbiosis International (Deemed University) Pune, Maharashtra, India*
[4]*Department of Electronics and Communication Department, Graphic Era Deemed to be University, Dehradun, Uttarakhand, India*
[5]*Department of Electronics and Communication Engineering, Manipal Institute of Technology, Manipal Academy of Higher Education, Manipal, India*

Abstract
In this chapter, the reliability issues of interface trap charges (ITCs) on 4H-SiC (4H–silicon carbide) Trench metal-oxide-semiconductor field-effect transistor (4H-SiC-T-MOSFET) have been discussed in terms of analog parameters. Reliability issues in 4H-SiC-T-MOSFET related to trap charges primarily stem from the interface between the gate oxide and the silicon carbide semiconductor. Trap charges at this interface can lead to threshold voltage shifts, degradation of device performance, and long-term reliability issues. The density of interface traps at the SiC/SiO_2 interface is a critical parameter affecting device reliability. High trap densities can lead to increased interface state charges, which can adversely affect device performance and reliability. Trap charges at the interface can cause threshold voltage shifts, leading to variations in device characteristics and performance. Threshold voltage instability can affect circuit operation and reliability over time. Therefore, while trap charges at the SiC/SiO_2 interface can pose

**Corresponding author*: ajay.kumar@jiit.ac.in

Ekta Goel, Archana Pandey, Shiromani Balmukund Rahi and Arun Samuel (eds.) Handbook of Advanced Semiconductor Field Effect Transistors, (465–478) © 2026 Scrivener Publishing LLC

reliability challenges for 4H-SiC-T-MOSFET, ongoing research and development efforts focus on mitigating these issues through improved materials, processing techniques, and device design strategies. By addressing interface trap–related reliability concerns, 4H-SiC-T-MOSFET can achieve enhanced long-term reliability and performance for various high-power and high-temperature applications.

***Keywords*:** 4H-SiC-T-MOSFET, analog, reliability, black phosphorus, trap charges

22.1 Introduction

Interface trap charges (ITCs) significantly impact the performance and reliability of field-effect transistors (FETs). These charges exist between the Si/SiO_2 interface in the FET structure. The presence of these traps alters the surface potential, requiring either more or less gate voltage to turn the FET on or off. This leads to unpredictable behavior, particularly in sensitive analog and low-power circuits. Trap traps can contribute to a higher leakage current in the subthreshold region by providing additional conduction paths through charge trapping and emission processes. This degrades the off-state behavior of the FET, impacting power consumption, especially in low-power devices. The charge carriers (electrons or holes) that move near the semiconductor-insulator interface are scattered by the trapped charges, which reduces the effective mobility of the carriers. This impacts the FET's drive current, leading to slower switching speeds and reduced overall performance. The process of charge trapping and de-trapping from the interface traps generates fluctuations in the current, contributing to the noise. This is especially relevant for analog and RF applications where low noise is critical. Prolonged stress, such as high electric fields or hot carrier injection, can create new interface traps, worsening the device's performance and potentially causing failure. This is a key concern in the long-term operation of integrated circuits (ICs). In short-channel devices, interface traps can interact with the drain field and contribute to drain-induced barrier lowering (DIBL), which lowers the threshold voltage as the drain voltage increases. This leads to degraded control of the channel by the gate. In some FETs, especially with thinner gate oxides, trapped charges may cause tunneling leakage through the gate dielectric. This negatively affects power efficiency and increases standby power consumption.

4H-SiC (4H–silicon carbide) is of great importance in the field of metal-oxide-semiconductor field-effect transistors (MOSFETs). Silicon carbide has a wide bandgap, making it capable of sustaining high electric fields without breakdown. This property allows 4H-SiC MOSFETs to operate at higher voltages compared to silicon-based devices [1]. High breakdown voltages are crucial for applications in power electronics, where devices must handle high voltages and currents efficiently [1, 2]. 4H-SiC MOSFETs can operate at much higher temperatures than traditional silicon MOSFETs. This makes them suitable for use in harsh environments where elevated temperatures are common, such as automotive, aerospace, and industrial applications. The ability to operate at high temperatures increases the reliability and lifetime of electronic systems. Silicon carbide exhibits high electron mobility compared to silicon, which results in faster switching speeds and reduced on-state resistance in MOSFET devices [3–6]. This translates to higher efficiency and improved performance in power conversion applications, where fast switching and low losses are critical. Due to its superior material properties, 4H-SiC MOSFETs have lower switching losses compared to silicon-based devices [7, 8]. This allows for higher frequency operation and enables more compact and efficient power converter designs. Lower switching losses also contribute to improved system efficiency and reduced cooling requirements. This makes it ideal for applications such as power distribution systems, electric vehicles, renewable energy systems, and military electronics.

4H-SiC is a polytype of silicon carbide that has gained significant importance in the development and application of FETs due to its superior material properties. 4H-SiC has a wide bandgap of approximately 3.26 eV, compared to silicon's 1.12 eV. This wide bandgap allows 4H-SiC FETs to operate at higher voltages, temperatures, and frequencies. It also reduces leakage currents and improves the breakdown voltage, making 4H-SiC FETs suitable for high-power and high-temperature applications [2, 9–17]. 4H-SiC has excellent thermal conductivity, around 3.7 W/cm·K, which is significantly higher than silicon. This property helps in efficiently dissipating heat generated during device operation, enhancing the reliability and performance of FETs in high-power applications. Although 4H-SiC's electron mobility is lower than that of silicon, it is still adequate for high-frequency applications. The electron mobility in 4H-SiC is around 800 $cm^2/V{\cdot}s$, supporting fast switching speeds and efficient RF and microwave applications performance. 4H-SiC can withstand electric fields up to 2.5-3 MV/cm, which is about ten times higher than silicon. This high-breakdown electric field allows 4H-SiC FETs to handle high voltages without undergoing breakdown, making them ideal for high-voltage power electronics. 4H-SiC exhibits strong resistance to radiation damage, which makes it suitable for applications in harsh environments, such as space and nuclear industries, where radiation exposure can degrade the performance of silicon-based devices. 4H-SiC FETs can operate at much higher temperatures than silicon devices. This is due to both the wide bandgap and the high thermal conductivity of 4H-SiC, which enable reliable performance in high-temperature environments without the need for extensive cooling mechanisms. 4H-SiC FETs are used in high-power converters, inverters, and motor drives, particularly in electric vehicles. The high-frequency performance of 4H-SiC makes it suitable for RF and microwave applications, including communication systems and radar [18–23]. In aerospace, automotive, and industrial sectors, where devices are subjected to high temperatures, 4H-SiC FETs are preferred due to their stability and reliability.

22.2 Device Design and Its Parameters

The device schematics for the 4H-SiC-recesses channel (RC)-MOSFET are displayed in Figure 22.1. Table 22.1 lists the parameters that were used. The specifications of a polysilicon-gated recessed channel MOSFET are identical, with the exception of the gate material (φ_{PS} = 4.17 eV).

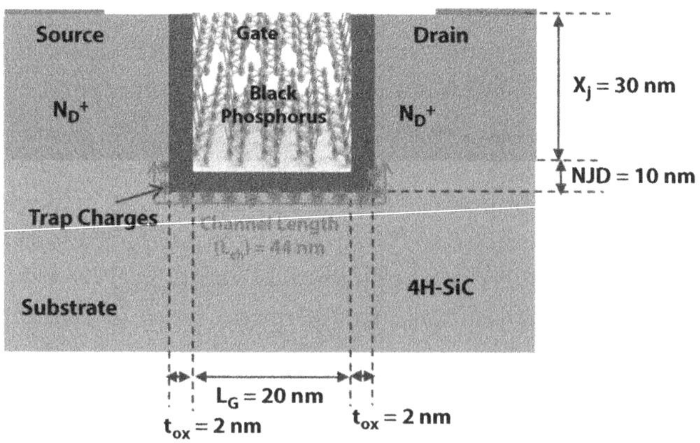

Figure 22.1 Device structure of 4H-SiC Trench MOSFET with the presence of trap charged at the interface of 4H-SiC/SiO_2.

Table 22.1 Device design parameters of 4H-SiC-T-MOSFET.

Device parameters	**4H-SiC-T-MOSFET**
Gate Length (L_g)	20 nm
Channel Length (L_{ch})	44 nm
Source and the Drain region Length	30 nm
Width of the Device (W)	200 nm
Groove Depth	38 nm
Negative Junction Depth (NJD)	10 nm
Source/Drain Doping (N_D^+)	$5x10^{16}cm^{-3}$ (N_D^+)
Physical Oxide Thickness (t_{ox})	2 nm
Gate metal work function (Φ)	5.16 eV
Gate to source voltage (V_{gs})	1.5 V
Drain to source voltage (V_{ds})	0.2 V

22.3 Methodology

Several models have been implemented for TCAD simulations [24–26]. For example, we have included the fldmob model to account for all the parameters influencing the mobility of carriers under changing electric fields, including the velocity saturation effect [27]. We have implemented the concentration-dependent SRH for carrier generation-recombination. It is a recombination model because it can incorporate a carrier lifetime that varies with concentration. The most precise way to simulate electron-hole pair generation-recombination, a crucial process in forecasting MOSFET behavior, is to use the Klasrh model. The models mentioned above take into consideration nearly all of the physics related to the MOSFET device, which leads to incredibly precise outcomes. The gate bias is swept from 0 to 5 V during the measurement.

22.4 Results and Discussion

The presence of trap charges in the gate oxide or at the oxide-semiconductor interface significantly impacts the transfer characteristics of a MOSFET. To enhance the reliability of MOSFETs in the presence of ITCs, several strategies such as material and process optimization, passivation techniques, device design, operational techniques, and circuit-level solutions can be employed. Among all these techniques, device design is most effective in pacifying the impact of ITCs. Figure 22.2 shows the impact of ITCs at the 4H-SiC/SiO_2 interface on the I_{ds}-V_{gs} of 4H-SiC Trench MOSFET (4H-SiC-T-MOSFET). Figure 22.2 shows that, in the presence of trap charges (negative and positive), off-current does not have any effect; however, on-current increases in the presence of positive (P) ITCs and reduces in the presence of negative (N) ITCs as compared to the absence of ITCs.

The reliability of trap charges in the conduction band is a critical factor in the performance and stability of various semiconductor devices, including transistors, sensors, and memory devices. Trap charges are defects or impurities that can capture and release charge carriers (electrons or holes) in the semiconductor material, influencing the device's electrical properties [21, 28–30]. Intrinsic traps are inherent to the semiconductor material, arising from imperfections in the crystal lattice. Extrinsic traps are due to impurities or defects introduced during the fabrication process. The position of the trap energy level relative to the conduction band edge determines how easily electrons can be captured or emitted. Figure 22.3 shows the impact of ITCs at the 4H-SiC/SiO_2 interface on the conduction band profile of 4H-SiC-T-MOSFET. When P-ITCs are present at the interface of 4H-SiC/SiO_2, the conduction band energy does not change much in the presence of P-N-ITCs compared to the absence of trap charges, so that the device shows reliability over the trap charges present at the interface of 4H-SiC/

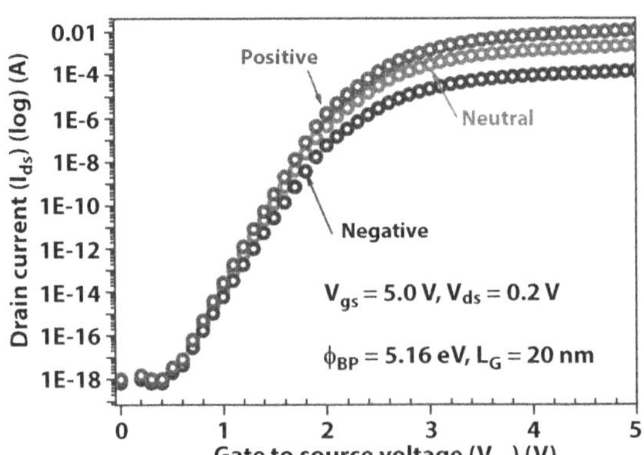

Figure 22.2 Impact of ITCs at the 4H-SiC/SiO_2 interface on I_{ds}-V_{gs} characteristics of 4H-SiC Trench MOSFET.

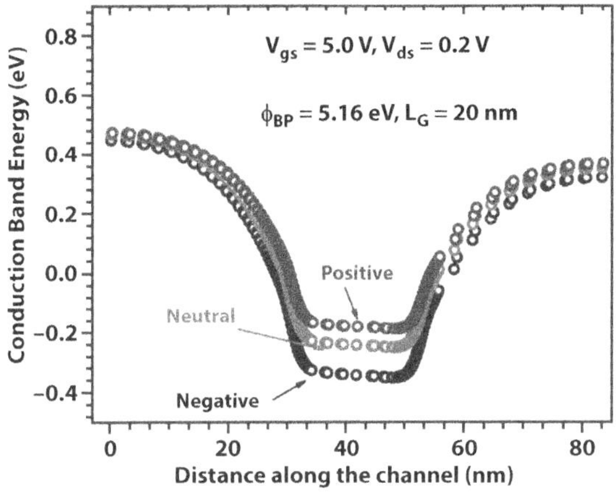

Figure 22.3 Impact of ITCs at the 4H-SiC/SiO_2 interface on the conduction band profile of 4H-SiC Trench MOSFET.

SiO_2 due to the trench/recessed architecture of the device [20, 21, 31–33]. On the other hand, in conventional MOSFET, ITCs significantly affect the on-current by altering the threshold voltage, mobility, and overall device behavior. Interface traps can capture or release charges, acting as fixed charges near the gate dielectric and semiconductor interface. Interface traps lead to the scattering of charge carriers as they move through the channel. This increases the Coulomb scattering, which reduces the carrier mobility. Lower mobility reduces the current for a given gate voltage because fewer carriers can contribute to conduction. In some cases, interface traps can increase gate leakage current due to additional pathways for tunneling. Although this does not directly affect on-current, it can contribute to higher overall power consumption, reducing the effective performance of the MOSFET.

Furthermore, Figure 22.4 shows the impact of ITCs at the 4H-SiC/SiO_2 interface on the valence band profile of 4H-SiC-T-MOSFET, as the reliability of trap charges in the valence band is as critical as those in the conduction band, especially for p-type semiconductors and devices where holes are the primary charge carriers. Trap charges in the valence band can significantly influence the performance and stability of these devices. Intrinsic traps are defects inherent in the semiconductor material, whereas extrinsic traps are introduced through impurities or during the fabrication process. The energy level of traps relative to the valence band edge determines the ease with which holes can be captured or emitted. Figure 22.4 shows that, when P-ITCs are at the interface of 4H-SiC/SiO_2, the valence band energy does not change much in the presence of P-N-ITCs compared

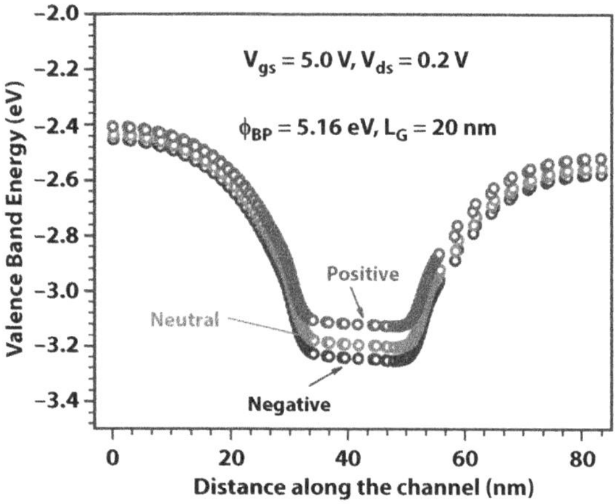

Figure 22.4 Impact of ITCs at the 4H-SiC/SiO_2 interface on valence band profile of 4H-SiC Trench MOSFET.

to the absence of trap charges, so that the device shows reliability over the trap charges present at the interface of 4H-SiC/SiO_2 due to the trench/recessed architecture of the device [25, 34, 35].

Figure 22.5 reflects the impact of ITCs at the 4H-SiC/SiO_2 interface on the surface potential of 4H-SiC-T-MOSFET as the reliability of trap charges in the surface potential of semiconductor devices is a crucial factor in ensuring their stable operation and longevity. Understanding the

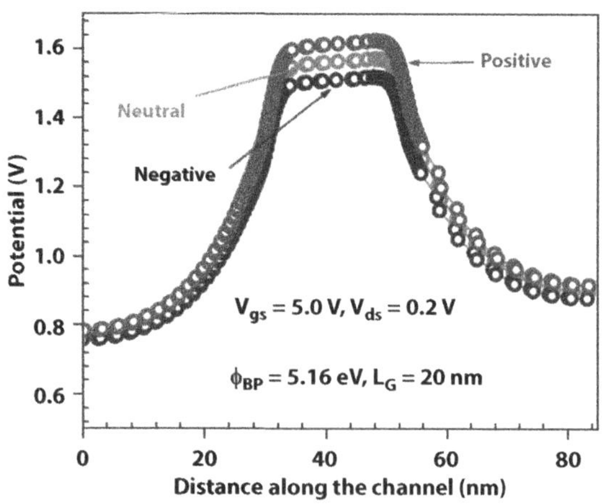

Figure 22.5 Impact of ITCs at the 4H-SiC/SiO_2 interface on surface potential of 4H-SiC Trench MOSFET.

nature and impact of these traps, along with effective mitigation strategies, is essential for improving device performance. Advanced characterization techniques, high-quality material growth, and careful fabrication processes are key to addressing the challenges posed by trap charges and enhancing the device's reliability. Figure 22.5 shows that, when P-ITCs are at the 4H-SiC/SiO_2 interface, the surface potential does not change much in the presence of positive and negative trap charges as compared to the absence of trap charges.

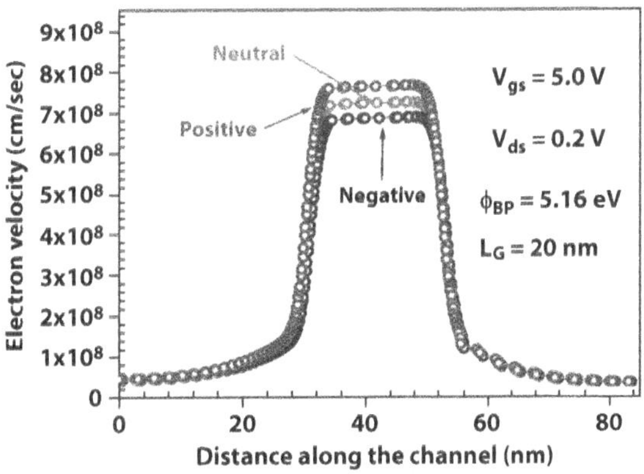

Figure 22.6 Impact of ITCs at the 4H-SiC/SiO_2 interface on electron velocity of 4H-SiC Trench MOSFET.

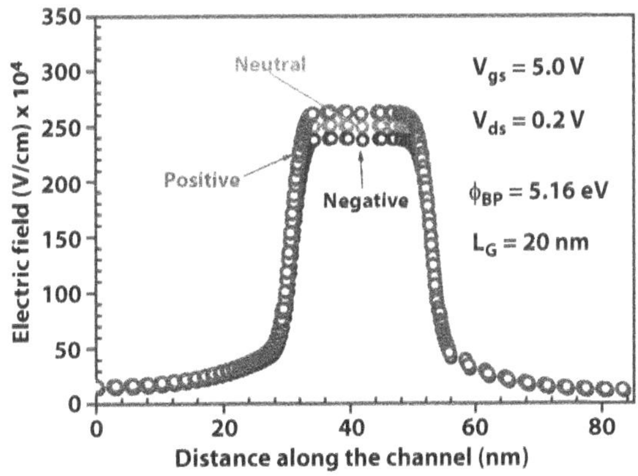

Figure 22.7 Impact of ITCs at the 4H-SiC/SiO_2 interface on the electric field of 4H-SiC Trench MOSFET.

Furthermore, electron velocity is observed when ITCs are present to show the device's reliability over trap charges. Figure 22.6 shows the impact of ITCs at the 4H-SiC/SiO_2 interface on the electron velocity of 4H-SiC-T-MOSFET. The impact of ITCs is negligible at the drain and source side. However, a very small effect is observed in the channel region. In the presence of P-ITCs, electron velocity increases and reduces when N-ITCs are present as compared to the absence of trap charges. Hence, 4H-SiC-T-MOSFET shows reliability over the trap charges. Moreover, the electric field is observed in 4H-SiC-T-MOSFET in the presence of ITCs to show the reliability of trap charges. Figure 22.7 shows the impact of ITCs at the 4H-SiC/SiO_2 interface on the electric field of 4H-SiC-T-MOSFET. The impact of positive and negative trap charges is negligible at the drain and source side. However, a very small effect is observed in the channel region. In the presence of positive trap charges, the electric field increases and reduces when negative trap charges are present as compared to the absence of trap charges due to a change in electron velocity. Hence, 4H-SiC-T-MOSFET shows reliability over the trap charges.

Electron mobility is evaluated in when ITCs are present to show the device's reliability over trap charges. Figure 22.8 shows the impact of positive, negative, and neutral trap charges at the 4H-SiC/SiO_2 interface on the electron mobility of 4H-SiC-T-MOSFET. The impact of ITCs on electron mobility is negligible at the drain and source side. However, a very small effect is observed in the channel region. When P-ITCs are present, electron mobility increases and reduces when N-ITCs are present. Hence, 4H-SiC-T-MOSFET shows reliability on electron mobility over the trap charges. Furthermore, the electron temperature is observed in 4H-SiC-T-MOSFET

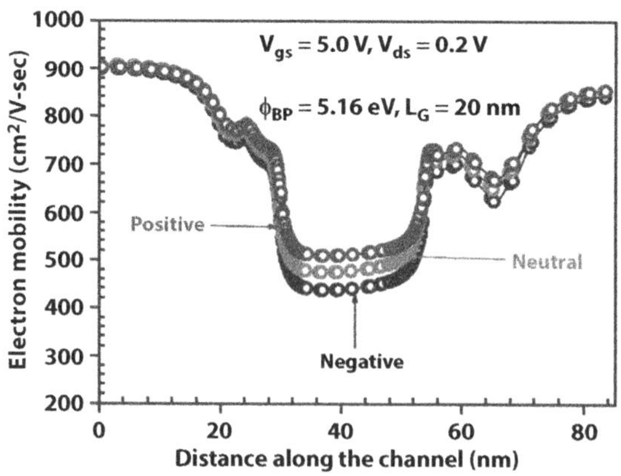

Figure 22.8 Impact of ITCs at the 4H-SiC/SiO_2 interface on electron mobility of 4H-SiC Trench MOSFET.

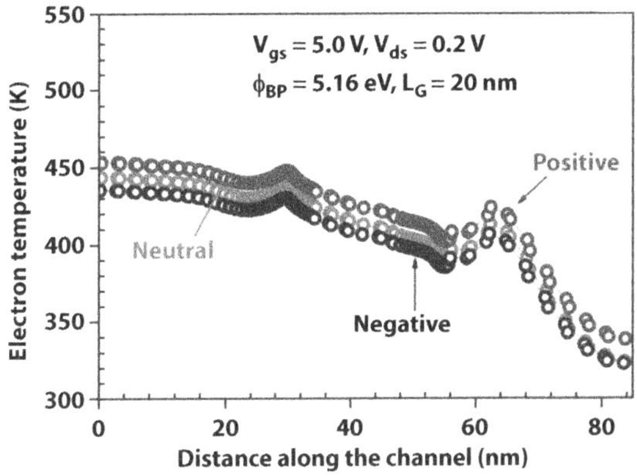

Figure 22.9 Impact of ITCs at the 4H-SiC/SiO_2 interface on electron temperature of 4H-SiC Trench MOSFET.

when ITCs are considered to show the reliability of ITCs. Figure 22.9 shows the impact of ITCs at the 4H-SiC/SiO_2 interface on the electron temperature of the 4H-SiC-T-MOSFET. The impact of P-N-ITCs is negligible at the drain, source, and entire channel region due to low leakage current. When P-ITCs are considered, the electron temperature increases and reduces when N-ITCs are present due to a change in the electron electric field. Hence, 4H-SiC-T-MOSFET shows reliability over the trap charges.

22.5 Conclusion

In terms of analog parameters, this chapter has addressed the reliability difficulties related to ITCs on 4H-SiC-T-MOSFET. Trap charge–related reliability problems in 4H-SiC-T-MOSFETs are mainly caused by the interface between the silicon carbide semiconductor and the gate oxide. Trap charges at this interface may cause long-term reliability problems, device performance deterioration, and threshold voltage shifts. Device reliability is significantly impacted by the density of interface traps at the 4H-SiC/SiO_2 interface. Elevated interface state charges brought on by high trap densities may negatively impact the device's dependability and performance. Device properties and performance may vary as a result of threshold voltage shifts brought on by trap charges at the interface. Over time, threshold voltage instability may have an impact on the dependability and operation of a circuit. Thus, even though trap charges at the 4H-SiC/SiO_2 interface may provide problems for 4H-SiC-T-MOSFET reliability, current research and development initiatives concentrate on resolving these concerns through enhanced materials, processing methods, and device design approaches. 4H-SiC-T-MOSFET can improve long-term reliability and performance for a variety of high-power and high-temperature applications by resolving interface trap–related reliability issues.

References

1. Chen, L., Guy, O., Jennings, M., Igic, P., Wilks, S., Mawby, P., Study of 4H-SiC trench MOSFET structures. *Solid State Electron.*, 49, 7, 1081–1085, 2005.
2. Mahabadi, S.J. and Moghadam, H.A., Comprehensive study of a 4H-SiC MES-MOSFET. *Physica E*, 74, 25–29, 2015.
3. Chen, Q.B., Compact modeling of multi-gate MOSFETs for RF designs, in: *2013 IEEE International Wireless Symposium (IWS)*, 2013.
4. Zareiee, M., High Performance Nano Device with Reduced Short Channel Effects in High Temperature Applications. *ECS J. Solid State Sci. Technol.*, 6, 7, M75–M78, 2017.
5. Zareiee, M. and Orouji, A.A., Superior electrical characteristics of novel nanoscale MOSFET with embedded tunnel diode. *Superlattices Microstruct.*, 101, 57–67, 2017.
6. Zareiee, M., A novel high performance nano-scale MOSFET by inserting Si3N4 layer in the channel. *Superlattices Microstruct.*, 88, 254–261, 2015.
7. Lam, K.-T., Dong, Z., Guo, J., Performance limits projection of black phosphorous field-effect transistors. *IEEE Electron Device Lett.*, 35, 9, 963–965, 2014.
8. Das, S., Zhang, W., Demarteau, M., Hoffmann, A., Dubey, M., Roelofs, A., Tunable transport gap in phosphorene. *Nano Lett.*, 14, 10, 5733–5739, 2014.
9. Hajjiah, A. and Huang, A., Novel SiC-Trench-MOSFET with reduced oxide electric field, in: *Solid-State and Integrated Circuits Technology, 2004. Proceedings. 7th International Conference on*, vol. 1, IEEE, pp. 340–344, 2004.
10. Han, K., Baliga, B., Sung, W., Split-Gate 1.2-kV 4H-SiC MOSFET: Analysis and Experimental Validation. *IEEE Electron Device Lett.*, 38, 10, 1437–1440, 2017.
11. Liangchun, C.Y., Dunne, G.T., Matocha, K.S., Cheung, K.P., Suehle, J.S., Sheng, K., Reliability issues of SiC MOSFETs: A technology for high-temperature environments. *IEEE Trans. Device Mater. Reliab.*, 10, 4, 418–426, 2010.
12. Maresca, L., Matacena, I., Riccio, M., Irace, A., Breglio, G., Daliento, S., Influence of the SiC/SiO2 SiC MOSFET interface traps distribution on CV measurements evaluated by TCAD simulations. *IEEE J. Emerging Sel. Top. Power Electron.*, 9, 2019.

13. Morkoc, H., Strite, S., Gao, G., Lin, M., Sverdlov, B., Burns, M., Large-band-gap SiC, III-V nitride, and II-VI ZnSe-based semiconductor device technologies. *J. Appl. Phys.*, 76, 3, 1363–1398, 1994.
14. Pengelly, R.S., Wood, S.M., Milligan, J.W., Sheppard, S.T., Pribble, W.L., A review of GaN on SiC high electron-mobility power transistors and MMICs. *IEEE Trans. Microwave Theory Tech.*, 60, 6, 1764–1783, 2012.
15. Shafiei, M., *et al.*, Platinum/graphene nanosheet/SiC contacts and their application for hydrogen gas sensing. *J. Phys. Chem. C*, 114, 32, 13796–13801, 2010.
16. Song, Q., Zhang, Y., Zhang, Y., Tang, X., 4H-SiC trench gate MOSFETs with field plate termination. *Sci. China Technol. Sci.*, 57, 10, 2044–2049, 2014.
17. Yadav, M., Pradhan, K., Sahu, P., A complete analytical potential based solution for a 4H-SiC MOSFET in nanoscale. *Adv. Nat. Sci.: Nanosci. Nanotechnol.*, 7, 2, 025011, 2016.
18. Kumar, A., Tripathi, M., Chaujar, R., Comprehensive analysis of sub-20 nm black phosphorus based junctionless-recessed channel MOSFET for analog/RF applications. *Superlattices Microstruct.*, 116, 171–180, 2018.
19. Kumar, A., Tripathi, M.M., Chaujar, R., Small-signal modeling of In2O5Sn based transparent gate recessed channel MOSFET for microwave/RF applications, in: *2017 IEEE Asia Pacific Microwave Conference (APMC)*, Kuala Lumpur, Malaysia, pp. 976–979, 2017, doi: 10.1109/APMC.2017.8251614.
20. Kumar, A., Tripathi, M.M., Chaujar, R., In2O5Sn based transparent gate recessed channel MOSFET: RF small-signal model for microwave applications. *AEU Int. J. Electron. Commun.*, 93, 233–241, 2018. 2018/09/01/.
21. Kumar, A., Tripathi, M.M., Chaujar, R., Reliability issues of In_2O_5Sn gate electrode recessed channel MOSFET: Impact of interface trap charges and temperature. *IEEE Trans. Electron Devices*, 65, 3, 860–866, 2018.
22. Kumar, A., Tripathi, S.K., Gupta, N., Tripathi, P.M., Chaujar, R., GaN Silicon-on-Insulator (SOI) N-Channel FinFET for High-Performance Low Power Applications, in: *2019 IEEE 14th Nanotechnology Materials and Devices Conference (NMDC)*, IEEE, pp. 1–4, 2019.
23. Kumar, B. and Chaujar, R., Analog and RF Performance Evaluation of Junctionless Accumulation Mode (JAM) Gate Stack Gate All Around (GS-GAA) FinFET. *Silicon*, 13, 3, 919–927, 2021.
24. Gupta, N., Kumar, A., Chaujar, R., TCAD analysis of frequency dependent intrinsic and extrinsic parameters of GEWE-SiNW MOSFET. *Nanoelectron. Mater. Devices*, 2, 185–188, 2015.
25. Kumar, A., Gupta, N., Chaujar, R., TCAD RF performance investigation of Transparent Gate Recessed Channel MOSFET. *Microelectron. J.*, 49, 36–42, 2016.
26. Kumar, B. and Chaujar, R., TCAD Temperature Analysis of Gate Stack Gate All Around (GS-GAA) FinFET for Improved RF and Wireless Performance. *Silicon*, 13, 1–13, 2021.

27. Silvaco, I., *ATLAS User's Manual*, vol. 5, SILVACO International, Santa Clara, CA, 2011.
28. Heiman, F. and Warfield, G., The effects of oxide traps on the MOS capacitance. *IEEE Trans. Electron Devices*, 12, 4, 167–178, 1965.
29. Kumar, A., Tiwari, B., Singh, S., Tripathi, M.M., Chaujar, R., Radiation Analysis of N-Channel TGRC-MOSFET: An X-Ray Dosimeter. *IEEE Trans. Electron Devices*, 65, 11, 5014–5020, 2018.
30. Kumar, A., Tripathi, M., Chaujar, R., Influence of interface trap charge density on reliability issues of transparent gate recessed channel (TGRC) MOSFET, in: *Microelectronic Devices, Circuits and Systems (ICMDCS), 2017 International conference on*, IEEE, pp. 1–4, 2017.
31. Kumar, A., Gupta, N., Singh, S., Tiwari, B., Tripathi, M.M., Chaujar, R., Carbon Nanotube Recessed Channel (CNT-RC) MOSFET for High Linearity/ULSI Applications, *TENCON 2019 - 2019 IEEE Region 10 Conference (TENCON)*, Kochi, India, pp. 2564–2567, 2019, doi: 10.1109/TENCON.2019.8929540.
32. Kumar, A., Tripathi, M., Chaujar, R., Small-signal modeling of In2O5Sn based transparent gate recessed channel MOSFET for microwave/RF applications, in: *2017 IEEE Asia Pacific Microwave Conference (APMC)*, IEEE, pp. 976–979, 2017.
33. Kumar, A., Tripathi, M.M., Chaujar, R., Sub-30nm In2O5Sn gate electrode recessed channel MOSFET: A biosensor for early stage diagnostics. *Vacuum*, 164, 46–52, 2019.
34. Kumar, A., Assessment of high-k gate stacked In2O5Sn gate recessed channel MOSFET for x-ray radiation reliability. *Eng. Res. Express*, 2, 3, 035017, 2020.
35. Kumar, A., Gupta, N., Chaujar, R., Power gain assessment of ITO based Transparent Gate Recessed Channel (TGRC) MOSFET for RF/wireless applications. *Superlattices Microstruct.*, 91, 290–301, 2016.

About the Editors

Dr. Ekta Goel received her B.Tech. degree in Electronics and Communication Engineering from Uttar Pradesh Technical University, Lucknow. She completed her M.Tech. and Ph.D. in Microelectronics Engg. from IIT (BHU), Varanasi in 2013 and 2018 respectively. She is working as an assistant professor in NIT Warangal (An Institute of National Importance) prior to that she has worked in Jaypee institute of information technology, Noida. Her area of research includes modeling and simulation of advanced nanoscale MOS devices, VLSI circuits, FET based biosensors, Tunnel FET, photodiodes and photovoltaic cells. She has published more than 50 research articles in peer reviewed journals like IEEE Transaction on Electron Devices, IEEE Photonics Journal, Physica Scripta, etc. She is a recipient of Outstanding Teaching Excellence Award 2023 by ARDORCOMM MEDIA, Young Women Researcher in Electronics Award 2023 by Venus International Women Awards, Bharat Vikas Award for outstanding research by INSTITUTE OF SELF RELIANCE Bhubaneswar, Odisha in 2019. She is working as the reviewers of many journals.

Dr. Archana Pandey is Assistant Professor (Senior Grade) in the Electronics & Communication Engineering department at JIIT, Noida: Dr. Pandey earned her Ph.D. from IIT Roorkee, focusing on FinFET parasitic effects in circuit design, and also holds an M.Tech in Solid-State Electronic Materials from IIT Roorkee and a B.Tech in Electronics & Communication Engineering. Since July 2017, she has been a senior-grade faculty member at JIIT, specializing in the design, modeling, and circuit-level integration of advanced semiconductor devices—such as FinFETs, GAAFETs, nanosheet FETs, negative-capacitance FETs, and FET-based sensors—as well as VLSI circuits and machine-learning integration. Her work has appeared in Q1 journals including IEEE Sensors Journal, IEEE Transactions on Electron Devices, with notable publications.

Shiromani Balmukund Rahi, PhD (IIT Kanpur), is an assistant professor at University School of Information and Communication Technology (SoICT) Gautam Buddha University Greater Noida, Uttar Pradesh, India. He has successfully published 26 international research publications, 4 conference proceedings and 40 book chapters, in addition to presenting his research at various international conferences and workshops. In addition to his original work, he has successfully published 11 edited books and 10 book projects on progress. He has also received various awards and recognitions for his work as an editor, reviewer and best paper award for several international journals and conferences. He has also worked as post-doctoral researcher in Korea Military Academy Seoul, Republic of Korea. He has also supervised more than 15 UG/PG students for various technical projects. He also serves as a reviewer for various national and international journals, conferences, and workshops. His research interest is simulation and modelling of various semiconductor devices like TFET, NCFETs, and Nanosheet FETs and AI and machine learning in semiconductor devices and circuit-based applications in research.

Arun Samuel T.S received a B.E degree in Electronics and Communication Engineering from Syed Ammal Engineering College (2004) and M.E degree in Computer and communication engineering from National Engineering College (2006). He has been awarded a Ph.D. in Nanoelectronic Devices (2014) from Thiagarajar College of Engineering, Tamilnadu, India, under Anna University Chennai. He is currently working in National Engineering College, Kovilpatti, India, as Professor. He has authored more than 70 research articles in International Journals (SCI and Scopus Indexed Journals), 20 articles in international conferences, 1 patent, 6 book chapters and editor of 3 books. He is a life member of the Institute of Engineering (IE), India, and a Senior member of IEEE. His research interest includes Modelling and Simulation of Multigate transistors and Tunnel Field-effect Transistors.

Index

Also of Interest

Check out these other related titles from Scrivener Publishing

Advanced Ultra Low-Power Semiconductor Devices: Design and Applications Edited by Shubham Tayal, Abhishek Kumar Upadhyay, Shiromani Balmukund Rahi, and Young Suh Song, ISBN: 9781394166411. Written and edited by a team of experts in the field, this important new volume broadly covers the design and applications of metal oxide semiconductor field effect transistors.

Field Effect Transistors, Edited by P. Suveetha Dhanaselvam, K. Srinivasa Rao, Shiromani Balmukund Rahi, and Dharmendra Singh Yadav, ISBN: 9781394248476. *Field Effect Transistors* is an essential read for anyone interested in the future of electronics, as it provides a comprehensive yet accessible exploration of innovative semiconductor devices and their applications, making it a perfect resource for both beginners and seasoned professionals in the field.

Integrated Devices for Artificial Intelligence and VLSI: VLSI Design, Simulation and Applications, Edited by Balwinder Raj, Suman Lata Tripathi, Tarun Chaudhary, K. Srinivasa Rao, and Mandeep Singh, ISBN: 9781394204359. With its in-depth exploration of the close connection between microelectronics, AI, and VLSI technology, this book offers valuable insights into the cutting-edge techniques and tools used in VLSI design automation, making it an essential resource for anyone seeking to stay ahead in the rapidly evolving field of VLSI design.

Controller Design for Industrial Applications, Edited by Arindam Mondal & Souvik Ganguli, ISBN: 9781394287079. *Controller Design for Industrial Applications* is essential for anyone looking to master the advanced techniques of intelligent controller design, enabling you to effectively tackle the complexities of modern industrial processes and optimize performance in an ever-evolving landscape.

Special Electrical Machinery, Edited by Dr. Jigneshkumar P. Desai, ISBN: 9781394193912. This book is a comprehensive guide to specialized motors, covering everything from switched reluctance motors to axial flux ironless PM motors, giving in-depth information on their operating principles, applications, and controls.

NANODEVICES FOR INTEGRATED CIRCUIT DESIGN, Edited by Suman Lata Tripathi, Abhishek Kumar, K. Srinivasa Rao, and Prasantha R. Mudimela, ISBN: 9781394185788. Written and edited by a team of experts in the field, this important new volume broadly covers the design of nano-devices and their integrated applications in digital and analog integrated circuits (IC) design.

Electrical and Electronic Devices, Circuits, and Materials: Technical Challenges and Solutions, Edited by Suman Lata Tripathi, Parvej Ahmad Alvi, and Umashankar Subramaniam, ISBN: 9781119750369. Covering every aspect of the design and improvement needed for solid-state electronic devices and circuit and their reliability issues, this new volume also includes overall system design for all kinds of analog and digital applications and developments in power systems.

Printed and bound by CPI Group (UK) Ltd, Croydon, CR0 4YY

17/09/2026

14964283-0008